SO-CFI-955

Marine Biology

Marine Biology

Peter Castro, Ph.D.

Professor
Biological Sciences Department
California State Polytechnic University, Pomona

Michael E. Huber, Ph.D.

Senior Lecturer in Biology and Head,
Motupore Island Research Center
University of Papua New Guinea

Artwork by
William Ober, M.D.
and
Claire Garrison, B.A.
Shoals Marine Laboratory, Cornell University

WCB **Wm. C. Brown Publishers**
Dubuque, Iowa • Melbourne, Australia • Oxford, England

Wm. C. Brown Publishers
A Division of Wm. C. Brown Communications, Inc.

Vice President and General Manager *Beverly Kolz*
Vice President, Publisher *Kevin Kane*
Vice President, Director of Sales and Marketing *Virginia S. Moffat*
Vice President, Director of Production *Colleen A. Yonda*
National Sales Manager *Douglas J. DiNardo*
Marketing Manager *Craig S. Marty*
Advertising Manager *Janelle Keeffer*
Production Editorial Manager *Renée Menne*
Publishing Services Manager *Karen J. Slaght*
Permissions/Records Manager *Connie Allendorf*

Wm. C. Brown Communications, Inc.

President and Chief Executive Officer *G. Franklin Lewis*
Corporate Senior Vice President, President of WCB Manufacturing *Roger Meyer*
Corporate Senior Vice President and Chief Financial Officer *Robert Chesterman*

Front cover photograph: © Animals Animals/James Watt
Back cover photograph: © Aninals Animals/Henry Asloos
Original maps by Pagecrafters, Inc.
Credits for all materials used by permission appear after the glossary.

Copyright © 1992 by Wm. C. Brown Communications, Inc. All
rights reserved

A Times Mirror Company

Library of Congress Catalog Card Number: 91–29217

ISBN 0–697–23480–0

No part of this publication may be reproduced, stored in a retrieval
system, or transmitted, in any form or by any means, electronic,
mechanical, photocopying, recording, or otherwise, without the
prior written permission of the publisher.

Printed in the United States of America by Wm. C. Brown Communications, Inc.,
2460 Kerper Boulevard, Dubuque, IA 52001

10   9   8   7   6   5

Preface

The study of marine life remains a popular subject in schools everywhere. Most colleges and universities offer an introductory, lower-division course in marine biology, one that usually requires no prerequisites. Many schools also offer a related course in oceanography. In addition, numerous high schools offer courses in marine biology for advanced students.

It is our experience that most college students enrolling in introductory marine biology are nonscience majors; many will take no other college science course. Some are technical or professional majors ranging from agriculture to chemistry, engineering to urban planning. Business and agriculture students may have a special interest in mariculture or in the exploitation of specific types of resources. Engineering students at some schools specialize in ocean engineering and are interested in topics such as fouling organisms and oceanographic instrumentation. Students with majors in political science or public policy may be particularly concerned with fisheries and the effects of pollution. Biology students, by the way, may also take the course. A great many students in all of the above categories enroll in the course because of such personal interests as scuba diving, sport fishing, or keeping marine aquaria at home. Many are interested in particular groups such as marine mammals or sharks. Students may enroll in the course to fulfill a general science requirement, although at a number of institutions introductory marine biology does not satisfy such requirements.

Our book is an introductory level text in marine biology. It is designed to provide a stimulating, up-to-date overview of marine biology while integrating the fundamental basic science background required for a general education course. It familiarizes students not only with the basic principles of biology, but with the physical sciences and the methods and assumptions of modern science as well. This approach demonstrates the relevance of the physical sciences to marine biology and thus renders the study of all sciences less intimidating to nonscience students. The book also seeks to satisfy the needs and expectations of a wide range of students, an approach that we hope guarantees a positive learning experience for all.

We have also tried to adopt a global, nonregional view of the world ocean that is integrated throughout the text. This global view of the oceans, a major development in marine science resulting from recent technological advances, forces us to consider not just the organisms living closest to our own shore, but all of those inhabiting the one ocean that so much influences our lives.

ORGANIZATION

Marine Biology is organized into four parts. The first (Chapters 1 to 3) will introduce students to marine biology and related fields of science. Chapter 1 describes the history of marine biology. It also describes the fundamentals of the scientific method, essential in understanding the workings of science. This feature, found in no other marine biology text, presents science as a process, an ongoing human endeavor. We feel that it is important for students to realize that science does have limitations and that there is still much to be learned. Chapters 2 and 3 present geological, physical, and chemical oceanography as they relate to marine biology. This link between the marine sciences is a central theme that will be repeated throughout the book. *Marine Biology* presents more information on the geology, chemistry, and physics of the marine environment than other texts, but we have kept Chapters 2 and 3 as short and basic as possible. Many physical aspects of the marine environment are discussed where they are biologically relevant. Upwelling, for example, is described in the chapter on the epipelagic (Chapter 14). Chapters 2 and 3 also include original world maps that were drawn using the Robinson projection to minimize distortion.

The exciting nature of life in the sea is the subject of the second part (Chapters 4 to 8). Chapter 4, "The Business of Life," is a brief introduction to the basic principles of biology. As with the fundamentals of geological, physical, and chemical oceanography, the basic concepts of biology are reviewed throughout the book by means of short **Marginal Review Notes.** Because the most important material from Chapter 4 is reviewed in these notes, students may proceed directly to Chapter 5 from Chapter 3 if introductory biology is a prerequisite for the course. The major groups of marine organisms are surveyed in Chapters 5 to 8. As in the first part of the book, we seek to provide introductory information that will be reviewed and expanded on in future chapters. In discussing the various groups, we emphasize functional morphology, outstanding ecological and physiological adaptations, and economic importance or significance to humankind. Classification and phylogeny are not stressed. A classification scheme for the major groups, however, is given at the end of each of the four chapters. Terminology has been limited to those terms that best illustrate significant concepts. Here and throughout the book we have selected organisms from around the world for illustration in photographs, line drawings, and color paintings, but we have emphasized those inhabiting the Atlantic, Gulf, and Pacific coasts of North America. Organisms are referred to by their most widely accepted common names; one or two common or important genera are listed the first time a group is mentioned in a chapter.

The third and most extensive part (Chapters 9 to 15) constitutes the heart of the book. The first chapter (Chapter 9) introduces students to the fundamental principles of ecology. As in Chapter 4, important concepts presented here will be reviewed in other chapters using marginal review notes. In the remaining six chapters of the third part, we have presented the major environments of the world ocean, proceeding from nearshore to offshore and from shallow to deep water. Chapters, however, are designed so that they can be covered in any sequence, depending on the instructor's preference and needs. Basic themes are adaptations to the physical constraints of the environment and the interactions of organisms within the environment. The chapters incorporate some of the most recent and exciting discoveries, as in the case of coral reefs and deep water.

The fourth part looks at the many ways humans interact with the world ocean: the use of resources, our impact on the marine environment, and the influence of the ocean on culture and other human experiences. These chapters present an up-to-date and comprehensive view of issues and concerns that are shared by many students. The use of resources (Chapter 16) looks not only at traditional uses such as fisheries and mariculture, but also at more modern aspects such as the pharmacological uses of marine natural products and the application of genetic engineering to mariculture. In Chapter 17, the discussion of marine pollution is followed by an examination of the conservation and enhancement of the marine environment. The book closes with a chapter on the impact of the ocean on human affairs (Chapter 18), one that the authors hope will stimulate students to reflect on the past and future significance of the world ocean.

TEACHING AND LEARNING AIDS

Marine Biology has been carefully written and designed to maximize its efficiency as a teaching aid. By using numerous and relatively short chapters we present information in readily absorbed units and at the same time increase the instructor's flexibility.

Key terms and concepts are given in **boldface** throughout each chapter and are defined in a **Glossary.** The extensive glossary not only provides complete definitions, but often refers to text illustrations that help explain the concept and to other key terms that are also defined in the glossary. **Key Concept Summaries** placed within the text highlight the most important concepts and ideas presented. **Marginal Review Notes** define key terms and concepts that are explained in previous chapters. Each note refers to the chapter and page where the concept was explained if additional review is needed.

The text is accompanied by a superb collection of photographs and illustrations, many in color, that we have carefully selected and designed. Many are used to further explain, clarify, and reinforce key concepts. Some of their captions contain questions that seek to further stimulate the student. Numerous maps especially created for *Marine Biology* to our specifications are found throughout the book and in Appendix C.

All chapters contain short **Boxed Essays** that present interesting, particularly relevant, or simply fascinating supplementary information—material as varied as experimental setups, John Steinbeck, intelligence among dolphins, and marine archeology.

Each chapter concludes with an enumerated **Do-It-Yourself Summary** that allows students to test their comprehension and retention of the most relevant information presented in the chapter. Key terms are omitted and the answers provided in Appendix A. The **Thought questions** that are given at the end of each chapter aim to challenge students and stimulate class discussion. Many of these have no "right" answer—that is often the point. A short annotated list of recent references, **For further reading,** is also provided in each chapter. It lists general articles of a level that is appropriate to students with a limited background in science. We have relied heavily on richly illustrated articles in such accessible publications as *National Geographic Magazine, Natural History, Oceanus, Scientific American,* and *Smithsonian.*

SUPPLEMENTS

To help students and professors alike, *Marine Biology* is accompanied by a number of useful supplements. An Instructor's Resource Manual/Test Bank is an invaluable aid for the instructor. For each chapter, the manual presents an outline, summary, and suggestions on how to introduce the chapter concepts to the students. The test bank contains questions for each chapter that will assist the instructor in compiling exams and quizzes. A computerized version of the test bank is also available in IBM and Apple formats. Finally, the manual contains 75 transparency masters of key illustrations in the text for use during lecture.

Transparency acetates of 40 other key illustrations provide additional visual support in the classroom. Thirty of these transparency acetates are in full color.

ACKNOWLEDGMENTS

The authors thank the many friends, colleagues, and former teachers who generously contributed photographs that have added so much to the book. Their names are acknowledged in the credits section of the book, but we extend special thanks to A. Charles Arneson, who provided many excellent photos. We are also grateful for the outstanding artwork of Bill Ober and Claire Garrison—their patience in sending sketches halfway around the world is very much appreciated. Mary Temperelli provided expert services in photo research. Mike Huber also thanks Kerry McGregor for her help and patience.

We also thank the exceptional staff at Mosby–Year Book, particularly Kathleen Scogna, Developmental Editor, who patiently and efficiently kept up with an enormous amount of detail while the authors worked in their separate corners of the world.

Peter Castro
Michael E. Huber

Reviewers

Keith E. Arnold, *California State Polytechnic University, Pomona*
Hans Bertsch, *National University* and *El Camino Community College*
Anthony L. Brundage, *California State Polytechnic University, Pomona*
Susan M. Cormier, *University of Louisville*
Harold Hirth, *University of Utah*
Ralph A. Lewin, *Scripps Institution of Oceanography*
David A. McKee, *Corpus Christi State University*
Edward K. Mercer, *California State Polytechnic University, Pomona*
Thomas M. Niesen, *San Francisco State University*
John S. Pearse, *University of California, Santa Cruz*
John E. Randall, *Bernice P. Bishop Museum, Honolulu*
Roger A. Rulifson, *East Carolina University*
Leslie J. Snider, *Mira Costa College*
Patrick J. Walsh, *Rosenstiel School of Marine and Atmospheric Science, University of Miami*
A. Quinton White, *Jacksonville University*

Contents

Part Three Structure and Function of the Marine Environment

Appendixes

Part One

Principles of
Marine Science

I The Science of Marine Biology

Marine biology is the scientific study of the plants, animals, and other organisms that live in the ocean. The ocean is a vast realm that contains many strange and wonderful creatures. It is often the beauty, mystery, and variety of life in the sea that attracts students to a course in marine biology. Even professional marine biologists feel a sense of adventure and wonder in their studies (Figure 1-1).

There are also many practical reasons to study marine biology. Marine life represents a vast source of human wealth. It provides us with food, medicine, and raw materials, as well as offering recreation to millions and supporting tourism all over the world. On the other hand, marine organisms can also create problems for humans. For example, some organisms harm humans directly by causing disease or attacking people. Others may harm us indirectly by injuring or killing marine organisms that we value for food or other purposes. Marine organisms may erode the piers, walls, and other structures we build in the ocean; foul the bottoms of ships; and clog pipes. They may even interfere with our weapons of war, for better or for worse.

At a much more fundamental level, marine life helps determine the very nature of our planet. Marine organisms produce much of the oxygen we breathe and probably help regulate the earth's climate. Our shorelines are shaped and protected by marine life, at least in part, and some marine organisms even create new land.

To make both full and wise use of the sea's living resources, to solve the problems marine organisms create, or to predict the effects of human activities on the life of the sea, we must learn all we can about marine life. In addition, marine organisms provide clues to the earth's past and to our own bodies that we must learn to understand. This is the challenge, the adventure, of marine biology.

Figure 1-1 Life below the surface—a world of beauty, puzzling mysteries, and wonderful surprises.

THE SCIENCE OF MARINE BIOLOGY

Marine biology is not really a separate science, but rather the more general science of biology applied to the sea. Nearly all the disciplines of biology in general are represented in marine biology. There are marine biologists who study the basic chemistry of living things, for example. Others are interested in plants and animals as whole organisms: the way they behave, where they live and why, and so on. Other marine biologists adopt a global perspective and look at the way entire oceans function as systems. Marine biology is thus both part of a broader science and itself made up of many different disciplines and viewpoints.

A marine biologist's interests may overlap broadly with those of biologists who study land organisms. The basic ways by which living things make use of energy, for example, are similar whether the organism lives on land or in the sea. Nevertheless, marine biology does have a flavor all of its own. Part of that flavor is due to its history.

The History of Marine Biology

People probably started learning about marine life from the first time they saw the ocean. After all, the sea is full of good things to eat. Archeologists have found piles of shells, the remains of ancient "clambakes," dating back to the Stone Age (Figure 1-2). Ancient harpoons and simple fishhooks of bone or shell have also been found. While they gathered food, people learned through experience which things were good to eat and which were bad tasting or harmful. The tomb of an Egyptian pharaoh bears a warning against eating a puffer, a kind of poisonous fish (Figure 1-3). Coastal peoples in virtually every culture developed a store of practical knowledge about marine life.

Figure 1-2   Ancient piles of shell and other debris, called middens, provide clues to the use of marine resources in ancient human cultures. They have been found by archeologists all over the world, sometimes dating back to the Stone Age. This midden from Baja California, Mexico, consists mostly of the shells of marine animals such as black abalone, giant owl limpet, and California mussel.

Figure 1-3   Inscription from the tomb of the Egyptian pharaoh Ti. It shows the puffer fish that the Egyptians knew contained a deadly toxin.

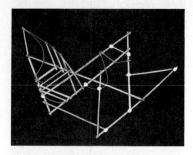

Figure 1-4 A Micronesian stick map from the Marshall Islands. These constructions were used by Pacific islanders to navigate between islands.

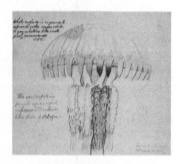

Figure 1-5 This Greek plate from around 330 BC reflects a considerable knowledge of marine life. An electric ray (*Torpedo*) can be identified.

Figure 1-6 A sketch of a jellyfish made by Joseph Drayton, an artist in the U.S. Exploring Expedition that circled the world between 1838 and 1842.

Knowledge of the ocean and its organisms developed as people gained skills in seamanship and navigation. Ancient Pacific Islanders, for example, were consummate seamen, navigating with the aid of unusual three-dimensional maps made of sticks and shells (Figure 1-4). The Phoenicians were the first accomplished Western navigators. By 2000 B.C. they were sailing around the Mediterranean Sea, Red Sea, eastern Atlantic Ocean, and Indian Ocean.

By the time of the ancient Greeks, a fair amount was known about the things that live near shore (Figure 1-5). The Greek philosopher Aristotle, who lived in the fourth century B.C., is considered by many to be the first marine biologist. He described many forms of marine life, and many of his descriptions are considered valid to this day. Aristotle made other studies as well. He recognized, for example, that gills are the breathing apparatus of fish.

During the centuries popularly known as the Dark Ages, scientific inquiry in most of Europe came to a grinding halt. Progress in the study of marine biology ceased. Indeed, much of the knowledge of the ancient Greeks was lost or distorted. Not all exploration of the ocean stopped, however. During the ninth and tenth centuries the Vikings continued to explore the northern Atlantic. In 995 A.D. a Viking party led by Leif Erikson discovered "Vinland"—what we now call North America. Arab traders were also active during the Middle Ages, voyaging to eastern Africa, southeast Asia, and India. In the process they learned about wind and current patterns, including the monsoons—strong winds that reverse direction with the seasons. In the Far East and the Pacific, people also continued to explore the sea.

In the Renaissance, spurred in part by the rediscovery of ancient knowledge preserved by the Arabs, Europeans again began to investigate the world around them. At first these investigations were mainly voyages of exploration. Christopher Columbus rediscovered the new world in 1492—word of the Vikings' find had never reached the rest of Europe. In 1519 Ferdinand Magellan embarked on the first expedition to sail around the globe. Many other epic voyages contributed to our knowledge of the oceans. Fairly accurate maps, especially of areas outside Europe, began to appear for the first time (see the section in Chapter 18, "Oceans as Barriers and Avenues").

Before long, explorers became curious about the ocean they sailed and the things that lived in it. An English sea captain, James Cook, was one of the first to make scientific observations along the way and to include a full-time naturalist among his crew. In a series of three great voyages, beginning in 1768, he explored all the oceans. He was the first European to see the Antarctic ice fields and to land in Hawaii, New Zealand, Tahiti, and a host of other Pacific islands. Cook was the first to make use of the chronometer, an accurate timepiece. This new technology enabled him to accurately determine his position, and therefore prepare accurate charts. From the Arctic to the Antarctic, from Alaska to Australia, Cook extended and reshaped the European conception of the world, bringing back specimens of plants and animals and tales of strange new lands. Though Cook was generally re-

PRINCIPLES OF MARINE SCIENCE

spectful and appreciative of indigenous cultures, he was killed in 1779 in a fight with native Hawaiians at Kealakekua Bay, Hawaii.

By the nineteenth century it was common for vessels to take along a naturalist to collect and study the life forms that were encountered (Figure 1-6). Perhaps the most famous of these shipboard naturalists was Charles Darwin (Figure 1-7), another Englishman. Beginning in 1831, Darwin sailed around the world on H.M.S. *Beagle* for 5 years, horribly seasick most of the time. Darwin, best known for proposing the modern theory of evolution, also made important contributions to marine biology. For example, he proposed an explanation for the formation of atolls, distinctive rings of coral reef. Though it wasn't until the 1950's that supporting evidence was obtained, his explanation is now almost universally accepted (see the section in Chapter 13, "How Atolls Are Formed"). He also used nets to capture the tiny drifting creatures known as plankton. Specialists still refer to his treatise on a group of animals called barnacles (Figure 1-8).

The Challenger Expedition.   By the middle of the nineteenth century a few scientists were able to undertake voyages for the specific purpose of studying the oceans. One was Edward Forbes, who in the 1840's and 1850's carried out extensive dredging of the sea floor, mostly around his native Britain but also in the Aegean Sea and other places. Forbes died prematurely in 1854, at the age of 39, but was the most influential marine biologist of his day. He discovered many previously unknown organisms and recognized that the life of the sea floor changes at different depths (see Chapter 15). His most important contribution, however, may have been to inspire interest in the life of the sea floor (Figure 1-9).

Many of Forbes' contemporaries and successors—especially from Britain, Germany, Scandinavia, and France—carried on his studies of sea floor life. Though their ships were poorly equipped and the voyages short, their studies yielded many interesting results. In fact, they were so successful that in 1872 British scientists were able to convince their government to fund the first major oceanographic expedition, under the scientific leadership of Charles Wyville Thompson. The British navy supplied a light warship to be fitted out for the purpose. The ship was named H.M.S. *Challenger*.

Figure 1-7 Charles Darwin (1809-1882), most often associated with the theory of evolution by natural selection, was also an able marine biologist.

Figure 1-8 Two of the drawings that accompanied Charles Darwin's original descriptions of barnacles.

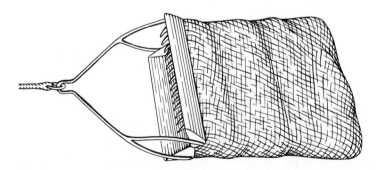

Figure 1-9 A typical dredge used by Edward Forbes and other early naturalists. The dredge was dragged behind a ship along the bottom, catching organisms in the mesh bag. Similar dredges are still used.

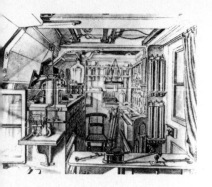

Figure 1-10 The naturalists' laboratory aboard the H.M.S. *Challenger.*

The *Challenger* underwent extensive renovations in preparation for the voyage. Most of her guns were removed—two were left, probably for moral support more than anything else. Laboratories and quarters for the scientific crew were added, and gear for dredging and taking water samples in deep water was installed. Though primitive by modern standards, the scientific equipment on board was the best of its day (Figure 1-10). Finally, in December of 1872, the *Challenger* set off.

During the 3½ years that followed, the *Challenger* and her crew sailed around the world gathering information and collecting samples (Figure 1-11). The sheer volume of the data was enormous. After the *Challenger* expedition returned to port, it took 19 years to publish the results, which fill 50 thick volumes. The *Challenger* brought back more information about the ocean than had been learned in all previous human history.

It was not just the duration of the voyage or the amount of information gathered that set the *Challenger* expedition apart from earlier efforts. More importantly, the expedition set new standards for the way that people studied the ocean. Measurements were made systematically and carefully, and meticulous records were kept. The crew worked with great efficiency and dedication to the task. For the first time, scientists began to get a coherent picture of what the ocean was like. The *Challenger* expedition laid the foundations of modern marine science.

In the years that followed, a number of other expeditions continued the work begun by the *Challenger*. Major oceanographic cruises continue to this day. In many ways, though, the voyage of the *Challenger* remains one of the most important in the history of oceanography.

Figure 1-11 The route taken by the *Challenger* expedition.

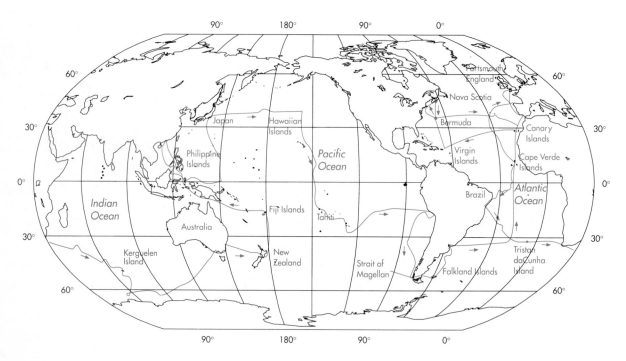

PRINCIPLES OF MARINE SCIENCE

The Growth of Marine Labs. Even before the *Challenger* set off, biologists were excited about the organisms brought back by ocean expeditions. Unfortunately, oceanographic vessels had quarters for only a limited number of scientists. Most biologists just got to see the dead, preserved specimens that the ships brought back to port. This was fine as long as biologists were content with simply describing the structure of new forms of marine life. They soon became curious, however, about how the organisms actually *lived:* how they functioned and what they did. Living specimens were essential for the study of these aspects of biology, but ships usually stayed in one place only for a short time, making long-term observations and experiments impossible.

As an alternative to studying the ocean from ships, some biologists decided to conduct their studies at the seashore. Among the first of these were two Frenchmen, Henri Milne-Edwards and Victor Andouin, who in about 1826 began making regular visits to the shore to study the life there. Other biologists soon followed suit. These excursions offered the opportunity to study live organisms, but there were no permanent facilities and only a limited amount of equipment could be taken along. This restricted the scope of the investigations. Eventually, permanent laboratories dedicated to the study of marine life were established on shore. These labs allowed marine biologists to keep organisms alive and to conduct their studies over long periods. The first such laboratory was the Stazione Zoologica (Figure 1-12), founded in Naples, Italy, by German biologists in 1872—the same year the *Challenger* embarked. The laboratory of the Marine Biological Society of the United Kingdom was founded at Plymouth, England, in 1879.

The first major American marine laboratory was the Marine Biological Laboratory at Woods Hole, Massachusetts (Figure 1-13). It is hard to pinpoint the exact date when this laboratory was established. The first marine laboratory at Woods Hole was initiated by the United States Fish Commission in 1871, but it did not flourish. Several other short-lived laboratories were subsequently established in the area around Woods Hole. In 1888 one of these, originally located on Cape Ann, Massachusetts, moved to Woods Hole and opened its doors as the Marine Biological Laboratory, which to this day is one of the world's most prestigious marine stations.

After these early beginnings, other marine laboratories were established. Among the earliest surviving laboratories to be founded in the United States were the Hopkins Marine Station, in Pacific Grove, California; Scripps Institution of Oceanography, in La Jolla, California; and the Friday Harbor Marine Laboratory in Friday Harbor, Washington. In the ensuing years, more laboratories appeared all over the world. These labs played vital roles in the growth of marine biology and remain active to this day.

The onset of World War II had a major effect on the development of marine biology. A new technology, **sonar** or sound **na**vigation **r**anging, was developed in response to the growing importance of submarine warfare. Sonar is based on the detection of underwater echoes—"listening" to the sea (Figure 1-14). The ocean, long thought of as a silent realm, was suddenly found to be full of sound—much of it made by animals! Learning about these animals was no longer the casual pursuit of a few interested marine

Figure 1-12 The Stazione Zoologica at Naples, Italy. Founded in 1872, the Stazione continues to attract marine biologists from around the world.

Figure 1-13 The Marine Biological Laboratory at Woods Hole, Massachusetts. The oldest American marine lab, it continues to be one of the most prestigious in the world.

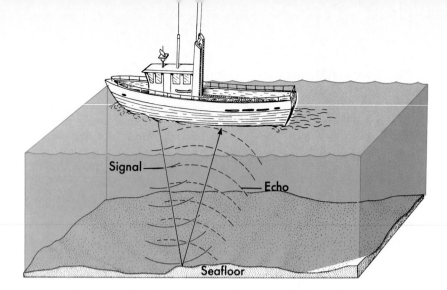

Signal

Echo

Seafloor

Figure 1-14 A ship uses sonar by "pinging," or emitting a loud pulse of sound and timing how long it takes the echo to return from the bottom. From this the depth of the water can be determined.

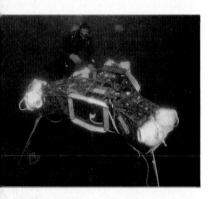

Figure 1-15 Scuba is a most important tool in the research of many marine scientists.

Figure 1-16 These marine scientists are hauling in a net used to capture minute marine organisms. One scientist is signalling instructions to the winch operator.

biologists, but a matter of national concern. As a result of this urgency, several marine laboratories, such as Scripps and Woods Hole, underwent rapid growth. When the war ended, these labs not only remained as vital research centers, but continued to grow.

The years immediately after World War II saw the refinement of the first really practical **scuba,** or self-contained **u**nderwater **b**reathing **a**pparatus. The basic technology used in scuba was developed in occupied France by the engineer Emile Gagnan to allow automobiles to run on compressed natural gas. After the war, Gagnan and fellow Frenchman Jacques Cousteau modified the apparatus, using it to breathe compressed air under water. Using scuba, marine biologists could for the first time descend below the surface of the water to observe marine organisms in their natural environment (Figure 1-15). Furthermore, marine biologists could work comfortably in the ocean, collecting specimens and performing experiments, though they were still limited to relatively shallow water.

Marine Biology Today

Oceanographic ships and shore-based laboratories are as important to marine biology now as they were in the past (Figure 1-16). Today many universities and other institutions operate research vessels. Modern ships are equipped with the latest equipment for navigation, sampling, and studying the creatures that are collected (Figure 1-17). Many of these, like the *Challenger,* were originally built for other uses, but a growing number of vessels are being designed expressly to conduct scientific research at sea.

Since almost everything encountered was previously unknown, the early oceanographic expeditions traveled long distances, often around the world, describing broad geographic patterns and new forms of life. Though new discoveries are still being made, the general picture is now fairly well known. Rather than making long voyages describing general patterns, seagoing scientists today are more likely to focus on specific questions, often in a relatively small geographic area (Figure 1-18). Because of the descriptive

PRINCIPLES OF MARINE SCIENCE

work performed in the early expeditions, modern oceanographers are free to perform detailed studies, conducting experiments and testing hypotheses about the sea.

In addition to ships as we normally think of them, some remarkable craft are used to study the marine world. High-tech submarines can descend to the deepest parts of the ocean, revealing a world that was once inaccessible (Figure 1-19). A variety of odd-looking vessels ply the oceans, providing specialized facilities to marine scientists (Figures 1-20 and 1-21).

Marine laboratories, too, have come a long way since the early days. Today such labs dot coastlines around the world and are used by an international community of scientists (Figure 1-22). All 21 coastal states in the United States have marine labs. Some labs are equipped with the most up-to-date facilities available. Others are simple field stations, providing a base for scientists to work in remote areas. There are even undersea habitats where scientists can live for weeks at a time, literally immersed in their work (Figure 1-23).

In addition to their role in research, marine laboratories have become important centers of education as well. Many labs offer summer courses that allow students to learn about marine biology first hand (Figure 1-24). Most labs provide facilities for graduate students from various universities to pursue their studies, and some offer graduate degrees in marine biology on their own. Thus, in addition to furthering today's research, marine laboratories are busy training the professional marine biologists of tomorrow.

New technology offers exciting opportunities for the study of the oceans. Computers have had a tremendous impact, because they allow scientists to rapidly analyze huge amounts of information (Figures 1-25 and 1-26) Space technology has also aided the study of the sea. Satellites now orbit the earth, peering down at the ocean below. Because they are so far away, these satellites can see wide areas of the ocean all at once, giving a

Figure 1-17 The R/V *Seward Johnson* is a modern research vessel operated by the Harbor Branch Oceanographic Institution in Florida.

Figure 1-18 Marine scientists used a floating enclosure like this to investigate the effects of pollutants on open-water organisms. The experiment, called CEPEX (or Controlled Ecosystem Pollution Experiment), was carried out off the coast of British Columbia, Canada.

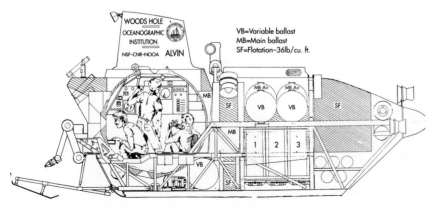

Figure 1-19 Cutaway section of *Alvin,* a deep submersible operated by the Woods Hole Oceanographic Institution. Notice the mechanical arms that are used to collect specimens.

Figure 1-20 R/V *FLIP*, short for "*floating instrument platform*," operated by Scripps Institution of Oceanography, provides a stable platform for research at sea. It is towed into position, then one end of the hull is flooded and swings into position.

Figure 1-21 A research barge operated by marine scientists from the Rosentiel School of Marine and Atmospheric Science of the University of Miami.

Figure 1-22 The Smithsonian Tropical Research Institute's marine laboratory on the Pacific coast of Panamá.

Figure 1-23 This scientist is working inside Hydrolab, one of several undersea habitats that have been used to help marine biologists pursue their studies. These cramped quarters serve as laboratory, living room, bedroom, and kitchen. The "office," however, is much larger: scientists just have to go out the front door to begin their underwater work.

Figure 1-24 A marine biology class at Duke University's marine laboratory in Beaufort, North Carolina.

much broader perspective than is possible at the earth's surface (Figures 1-27 and 1-28). Much of our knowledge of the ocean, especially of large-scale features like ocean current systems, could not have been obtained without the use of computers and satellites. With the aid of computers, scientists are learning to use the information gathered by the satellites to measure the temperature of the sea surface, track ocean currents, and determine the abundance and kinds of organisms present.

Marine biologists today use every available tool in their study of the sea. Information about the ocean pours in at an ever-increasing pace (Figure 1-29). Much remains to be learned, however, and the oceans remain a realm of great mystery and excitement.

Figure 1-25 A marine scientist working at a computer facility.

Figure 1-26 This computer-generated image represents a series of flashes of light from a krill, a shrimplike animal that produces light much like fireflies do. The plot shows the brightness, or intensity, of the light and its color, or wavelength, at different times. A millisecond (msec) equals one one-thousandth of a second: each flash lasted less than a second! This plot could not have been produced without modern computers and light-sensing technology that was originally developed to study distant stars.

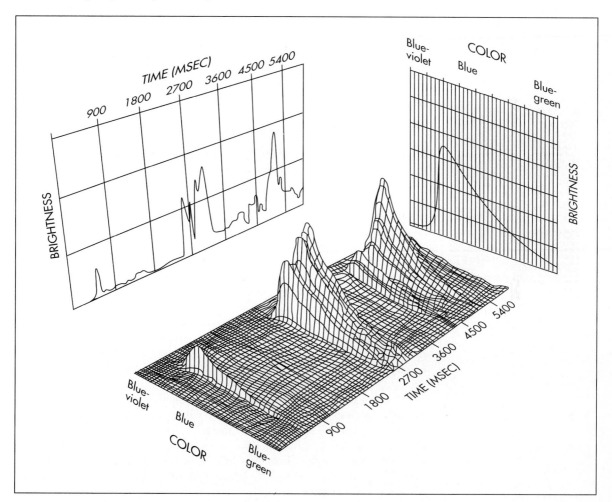

Figure 1-27 The Nimbus-G environmental satellite was the first global monitor of atmospheric pollutants. Its remote sensors obtain continuous information on the oceans, the atmosphere, and our planet's heat balance.

Figure 1-28 A satellite image showing the amount of plant life on earth. This image is a composite of photos taken over a 3-year period. Data for the ocean were obtained from the Coastal Zone Color Scanner (CZCS), which was mounted on the Nimbus-7 satellite. Advances in computer and space technology made this image possible.

Figure 1-29 A marine biologist at work.

THE SCIENTIFIC METHOD

Marine biology is an adventure, to be sure, but it is also a science. Scientists, including marine biologists, share a certain way of looking at the world. Students of marine biology should have some familiarity with this point of view and how it affects our understanding of the natural world, including the ocean.

We live in an age of science. Advertisers constantly boast new "scientific" improvements to their products. Newspapers regularly report new breakthroughs, and many television stations have special science reporters. Governments and private companies spend billions of dollars every year on scientific research and science education. Why has science come to occupy a position of such prestige in our society? The answer, quite simply, is that it works! Science has been among the most successful of human endeavors. Modern society would be impossible without the knowledge and technology produced by science. The lives of almost everyone have been enriched by scientific advances in medicine, agriculture, communication, transportation, and countless other fields.

Much of the practical success of science results from the way it is done. Scientists do not see the world as a place where things happen haphazardly or for no reason. They are convinced instead that all events in the universe can be explained by physical laws. Scientists don't go about discovering these laws haphazardly, either, but proceed according to time-tested procedures. The set of procedures by which scientists learn about the world is known as the **scientific method.**

PRINCIPLES OF MARINE SCIENCE

Scientists sometimes disagree over the fine points of exactly what constitutes the scientific method. Because of this, they often apply the method in slightly different ways. In spite of these minor differences, most scientists do agree on the basic principles of the scientific method. The scientific method should not be seen as a fixed set of rules to be rigidly followed, but rather as a flexible framework that guides the study of nature.

Figure 1-30 The sailfish (*Istiophorus platypterus*).

Observation, the Currency of Science

The goal of science is to discover facts about the natural world and the laws that explain these facts. At the heart of the scientific method is the belief that we can learn about the outside world only through our senses. Scientists concern themselves solely with what they can see, hear, feel, taste, and smell, though they may use tools to extend and improve the senses, similar to the way that microscopes improve vision. In other words, scientific knowledge is based only on what can be directly observed. Anything that cannot be observed is outside the realm of science.

One of the advantages of relying on observations is that these findings are accessible to others. A person's thoughts, feelings, and beliefs are internal. No one really knows what goes on in the mind of another. On the other hand, the world studied by scientists is external to any single person. Many people can look at the same object. Even though sensory perception may be imperfect and the scientist, like anyone else, may not be completely impartial, the object is there for all to see. Thus there is a way to check and verify any one person's observations.

Figure 1-31 The grey reef shark (*Carcharhinus amblyrhynchos*). The vertical bars just behind the mouth are the gill slits, where water exits after passing over the gills.

Two Ways of Thinking

Scientists don't always agree on the best way to do science. In the past there were serious disputes over which methods of scientific reasoning were acceptable. Some people thought that the only truly scientific form of thinking was **induction,** in which one starts with a number of separate observations and then arrives at general principles. Others felt that scientists should use **deduction,** and reason from general principles to specific conclusions. Most scientists now agree that both ways of thinking are indispensable to the scientific method. It is still useful, however, to consider them separately.

Induction. When using induction, a scientist starts by making a series of individual observations. Ideally, he or she has no goal or hunch about the outcome and is completely objective. The combination of these observations suggests a general conclusion. For example, suppose a particular marine biologist examined a sailfish (Figure 1-30), a shark (Figure 1-31), and a tuna (Figure 1-32), and found that they all had gills. Since sailfish, sharks, and tuna are all fishes, he might draw the general conclusion *"All fishes have gills."* This is an example of induction.

Figure 1-32 The bluefin tuna (*Thunnus thynnus*).

In the process of induction, general conclusions are made on the basis of specific observations.

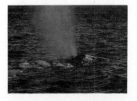

Figure 1-33 The gray whale (*Eschrichtius robustus*) breathes air with lungs, rather than using gills to get oxygen from water as fishes do. The spout is mostly water vapor, like steam, rather than liquid water.

John Steinbeck and Ed Ricketts

Most people know the American writer John Steinbeck as the author of such beloved works as *The Grapes of Wrath*, *Of Mice and Men*, and *East of Eden*. Less well known are Steinbeck's contributions to marine biology, which resulted largely from his close friendship with a man named Ed Ricketts.

Steinbeck and Ricketts first met in 1930, by Steinbeck's account, in a dentist's office in Pacific Grove, California. Steinbeck had a long-standing interest in marine biology and had even taken college courses in the subject at the nearby Hopkins Marine Station. He had wanted to meet Ricketts for some time. Ricketts owned and operated the Pacific Biological Laboratory, located near the Hopkins station and the present site of the Monterey Bay Aquarium. Working out of his laboratory, Ricketts collected specimens of marine life along the Pacific coast and sold them to universities and museums. He was immensely popular in the area and knew more about marine biology than anyone around.

The two men became close friends almost immediately. Before long Steinbeck, then struggling as a writer, was spending a lot of time hanging around his friend's laboratory. He gave his interest in marine biology free rein and soon was helping Ricketts in his business, going on collecting trips and assisting in day-to-day operations. Steinbeck got so involved in this work that he could even get excited about a microscope:

My dream for some time in the future is a research scope with an oil immersion lens, but that costs about 600 dollars and I'm not getting it right now. . . . that research model, Oh boy! Oh boy! Sometime I'll have one. (From Steinbeck, E. and R. Wallsten: *Steinbeck: A Life in Letters*, New York, 1975, Viking Press.)

John Steinbeck would eventually credit Ed Ricketts with shaping his views of man and the world. Steinbeck was so taken with his friend that he included Ricketts in his writing. Characters in at least six of Steinbeck's novels were based on Ricketts. The most famous is "Doc," the main character of *Cannery Row*, who runs the "Western Biological Laboratory":

It sells the lovely animals of the sea, the sponges, tunicates, anemones, the stars and buttlestars [sic], the sunstars, the bivalves, barnacles, the worms and shells, the fabulous and multiform little brothers, the living moving flowers of the sea, nudibranchs and tectibranchs, the spiked and nobbed and needly urchins, the crabs and demi-crabs, the little dragons, the snapping shrimps, and ghost shrimps so transparent that they hardly throw a shadow. . . . You can order anything living from Western Biological and sooner or later you will get it. (From Steinbeck, J.: *Cannery Row*, New York, 1945, Viking Press.)

The scientist must be careful in making inductions. The step from isolated observation to general statement depends on the number and quality of the observations. If the biologist had stopped after examining the sailfish, which happens to have a bill, he might use induction to make the false conclusion *"All fishes have bills."* Even after examining all three fishes, he might have concluded *"All marine animals have gills"* instead of just *"All fishes have gills."* This is where deduction comes into play.

Deduction. In deductive reasoning, scientists start with a general statement about nature and predict what specific consequences would be if

BOX 1-1

The friendship of Ed Ricketts and John Steinbeck was beneficial to marine biology as well as to literature. One long trip to Mexico resulted in their collaboration on *The Sea of Cortez*. This book is partly literature, partly travelogue, and partly a scientific report. It contains lists and descriptions of the marine life the two men encountered on their expedition.

Ed Ricketts' most enduring contribution to marine biology was the publication in 1939 of *Between Pacific Tides*. Written with Jack Calvin, a mutual friend of Ricketts and Steinbeck, *Between Pacific Tides* is a comprehensive guide to the seashore life of the Pacific coast of North America. Revised and updated—it's now in its fifth edition—it is still used by amateurs and professionals alike.

Though Ricketts was an able biologist and was largely responsible for the content of *Between Pacific Tides*, he had difficulty getting his observations and ideas down on paper. Steinbeck almost certainly helped him to write the book. Steinbeck, who was beginning to achieve success as a writer, also helped Ricketts get the book published. At one point, Ricketts felt that the publisher, Stanford University Press, was dragging its feet, and Steinbeck fired off this sarcastic letter:

Gentlemen:

May we withdraw certain selected parts of *Between Pacific Tides* which with the passing years badly need revision. Science advances but Stanford Press does not.

There is the problem also of the impending New Ice Age.

Sometime in the near future we should like to place our order for one (1) copy of the forthcoming (1948, no doubt) publication, The Internal Combustion Engine, Will it Work?

Sincerely,
John Steinbeck
Ed Ricketts

P.S. Good Luck with A Brief Anatomy of the Turtle. (From Steinbeck, E. and R. Wallsten: *Steinbeck: A Life in Letters*, New York, 1975, Viking Press.)

Ed Ricketts was killed in a train accident in 1948. Steinbeck, saddened at the death of his friend, wrote "There died the greatest man I have ever known and the best teacher" (*ibid.*). The memory of Ed Ricketts, and the love of marine biology he inspired, would remain with John Steinbeck for the rest of his life.

Ed Ricketts.

the general statement is true. They might arrive at the general statement by hunch or intuition, but often the statement is the result of induction. Suppose our marine biologist used induction to make the general statement *"All marine animals have gills."* He might then reason that if all marine animals have gills and whales are marine animals, then whales must have gills. The biologist has used a general statement about all marine animals to make a statement about a particular kind of marine animal.

In the process of deduction, specific predictions are made by applying a general principle.

Testing Ideas

Scientists are never content to simply make statements about the world and let it go at that. Instead, they are obsessed with testing the statements to see whether they are, in fact, true. Both induction and deduction lead the scientist to make statements that *might* be true. A statement that might be true is called a **hypothesis.** A crucial feature of the scientific method is that all hypotheses are tested, usually over and over. This insistence on testing is one of the great strengths of the scientific method. Incorrect hypotheses are quickly weeded out and discarded.

Constructing the Hypothesis. Scientific hypotheses must be stated in a way that allows them to be tested. What this means is that it must be possible, at least potentially, to *prove* that the hypothesis is false if it really is false. Sometimes this is simple. For example, the hypothesis that whales have gills is easy to test. All the biologist has to do is to examine a whale to see whether it has gills. By doing so, he would find that whales have lungs, not gills (Figure 1-33)! He would have proven the hypothesis "*Whales have gills*" false. He would also have disproved the more general hypothesis "*All marine animals have gills.*" The steps our marine biologist used to construct and test his hypotheses are illustrated in Figure 1-34.

People sometimes make the mistake of proposing hypotheses that cannot be fairly tested. Someone who believes in mermaids might say "*Somewhere in the ocean there are mermaids.*" The problem with this hypothesis is that it could never be *proven* to be false. An army of marine biologists could spend their lives looking for a mermaid without success, but the believer could always say "*The mermaids are there, you just didn't find them.*" No matter how hard they look, the biologists could never prove that there are no mermaids. Therefore the statement "*There are mermaids in the ocean*" is not a valid scientific hypothesis.

> **A scientific hypothesis is a statement about the world that might be true and is testable. A testable hypothesis is one that at least potentially can be proven false.**

The Nature of Scientific Proof. It must be at least possible to disprove a hypothesis before the hypothesis can be considered scientific. But how can a hypothesis be proven to be true? This question has always troubled scientists, and the answer may trouble you too. In general, no scientific hypothesis can be proven absolutely true. For example, consider the hypothesis that all fish have gills. It is easy to see that this hypothesis can be proven false by finding a fish that lacks gills. But even though every fish so far examined possesses gills, this does not prove that *all* fish have gills. Somewhere out there may lurk a fish without them. Just as it cannot be proven that there are no mermaids, it can never be proven that all fish have gills.

In science, then, there are no absolute truths. Knowing this, the scientist could throw up his hands and look for another line of work. Fortunately, most scientists have learned to accept and deal with the lack of absolute certainty that is inherent in science by making the best of the available evidence. Any scientific hypothesis is examined and tested, poked and prodded, to see if it agrees with actual observations of the world. When a

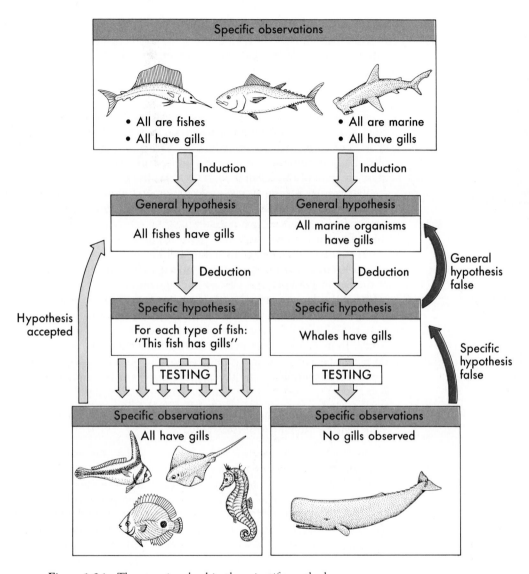

Figure 1-34 The steps involved in the scientific method.

hypothesis withstands all these tests, it is provisionally accepted as "true" in the sense that it is consistent with the information at hand. Scientists speak of *accepting* hypotheses, not *proving* them. They accept the hypothesis that all fish have gills because every attempt to reject it has failed. At least for the time being, the hypothesis works well. The good scientist never quite forgets, however, that any hypothesis, even one of his or her favorites, could suddenly be thrown out the window by new information. No hypothesis in science is exempt from testing, and none is immune to being discarded if it conflicts with the evidence. The bottom line in science is observation of the world, *not* preconceived human ideas or beliefs.

No hypothesis can be scientifically *proven* to be true. Instead, hypotheses are *accepted* for as long as they are supported by the available evidence.

Testing the Hypothesis. Because scientific hypotheses generally can't be proven true, scientists surprisingly spend most of their time trying to disprove, not prove, hypotheses. The basic idea is that more confidence can be placed in a hypothesis that has stood up to hard testing than in an untested one. The role of the scientist is to be a skeptic.

Often scientists are trying to decide among more than one competing hypothesis. After looking at the sailfish, shark, and tuna, our imaginary marine biologist advanced two possible hypotheses: that all fish have gills and that all marine animals have gills. Both hypotheses were consistent with his observations to that point. Examining a whale, he was able to reject the second hypothesis and, in doing so, strengthened the first one. He arrived at the best hypothesis by a process of elimination.

Scientific knowledge is based on observation, and sometimes simple observation is the best way to test a hypothesis. The marine biologist was able to test the hypothesis that all marine animals have gills just by examining a whale. In many cases, however, the conditions required to test a hypothesis do not occur naturally and cannot be merely observed. The scientist must perform an **experiment,** and thereby artificially create a situation to test the hypothesis.

In experiments, scientists create situations to test hypotheses instead of relying on naturally occurring events.

Suppose another marine biologist decides that she wants to find out whether oysters grow faster in warm or cold water. One approach would be to find two places, one warm and one cold, and measure how fast oysters grow at each place. The temperature at any given place will change all the time, however, and the biologist would have difficulty finding two locations where one is always warmer than the other. Even if she does, there might be many differences between the two places besides temperature. The oysters might be of different types at the two places, for example. They might be eating different foods or different amounts of food. There might be pollution or an outbreak of disease at one of the sites. In any natural situation, there will be countless factors other than temperature that might affect how fast the oysters grow. Factors that might affect observations are called **variables.**

Faced with all these variables, the biologist decides to perform an experiment. She collects oysters from one place and divides them at random into two groups. Now she knows that the oysters in the two groups are pretty much the same. She places the two groups in holding tanks, where she can regulate the water temperature with thermostats, and grows one group in warm water, the other in cold. She feeds all the oysters the same amounts of the same food at the same time, protects the oysters from pollution and

PRINCIPLES OF MARINE SCIENCE

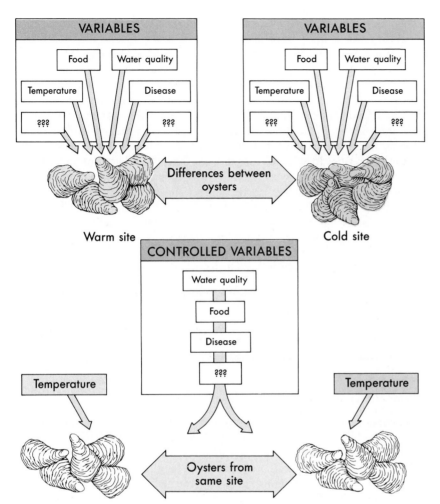

Figure 1-35 The steps involved in the experimental process.

disease, and keeps all the other living conditions exactly the same for both groups. Because all these variables are the same for both groups, the biologist knows that they cannot be responsible for any differences she observes in the growth of the oysters. When a scientist artificially keeps a variable from changing to prevent it from affecting an experiment, the variable is said to be **controlled,** and the experiment is called a **controlled experiment** (Figure 1-35). Since the biologist has controlled the effects of other variables while growing the oysters at different temperatures, she can be more confident that any differences in growth rate between the two groups are due to temperature.

Similarly, the biologist could study how food supply affects oyster growth by keeping the oysters at the same temperature but giving them different amounts of food. Experiments thus allow the effects of different variables to be separated. The way that variables interact can also be studied. The oysters could be maintained in different *combinations* of temperature and food supply, for example, to see whether the temperature at which they grow faster depends on how much food they get.

The Scientific Theory. Many people think of a theory as a rather shaky proposition, and most of us have heard people ridicule some idea or other because it was only theoretic. The public usually reserves such scorn for controversial or unpopular theories. The theory of gravity, for instance, is rarely criticized for being "just a theory." In reality, a scientific theory is on much firmer ground than most people realize. To the scientist, the term **theory** means a hypothesis that has passed so many tests that it is generally regarded as true. In science, the status of theory is a lofty one indeed. Because it has been extensively tested, scientists place considerable confidence in it. They use it as a framework to order their thinking and move on to new discoveries.

It must be remembered, however, that a theory is still a hypothesis, though a well-tested one. As with other hypotheses, theories cannot be absolutely proven, and are accepted as true only as long as they are supported by evidence. Good scientists accept theories for the time being because the best available evidence supports them. They also recognize that any theory may be overturned at any time by new evidence.

> A scientific theory is a hypothesis that has been so extensively tested that it is generally regarded as true. Like any hypothesis, however, it is subject to rejection if enough evidence accumulates against it.

Limitations of the Scientific Method

No human enterprise, including science, is perfect. Just as it is important to understand how and why the scientific method works, it is important to understand the limitations of the scientific method. For one thing, remember that scientists are people too and are prone to the same human shortcomings as anyone else. Scientists may be attached to favorite theories even when confronted with contradictory evidence. Like anyone else, they may let their personal biases affect their thinking. No one can be completely objective all the time. Fortunately, factual errors are usually corrected, because hypotheses are tested not just by one person but by many. The practical success of science is evidence that the self-checking nature of the scientific method does work most of the time.

Science also has some built-in limitations. Ironically, these limitations arise from the same features that give the scientific method its power: the insistence on direct observation and testable hypotheses. This means that science cannot make judgments about values, ethics, or morality. Science can reveal how the world is, but not how it should be. Science cannot dictate what is beautiful. Science can't even tell humanity how to use the knowledge and technology it produces. These things all depend on values, feelings, and beliefs, which are beyond the scope of science.

PRINCIPLES OF MARINE SCIENCE

Do It Yourself

SUMMARY

1. Science deals only with phenomena that can be _____.
2. Reasoning from isolated observations to general principles is called _____. _____ means using general principles to make specific conclusions or predictions.
3. A _____ is a statement about the natural world that might be true. To be considered scientific, the statement must be _____. This means that it must at least be possible to prove the statement _____.
4. In _____, scientists set up conditions in which hypotheses can be tested.
5. Such factors as temperature, food supply, depth, and so on that could have an effect on natural processes are called _____.
6. A _____ is a scientific hypothesis that has survived extensive testing and is generally accepted.

THOUGHT QUESTIONS

1. Nearly all of the major advances in marine biology have come in the last 200 years. What do you think are the reasons for this?
2. In this chapter it was explained that the statement "There are mermaids in the ocean" is not a valid scientific hypothesis. Is the statement "There are *no* mermaids in the ocean" one? Why?
3. Imagine that you are a marine biologist, and you notice that a certain type of crab tends to be considerably larger in a local bay than the same type of crab is in the waters outside the bay. What hypotheses might account for this difference? How would you go about testing these hypotheses?
4. Many species of whale have been hunted to the brink of extinction. Many people think that we do not have the right to kill whales and that all whaling should cease. On the other hand, in many cultures whales have been hunted for centuries and still have great cultural importance. People from such cultures argue that limited whaling should be allowed to continue. What is the role that science can play in deciding who is right? What questions cannot be answered by science?

FOR FURTHER READING

Judge, J.: "Where Columbus Found the New World." *National Geographic Magazine*, vol. 170, no. 5, November 1986, pp. 566-599. *Description of Columbus' voyage and of the island in the Bahamas where he first saw the Americas, both as it was then and today.*

Lenhoff, H.M. and S.G. Lenhoff: "Trembley's Polyps." *Scientific American*, vol. 258, no.4, April 1988, pp. 108-113. *The story of Abraham Trembley, one of the first experimental biologists, and the work he did in the 1740's.*

2 The Sea Floor

*T*he oceans are not simply areas where the land happens to be covered by water. The sea floor is geologically distinct from the continents. It is locked in a perpetual cycle of birth and destruction (Figure 2-1) that shapes the oceans and controls much of the geology and geologic history of the continents. Geologic processes that occur beneath the waters of the sea affect not only marine life, but dry land as well.

The processes that mold ocean basins occur slowly, over hundreds of millions of years. On these time scales, where a human lifetime is but the blink of an eye, solid rocks can flow like liquid, entire continents can move across the face of the earth, and mountains can grow from flat plains. To understand the sea floor, then, we must learn to adopt the unfamiliar point of view of geologic time.

It may seem odd to devote an entire chapter in a marine biology book to geology, but geology is important to the marine biologist. **Habitats,** or places in which organisms live, are directly shaped by geologic processes. The form of coastlines; the depth of the water; whether the bottom is muddy, sandy, or rocky; and many other features of a marine habitat are determined by its geology. Even the history of marine life is related to geology.

Figure 2-1 Underwater volcanic activity at Loihi, a submarine volcano off the coast of the island of Hawaii.

THE WATER PLANET

The presence of large amounts of liquid water makes our planet unique. Most other planets have very little water, and on those that do, the water exists only as perpetually frozen ice or as vapor in the atmosphere. The earth, on the other hand, is very much a water planet. The oceans cover most of the globe and play crucial roles in regulating our climate and atmosphere. Without water, life itself would be impossible.

The Geography of the Ocean Basins

The oceans occupy 71% of the earth's surface. They are not distributed equally with respect to the Equator. About two thirds of the land area is found in the Northern Hemisphere, which is only 61% ocean. The oceans cover about 80% of the Southern Hemisphere.

The oceans are traditionally divided into four large basins (Figures 2-2 and 2-3). The **Pacific** is the deepest and largest ocean, almost as large as all the others combined (Table 2-1). The **Atlantic** Ocean is a little larger than the **Indian** Ocean, but the two are similar in average depth. The **Arctic** is

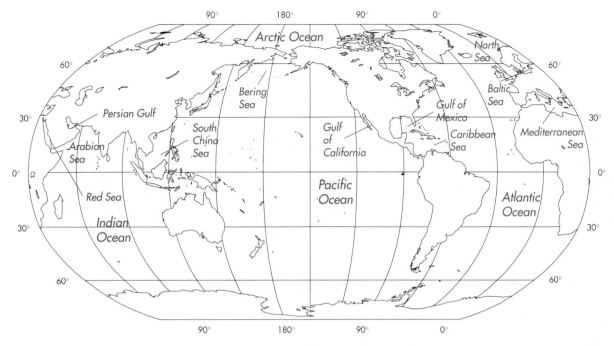

Figure 2-2 The major ocean basins of the world.

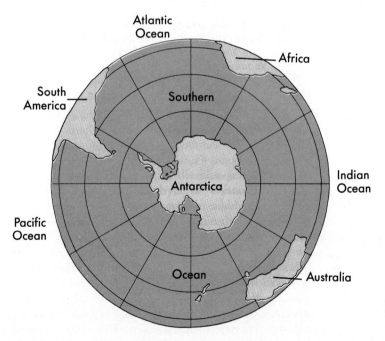

Figure 2-3 South polar view of the world. The major ocean basins can be seen as extensions of one interconnected world ocean. The ocean that surrounds Antarctica is sometimes called the Southern Ocean.

TABLE 2-1

Average depths and total areas of the four major ocean basins

OCEAN	AREA (millions of km²)	AVERAGE DEPTH (m)
Pacific	166.2	4,188
Atlantic	86.5	3,736
Indian	73.4	3,872
Arctic	9.5	1,330

the smallest and shallowest ocean. Connected or marginal to the main ocean basins are a number of shallow seas, such as the Mediterranean Sea, the Gulf of Mexico, and the South China Sea.

Though we usually treat the oceans as four separate entities, they are actually interconnected. This can be seen most easily by looking at a map of the world as seen from the South Pole (see Figure 2-3). In this view it is clear that the Pacific, Atlantic, and Indian oceans are large branches of one vast system. The connections among the major basins allow seawater, materials, and some organisms to move from one ocean to another. Oceanographers often refer to all the oceans together as the **world ocean** to indicate that the oceans are actually one great interconnected system. They may also refer to the continuous body of water that surrounds Antarctica as the Southern Ocean (see Figure 2-3).

The world ocean, which covers 71% of the planet, is divided into four major basins: the Pacific, Atlantic, Indian, and Arctic oceans.

The Structure of the Earth

The earth and the rest of the solar system are thought to have originated about 4.5 billion years ago from a cloud or clouds of dust and gas, the debris remaining from a huge cosmic explosion that is often called the **big bang.** The particles in the cloud were attracted to each other by the force of gravity and began to aggregate into denser clouds within the original cloud. Gravitational attraction was even stronger in these smaller clouds, which caused further condensation into even more compact masses, as well as causing temperature increases. Because of their stronger gravity, these masses continued to attract particles and grow. Eventually they formed the sun and the planets of the solar system.

So much heat was produced as the early earth condensed from the dust cloud that the planet was probably molten. This allowed materials to settle within the planet according to their **density.** Density is the weight, or more correctly the **mass,** of a given volume of a substance. Obviously, a pound of styrofoam weighs more than an ounce of lead, but most people think of lead as being "heavier" than styrofoam. This is because lead weighs more than styrofoam if *equal volumes* of the two are compared. In other words, lead is more dense than styrofoam. The density of a substance is calculated by dividing its mass by its volume. If two substances are mixed, the denser material will tend to sink and the less dense to float.

$$\text{density} = \frac{\text{mass}}{\text{volume}}$$

Density is the mass of a substance per unit volume. Substances of low density will float on substances of higher density.

During the time that the young earth was molten, the densest material tended to flow toward the center of the planet, while lighter materials floated toward the surface. The lighter material at the surface cooled to make a thin crust. Eventually, the atmosphere and oceans began to form. If the earth had settled into an orbit only slightly closer to the sun, the planet would have been too hot and the water would all have evaporated into the atmosphere. With an orbit only slightly farther from the sun, all the water would be perpetually frozen. Fortunately for us, our planet orbits the sun in the narrow zone in which water can exist. Without liquid water, there could be no life on earth.

Internal Structure. The internal structure of the earth reflects the planet's early beginnings. As materials sank or floated according to their density, they formed concentric layers like those of an onion (Figure 2-4). The innermost layer, the **core,** is composed mostly of mixtures, or alloys, of iron. The pressure in the core is more than a million times that at the earth's surface, and the temperature is estimated to be over 4,000° C (7,200° F). The core is composed of a solid inner core and a liquid outer

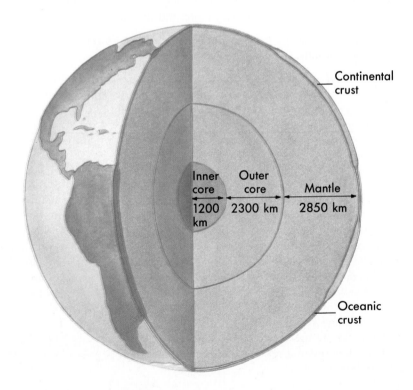

Figure 2-4 The interior of the earth in divided into the core, mantle, and crust. The core is subdivided into the solid inner core and the liquid outer core. The mantle also has several layers, which are not shown here. The thickness of the crust is exaggerated.

TABLE 2-2

Comparison of the characteristics of continental and oceanic crust

OCEANIC CRUST	CONTINENTAL CRUST
Density about 3.0 g/cm^3	Density about 2.7 g/cm^3
Only about 5 km thick	20 to 50 km thick
Geologically young	Can be very old
Dark in color	Light in color
Rich in iron and magnesium	Rich in sodium, potassium, calcium, and aluminum

core. It is thought that the motion of the liquid material in the iron-rich outer core produces the earth's magnetic field.

The layer outside the earth's core is the **mantle.** Though most of the mantle is thought to be solid, it is very hot— near the melting point of rocks. Because of this, the mantle actually flows almost like a liquid, though much slower. Over hundreds of millions of years, the mantle swirls and mixes much like very thick soup heating in a saucepan. This heat-driven motion is called **convection.**

The **crust** is the outermost, and therefore most familiar, layer of the earth. Compared with the deeper layers, it is extremely thin, like a rigid skin floating on top of the mantle. The composition and characteristics of the crust differ greatly between the oceans and the continents.

The earth is composed of three main layers: the iron-rich core, the semiplastic mantle, and the thin outer crust.

Continental and Oceanic Crust. The geologic distinction between ocean and continents is the result of clear-cut differences in the crust itself (Table 2-2). The crust under the oceans is much thinner than that under the continents. Oceanic crust is only about 5 km (3 miles) thick. Under the continents the crust generally varies between 20 to 50 km (10 to 30 miles) in thickness.

Oceanic crust, the sea floor, is made of a mineral called **basalt** that has a dark color. Most continental rocks are of a general type called **granite,** which has a different chemical composition than basalt and is lighter in color. Oceanic crust is denser than continental crust, though both are less dense than the underlying mantle. Oceanic crust has a density of about 3.0 g/cm^3, compared with a density averaging about 2.7 g/cm^3 for continental crust. The continents can be thought of as thick blocks of crust "floating" on the mantle much as icebergs float in water. Oceanic crust floats on the mantle too, but since it is denser it doesn't float as high. This is why the continents lie above sea level and oceanic crust lies below sea level.

Oceanic and continental crusts also differ in geologic age. The oldest oceanic rocks are less than 200 million years old, quite young by geologic standards. Continental rocks, on the other hand, can be very old—as old as 3.8 billion years.

THE ORIGIN AND STRUCTURE OF THE OCEAN BASINS

For centuries people viewed the earth as static and unchanging. Evidence of geologic change was all around, however, from catastrophic earthquakes and volcanic eruptions to the slow erosion of river valleys. People began to see that the face of the earth did indeed change. Today scientists recognize the earth as a world of constant change where even the continents move.

Early Evidence of Continental Drift

As early as 1620 the English philosopher, writer, and statesman Sir Francis Bacon noted that the coasts of the continents on opposite sides of the Atlantic fit together like pieces of a puzzle (Figure 2-5). It was later suggested that the Western Hemisphere might once have been joined to Europe and Africa. Evidence that the continents were once merged slowly accumulated over the centuries. Coal deposits and other geologic formations, for example, match up on opposite sides of the Atlantic. Fossils collected on opposing coasts are also similar.

In 1912 a German geophysicist named Alfred Wegener combined the evidence available at the time and proposed the first detailed hypothesis of **continental drift.** Wegener suggested that all the continents had once been joined in a single "supercontinent," which he called **Pangaea.** He thought

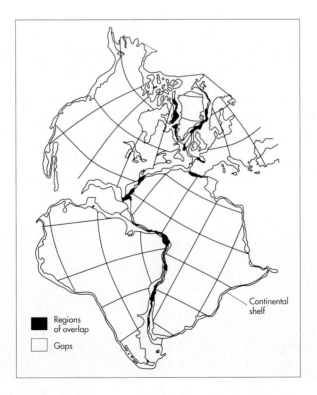

Figure 2-5 When the continents bordering the Atlantic Ocean are fit together like a jigsaw puzzle, their coastlines match up, indicating that the continents were once joined.

that Pangaea began breaking up into the continents we know today about 180 million years ago.

The Theory of Plate Tectonics

Wegener's hypothesis was not widely accepted because he could not supply a plausible mechanism to account for the motion of the continents. We now know that the mechanism he did propose is false. Later proposals of continental drift also failed to provide a workable mechanism, though evidence continued to accumulate. In the late 1950's and 1960's scientists were able to put all the evidence together. They concluded that the continents *did* drift, as part of a process that involves the entire surface of our planet. This process is called **plate tectonics.**

Discovery of the Mid-Ocean Ridge.   In the years after World War II, sonar allowed the first detailed surveys of large areas of the sea floor. These surveys resulted in the discovery of the **mid-ocean ridge** system, a continuous chain of volcanic submarine mountains that encircles the globe like the seams on a baseball (Figures 2-6 and 2-7). The mid-ocean ridge system is the largest geologic feature on the planet. It is interrupted at regular inter-

Figure 2-6   The sea floor.

PRINCIPLES OF MARINE SCIENCE

vals by large geologic faults called **transform faults.** Occasionally the submarine mountains of the ridge rise so high that they break the surface to form islands, such as Iceland and the Azores.

The portion of the mid-ocean ridge in the Atlantic, known as the **Mid-Atlantic Ridge,** runs right down the center of the Atlantic Ocean, closely following the curves of the opposing coastlines. The ridge forms an inverted "**Y**" in the Indian Ocean and runs up the eastern side of the Pacific (see Figure 2-7). The main section of ridge in the eastern Pacific is called the **East Pacific Rise.**

> **The mid-ocean ridge system is a continuous chain of submarine volcanic mountains that runs through all of the ocean basins.**

Surveys of the sea floor also revealed the existence of a system of deep depressions in the sea floor called **trenches** (see Figures 2-6 and 2-7). Trenches are especially common in the Pacific. How they are formed will be discussed later in this chapter (see the section "Sea Floor Spreading and Plate Tectonics").

Significance of the Mid-Ocean Ridge. When the mid-ocean ridge system was discovered, geologists wanted to know how it was formed and began studying it intensively. They found that there is a great deal of geologic activity around the ridge. Earthquakes are clustered at the ridge, for example (Figure 2-8), and volcanoes are common (Figure 2-9).

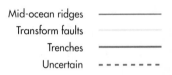

Mid-ocean ridges	——————
Transform faults	——————
Trenches	——————
Uncertain	- - - - - - -

Figure 2-7 The major features of the sea floor. Compare this map to that in Figure 2-6.

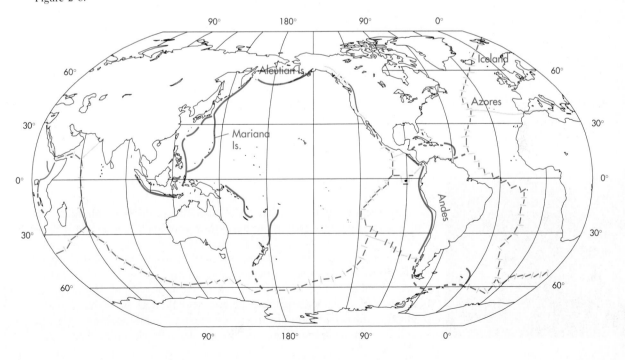

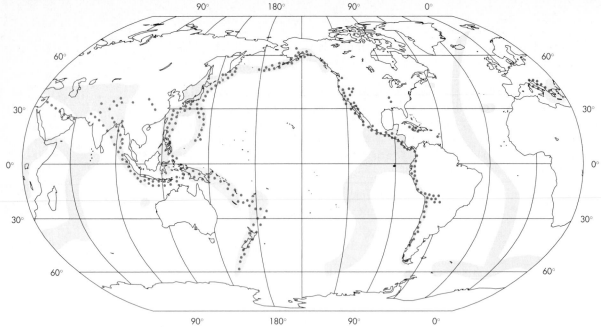

Shallow focus earthquakes ☐
Deep focus earthquakes ☐ (dots)

Figure 2-8 World distribution of earthquakes. Compare the distribution of earthquakes with the locations of mid-ocean ridges and trenches as shown in Figure 2-7.

Figure 2-9 The island of Surtsey was born in 1963 when a volcano on the Mid-Atlantic Ridge southwest of Iceland built up above sea level. Iceland itself is also part of the Mid-Atlantic Ridge.

The characteristics of the sea floor are also related to the mid-ocean ridge. The layer of **sediment,** the loose material such as mud or sand that settles to the bottom, gets thicker as one moves away from the crest of the mid-ocean ridge. Beginning in 1968 a deep-sea drilling ship, the *Glomar Challenger,* obtained samples of the actual sea floor rock (Figure 2-10). It was found that the rocks are older the farther they are from the ridge crest.

One of the most important findings came from studying the magnetism of rocks on the sea floor. Geologists already knew that the earth's magnetic field reverses direction a few times every million years. During magnetic reversal periods, the earth's magnetic north pole is opposite to where it is now. During a reversal, a compass would point toward what is now the South Pole! The cause of these reversals is unknown but is thought to be related to changes in the movement of material in the molten outer core.

Many rocks contain tiny particles of **magnetite,** a magnetic mineral containing iron. When a rock is molten the magnetite particles are free to move and act like tiny compasses, pointing toward the magnetic north pole, wherever it happens to be. Thus the magnetite particles will point in opposite directions during times of normal and reversed magnetic fields. When the rocks cool the magnetite particles are frozen in place, and they maintain their orientation even when the magnetic field changes. It is thus possible to tell what the orientation of the earth's magnetic field was when the rocks cooled. By measuring the magnetization of volcanic rocks on land, it is pos-

PRINCIPLES OF MARINE SCIENCE

sible to determine the earth's magnetic history over the last several hundred million years.

Geologists studying the mid-ocean ridge system discovered a pattern of magnetic bands or "stripes" in the sea floor (Figure 2-11). The stripes run parallel to the mid-ocean ridge and represent zones in which the rocks on the sea floor alternate between normal and reversed magnetization. The bands are symmetric around the ridge, so that the pattern on one side of the ridge is a mirror image of the pattern on the other side. The discovery of these magnetic bands, called **magnetic anomalies,** was important because

Figure 2-10 These geologists are cutting up a core of rock drilled from the deep sea floor by the *Glomar Challenger.*

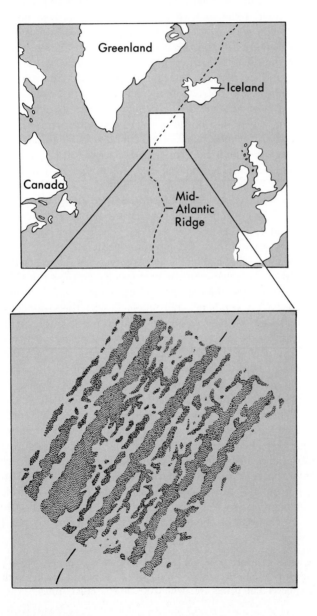

Figure 2-11 Rocks with normal magnetization cooled from molten rocks at a time when the earth's magnetic field was normal, that is, the same as it is today. If the positions of sea floor rocks with normal magnetization (*dark*) and reversed magnetization (*light*) are drawn on a map, they forms bands running parallel to the mid-ocean ridge. Note that the rocks at the ridge crest have normal magnetization.

they offer evidence to dispute the idea that the sea floor was formed all at once as molten material cooled. The pattern of magnetic anomalies suggests that different sections of the sea floor were created at different times, sometimes during periods of normal magnetism and sometimes during reversed magnetism.

Earthquakes and volcanoes are associated with the mid-ocean ridge. The sediments get thicker and the rock of the actual sea floor is older the farther they are from the ridge. Bands of rock alternating between normal and reversed magnetism parallel the ridge.

Creation of the Sea Floor. It was the discovery of magnetic anomalies on the sea floor, together with the other evidence, that finally led to an understanding of plate tectonics. The jump from the various observations of the sea floor and mid-ocean ridges to the theory of plate tectonics is a good example of the use of induction in science.

At the mid-ocean ridges, huge pieces of oceanic crust are separating. As the pieces move apart they create a crack in the crust called a **rift.** When a rift occurs it releases some of the pressure on the underlying mantle, like removing the cap from a bottle of soda pop. Because of the reduced pressure, some of the mantle material melts and rises up through the rift. The ascending mantle material pushes up the oceanic crust around the rift to form the mid-ocean ridge (Figure 2-12). When this molten material reaches the earth's surface it cools and solidifies to form new oceanic crust. The process repeats itself as the sea floor continues to move away from the mid-ocean ridge. The entire process by which the sea floor moves away from the mid-ocean ridges to create new sea floor is called **sea floor spreading,** and the ridges are sometimes called **spreading centers.**

Sea floor spreading explains many observations related to the mid-ocean ridge. Right at the ridge crest the crust is new and has not had time to accumulate a layer of sediment. As the crust moves away from the ridge it ages and sediment builds up. This explains why the sediment gets thicker and the rocks get older as one moves away from the ridge. Sea floor spreading also explains the pattern of magnetic stripes. As new sea floor is created, it freezes the magnetic field prevailing at the time and preserves that magnetization as it moves away from the ridge. Eventually the earth's magnetic field reverses, starting a new stripe.

Sea Floor Spreading and Plate Tectonics. Sea floor spreading is only part of the story of plate tectonics. We now know that the earth's surface is covered by a fairly rigid layer composed of the crust and the uppermost part of the mantle. This layer, approximately 100 km (60 miles) thick, is called the **lithosphere,** meaning "rock sphere." The lithosphere is broken into a number of plates called **lithospheric plates** (Figure 2-13). The plates can contain continental crust, oceanic crust, or both. They float on the rest of the upper mantle, which is denser and more plastic. It is thought that slow, swirling movements of the mantle drive the motion of the lithospheric plates.

The earth's surface is broken up into a number of plates. These plates, composed of the crust and the top part of the mantle, make up the lithosphere. The plates are about 100 km (60 miles) thick.

induction the development of a generalized conclusion from a series of isolated observations
Chapter 1, p. 13

PRINCIPLES OF MARINE SCIENCE

Mid-ocean ridge

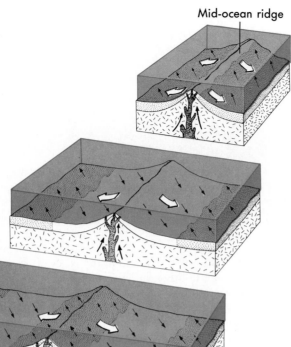

Figure 2-12 Cross-section of the sea floor at the mid-oceanic ridge showing the mechanism of sea-floor spreading. As the sea floor moves away from the rift, molten material rises from the mantle and cools to form new sea floor. When the rocks cool, they "freeze" whatever magnetic orientation, normal or reversed, is present at the time. The entire floor of the ocean was created at the mid-ocean ridge in this manner.

Figure 2-13 The division of the earth's surface into lithospheric plates. Some areas are not fully understood, and there may be more plates than are shown here. Compare this map with those shown in Figures 2-7 to 2-8.

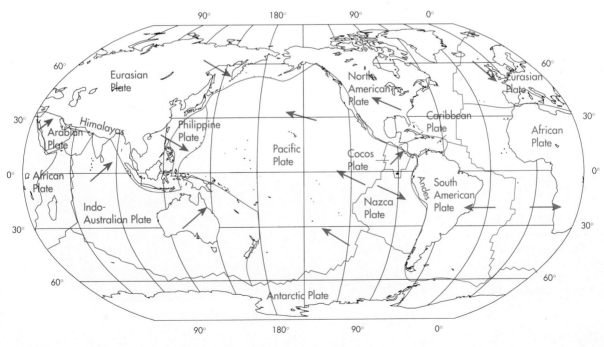

⟶ Direction of plate movement

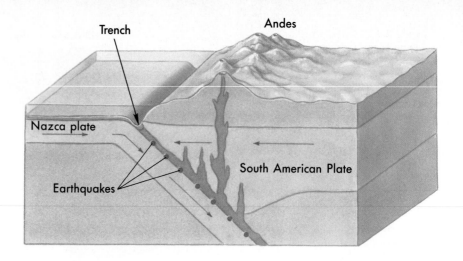

Figure 2-14 Some trenches are formed by the collision of continental and oceanic plates. In this example the Pacific Plate dipped beneath the South American Plate. Earthquakes are produced as the Pacific Plate descends into the mantle. The lighter materials from the sinking plate rise back to the surface as they melt, creating the volcanic Andes Mountains (see Figure 2-6).

The mid-ocean ridges form the edges of many of the plates. It is at the ridges that the lithospheric plates move apart and new sea floor—that is, new oceanic lithosphere—is created by sea floor spreading. If the plate includes a block of continental crust, the continent is carried along with the plate as it moves away from the ridge. This is the mechanism of continental drift. The plates move apart at between 2 and 18 cm (0.8 to 7 inches) per year. For comparison, human fingernails grow at about 6 cm (2.4 inches) per year.

As new lithosphere is created, old lithosphere is destroyed somewhere else. Otherwise the earth would have to constantly expand to accommodate the new lithosphere. Lithosphere is destroyed at trenches, which are another important type of boundary between lithospheric plates. A trench is formed when two plates collide and one of the plates dips below the other and slides back down into the mantle (Figures 2-14 and 2-15). This downward movement of the plate into the mantle is called **subduction.** Because

Figure 2-15 Trenches can also be caused by the collision of two oceanic plates. The oceanic portion of the North American Plate has collided with the Pacific Plate. In this case, the Pacific Plate dipped below the North American Plate, but either of the two plates could have done so. Earthquakes are produced by the descending plate. The volcanoes associated with the trench have produced the Aleutian Island arc (see Figures 2-6 and 2-16).

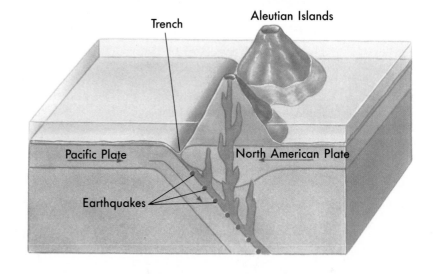

subduction occurs at trenches, trenches are sometimes called **subduction zones.**

As the plate moves down, it weakens under the heat and pressure of the mantle and begins to break up. The movements produced as the plate breaks up cause earthquakes. Eventually the plate gets so hot that it melts. Some of the molten material rises back to the surface to form volcanoes. The rest of the molten rock continues to sink into the mantle. Some of this material may eventually be recycled, rising again at another mid-ocean ridge several hundred million years later.

> **At trenches a lithospheric plate descends into the mantle, where it breaks up and melts. This process, called subduction, produces earthquakes and volcanoes.**

The collision that produces a trench can be either between an oceanic plate and a continent or between two oceanic plates. When an oceanic plate collides with a continent it is always the oceanic plate that descends into the mantle. This is because the continental block is less dense than the oceanic plate and floats on top. This explains why very old rocks are found only on the continents: because oceanic crust is destroyed at trenches, it never gets very old by geologic standards. Continental crust, on the other hand, is not destroyed in trenches and can therefore last for billions of years.

A collision between oceanic and continental plates causes the development of continental volcanoes associated with the trench. These volcanoes may form coastal mountain ranges. A good example of this can be seen on the Pacific coast of South America (Figure 2-14).

When two oceanic plates collide, one of the plates dips beneath the

Figure 2-16 Mt. Veniaminof, an active volcano on the Alaska Peninsula. The Alaskan Peninsula is geologically part of the Aleutian Island chain that is associated with the Aleutian Trench.

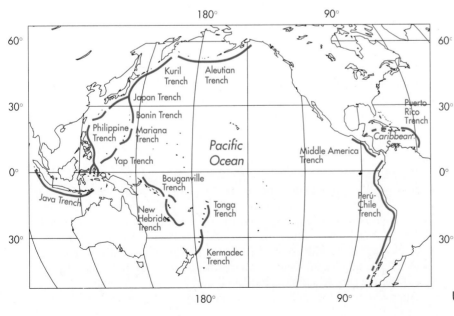

Figure 2-17 Trenches and island arcs in the Pacific Basin and the western Atlantic.

Trenches ————
Uncertain trenches - - - - - -

other to form the trench. Again the trench is associated with earthquakes and volcanoes (Figures 2-15 and 2-16). The volcanoes may rise from the sea floor to create chains of volcanic islands. As viewed on a map, trenches are curved because of the earth's spheric shape. The volcanic island chains associated with the trenches follow the trenches' curvature and are called **island arcs.** Examples include the Aleutian and Mariana islands (Figure 2-17).

Occasionally two continental plates collide. Because of the relatively low density of continental crust, both plates tend to float and neither is subducted. Therefore no trench is formed. Instead, the two continental blocks push against each other with such tremendous force that the two continents become "welded" together. The force is eventually too much for the rocks, which buckle and fold like an accordion. The huge folds form mountain ranges. The Himalaya mountains (Figure 2-18), for example, were formed when India collided with the rest of Asia (see the upcoming section "Continental Drift and the Changing Oceans").

There is a third type of plate boundary in addition to trenches and mid-ocean ridges. Sometimes two plates are moving in such a way that they slide past each other, neither creating nor destroying lithosphere. This type of plate boundary is called a **shear boundary.** In the zone where the two plates are moving past each other, called a **fault,** there is a great deal of friction between the plates. This friction prevents the plates from sliding smoothly. Instead they "lock," and stress builds up until the plates break free and slip all at once, causing an earthquake. The San Andreas Fault in California is the largest and most famous example of a shear boundary (Figures 2-19 and 2-20).

Geologic History of the Earth

We now realize that the earth's surface has undergone dramatic alterations. The continents have been carried long distances by the moving sea floor, and the ocean basins have changed in size and shape. In fact, new oceans have been born. Knowledge of the process of plate tectonics has allowed scientists to reconstruct much of the history of these changes.

Continental Drift and the Changing Oceans. About 200 million years ago all the continents were joined in the supercontinent Pangaea, just as Wegener proposed. Antarctica was in approximately the same place it is today, but all the other continents were in different positions (Figure 2-21, A). India was attached to Antarctica and Africa, rather than to Eurasia as it is now.

Pangaea was surrounded by a single vast ocean called **Panthalassa.** Panthalassa, which covered all of the rest of the planet, was the ancestor of the modern Pacific Ocean. A relatively shallow sea, the **Tethys Sea,** separated Eurasia from Africa. The Tethys Sea, the precursor of the present-day Mediterranean, was home to many of the world's shallow-water organisms. Another indentation in the coast of Pangaea, the **Sinus Borealis,** was to become the Arctic Ocean. Before Pangaea began to break up, there was no sign of the modern Atlantic or Indian oceans.

Approximately 180 million years ago, Pangaea began to break up. A

Figure 2-18 The Himalayas, the world's highest mountain range on land, were formed when India collided with Asia.

PRINCIPLES OF MARINE SCIENCE

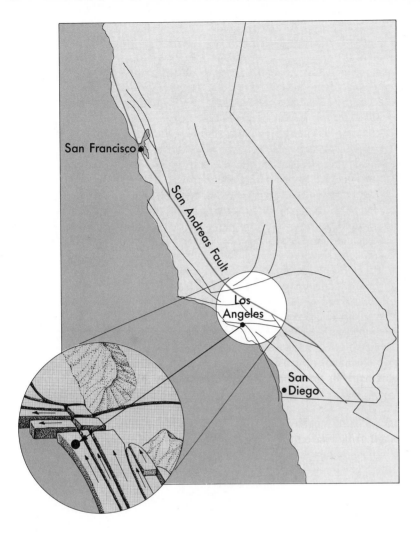

Figure 2-19 California's San Andreas Fault marks the boundary between the Pacific Plate to the west and the North American Plate to the east. The San Andreas Fault is actually part of a complex system of faults that separates the two plates.

Figure 2-20 Aerial view of the San Andreas Fault near Bakersfield, California.

new rift appeared between North America and the combined continents of South America and Africa (Figure 2-21, B). This rift was the beginning of the Mid-Atlantic Ridge, and its formation marked the birth of the northern Atlantic Ocean. Pangaea was now separated into two large continents. One was **Laurasia,** composed of North America and Eurasia. South America, Africa, Antarctica, India, and Australia made up the southern continent of **Gondwana.**

At around the same time another rift split up Gondwana (see Figure 2-21, B) and marked the beginning of the Indian Ocean. South America and Africa began to move to the northeast, and India—separated from the other continents—began to move north.

About 135 million years ago the south Atlantic was born when a new rift between South America and Africa occurred, one that eventually joined the mid-ocean ridge in the north Atlantic to form a single mid-ocean ridge (Figure 2-21, C). As the Atlantic Ocean grew, the Americas were carried away from Eurasia and Africa. To accommodate the new sea floor produced

Figure 2-21 The breakup of Pangaea and the history of continental drift. Dense shading indicates mountains. **(A)** The supercontinent Pangaea as it probably looked 190 million years ago. **(B)** The world 150 million years ago, some 30 million after Pangaea began to break up. The northern Atlantic Ocean was born when one rift separated Laurasia from Gondwana. A second rift that appeared between Africa and Antarctica began the breakup of Gondwana and formed the young Indian Ocean. **(C)** The world 95 million years ago after the southern Atlantic Ocean was born as a new rift between South America and Africa.

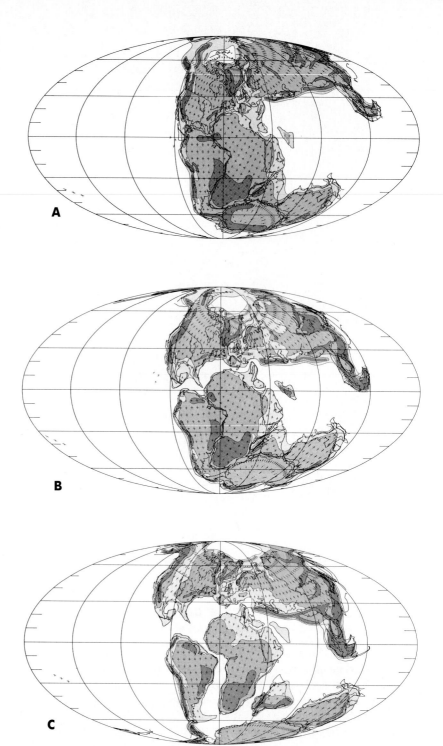

PRINCIPLES OF MARINE SCIENCE

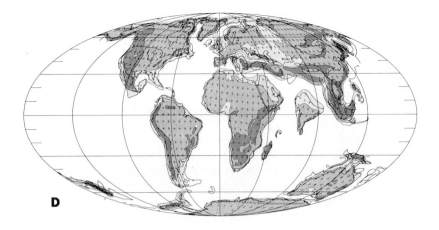

Figure 2-21, cont'd **(D)**
The world 45 million years
ago, 15 million years after
the rift in the Indian
Ocean extended between
Antarctica and Australia,
separating the two conti-
nents. **(E)** The world
about 15 million years ago.

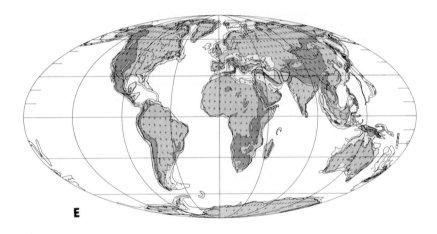

in the Atlantic, the Pacific Ocean—the descendant of Panthalassa—
steadily shrank. The Atlantic is still growing, and the Pacific shrinking.

The Y-shaped ridge that produced the Indian Ocean gradually extended
to separate Australia from Antarctica (Figure 2-21, *D*). The base of the inverted "Y" extended into the African continent, forming the Red Sea,
which is actually a young ocean. India continued to move to the north until
it collided with Asia (Figure 2-21, *E*) to create the Himalayas.

> The continents were once united in a single supercontinent called Pangaea
> that began to break up about 180 million years ago. The continents have
> since moved to their present positions.

The Record in the Sediments. We have already seen how the increase
in sediment thickness moving away from mid-ocean ridges gave a clue to
the workings of plate tectonics. Marine sediments also provide a great deal
of other information about the earth's past. The type of sediment laid down
on the sea floor often reflects the prevailing conditions in the ocean above.

By studying sediments that were deposited in the past, oceanographers have learned a great deal about the history of the world ocean.

Two major types make up most of the sediments found in the sea. The first of these is **lithogenous** sediment, which is derived from the physical and chemical breakdown—or **weathering**—of rocks, mostly on the continents. Coarse sediments, which consist of relatively large particles, tend to sink to the bottom rapidly rather than being carried out to sea by ocean currents (see the section in Chapter 10, "The Shifting Sediments"). Most coarse lithogenous sediments, therefore, are deposited near the edges of the continents. Finer material sinks much more slowly and is carried far out to sea by ocean currents. Some fine material is even transported as dust by wind. The most common kind of lithogenous sediment on the open ocean floor is a fine sediment called **red clay.**

The second major type of marine sediment, **biogenous** sediment, is made up of the skeletons and shells of marine organisms. Some biogenous sediments are composed of the mineral **calcium carbonate ($CaCO_3$).** This type of sediment is called **calcareous ooze.** The other type of biogenous sediment is made of **silica (SiO_2),** which is similar to glass. It is called **siliceous ooze.**

> **The two most abundant types of marine sediment are lithogenous sediment, which comes from the weathering of rocks on land, and biogenous sediment, which is composed of the tiny shells of marine organisms. Biogenous sediment may be made of calcium carbonate or silica.**

Most of the organisms that produce biogenous sediments are microscopic or nearly so. The sediment particles are sometimes called **microfossils,** since each particle represents the preserved remains of a dead organism. Microfossils tell scientists what organisms lived in the ocean in the past. Since some of these organisms are known to prefer cold or warm water, microfossils also give clues to ancient ocean temperatures. Ocean temperatures are determined by the earth's climate and ocean currents.

The climate of the earth in the past can also be determined by the chemical composition of the microfossils. By various methods including **carbon dating,** a procedure in which the relative amounts of different forms of carbon are measured, the age of the microfossils can be determined. It is also possible to tell the temperature of the water in which the organisms lived by studying oxygen atoms and other substances in the microfossils. Microfossils thus give a great deal of information about the earth's past climate.

Climate and Changes in Sea Level.   The climate of the earth has fluctuated through much of its history. During cold periods, or **ice ages,** great glaciers build up on the continents. Since large amounts of water are trapped as ice instead of flowing to the sea in rivers, there is less water in the ocean. Thus sea level falls during periods of glaciation.

The **Pleistocene,** which began a little less than 2 million years ago, was the last major period of glaciation. During the Pleistocene there were a series of ice ages interspersed by warm periods of melting. The peak of the last ice age occurred about 18,000 years ago. At that time, vast ice sheets as thick as 3 km (2 miles) covered much of North America and Europe. Sea level was about 130 m (425 feet) lower than it is today.

Sea level falls during ice ages, because water is trapped in glaciers on the continents. The last major ice age occurred about 18,000 years ago.

Sea level continues to rise, though the rate of melting has slowed in the past 3,000 years. Some scientists think that the natural conditions of the earth would eventually produce another ice age. Human impact on the atmosphere, however, has caused the **greenhouse effect** (see Box 17-1). Global temperatures are expected to increase dramatically, producing a sharp rise in sea level.

THE GEOLOGIC PROVINCES OF THE OCEAN

The structure of the ocean floor is dominated by the workings of plate tectonics. Because plate tectonics is a global process, the major features of the sea floor are quite similar from place to place around the world. The sea floor is divided into two main regions: the **continental margins,** which represent the edges of the continents, and the deep sea floor itself.

Continental Margins

The continental margins are the boundaries between continental crust and oceanic crust. Most of the sediment from the continents settles to the bottom soon after reaching the sea and accumulates on the continental margins. Sediment deposits on the continental margins may be as thick as 10 km (6 miles). Continental margins generally consist of a shallow, gently sloping **continental shelf,** a steeper **continental slope** seaward of the continental shelf, and another gently sloping region—the **continental rise**—at the base of the continental slope. (Figure 2-22).

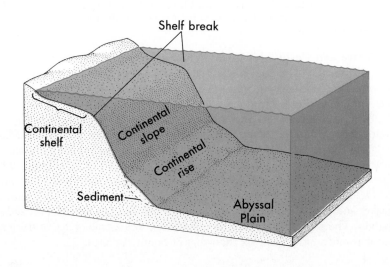

Figure 2-22 A "typical" continental margin consists of a continental shelf, continental slope, and a continental rise. Seaward of the continental rise lies the deep sea floor, or abyssal plain. These basic features vary considerably from place to place.

The Continental Shelf. The shallowest part of the continental margin is the continental shelf. Though they make up only about 8% of the ocean's surface area, continental shelves are the biologically richest part of the ocean, with the most life and the best fishing. The shelf is composed of continental crust and is really just part of the continent that presently happens to be under water. During instances of low sea levels in the past, in fact, most of the continental shelves were exposed. At these times, rivers and glaciers flowed across the continental shelves and eroded deep canyons. When the sea level rose, these canyons were submerged to become **submarine canyons** (Figure 2-23).

The continental shelf extends outward at a gentle slope that in most places is too gradual to see with the naked eye. The shelf varies in width from less than 1 km (0.6 mile) on the Pacific coast of South America and other places to more than 750 km (470 miles) on the Arctic coast of Siberia. The continental shelf ends at the **shelf break,** where the slope abruptly gets steeper. The shelf break usually occurs at depths of 120 to 200 m (400 to 600 feet) but can be as deep as 400 m (1,300 feet).

The Continental Slope. The continental slope is the closest thing to the exact edge of the continent. It begins at the shelf break and descends downward to the deep sea floor. Submarine canyons beginning on the continental shelf cut across the continental slope to its base at 3,000 to 5,000 m (10,000 to 16,500 feet). These canyons channel sediments from the continental shelf to the sea floor.

The Continental Rise. Sediment moving down a submarine canyon accumulates at the canyon's base in a deposit called a **deep-sea fan,** similar to a river delta. Adjacent deep sea fans may merge to form the continental rise. The rise consists of a thick layer of sediment piled up on the sea floor. Sediment may also be carried along the base of the slope by currents, extending the continental rise away from the deep-sea fans.

Continental margins have three main parts. The continental shelf is the submerged part of the continent and is almost flat. The relatively steep continental slope is the actual edge of the continent. The continental rise is formed by sediments building up on the sea floor at the base of the continental slope.

Active and Passive Margins. The nature of the continental margin depends to a large extent on the plate tectonic processes occurring in the region. The continent of South America provides a good example of the relationship between the continental margin and plate tectonics (Figure 2-24). The South American Plate (see Figure 2-13) consists of both the continent itself and the part of the Atlantic sea floor created by the Mid-Atlantic Ridge. South America is carried westward along with the plate as new sea floor is created. The west coast of South America is colliding with the Nazca Plate, leading to the creation of trenches (see Figures 2-14 and 2-17). Trenches are zones of intense geologic activity, including earthquakes and volcanoes, so this type of continental margin is called an **active margin.** The west coast of North America also has a type of active margin, but it is much more complex than that of South America.

PRINCIPLES OF MARINE SCIENCE

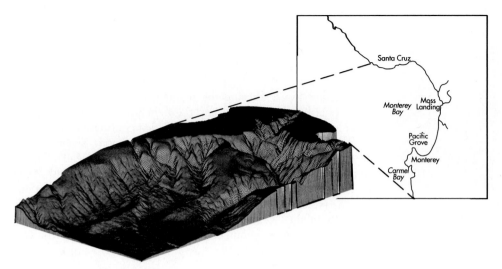

Figure 2-23 The Monterey Canyon originates less than 1 km (0.6 mile) off Moss Landing in Monterey Bay, California, and plummets for about 175 km (110 miles) into the open ocean.

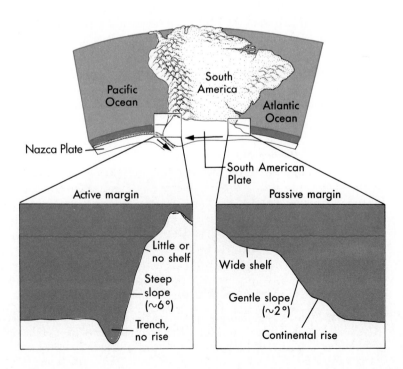

Figure 2-24 The opposite sides of South America have very different continental margins. The "leading edge," or west coast, is colliding with the Nazca Plate. It has a narrow shelf and steep slope, and has a trench rather than a continental rise. On the "trailing edge"—that is, the Atlantic Coast—there is a wide shelf, a relatively gentle continental slope, and a well-developed continental rise. Compare this with the map in Figure 2-7.

Figure 2-25 Steep, rocky shorelines like this one in Monterey Bay, California, are typical of active margins.

As the colliding plate descends into the trench, some of the sediment gets scraped off, folded, and "plastered" onto the continental margin. The edge of the continent is lifted slightly by the oceanic plate passing below, and the coast is built up by volcanoes. These processes give active margins steep, rocky shorelines (Figure 2-25), narrow continental shelves, and steep continental slopes. Because the sediments at the base of the continental slope are either carried down into the trench or scraped onto the continent, active margins usually lack a well-developed continental rise.

South America's east coast, on the other hand, is not a boundary between plates and is therefore relatively inactive geologically. The continental margin here can be thought of as the edge left when South America separated from Africa. This type of margin is called a **passive margin.** Passive margins typically have flat coastal plains (Figure 2-26), wide shelves, and relatively gradual continental slopes. Because there are no tectonic processes to remove it, sediment accumulates at the base of the continental slope. Passive margins therefore usually have a thick continental rise.

> Active continental margins have narrow shelves, steep slopes, and little or no continental rise. Passive margins have wide shelves, relatively gentle slopes, and a well-developed rise.

Deep Ocean Basins

Most of the deep sea floor lies at a depth of 3,000 to 5,000 m (10,000 to 16,500 feet), averaging about 4,000 m (13,000 feet). The sea floor is almost flat and is called the **abyssal plain.** The abyssal plain is not perfectly flat, however, and rises toward the mid-ocean ridges at a very gentle slope of less than 1 degree. The abyssal plain is dotted with submarine volcanoes called **seamounts** and volcanic islands. Distinctive flat-topped seamounts called **guyots** are common in parts of the Pacific Ocean.

Figure 2-26 The Atlantic coast of North America has a passive margin. Because of the lack of geological activity and the buildup of sediments, most of the eastern seaboard has a broad coastal plain, with marshes, lagoons, and estuaries.

At trenches, where the plate descends into the mantle, the sea floor slopes steeply downward. Trenches are the deepest parts of the world ocean. The Mariana Trench in the western Pacific is the deepest place of all—11,022 m (36,163 feet) deep.

The Mid-Ocean Ridge and Hydrothermal Vents

The mid-ocean ridge itself is an environment that is unique in the ocean. As noted previously, the ridge is formed when material rising from the mantle pushes up the oceanic crust. Right at the center of the ridge, however, the plates are pulling apart. This leaves a great gap or depression known as the **central rift valley**, which has been called a "wound in the earth's crust." The floor and sides of the valley are riddled with crevices and fractures. Seawater seeps down through these cracks until it gets heated to very high temperatures by the hot mantle material (Figure 2-27). The heated water then forces its way back up through the crust and emerges in **hydrothermal vents,** or **deep-sea hot springs.**

The water coming from many hydrothermal vents is warm, perhaps 10° to 20° C (50° to 68° F), much warmer than the surrounding water (see the

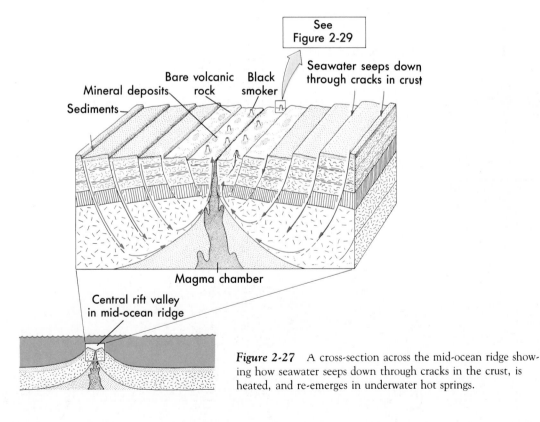

Figure 2-27 A cross-section across the mid-ocean ridge showing how seawater seeps down through cracks in the crust, is heated, and re-emerges in underwater hot springs.

BOX 2-1

"Hot Spots" and the Creation of the Hawaiian Islands

Eruption on Hawaii.

Kauai.

Deep in the lower mantle lie "hot spots," which are sources of hot, molten rock called magma. Hot spots remain stationary under the lithospheric plates moving above. From time to time a plume of hot magma forces its way up through the crust to create an undersea volcano. Sometimes these volcanoes get high enough to become islands.

As the plate moves over the hot spot, each new eruption of magma breaks out at a slightly different place. The result is a linear chain of volcanoes. The Hawaiian Islands are part of one such chain called the Hawaiian Ridge. The Hawaiian Ridge is connected to the Emperor Seamount chain, a string of seamounts stretching to the northwest. Each seamount and island in the chain was produced when magma erupted while the plate was over the hot spot. The volcanoes therefore get older moving away from the hot spot. The island of Hawaii, the youngest of the Hawaiian Islands, began forming less than a million years ago. It is still erupting today, and it grows as the lava from the eruptions solidifies into new rock. Much of the island is smooth, bare volcanic rock, because there has not been enough time

for erosion or for vegetation to grow. Kauai, at more than 5 million years old, is the oldest of the main Hawaiian Islands. It is densely vegetated, and erosion has produced steep, jagged cliffs.

The remaining islands and seamounts of the Hawaiian Ridge and the seamounts of the Emperor chain get progressively older moving to the northwest. For example, Midway Island, about two thirds of the way up the Hawaiian Ridge, is about 25 million years old. The bend that separates the Emperor chain from the Hawaiian Ridge was caused when the Pacific plate changed its direction of movement a little less than 40 million years ago. Meiji Seamount, at the northernmost end of the chain, is 70 million years old.

The formation of the Hawaiian Islands by a hot spot provides the solution to a mystery that once puzzled biologists. The Hawaiian Islands contain more than 800 different kinds of fruit flies, most of which are found nowhere else. And yet, the islands are geologically very young. How could so many different kinds of fruit fly have developed in such a short time?

The answer is that they didn't. The fruit flies have been around a lot longer than today's Hawaiian Islands. They once lived on older islands in the chain, most of which have sunk below sea level and are now seamounts. The flies traveled to the new islands that appeared over the hot spot.

A new Hawaiian island may be on the way. A young submarine volcano called Loihi has been found southeast of the island of Hawaii. It will probably be several million years before Loihi breaks the surface, so don't get your surfboards ready yet.

The Emperor Seamounts and the Hawaiian Island chain.

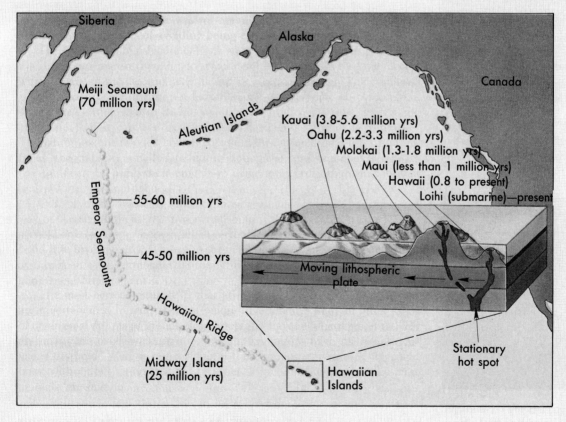

Siberia

Alaska

Canada

Meiji Seamount
(70 million yrs)

Aleutian Islands

Kauai (3.8-5.6 million yrs)
Oahu (2.2-3.3 million yrs)
Molokai (1.3-1.8 million yrs)
Maui (less than 1 million yrs)
Hawaii (0.8 to present)
Loihi (submarine)—present

Emperor Seamounts

55-60 million yrs

45-50 million yrs

Moving lithospheric plate

Hawaiian Ridge

Midway Island
(25 million yrs)

Hawaiian Islands

Stationary hot spot

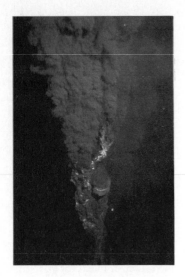

Figure 2-28 "Black smokers" are commonly found in hydrothermal vent areas. This black smoker is at a depth of 2,600 m (8,500 feet) in the East Pacific Rise.

section in Chapter 3, "The Three-Layer Ocean"). At some vents, however, the water is blisteringly hot, up to 350° C (662° F). The water is so hot that when scientists first tried to measure its temperature the thermometer they were using started to melt! To take accurate readings, they had to return with a specially-designed thermometer.

As the hot water seeps through cracks in the earth's crust, it dissolves a variety of minerals, mainly those known as **sulfides.** When the mineral-laden hot water emerges at the vent it mixes with the surrounding cold water and is rapidly cooled. This causes many of the minerals to solidify, forming mineral deposits around the vents (see the section in Chapter 16, "Ocean Mining"). **Black smokers** (Figures 2-28 and 2-29) are one type of mineral deposit found at hydrothermal vents. These are chimney-like structures that progressively build up around a vent as the minerals solidify. The "smoke" is actually a dense cloud of mineral particles.

Deep-sea hot springs are of great interest not only to geologists, but to biologists. One of the most exciting developments *ever* in marine biology has been the discovery of unexpectedly rich marine life around hydrothermal vents. These organisms are discussed in the section in Chapter 15 titled "Deep-Sea Hot Springs."

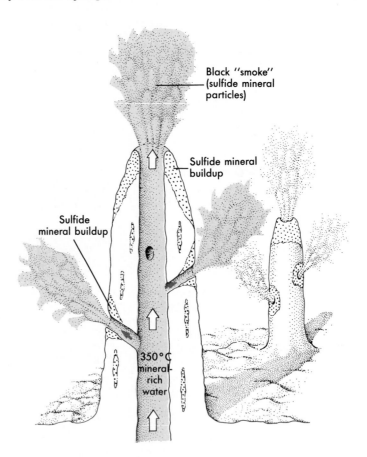

Black "smoke" (sulfide mineral particles)

Sulfide mineral buildup

Sulfide mineral buildup

350°C mineral-rich water

Figure 2-29 A cross-section of a black smoker illustrating how it builds up.

PRINCIPLES OF MARINE SCIENCE

SUMMARY

1. The major basins of the world ocean are the _____, _____, _____, and _____. The ocean around Antarctica is sometimes called the _____.
2. The _____ of a substance is calculated by dividing its mass by its volume.
3. The three main divisions of the earth's interior are the _____, the _____, and the crust.
4. A long chain of submarine volcanic mountains that runs through the world ocean is called the _____ system.
5. Sea floor is created at _____ and destroyed at _____.
6. In the process of _____ a lithospheric plate descends into the _____.
7. A single supercontinent called _____ began to break up into the present day continents about _____ years ago.
8. The two most abundant types of marine sediment are _____ sediment, which comes from rocks, and _____ sediment, which is produced by organisms.
9. The continental margins consist of the _____, the continental slope, and the _____.
10. Most of the deep sea floor, or _____, is between _____ and _____ meters deep.
11. Deep-sea hot springs, or _____, are places at the mid-ocean ridge from which _____ emerges.

THOUGHT QUESTIONS

1. The process of plate tectonics is occurring today in the same way as in the past. Can you project the future positions of the continents by looking at a map of their present positions and the positions of the mid-ocean ridges (Figure 2-7)?
2. Why are most oceanic trenches found in the Pacific Ocean?
3. Scientists who study forms of marine life that lived more than approximately 200 million years ago usually have to obtain fossils not from the sea floor, but from areas that were once undersea and have been uplifted onto the continents. Why do you think this is?
4. What are some of the major pieces of evidence for the theory of plate tectonics? How does the theory explain these observations?

FOR FURTHER READING

Gore, R. and J.A. Sugar: "Our Restless Planet Earth." *National Geographic Magazine*, vol. 168, no. 2, August 1985, pp. 142-181. *A description of plate tectonics and continental drift and some of the geological and biological implications of these processes.*

Macdonald, K.C. and P.J. Fox: "The Mid-Ocean Ridge." *Scientific American*, vol. 262, no. 6, June 1990, pp. 72-79. *Description of new findings about the mid-ocean ridge system discovered with high-tech sonar.*

Propp, M.V. and V.G. Tarasov: "Caldron in the Sea." *Natural History*, August 1989, pp. 28-33. *The Soviet authors describe the marine life around the hot springs of a partially submerged volcanic crater in the Kuril Islands, which lie northeast of Japan.*

Chemical and Physical Features of the World Ocean 3

"E" verybody talks about the weather, but nobody does anything about it." Mark Twain's famous quotation expresses the plight of the organisms that live in the sea as well it does that of human beings. The mussels in Figure 3-1 are being buffeted by waves that were generated by a distant storm. They may be immersed in seawater at high tide or exposed to the air at low tide. The saltiness or temperature of the water may vary with the season or with fluctuations in rainfall and ocean currents. From the point of view of the things that live in the sea, wind and waves, tides and currents, temperature and salt all make up the "weather." All are beyond the control of the mussels or any other marine organism.

Since marine organisms can't control the physical and chemical nature of their environment, they simply have to "grin and bear it;" that is, they have to *adapt* to the place where they live, or live elsewhere. The types of organisms found at a given place in the ocean and the way those organisms live is controlled to a large extent by chemical and physical factors. To understand the biology of marine organisms, therefore, we must know something about the environment in which they live. This chapter describes the chemistry and physics of the oceans in relation to life in the sea.

THE WATERS OF THE OCEAN

Everyone knows that the ocean is full of water. Most people think of water as commonplace, because there is so much of it around. From a cosmic perspective, though, water is not so common after all. Earth is the only known planet that has liquid water on its surface.

Even so, most of us never give water a second thought—unless we happen to be hot or thirsty. Water quenches our thirst because it makes up most of our bodies. Marine organisms, too, are mostly water—80% or more in most cases, in jellyfishes over 95%. Water not only fills the ocean, it makes life itself possible.

Figure 3-1 Blue mussels (*Mytilus edulis*) being buffeted by storm waves.

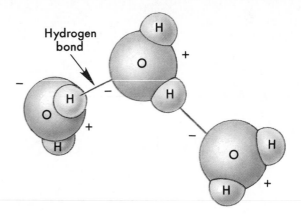

Hydrogen
bond

Figure 3-2 The two ends of water molecules have different electrical charges. The oxygen end of the molecule has a weak negative charge; the hydrogen end, a slight positive charge. Opposite charges attract each other much as the opposite poles of a magnet attract each other, so the oxygen end of one molecule is attracted to the hydrogen ends of neighboring molecules. These weak attractions among water molecules are known as hydrogen bonds.

The Unique Nature of Pure Water

All matter is made of tiny particles called **atoms.** The basic **elements,** of which 104 are thus far known, are composed entirely of individual atoms of the same kind. In all other substances the atoms are combined into large particles called **molecules.** Water molecules are made up of one oxygen atom, which is relatively large, combined with two small hydrogen atoms. Water molecules have an unusual structure that causes them to "stick" together, or more properly, to attract each other (Figure 3-2). The attractions between liquid water molecules are called **hydrogen bonds.** Hydrogen bonds are not very strong, but they make water different from any other substance on earth.

The States of Water. Any substance can exist in three different physical forms, called **states** or **phases.** A substance can occur as a solid, a liquid, or a gas. Water is the only substance that naturally occurs in all three phases on earth.

The molecules in liquid water move constantly. Hydrogen bonds hold small groups of molecules together (Figure 3-3), but the bonds are weak and the groups continually form and break apart. Any particular molecule in liquid water spends most of its time as part of a group, connected to other molecules by hydrogen bonds.

Temperature is a reflection of the average speed of the molecules—the faster they move, the higher the temperature. If a water molecule moves fast enough, it may break free of all the hydrogen bonds. In this process, called **evaporation,** the molecules go from the liquid phase to the gaseous or vapor phase. The molecules in water vapor are not held together by hydrogen bonds. The molecules are separate and are much farther apart than in the liquid (see Figure 3-3). As the temperature rises so does the rate of evaporation, since the molecules move faster and more of them escape the hydrogen bonds. If water gets hot enough, nearly all of the hydrogen bonds are broken and the molecules try to enter the vapor state all at once; in other words, the water boils.

When liquid water cools, the molecules not only move slower, they pack closer together, so the volume of the water decreases. Since the volume decreases without changing the weight, the water becomes denser as it cools. Fresh water gets denser as the water gets colder down to 4° C (39° F),

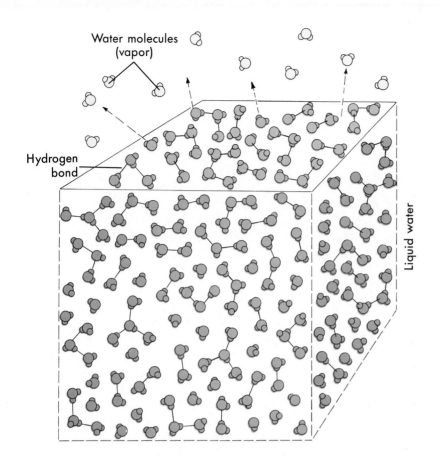

Water molecules (vapor)

Hydrogen bond

Liquid water

Figure 3-3 The molecules in liquid water form groups of various sizes held together by hydrogen bonds. The molecules are moving too rapidly to be held permanently in place, so the groups constantly break up and re-form. Evaporation occurs when molecules break free of all hydrogen bonds and enter the gaseous state. Molecules of water vapor are much farther apart than in the liquid state.

at which point it starts getting less dense. Seawater just keeps getting denser as it gets colder.

Seawater gets denser as it gets colder.

As the water continues to cool the molecules move slower and more weakly. Eventually the hydrogen bonds overcome the motion of the molecules. The bonds hold the molecules in a fixed three-dimensional pattern (Figure 3-4), and the water changes from the liquid to the solid state. Solids that consist of such regular patterns of molecules are called **crystals.** The solid form of water, of course, is ice.

In ice crystals, the water molecules are spaced farther apart than in liquid water (see Figure 3-4). This means that water expands as it freezes, as anyone who has made ice cubes knows. Since the same weight of water occupies more volume in ice than in liquid water, ice is less dense than liquid water. That is why ice floats.

It is extremely unusual for a substance's solid phase to be less dense than its liquid phase. This unique characteristic of water has important implications for aquatic organisms, both marine and freshwater. Because ice floats, it acts as an insulating "blanket" that helps keep the water below from rapidly cooling off in cold weather. Organisms can live below surface ice. If ice were denser than liquid water, it would sink. In cold regions, ice would

density   the weight of a given volume of a substance.
Density = weight/volume
Chapter 2, p. 24

Figure 3-4 The molecular structure of water changes with temperature. In ice, hydrogen bonds hold the molecules in a hexagonal, or six-sided, pattern. As heat is added, the ice warms up and the molecules vibrate more rapidly until they break free of the crystal structure. When this happens, the ice melts. While the ice is melting, heat that is added is absorbed by breaking hydrogen bonds, not by increasing the temperature. When the ice is completely melted, additional heat again causes the temperature to rise. Some of the molecules gain enough speed to break free of all the bonds and evaporate. At 100° C (212° F) nearly all the hydrogen bonds are broken and the water boils.

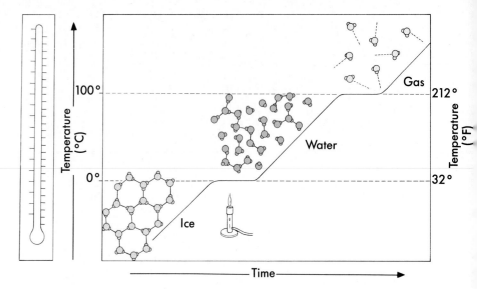

build up at the bottom and new ice would form at the surface. Lakes and some parts of the ocean would freeze solidly, and only a thin surface layer would melt in the summer. Lakes and the ocean would be much less hospitable to life.

Heat and Water. It takes a large amount of heat to melt ice. Water molecules in ice vibrate but do not move from place to place within the crystal. As heat energy is added and the temperature of the ice rises, the molecules vibrate faster, eventually breaking some of the hydrogen bonds that hold the crystal together (see Figure 3-4). The molecules then start to move around as well as vibrate. As a result, the ice begins to melt. Because hydrogen bonds must be broken, ice melts—or, moving in the opposite direction, freezes—at a much higher temperature than similar substances that do not form hydrogen bonds. If not for the hydrogen bonds, ice would melt at about −90° C (−130° F) instead of 0° C (32° F)!

Not only does ice melt at a comparatively high temperature, it absorbs a lot of heat in the process of melting. The amount of heat required to melt a substance is called the **latent heat of melting.** Water has a higher latent heat of melting than any other commonly occurring substance. The reverse is also true: a great deal of heat must be removed from liquid water to freeze it. It thus takes a long period of very cold weather before a body of water will freeze.

Once ice begins to melt, added heat energy goes into breaking more hydrogen bonds rather than increasing the speed of molecular motion. As long as there is some ice left to be melted, the temperature of the ice-water mixture remains at a constant 0° C (32° F). This is why ice works so well as a refrigerant in a drink or insulated cooler: heat that enters the glass or cooler goes into melting the ice, not into raising the temperature.

Once all the ice has melted, adding heat causes the molecules to move more rapidly, thereby raising the temperature. Some of the heat energy,

PRINCIPLES OF MARINE SCIENCE

however, goes into breaking the hydrogen bonds that hold groups of molecules together. Since not all the energy goes into increasing the speed of molecular motion, it takes a large amount of heat to raise the temperature. Thus the ability of water to hold heat energy, or its **heat capacity,** is high. In fact, water has one of the highest heat capacities of any naturally occurring substance. Water is used in the cooling systems of cars for this reason: It can absorb a lot of heat without greatly increasing in temperature. The high heat capacity of water also means that marine organisms are not subjected to the rapid and sometimes drastic changes in temperature that are experienced by organisms on land (Figure 3-5).

Figure 3-5 In the heat of the sun at low tide, this seaweed (*Hermosira banksii*) is beginning to shrivel. Intertidal organisms are subjected to much more extreme temperatures than organisms that are always submerged.

Water also absorbs a great deal of heat in the process of evaporating, that is, it has a high **latent heat of evaporation.** Again this is due to hydrogen bonding: only the fastest moving molecules, those with the most energy, can break free of the hydrogen bonds and enter the gaseous phase. Because the fastest molecules leave the liquid phase, those left behind have a lower average velocity and therefore a lower temperature. This is known as **evaporative cooling.** Our bodies take advantage of evaporative cooling: When perspiration evaporates, it cools our skin.

> Water has the highest latent heats of melting and evaporation and one of the highest heat capacities of any natural substance.

Water as a Solvent. Because it can dissolve more things than any other naturally occurring substance, water is often called the "universal solvent." Water is especially good at dissolving a general class of molecules called **salts.** Salts are made of combinations of particles that have opposite electrical charges. Such electrically charged particles, which can be either single atoms or groups of atoms, are known as **ions.** For example, ordinary table salt—or sodium chloride (NaCl)—is composed of a positively charged sodium ion (Na^+) combined with a negatively charged chloride ion (Cl^-). Recall that hydrogen bonds form because the hydrogen end of the water molecule has a slight positive charge and is attracted to the slightly negative oxygen end of other water molecules. Ions have much stronger electric charges than the opposite ends of the water molecule, so the electrical attraction between ions is much stronger than the hydrogen bonds that form between water molecules. If no water is present, the ions bind strongly together to form salt crystals (Figure 3-6).

When a salt crystal is placed in water, the strongly charged ions attract the water molecules—with their weak charges—much like iron filings are attracted to a magnet. A layer of water molecules surrounds each ion, insulating it from the surrounding ions (Figure 3-7). This greatly weakens the electrical attractions that hold the salt crystal together. The crystal pulls apart, or **dissociates,** and the salt dissolves.

Seawater

The characteristics of seawater are due both to the nature of pure water and to the materials dissolved in it. The solids dissolved in seawater come from two main sources. Some are produced by the chemical weathering of rocks on land and are carried to the sea by rivers. Other materials come from the

weathering the physical or chemical breakdown of rocks
Chapter 2, p. 40

SALT CRYSTAL

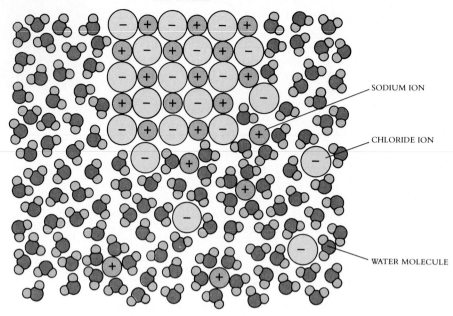

SODIUM ION

CHLORIDE ION

WATER MOLECULE

Figure 3-6 The arrangement of ions in a crystal of sodium chloride. Since the charges on the ions are much stronger than the charges on a water molecule, the bonds between ions are much stronger than hydrogen bonds.

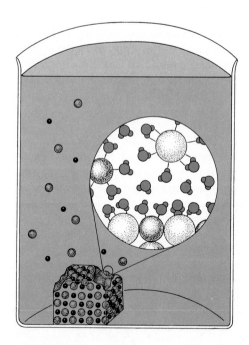

Figure 3-7 Because of their slight electric charges, water molecules cluster around the strongly charged ions in salt. This weakens the bonds between the ions, which then separate—or dissociate.

PRINCIPLES OF MARINE SCIENCE

TABLE 3-1

The composition of seawater of 35‰ salinity

ION	CONCENTRATION ‰	PERCENTAGE OF TOTAL SALINITY
Chloride (Cl^-)	19.345	55.03
Sodium (Na^+)	10.752	30.59
Sulfate (SO_4^{-2})	2.701	7.68
Magnesium (Mg^{+2})	1.295	3.68
Calcium (Ca^{+2})	0.416	1.18
Potassium (K^+)	0.390	1.11
Bicarbonate (HCO_3^-)	0.145	0.41
Bromide (Br^-)	0.066	0.19
Borate ($H_2BO_3^-$)	0.027	0.08
Strontium (Sr^{+2})	0.013	0.04
Fluoride (F^-)	0.001	0.003
Everything else	<0.001	<0.001

earth's interior and are released directly into the ocean at hydrothermal vents.

Salt Composition. Seawater contains at least a little of almost everything, but most of the **solutes**—or dissolved materials—are made up of a surprisingly small group of ions. In fact, only six ions compose over 98% of the solids in seawater (Table 3-1). Sodium and chloride account for about 85% of the solids, which is why seawater tastes like table salt.

If seawater is evaporated, the ions in it are left behind and combine to form various salts. **Salinity** is defined as the total amount of salt dissolved in seawater. Salinity is usually expressed as the number of grams of salt left behind when 1000 grams of seawater are evaporated. If we evaporated 1000 grams of seawater and were left with 35 grams of salt, we would say the seawater had a salinity of 35 parts per thousand, or 35‰. The salinity of water strongly affects the organisms that live in it. Most marine organisms, for instance, die if they are placed in fresh water. Even slight changes in salinity will harm some organisms. Many have specific mechanisms that allow them to cope with changes in salinity (see the section in Chapter 4, "Regulation of Salt and Water Balance").

Organisms may be affected not only by the total amount of salt, but by the kinds of ions in the water. The composition of the ions in seawater can be determined by analyzing the salts left after evaporation. In fact, that is how the information in Table 3-1 was obtained.

The chemist William Dittmar analyzed 77 seawater samples brought back by the *Challenger* expedition and found that the relative amounts of the various ions were constant. Thus the percentage accounted for by each ion is always the same, although the total amount of solid material dissolved in seawater varies from place to place. Chloride ion, for example, always makes up 55.03% of however much salt is present. This principle is called the **rule of constant proportions.**

> The rule of constant proportions says that the relative amounts of the various ions in seawater are always the same.

hydrothermal vents undersea "hot springs" associated with mid-ocean ridges
Chapter 2, p. 45

The Challenger expedition (1872-1876) marked the birth of modern oceanography
Chapter 1, p. 5

Figure 3-8 Not all the ions in seawater enter the ocean at the same place. Positive ions such as sodium and magnesium mostly come from the weathering of rocks and are carried to the sea by rivers. Negative ions such as chloride and sulfate enter the ocean mainly at hydrothermal vents. If the ocean was not thoroughly mixed, coastal water would have a relatively high proportion of sodium and magnesium. Deeper water, influenced by hydrothermal input, would be rich in chloride and sulfate. Actually, the proportions of ions *do* vary right near river mouths and hydrothermal vents, but most of the ocean is well mixed and the rule of constant proportions holds.

The rule of constant proportions has important implications. For one thing, it means that though marine organisms may be exposed to changes in total salinity, they do not have to deal with changes in the ratios of the various ions. This makes it much easier for them to control their internal salt and water balance. The rule of constant proportions also indicates that the oceans are thoroughly mixed (Figure 3-8), and it implies that ocean salinity varies only as a result of the addition or removal of pure water, not the addition or removal of salt. If salinity varied by adding or removing any particular salt, then the relative amounts of the ions in the seawater would change. If magnesium chloride ($MgCl_2$) were added, for example, then the proportion of magnesium in the water would go up.

Changes in ocean salinity, then, are controlled by the addition and removal of pure water. Water is added to the ocean by **precipitation**—rain and snow. This includes not only direct precipitation into the ocean but precipitation on land that is carried to the ocean by rivers. Water is removed from the ocean primarily by evaporation, but it may also be removed by freezing. When salt water freezes the ions are not included in the form-

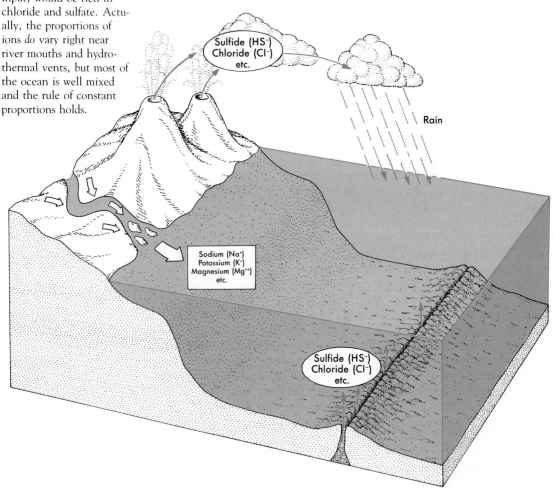

ing ice. They are left behind in the unfrozen water, increasing its salinity, and the ice is made of almost pure water.

The average salinity of the ocean is about 35‰. The open ocean varies relatively little, between about 33‰ and 37‰, depending mostly on the balance between evaporation and precipitation. Partially enclosed seas may have much more extreme salinities. The Red Sea, for instance, is in a hot, dry region where evaporation predominates over precipitation. As a result the Red Sea is very salty, about 40‰. Near coasts or in enclosed basins, runoff from rivers may have a strong effect on salinity. The Baltic Sea gets a lot of river runoff, for example, and has a typical salinity of only about 7‰ at the surface.

Figure 3-9 The Nansen bottle is one type of container that is used to collect samples of seawater from various depths in the ocean; it is fitted with a rack of reversing thermometers to record water temperatures.

Salinity, Temperature, and Density.   We have already seen that temperature greatly affects the density of water. Salinity also influences the density of seawater: The saltier the water, the denser it is. The density of seawater therefore depends on *both* the temperature and the salinity of the water.

> The temperature and salinity of seawater determine its density: it gets
> denser as it gets saltier and colder.

Temperature in the open ocean varies between about −2° and +30° C (28° to 86° F). Temperatures below 0° C (32° F) are possible because salt water freezes at a colder temperature than pure water. This is one reason that the ocean is less prone to freezing than are lakes and rivers.

Temperature in the ocean varies considerably more than salinity, so as a practical matter density is controlled more by temperature than salinity. Exceptions do occur, however, and both the temperature and salinity of ocean water still need to be measured to determine the density.

Temperature and salinity can be measured by lowering specially designed bottles fitted with thermometers into the ocean to collect water samples that are analyzed to determine salinity (Figures 3-9 and 3-10). Ordinary thermometers cannot be used to measure the temperature at depth, since the temperature reading would change as the thermometer was brought back to the surface. To overcome this problem, oceanographers use special **reversing thermometers** to measure temperature in the ocean (see Figure 3-9). Reversing thermometers are made so that the mercury column breaks

Figure 3-10 Niskin bottles, another type of sampling bottle, are clamped to a cable and lowered to the desired depth with both ends open. A weight called a messenger sent down from the surface causes the spring-loaded end caps to snap shut, trapping a sample of seawater in the bottle.

when the thermometer is turned upside down, preventing the temperature reading from changing thereafter. The thermometers are mounted onto spring-loaded racks that flip the thermometers over when a trigger is activated. A weight called a **messenger** slides down the wire to trigger the thermometer rack.

A series of bottles can be attached to the cable, allowing the user to determine temperature and salinity, and therefore density, at several depths at the same time (Figure 3-11). A graph that shows the temperature at different depths in the ocean is called a temperature **profile.** A temperature profile obtained from a given location can be thought of as showing the temperature in a vertical shaft of water, or **water column,** extending down from the surface. Profiles can also be plotted for salinity, density, or any other characteristic. Profiles are drawn with the axes "upside down" as shown in Figure 3-11.

A profile is a plot that shows temperature, salinity, or any other characteristic of seawater at various depths in the water column.

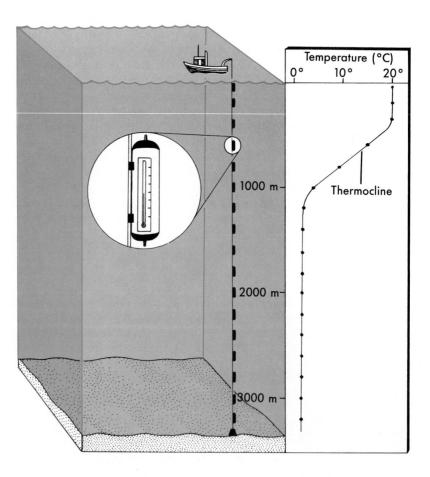

Figure 3-11 A series of bottles attached to the same wire is used to measure the temperature and salinity at several depths at once. Such information can be used to plot a profile or a graph showing how temperature, salinity, or any other property varies with depth. In this case, temperature measurements were used to plot a temperature profile. The zone where the temperature drops rapidly as it gets deeper is called a thermocline.

PRINCIPLES OF MARINE SCIENCE

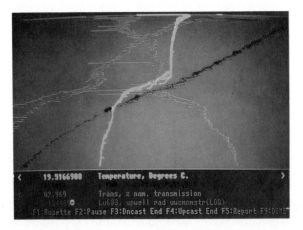

Figure 3-12 This computer screen is displaying information received from an instrument that simultaneously measures temperature, depth, and other information as it is lowered from a ship.

Using bottles and reversing thermometers is time consuming and expensive. Today oceanographers usually use electronic sensors that quickly and accurately measure the salinity, temperature, and depth as they are lowered through the water (Figure 3-12). Some of these are inexpensive enough to be disposable. This saves the time required to pull the instrument back to the surface. Oceanographic ships are very expensive to operate, and every savings of time also saves money. Another advantage of such instruments is that they take measurements at every depth, rather than just at intervals (Figure 3-12).

Though temperature, salinity, and various other things can be measured electronically, many other measurements still require a water sample. Niskin bottles and similar containers are still widely used. These days, oceanographers usually mount all the bottles to a single rack (Figure 3-13) rather than stringing them along the wire. As the rack is lowered, each bottle is closed at a different depth to obtain the profile. The rack is easier to use than a series of individual bottles. Various instruments, including temperature and salinity sensors, can also be mounted on the rack.

A

B

C

Figure 3-13 Today, water sampling bottles can be mounted in a rosette to a single frame **(A),** which is then lowered through the water column. **(B).** Many of these provide instantaneous information to a computer console on deck **(C).**

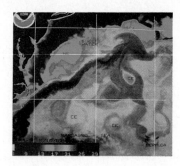

Figure 3-14 Satellite image showing the surface temperatures in the northwestern Atlantic Ocean. The warm Gulf Stream (*dark*) contrasts sharply with the colder water to the north (*light*).

Figure 3-15 This simple instrument, called a Secchi disk, is often used to measure water clarity. The disk is slowly lowered, and the exact depth where it is no longer visible is recorded. Electronic instruments that measure water transparency are now available, but the Secchi disk still gives remarkably reliable results.

Even with electronic instruments, measurements from ships can be made in only one place at a time. To gather information over a large area, the ship has to steam from place to place. In the meantime, conditions might change because of currents or the weather. As an alternative to ships, satellites have been used in recent years to study the ocean from space. Equipped with sensors that detect the amount of heat given off by the ocean, the satellites can determine the temperature of the sea surface.

Satellites can measure conditions only at the surface, but they give instantaneous coverage of a large area (Figure 3-14). Furthermore, a series of measurements can be made in a short time. This makes it possible to follow rapid changes in surface conditions that result from currents, weather, and so forth.

Oceanographers are not only using satellites to measure temperature, but they are also learning to use satellites to tell how much and what kind of plant life is in the water, as well as other characteristics of the water. Such information is greatly increasing our understanding of the ocean.

Dissolved Gases. There are gases as well as solid materials dissolved in seawater. The three most important gases for life in the ocean are **oxygen** (O_2), **carbon dioxide** (CO_2) and **nitrogen** (N_2). All three of these gases are found in the earth's atmosphere and dissolve in seawater at the boundary between the atmosphere and the sea surface. Sometimes the reverse occurs and the sea surface releases gases to the atmosphere. This process is known as **gas exchange** between the ocean and atmosphere. The amounts of all three gases in seawater are also affected by the activities of living things.

Transparency. One of the most biologically important properties of seawater is that it is relatively transparent, so sunlight can penetrate fairly deeply into the ocean. This is vital because all plants need light to grow. If seawater were not transparent, there would be very little plant life in the sea, and then only at the surface.

The transparency of water depends to a large extent on how much and what kind of material is suspended and dissolved in the water (Figure 3-15).

PRINCIPLES OF MARINE SCIENCE

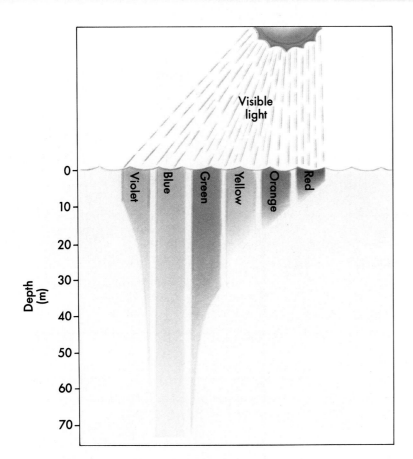

Figure 3-16 Different colors of light penetrate to different depths in the ocean. Blue light penetrates the deepest; red light, the least.

Obviously, muddy water is not as clear as clean water. Large quantities of plant life also reduce the transparency of water, much as the trees in a forest shade the ground. Water near coasts often contains a lot of material brought in by rivers. This material gives coastal waters a greenish tint and makes them less transparent than the blue waters of the open ocean.

Sunlight contains all the colors of the visible spectrum, but not all colors of light penetrate equally well. The ocean is most transparent to blue light. Other colors are absorbed more than blue, so as depth increases more and more of these other colors are filtered out. Before long, only blue light remains (Figures 3-16). Things that appear red on the surface look grey or black at depth because there is no red light to reflect off of them and be seen. Eventually even the blue light gets absorbed, and there is total darkness.

Pressure. **Pressure** is another factor that changes dramatically with depth in the ocean. Organisms on land are under 1 **atmosphere** (14.7 pounds per square inch, or psi, at sea level) of pressure—the weight of all the air above them. Marine organisms, however, are under the weight of not only the atmosphere, but of the water above them. Since water is much heavier than air, marine organisms are under much more pressure than those on land. The pressure increases dramatically with depth because the

Figure 3-17 The pressure at any place depends on the weight pressing down from above. At the sea surface or on land, the atmosphere is the only thing above. Divers or marine organisms, however, are also under the weight of the water column, which obviously gets greater going deeper.

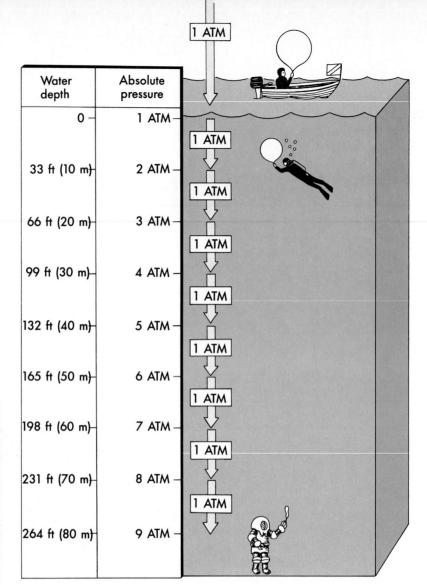

Water depth	Absolute pressure
0	1 ATM
33 ft (10 m)	2 ATM
66 ft (20 m)	3 ATM
99 ft (30 m)	4 ATM
132 ft (40 m)	5 ATM
165 ft (50 m)	6 ATM
198 ft (60 m)	7 ATM
231 ft (70 m)	8 ATM
264 ft (80 m)	9 ATM

Figure 3-18 This hatchetfish (*Argyropelecus pacificus*) has an internal air sac called a swim bladder that expanded when it was brought up from the deep. The gas bladder has expanded so much that it is protruding from the mouth. Fishes like this sometimes survive if the bladder is carefully punctured with a needle.

amount of water above is greater (Figure 3-17). With each 10 m (33 feet) of increased depth, another atmosphere of pressure is added.

As the pressure increases, gases are compressed. Gas-filled structures such as air bladders, floats, and lungs shrink or collapse. This limits the depth range of many marine organisms. It also means that submarines and housings for scientific instruments must be specially engineered to withstand pressure. This greatly increases the difficulty, expense, and sometimes the danger of studying the sea. The reverse effect also causes problems: organisms that contain gas-filled structures are often injured when brought up from the deep (Figure 3-18).

MOTION IN THE OCEAN

The ocean never rests. Anyone who has gone swimming in, sailed on, or simply walked beside the sea knows that it is in constant motion. Waves, currents, and tides all move and mix ocean waters, and all have important effects on life in the sea.

Surface Circulation

The most intense motion of the ocean occurs at the surface in the form of surface currents and waves. Both currents and waves are driven by the wind, which in turn is driven by heat from the sun. Winds and currents are also strongly influenced by what is known as the Coriolis effect.

The Coriolis Effect. Because the earth is rotating, anything that moves over its surface tends to turn a little to one side rather than moving in a straight line. One way to understand this bending, called the **Coriolis effect,** is to try a simple experiment. Put a piece of paper on a record player or some other rotating surface. Try to draw a straight line. Even if you use a ruler the line will come out curved because the paper is spinning, moving under the pen. Figure 3-19 shows some other ways to understand the Coriolis effect.

The Coriolis effect is too slight to notice when you're walking along or driving in a car, so most people aren't aware of it. It is very important, however, for things like winds and ocean currents that move over large dis-

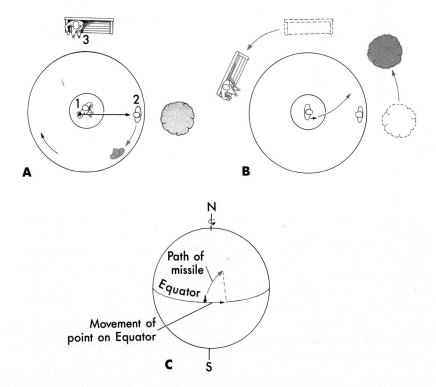

Figure 3-19 One way to understand the Coriolis effect is to think of a merry-go-round. **(A)** Imagine that someone at the center (*1*) throws a ball at someone on the outside (*2*). An observer on the ground (*3*) can see that the rider on the outside moved to a new position (shown in blue) while the ball was in the air. **(B)** To the people on the merry-go-round, however, it looks like the ball veers off to the side. Try it for yourself! **(C)** Places at the Equator travel faster than those near the poles: both places make one full revolution every day, but the place at the Equator has to travel a lot further. If a missile is fired from the Equator, it continues to move at the speed it had when fired. This means that as it moves toward the pole, it moves faster than the ground below. Thus it appears to curve relative to the earth's surface.

Figure 3-20 Air near the Equator is warmed by solar heating and rises. Air from higher latitudes rushes in over the earth's surface to replace the rising air, creating wind. These winds, the trades, are deflected by the Coriolis effect and approach the Equator at an angle of about 45 degrees.

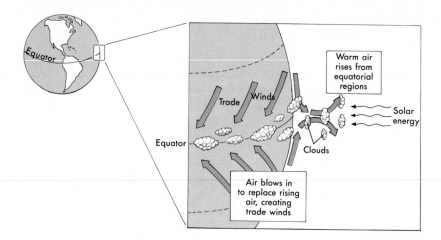

tances. In the Northern Hemisphere the Coriolis effect always pulls things to the right. In the Southern Hemisphere, things turn to the left.

The Coriolis effect deflects large-scale motions like winds and currents to the right in the Northern Nemisphere and to the left in the Southern Hemisphere.

Wind Patterns.   The winds in our atmosphere are driven by heat energy from the sun. Most of this solar energy is absorbed near the Equator. This makes sense: it is generally warm at the Equator and cold at the poles. The warm air at the Equator rises. Air from adjacent areas gets sucked in to replace the rising equatorial air, creating wind (Figure 3-20). The winds do not move straight toward the Equator, but are bent by the Coriolis effect. These winds, called the **trade winds,** approach the Equator at an angle of about 45 degrees. Over the ocean, where they are unaffected by land, the trade winds are the steadiest winds on earth.

Other winds are also driven by solar energy but tend to be more variable than the trades. At middle latitudes lie the **westerlies** (Figure 3-21), which move opposite the trade winds. At high latitudes are the **polar easterlies,** the most variable winds of all.

Surface Currents.   The major wind fields of the atmosphere push the sea surface, creating currents. All the major currents of the open ocean, in fact, are driven by the wind.

When pushed by the wind, surface water does not move in the same direction as the wind but, because of the Coriolis effect, moves off at an angle of 45 degrees (Figure 3-22). The surface currents driven by the trade winds, for example, do not move toward the Equator at all. Instead, these **equatorial currents** move parallel to the Equator (Figure 3-23). Under the influence of the Coriolis effect, the wind-driven surface currents combine into huge, more or less circular systems called **gyres.**

Global wind patterns and the Coriolis effect produce gyres, large circular systems of surface currents.

PRINCIPLES OF MARINE SCIENCE

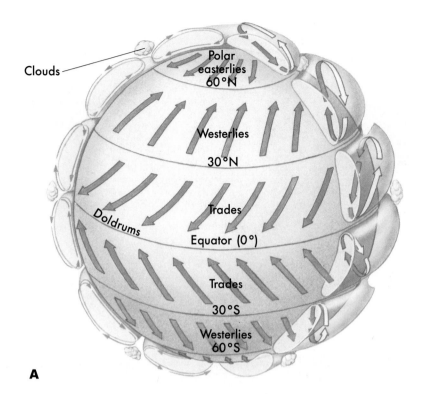

Clouds

Polar
easterlies
60°N

Westerlies

30°N

Trades

Doldrums

Equator (0°)

Trades

30°S

Westerlies
60°S

A

Figure 3-21 The major wind fields on earth are created by the rising of sun-warmed air and the sinking of cold air. The wind fields as they would look on an imaginary water-covered earth **(A)** are modified somewhat by the influence of the continents **(B)**. The trade winds lie between about 30 degrees north and south latitude and are the steadiest of all winds. From about 30 to 60 degrees are the westerlies, and above 60 degrees are the most variable winds, the polar easterlies. The transition zones or boundaries between these major wind belts have very light and changeable winds (see Box 3-1).

B

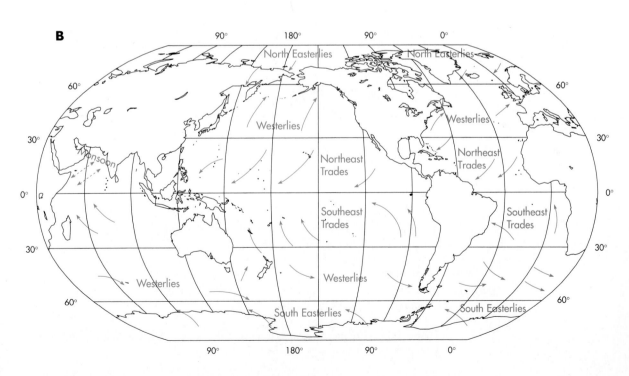

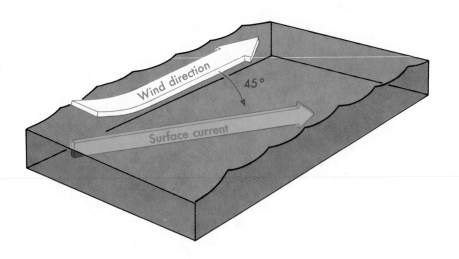

Figure 3-22 Because of the Coriolis effect, surface currents do not move parallel to the wind but at an angle of 45 degrees from the wind direction.

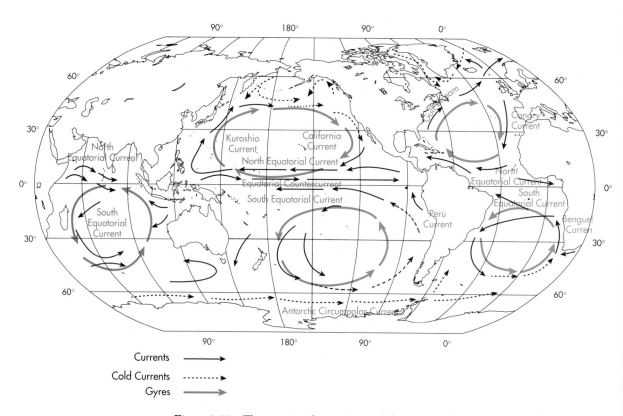

Figure 3-23 The major surface currents of the oceans. These currents combine to form major circulation systems called gyres.

BOX 3-1

Tall Ships and Surface Currents

Winds and surface currents were among the first oceanic phenomena to be observed and documented. For centuries, ships were at the mercy of the wind, and the names sailors gave to various areas reflected their knowledge of global wind patterns. Many of these names are still used. The "trade" winds were so named because traders relied on them during their voyages. The equatorial region where the winds are light and variable because of the rising air masses there (see Figure 3-20) is called the "doldrums." Winds are also variable at 30 degrees north and south latitudes, where the trade winds and westerlies are moving apart (see Figure 3-21). At these latitudes, sailors, becalmed and short of drinking water, sometimes had to throw their dying horses overboard. To this day these are known as the "horse latitudes."

Sailors also knew about surface currents. A clever navigator could shorten a passage by weeks or months by riding favorable currents and avoiding unfavorable ones. In the fifteenth century, Portuguese sailors under the guidance of Prince Henry the Navigator made careful observations of currents along the west coast of Africa. Soon they were using their knowledge of these currents on their trading voyages. On the southbound journey, the ships sailed close to shore while in the Northern Hemisphere, riding the Canary Current. When they crossed the Equator they swung to the west to avoid the Benguela Current. On the voyage home, they took the opposite path, completing a "figure 8."

Other currents were known to early mariners. Christopher Columbus noted the existence of the North Atlantic Equatorial Current on his third voyage to the New World. While searching for the "Fountain of Youth" the Spaniard Juan Ponce de León described the Florida Current. In the Pacific, fisherman recorded their knowledge of the Perú Current and the Kuroshio.

Even Ben Franklin has a place in this story. While serving as colonial postmaster he noticed that mail ships routinely made the trip to Europe 2 weeks faster than they returned. He began to question sea captains, particularly those of the Nantucket whaling fleet, about their knowledge of surface currents. Using this information, he published the first chart of the mighty Gulf Stream and issued instructions on how to avoid it when returning from Europe.

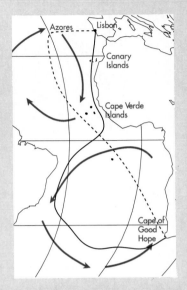

Route of Portuguese ships on trading voyages to the west coast of Africa. The Southbound journey is shown by a solid line; northbound journey by a dotted line. Prevailing currants are indicated by blue arrows.

Recall that water is particularly good at transporting heat because of its high heat capacity. The warm currents on the western sides of the gyres carry vast amounts of solar heat from the Equator to higher latitudes. Cold currents flow in the opposite direction on the eastern sides. Ocean currents thus act like a giant "thermostat," warming the poles, cooling the tropics, and regulating the climate of our planet. Indeed, recent studies indicate that the onset of an ice age is directly related to a major change in ocean circulation.

The role of surface currents in transporting heat is reflected in the temperature of the sea surface (Figure 3-24). Surface temperature is higher on the western sides of the oceans, where currents carry warm water away from

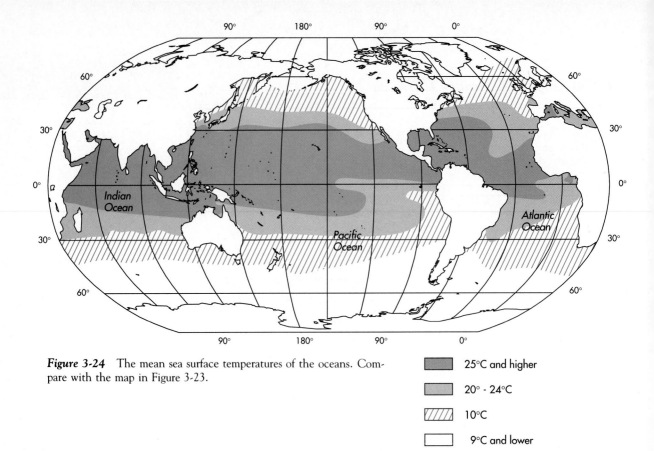

Figure 3-24 The mean sea surface temperatures of the oceans. Compare with the map in Figure 3-23.

▓	25°C and higher
░	20° - 24°C
▨	10°C
☐	9°C and lower

the Equator, than on the eastern sides, where cold currents flow toward the Equator. Because of this, tropical organisms like corals tend to extend into high latitudes on the west sides of the oceans (see the section in Chapter 13, "Conditions for Reef Growth" and Figure 13-13). Cold-loving organisms like kelps, on the other hand, occur closest to the Equator on the eastern shores of the oceans (see the section in Chapter 12, "Kelp Communities," and Figure 12-23).

Figure 3-23 shows the average pattern of currents over large distances and a long time span. At a given place on a given day the current might be different: currents shift with the season and the weather. On the continental shelf, currents are strongly affected by the bottom, the shape of the coastline, and the tides (see the section in Chapter 12, "Physical Characteristics of the Subtidal Environment").

Waves

The wind not only drives surface currents, it causes **waves.** Waves are probably the most familiar of all ocean phenomena, easily visible at the shore. They also affect the organisms that live on the shore.

The highest part of a wave is called the **crest** and the lowest part the

PRINCIPLES OF MARINE SCIENCE

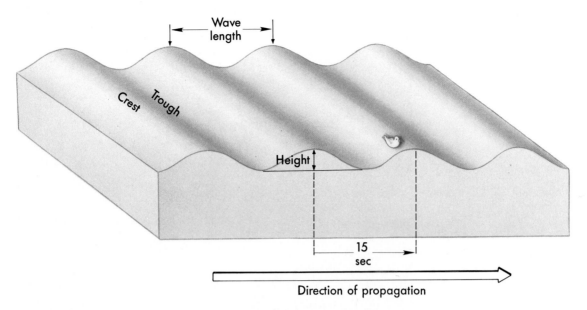

Figure 3-25 An idealized wave train. The highest point of a wave is called the crest; the lowest point, the trough. The wavelength is the distance between crests. The wave period refers to how long it takes the wave to go by. In this example the period is 15 seconds, which is how long it will take for the next crest to reach the bird.

trough (Figure 3-25). The size of an ocean wave is usually expressed as the wave **height,** which is the vertical distance between the trough and the crest. Wave crests, or troughs, can be close together or far apart: The distance between them is called the **wavelength.** The time a wave takes to go by any given point is called the **period** of the wave.

When under a wave crest the water moves up and forward; under the troughs it moves down and back. On the whole the water particles don't go anywhere at all as the wave passes; they just move in circles (Figure 3-26).

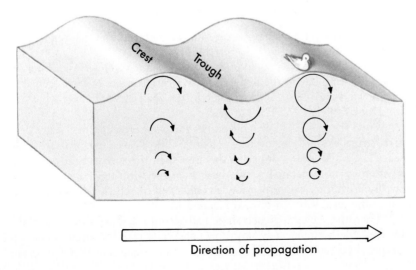

Figure 3-26 Water particles do not move along with a wave, but instead move in circles. When under the crest they move up and forward with the wave, then they get sucked back down. As wave after wave passes, the water and anything floating in or on it moves in circles.

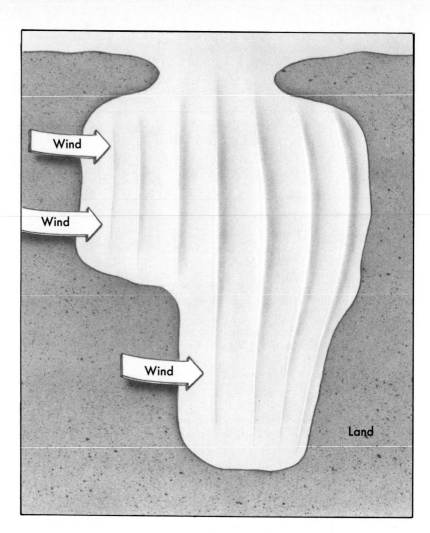

Figure 3-27 The wind is blowing at the same speed and has been blowing for the same length of time over this entire imaginary bay. On the top side of the bay, however, the wind is blowing over a much longer stretch of water. In other words, the fetch is longer on the top side, and so the waves are bigger there.

Though waves carry energy across the sea surface, they do not actually transport water.

Waves begin to form as soon as the wind starts to blow. The faster and longer the wind blows, the larger the waves get. The size of waves generated by the wind also depends on the **fetch,** the span of water over which the wind blows (Figure 3-27).

While the wind is blowing it pushes the wave crests up into sharp peaks and "stretches out" the troughs (Figure 3-28). Waves like this are called **seas.** The waves move away slightly faster than the speed of the wind from where they are generated. Once away from the wind the waves settle into **swells.** With their smoothly rounded crests and troughs, swells are very similar to the ideal waves shown in Figure 3-25.

When the waves approach shore and get into shallow water, they begin to "feel" the bottom. They slow down and "pile up," becoming higher and steeper. Eventually they become so high and so steep that they fall forward

PRINCIPLES OF MARINE SCIENCE

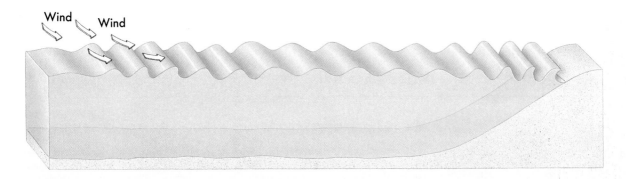

Wind Wind

and **break,** creating **surf.** The energy that was put into the wave by the wind is expended on the shoreline as the wave breaks.

The sea surface is usually a confused jumble instead of a series of nice, regular waves moving in one direction (Figure 3-29). This is because the surface at any given place is affected by a mixture of waves from many different places, generated by winds of different speeds blowing in different directions for different lengths of time. When the crests of two waves collide, they add together, producing a higher wave (see Box 3-2). This is called **wave reinforcement.** The crests pass through each other, however, and the high peak disappears. Similarly, when a crest and trough intersect they tend to **cancel** each other. The complex surface of the ocean results from many such interactions.

Tides

The sea surface has been rising and falling in the rhythmic pattern known as the **tides** for billions of years. Like wind waves, the tides are easily observable. In fact, tides can be thought of as waves that have a very long wavelength. The tides, however, are not caused by the wind.

Why Are There Tides? The tides are caused by the gravitational pull of the moon and sun and by the rotations of the earth, moon, and sun. Strictly speaking, the moon does not rotate around the earth. Instead the earth and moon both rotate around a common point, their combined center of mass (Figure 3-30). This rotation produces **centrifugal force,** which is the force that pushes you outward when you ride on a merry-go-round. The centrifugal force just balances the gravitational attraction between the earth and moon—otherwise the two would either fly away from each other or crash together.

Centrifugal force and the moon's gravity are not in perfect balance everywhere on the earth's surface, however. On the side of the earth nearest the moon, the moon's gravity is strongest and pulls the water toward the moon. On the side away from the moon, centrifugal force predominates, pushing the water away from the moon. If the earth were completely covered with water, the water would form two bulges on opposite sides of the planet. The water would be relatively deep under the bulges and shallow away from the bulges.

In addition to the rotation of the earth and moon illustrated in Figure 3-30, the earth is spinning like a top on its own axis. As it does so, any

Figure 3-28 Storm winds generate seas, peaked waves with relatively flat troughs. The waves move out of the storm area, carrying energy away, and become swells, with rounded crests and troughs. When the waves reach shallow water they get higher and shorter, that is, closer together. Eventually they become unstable and break, expending their energy on the shoreline.

Figure 3-29 The real sea surface is usually much more irregular than an "ideal" wave, because it is made up of a combination of many waves.

BOX 3-2

Waves That Kill

The Great Wave at Kanagawa.

Most of the time, waves are benign, providing entertainment for coast-watchers and thrills for surfers. Sure, they may sometimes cause seasickness or give the unwary beachgoer a tumble, but by and large they are relatively harmless. On rare occasions, however, waves can unleash all the awesome power of the sea.

So-called rogue waves are one type of wave that is said to be dangerous. Rogue waves are not really separate waves at all, but are formed by the interactions among different waves. When the crests of two waves happen to meet at the same place and time, they reinforce each other, producing a crest that is larger than either of the two individual waves. If by rare coincidence enough crests converge, the result is a wave crest of giant proportions—large enough to sink ships. Because rogue waves are not individual waves, but a combination of waves, they appear seemingly out of nowhere, which makes them especially dangerous.

It is not certain how often rogue waves occur, if they really exist at all. Oceanographers have never been able to document the occurrence of a rogue wave, so they can't be too common. On the other hand, mariners have attributed the loss of their ships to rogue waves on a number of occasions. Of course, a "rogue wave" might provide a convenient excuse after making a mistake at sea—no one can prove that the wave wasn't there!

There is no doubt about the existence of another deadly type of wave, **tsunami.** Tsunami were once called "tidal waves," but they have nothing to do with the tides. They are produced instead by earthquakes and other seismic disturbances of the sea floor, so they are more properly called **seismic sea waves.** When such a disturbance occurs, it can produce very long, fast-moving waves. Tsunami may have wavelengths of 240 km (150 miles) and travel at over 700 km/hr (435 miles/hr)—as fast as a jet airplane. In the open ocean, tsunami are not very high, usually less than 1 m (40 inches). Ships at sea usually don't even notice the passing of a tsunami. Like all waves, however, tsunami get higher—much higher—when they reach shallow water. The increase in the height of the wave can be dramatic, and when a tsunami strikes the shore it can cause vast damage.

Perhaps the most famous tsunami were caused not by an earthquake, but by the explosion in 1883 of the volcanic island of Krakatoa in the Indian Ocean. These tsunami caused destruction over half of the globe and killed more than 35,000 people. In 1960 an earthquake in Chile produced tsunami that caused damage around the Pacific. Hilo, Hawaii, suffered tremendous damage, and 61 people lost their lives. Destruction occurred along the west coast of the United States as a result of both the 1960 Chile earthquake and the great Alaska earthquake of 1964.

Today there is a world-wide network of seismic monitoring stations that provides instant notice of an earthquake or other seismic disturbance. Although the destruction of property cannot be prevented, it is hoped that tsunami will never again take their terrible toll of human life.

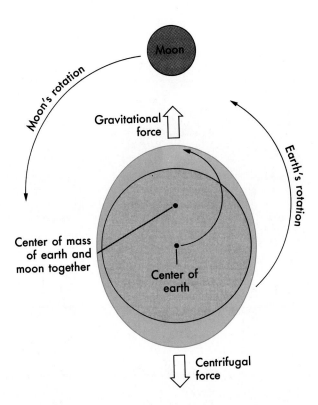

Figure 3-30 The moon and earth both rotate around their common center of mass. To visualize this motion, flip the pages starting with the next page. Centrifugal force produced by the earth's motion causes the water to bulge outward, away from the moon. On the side of the earth closest to the moon, however, the moon's gravitational pull overcomes the centrifugal force and pulls the water into a bulge *toward* the moon.

given point on the planet's surface will first be under a bulge and then away from the bulge (Figure 3-31). High tide will occur when the point is under a bulge. Since the earth takes 24 hours to complete a rotation, the point will have two high tides and two low tides every day. Actually, the moon advances a little in its own orbit in the course of a day. It takes the point on earth an extra 50 minutes to "catch up" and come directly in line with the moon again. A full tidal cycle therefore takes 24 hours and 50 minutes.

The sun produces tidal bulges in the same way as the moon. Though the sun is much larger than the moon, it is 400 times farther away, so the effect of the sun on the tides is only about half as strong as the moon's. When the sun and moon are in line with each other, which happens at the full and new moons (Figure 3-32), their effects add together. At these times the **tidal range,** or difference in water level between successive high and low tides, is large. Such tides are called **spring tides.** This name is a bit confusing because spring tides occur throughout the year, about once every 2 weeks.

When the sun and moon are at right angles their effects partially cancel each other. During these **neap tides** the tidal range is small. Neap tides occur when the moon is in the first and third quarters.

> **The tides are caused by a combination of the gravity of the sun and moon and the centrifugal force that results from the rotation of the earth, moon, and sun.**

Figure 3-31 As the earth spins on its axis, a given point on earth like the one marked by the flag alternates between being under a bulge, making it high tide, and away from the bulges, making it low tide.

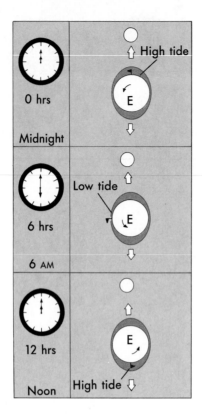

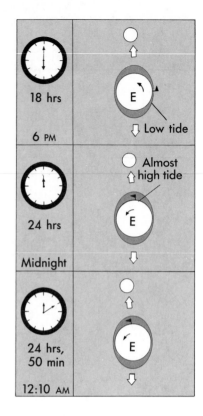

Tides in the Real World. Fortunately for us, the earth is not completely covered with water. Largely because of the continents, tides in the real world behave somewhat differently than they would on a water-covered earth. The tides vary from place to place, depending on the location and on the shape and depth of the basin.

As predicted, most places do have two high tides and two low tides a day (Figure 3-33, A and B); that is, they have **semidiurnal tides.** The east coast of North America and most of Europe and Africa have semidiurnal tides (Figure 3-34). Some places have a **mixed semidiurnal tide,** with successive high tides of different height (Figure 3-33, B). Mixed semidiurnal tides are characteristic of most of the west coast of the United States and Canada. **Diurnal tides** occur when there is only one high and one low tide every day (Figure 3-33, C). Diurnal tides are uncommon. They occur in the Gulf of Mexico, along the coasts of Antarctica, and parts of the Pacific Ocean (see Figure 3-34).

Tide tables that give the predicted time and height of high and low tides are available in most coastal areas. These tables give the values for one particular place. Other places in the area may have slightly different tides, depending on how far away they are and on the effects of channels, reefs, basins, and other local features. Weather patterns can also influence the

PRINCIPLES OF MARINE SCIENCE

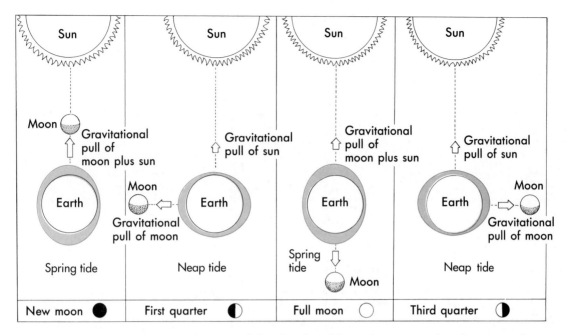

Figure 3-32 The tidal bulges are largest, and therefore the tidal range is greatest, when the moon and sun are in line and acting together. This happens at new and full moon. When the moon and sun are pulling in different directions, which occurs when the moon is in quarter, the bulges and tidal range are smallest.

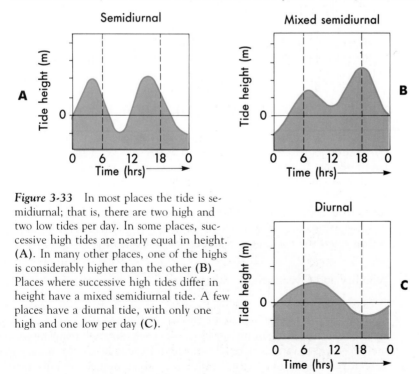

Figure 3-33 In most places the tide is semidiurnal; that is, there are two high and two low tides per day. In some places, successive high tides are nearly equal in height. (**A**). In many other places, one of the highs is considerably higher than the other (**B**). Places where successive high tides differ in height have a mixed semidiurnal tide. A few places have a diurnal tide, with only one high and one low per day (**C**).

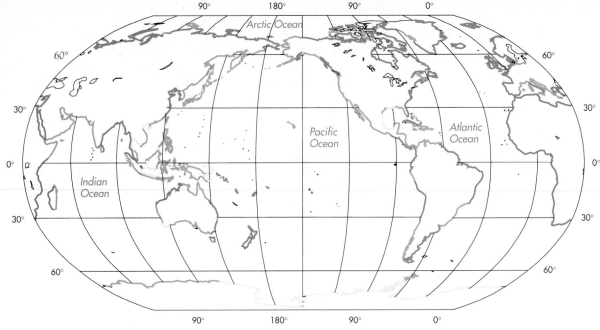

Semidiurnal tide ————————————

Mixed semidiurnal tide ————————————

Diurnal tide ————————————

Figure 3-34 The worldwide distribution of semidiurnal, mixed semidiurnal, and diurnal tides.

tides. Strong winds, for instance, can "pile up" water on shore, causing higher tides than predicted. On the whole, however, tide tables are remarkably accurate.

VERTICAL MOTION AND THE THREE-LAYER OCEAN

The oceans are a three-dimensional habitat. They vary not only horizontally—from north to south and east to west—but vertically, that is, at different depths. In many ways, changes in living conditions due to the third dimension—depth—are more important than changes due to geographic location. Much of the three-dimensional structure of the sea, especially in relation to depth, is controlled by the density of the water. That is why oceanographers take such pains to measure the temperature and salinity of ocean water, the two factors that determine its density.

Stability and Overturn

Because the densest water tends to sink, the ocean is usually layered—or **stratified**—with the densest water on the bottom and the least dense at the surface. This can be seen in typical profiles of density and temperature (Figure 3-35). Deep water is normally cold and dense, whereas the surface water is relatively warm and "light." Since the surface water naturally tends to

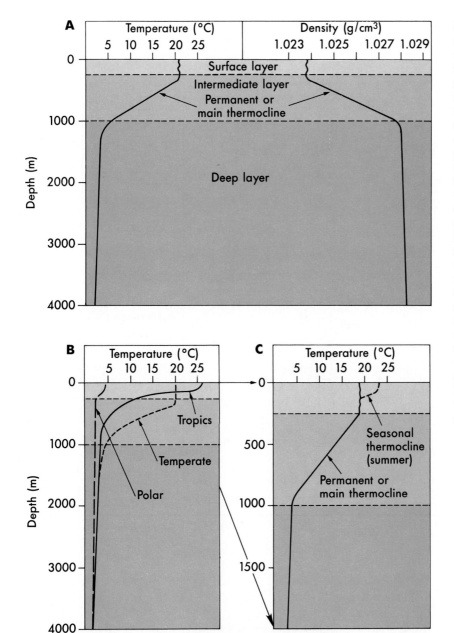

Figure 3-35 Typical profiles of temperature and density in the ocean away from the poles (**A**). Temperature and density profiles are usually mirror images of each other, because density in the ocean is controlled largely by temperature. Surface water is usually warmer and therefore less dense than the water below. The temperature of the surface, as you might expect, varies with latitude, with the highest surface temperature in the tropics (**B**). Deep-water temperatures are much more uniform. In temperate and polar waters, thermoclines may develop in the surface layer during the summer when the sun warms the uppermost part of the water column (**C**). Note that the depth scale in **C** is different than in **A** and **B**.

stay where it is and float on the denser water below, the water column is said to be **stable.** Unless some type of energy, such as wind or waves from a storm, stirs up the water column the surface water will not mix with the deep water.

The stability of the water column depends on the difference in density between the surface and deep water. If the surface water is only slightly

Figure 3-36 Overturn of the water column often occurs in temperate and polar waters. In summer **(A)** the sun warms up the surface layer, which therefore becomes much less dense than the water below. When fall comes **(B)** the surface water cools and storm winds mix colder deep water to the surface. If the surface water gets cold enough, usually during winter **(C)**, it gets denser than the deeper water and sinks.

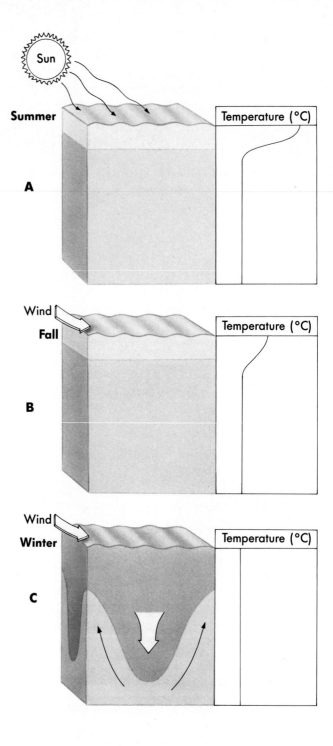

PRINCIPLES OF MARINE SCIENCE

more dense than the deep water, it will be relatively easy to mix the two. In this case the water column would be said to have low stability. A highly stable water column, on the other hand, would result from a large density difference between deep and shallow water.

> The water column in the ocean is usually stratified, with the densest water on the bottom. The greater the difference in density between surface and deep water, the more stable the water column is and the harder it is to mix.

Occasionally surface water becomes more dense than the water below. Such a water column is called **unstable** because the situation cannot last long. The surface water sinks, displacing the less dense water below in a process known as **overturn** (Figure 3-36). Since surface water, all with the same temperature and density, is descending through the water column, the temperature and density profiles are vertical straight lines. In fact, oceanographers identify conditions of overturn by looking for such straight-line profiles. Overturn usually occurs in temperate and polar regions, during the winter when the surface water cools. It is important in part because it brings oxygen-rich surface water to the deep sea (see the section in Chapter 15, "The Oxygen Minimum Layer"). Other important consequences of overturn will be discussed in Chapter 14 (in the section "Patterns of Production").

The Three-Layer Ocean

The processes that change salinity in the open ocean—precipitation, evaporation, and freezing—occur only at the surface. Temperature also changes primarily at the surface, through evaporative cooling, solar heating, or the exchange of heat with the atmosphere. Once surface water has sunk, therefore, the salinity and temperature cannot change. From then on the volume of water, or **water mass,** has a "fingerprint"—a characteristic combination of temperature and salinity. Oceanographers can therefore identify water masses by their temperature and salinity. This makes it possible to follow the movements, or **circulation,** of water masses over great distances (see the section in Chapter 15, "The Oxygen Minimum Layer").

After a water mass leaves the surface it descends to a depth determined by its density. If the water is very dense, it will sink all the way to the bottom. Water of intermediate density will descend only part way, "seeking" a level where it is denser than the water above but less dense than the water below. The stability of the water column is quickly restored as the layers of water sort themselves out.

Although there are actually many thin layers of water, each of slightly different density, it is not a bad approximation to view the ocean as being made of three principal layers (Figure 3-37). The **surface layer** is usually around 100 to 200 m (330 to 660 feet) thick. Much of the time the surface layer is mixed by wind, waves, and currents, so it is also known as the **mixed layer.** The surface layer is not always well mixed, however. Sometimes, usually in the spring and summer in temperate and polar water, the very uppermost part of the surface layer gets heated by the sun. The warm water floats in a shallow "lens" on top, and there is a sharp transition to the

Figure 3-37 Greatly simplified, the ocean can be thought of as having three main layers or water masses. The warm surface layer is relatively thin. The intermediate layer, the region of the main thermocline, is a transition between the surface layer and the cold and deep bottom layer.

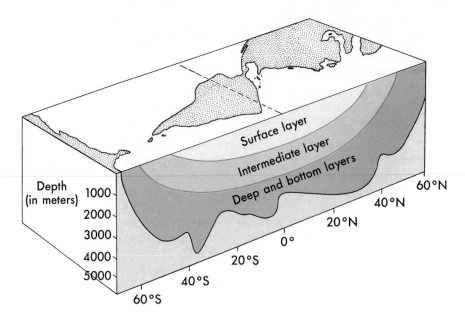

cooler water below. These sudden changes in temperature over small depth intervals, called **thermoclines,** are often noticed by divers. When the weather becomes colder, wind and waves again mix the water column and the thermocline breaks down.

The **intermediate layer** lies below the surface layer to a depth of around 1,500 m (5,000 feet). The **main thermocline,** a zone of transition between warm surface water and the cold water below, lies in the intermediate layer. The main thermocline should not be confused with the shallower, seasonal thermoclines described previously. The main thermocline rarely breaks down, and then in only a few places. The main thermocline is a feature of the open ocean. The waters over the continental shelf are not deep enough to have the main thermocline. Indeed, shelf waters are often mixed all the way to the bottom.

Below about 1,500 m (5,000 feet) lie the **deep** and **bottom layers.** Technically, deep and bottom water are different, but they are similar in being uniformly cold, typically less than 4° C (39° F).

The ocean has three main layers: the surface or mixed layer, the intermediate layer, and the deep layer.

PRINCIPLES OF MARINE SCIENCE

Do
It
Yourself

SUMMARY

1. Many of the unique properties of water result from the _____ that form between water molecules.
2. Substances can exist in three states or phases: _____, _____, and _____. _____ is the only substance that naturally occurs in all three phases on earth.
3. Water is capable of transporting large amounts of energy because of its high _____.
4. Water is called the _____ because it can dissolve so many things. Dissolved materials are known as _____.
5. Salts are composed of oppositely charged particles called _____.
6. The principle that the _____ is always the same is called the rule of constant proportions.
7. The _____ of seawater is determined by temperature and salinity.
8. Because other colors are _____ faster, _____ light penetrates the deepest in the ocean.
9. Most of the time the ocean is _____, or layered, according to the density of the water.
10. The _____ deflects all large-scale motions on the earth's surface.
11. The major surface currents form large circular systems called _____.
12. Waves transport _____ but not water.
13. The _____ affects the tides more than any other celestial body. Tides are also influenced by _____ and bottom features such as _____ and the shape of the basin of the body of water.

THOUGHT QUESTIONS

1. The winter of 1984-1985 was particularly cold in Europe. The northern part of the Black Sea, which lies between the U.S.S.R. and Turkey, froze, which is rare in the normally mild climate. The Adriatic Sea, located to the east, had just as cold a winter but never froze. The Black Sea has an unusually low salinity of about 18‰. What would you guess about the salinity of the Adriatic?
2. Just for the fun of it, someone walking along the shore in Beaufort, South Carolina, throws a bottle with a message in it into the sea. Some time later, someone in Perth, on the west coast of Australia, finds the bottle. Referring to Appendix C of this book or Figure 3-23, can you trace the path the bottle probably took?
3. If you owned a seaside home and a bad storm brought heavy winds and high surf to your coastline, would you prefer it to be during new moon or a quarter moon? Why?

FOR FURTHER READING

Broeker, W.S. and G.H. Denton: "What Drives Glacial Cycles?" *Scientific American*, vol. 262, no. 1, January 1990, pp. 48-56. *The authors of this somewhat technical article propose that changes in the circulation of the oceans are directly related to the onset of the ice ages described in Chapter 2.*

Linn, A.: "Oh, What a Spin We're in, Thanks to the Coriolis Effect." *Smithsonian*, vol. 13, no. 11, February 1983, pp. 66-73. *An interesting account of the Coriolis effect and some of its consequences.*

MacLeish, W.H.: "The Blue God." *Smithsonian*, vol. 19, no. 11, February 1989, pp. 44-59. *The first of a two-part series on the Gulf Stream: the history of its study; its effects on climate, marine life, and humans; and what still remains to be learned.*

MacLeish, W.H.: "Painting a Portrait of the Stream from Miles Above—and Below." *Smithsonian*, vol. 19, no. 12, March 1989, pp. 42-55. *The second of a two-part series describes the use of satellites, underwater electronics, and other modern technology in studying the Gulf Stream.*

Part Two

Life in the
Marine
Environment

4 The Business of Life

Figure 4-1 Although among the simplest of animals, this sea anemone performs all of the complex tasks needed to support life. Perhaps most important of these is reproduction. This anemone has released thousands of eggs.

Water is called the "universal solvent" because it can dissolve more different substances than any other liquid
Chapter 3, p. 55.

Now that we have discussed some of the major features of the marine environment, it is time to turn our attention to life in the sea. Perhaps the most basic question we could ask is "What *is* life?" Most people have a pretty good feel for what the word "living" means, but it is difficult to come up with a precise definition. About the best we can do is to describe the properties that living things have in common.

All living things use **energy,** the ability to do work, to maintain themselves and grow. This is accomplished by means of a vast number of chemical reactions that together are called **metabolism.** Living things also use energy to regulate their internal environments, that is, to maintain livable conditions inside themselves no matter what their surroundings are like. They are able to sense and react to the external environment. In addition, all life forms reproduce to perpetuate their kind, and they pass their characteristics on to their offspring (Figure 4-1).

All living things grow, metabolize, react to the external environment, and reproduce.

The most basic features of living things are shared by all organisms, not just those that live in the sea. However, here we will pay particular attention to the problems faced by marine organisms.

THE INGREDIENTS OF LIFE

The process of life involves an intricate series of interactions among an immense variety of chemicals. The most important of these chemicals is also one of the simplest: water. The bodies of most organisms are about two thirds water, and some, like jellyfishes, are more than 95% water. As the "universal solvent," water provides the medium in which all the other molecules dissolve and interact. Water is the base of a complex "chemical soup" inside all organisms. It is in this chemical soup that the chemical reactions of metabolism take place.

The Building Blocks

The processes that make life possible involve an enormous number of chemicals in addition to water. Most of these chemicals are **organic compounds,**

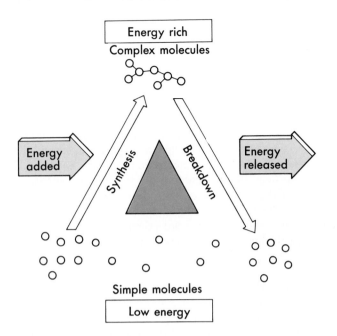

Energy rich
Complex molecules

Energy added →

Synthesis

Breakdown

← Energy released

Simple molecules
Low energy

Figure 4-2 Chemical reactions can be divided into those that require energy and those which release it. Organisms need energy to make complex molecules from simple ones, much as energy is required to push a rock uphill. The breakdown of energy-rich complex molecules is a "downhill" process that releases energy.

molecules that contain atoms of **carbon, hydrogen,** and usually **oxygen.** Organic compounds are high-energy molecules; that is, it takes energy to make them (Figure 4-2). The ability to use energy to make, or synthesize, organic compounds is an important characteristic of living things. Organic compounds tend to fall apart when left to themselves. When this happens they release the energy that went into their construction. Organisms have developed the ability to control the breakdown of organic molecules, as well as to synthesize them. Often the organism makes use of the energy that is released when the molecules break down.

Carbohydrates. Most organic molecules belong to one of four main groups. **Carbohydrates** are a group of organic compounds that are composed mainly of the basic elements of all organic molecules: carbon, hydrogen, and oxygen. The simplest carbohydrates are **glucose** and a few other **simple sugars.** More complex carbohydrates are formed by the combination of simple sugars into chains. Ordinary table sugar, for example, consists of two simple sugar molecules joined together. **Starches** and other complex carbohydrates consist of much longer chains, which may contain components other than simple sugars.

Carbohydrates serve a variety of functions. Simple sugars play vital roles in the most basic metabolic processes, as described below. Complex carbohydrates like starches are often used to store energy reserves. Other carbohydrates are important **structural molecules;** that is, they provide support and protection. Plants make use of a structural carbohydrate called **cellulose,** the main ingredient of wood and plant fibers (Figure 4-3). Some animals use a modified carbohydrate called **chitin** as a skeletal material.

Proteins. Another important group of organic molecules is the **proteins.** Proteins are the most varied and complex organic compounds. Like complex carbohydrates, proteins are composed of chains of smaller subunits.

molecules The smallest units of a chemical; they consist of two or more *atoms Chapter 3, p. 52*

Figure 4-3 Except for a dark clump in the center, this seaweed (*Enteromorpha*) is dead. The dead seaweed appears as a white cellulose "skeleton."

In proteins the subunits are **amino acids,** which contain atoms of **nitrogen,** carbon, hydrogen, and oxygen.

Since there are roughly 20 different amino acids, many different combinations are possible. Proteins therefore have an extraordinary variety of functions. Muscles are made largely of protein. **Enzymes** are proteins that speed up specific chemical reactions. Without enzymes, many metabolic reactions would proceed very slowly or not at all. There are also many structural proteins, such as those that help make up skin, hair, and the skeleton of some marine animals. Some proteins are **hormones,** chemicals that act as "messengers" to help different parts of the body work together. Still other proteins carry oxygen, act as poisons, produce light, help animals find mates, and have countless other uses.

Lipids. Fats, oils, and waxes are examples of **lipids,** another group of organic chemicals. Lipids are often used for energy storage (Figure 4-4). Another useful property of lipids is that they repel water. Many marine mammals and birds, for instance, use a coating of oil to keep their fur or feathers dry. Some marine plants that are exposed to the air at low tide have a coating of wax that helps keep them from drying out. Lipids are also useful in buoyancy and for insulation from the cold. Some hormones are lipids.

Nucleic Acids. Organisms use **nucleic acids** to store the basic genetic information they inherit from their parents and transmit it to the next generation. Nucleic acids are chains of repeating subunits called **nucleotides.** There are five different types of nucleotide, each consisting of a simple sugar joined to molecules that contain phosphorus and nitrogen. The order of the different nucleotides on the chain forms a kind of code, like a five-letter alphabet, that contains genetic information. Five letters may not sound like much, but Morse code can express any message with only two symbols: dots and dashes. Modern computers also use a two-letter alphabet.

One of the most important nucleic acids is **DNA** (deoxyribonucleic acid). An organism's DNA molecules specify the "recipe" for the organism—all the inherited instructions for its construction and maintenance.

> **Organic compounds are chemicals that contain carbon and hydrogen. The main types of organic compounds are carbohydrates, proteins, lipids, and nucleic acids.**

The Fuel of Life

The molecules that make up living things interact in many complex chemical reactions. The most basic of these chemical systems have to do with the capture, storage, and use of energy. More simply put, they deal with the production and use of food.

Photosynthesis: Making the Fuel. Almost all living things ultimately get their energy from the sun (Figure 4-5). In a process called **photosynthesis,** plants and a few other organisms capture the sun's energy and use it to make organic compounds, specifically simple sugars such as glucose. Most other organisms use these organic molecules as a source of energy. Organic material contains a tremendous amount of energy. As food it fuels our bodies and those of most other creatures. In such forms as oil, gas, and coal it heats our homes, runs our factories, and powers our cars.

Figure 4-4 The thick layers of blubber on this whale are composed almost completely of lipid. Whales can go without eating for long periods by using the energy stored in the blubber. The whale's blubber also helps protect it from the cold and gives it buoyancy. Humans once used the energy contained in blubber by burning whale oil for light and heat. These magnificent beasts have been driven nearly to extinction, largely for their blubber.

Photosynthesis (Figure 4-6) begins when solar energy, in the form of sunlight, is absorbed by chemicals called **photosynthetic pigments** that are contained within an organism. The most common photosynthetic pigment is **chlorophyll.** The bright green color that we think of as characteristic of plants is caused by chlorophyll. Most marine plants have additional pigments that may mask the green chlorophyll (see Table 5-1). Because of these pigments, marine plants come in many colors: brown, red, blue, even black.

In a long series of enzyme-controlled reactions, the solar energy captured by chlorophyll and other pigments is used to make simple sugars, with **carbon dioxide (CO_2)** and water (H_2O) as the raw materials. Carbon dioxide is one of very few carbon-containing molecules that are not considered organic compounds. Photosynthesis, then, converts carbon from an inorganic to an organic form. This is called **fixing** the carbon or **carbon fixation.** In this process, the solar energy that was absorbed by chlorophyll is stored as chemical energy in the form of simple sugars such as glucose. The glucose is then used to make other organic compounds (see the upcoming section, "Primary Production"). In addition, photosynthesis produces oxygen gas (O_2), which is released as a by-product (Figure 4-7). All the oxygen gas on earth, both in the atmosphere we breath and in the ocean, was produced by photosynthetic organisms. Photosynthesis constantly replenishes

Figure 4-5 The sunlight that streams through the water is captured by these kelps, seaweeds that inhabit temperate waters. The kelp uses the light energy to grow. Other organisms eat the kelp, making use of the energy that originally came from the sun.

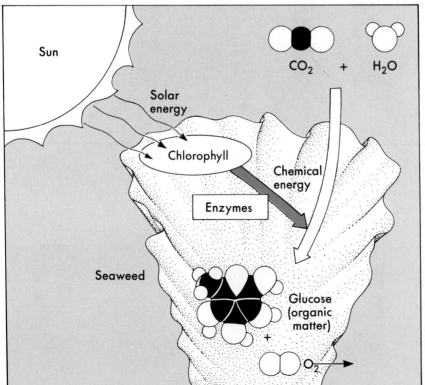

Figure 4-6 In photosynthesis, plants and other organisms combine carbon dioxide and water to make simple sugars like glucose. The energy for this comes from sunlight. Oxygen is given off as a by-product.

Figure 4-7 The mat of microscopic algae that covers this old seagrass blade is performing photosynthesis, releasing bubbles of oxygen as a by-product.

the earth's oxygen supply. Thus we rely on photosynthesis not only for food, but for the very air we breath.

Photosynthesis captures light energy from the sun to produce organic matter. Carbon dioxide and water are used to make the organic matter, and oxygen gas is liberated as a by-product:

$$H_2O + CO_2 \xrightarrow[\text{energy}]{\text{sun}} \text{organic matter} + O_2$$

Organisms that are capable of photosynthesis can obtain all the energy they need from sunlight and do not need to eat. Such organisms are called **autotrophs.** Plants are the most familiar autotrophs. Many other organisms cannot produce their own food and must obtain energy by eating organic matter that already exists. These organisms, which include all animals, are called **heterotrophs.**

Respiration: Burning the Fuel. The process of **respiration** allows all organisms, plants and animals alike, to make use of the solar energy that is trapped during photosynthesis. Though the chemical reactions are different, the net result of respiration is essentially the reverse of photosynthesis (Figure 4-8). Sugars are broken down using oxygen, and carbon dioxide and water are given off. This process is sometimes called "cellular respiration" to distinguish it from the physical act of breathing. Since breathing acts to provide oxygen for cellular respiration, the two processes are closely related.

Respiration, which occurs in all organisms, breaks down organic matter, releasing the energy it contains. Respiration consumes oxygen and produces carbon dioxide and water:

$$\text{organic matter} + O_2 \longrightarrow CO_2 + H_2O + \text{chemical energy}$$

Figure 4-8 Respiration is basically the opposite of photosynthesis. Respiration breaks down organic matter, using oxygen, and produces carbon dioxide and water. The energy that originally came from the sun is then made available to the organism.

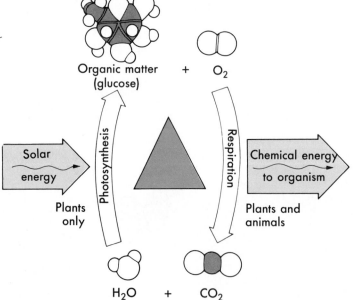

LIFE IN THE MARINE ENVIRONMENT

Respiration is somewhat similar to the burning of wood or oil in that organic material is broken down, oxygen is consumed, and the energy contained in the material is released. It is because of this that weight-watchers speak of "burning calories." In respiration, however, the energy stored in the organic material is not released in flames. Instead, much of the energy released by respiration is temporarily stored in yet another organic molecule (Figure 4-9). This molecule is **adenosine triphosphate,** or **ATP** for short.

The "backbone" of an ATP molecule is a compound called adenosine. A type of ion called **phosphate** (PO_4^{-3}) attaches to the adenosine, usually in a chain of two or three ions. When there are two phosphates the molecule is called **adenosine diphosphate (ADP).** When a third phosphate is added the molecule becomes adenosine *tri*phosphate, or ATP. A great deal of energy is needed to attach the third phosphate group, that is, to convert ADP to ATP. It is respiration that supplies this energy.

The energy that it took to make the ATP is easily liberated by breaking the third phosphate off again. When energy is needed, ATP is broken down into ADP. When energy is made available by respiration, ADP is converted back into ATP. This cycle goes on continuously. In an average day you go through about 57 kg (125 pounds) of ATP. Whales must use tons!

The advantage of using ATP is that most metabolic reactions can use the energy in ATP. Only respiration and a few other chemical reactions, on the other hand, can directly use the energy contained in glucose. Putting the energy into ATP makes it available to all the processes necessary for life. For this reason, ATP is often called the "energy currency" of life.

The energy released from organic matter by respiration is used to make ATP, the "energy currency" of life.

Primary Production. The sugars produced by photosynthesis supply both the raw material and the energy, via respiration, for the manufacture

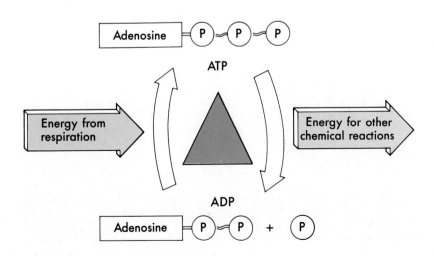

Figure 4-9 Respiration is linked to the production of ATP. When organic molecules are broken down, the released energy is used to attach a third phosphate group to molecules of ADP, converting them to ATP.

of other organic compounds. Through elaborate chemical processes, some of the glucose formed during photosynthesis is converted into other types of organic molecules—carbohydrates, proteins, lipids, and nucleic acids (Figure 4-10). The energy for these conversions, in the form of ATP, comes from burning up much of the rest of the glucose. Thus most of the sugar produced by photosynthesis is either converted to other types of organic matter or respired to fuel this conversion.

When plants and other autotrophs produce more organic matter than they use in respiration, there is an overall gain of organic matter. This net increase in organic matter is called **primary production.** Plants use the extra organic material to grow and reproduce, in other words, to produce more plant material—more food for animals and other heterotrophs. Plants and other organisms that perform the primary production of food are called **primary producers,** or sometimes just **producers.**

Primary production is the net gain in organic matter that occurs when autotrophs photosynthesize more than they respire.

Figure 4-10 In photosynthesis, plants capture energy from the sun and store it in chemical form in simple sugars. The sugars are then broken down in respiration, and the energy is transferred to ATP. The energy can now be used by the rest of the organism's metabolism, which includes making other organic matter from glucose. In the absence of light, plants can perform only respiration, like animals.

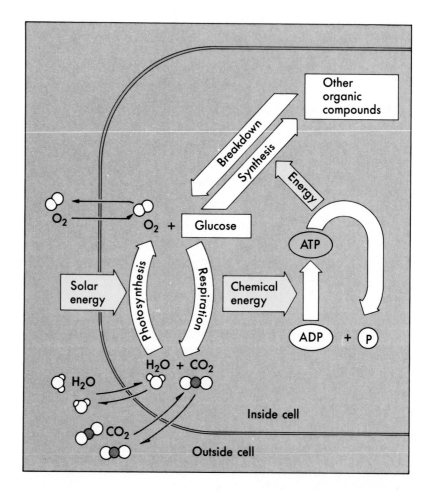

LIFE IN THE MARINE ENVIRONMENT

The Importance of Nutrients.   Only light, carbon dioxide, water, and sunlight are needed to make simple sugars by photosynthesis, but other materials are needed to convert the sugars into other organic compounds. These raw materials, called **nutrients,** include minerals, vitamins, and other substances. Plants need many different nutrients, but those needed in the largest amounts are nitrogen and phosphorus. Nitrogen is needed to make proteins; both nitrogen and phosphorus are needed to make nucleic acids. In one form or another, nitrogen and phosphorus are the main ingredients of most garden fertilizers. They are important "fertilizers" in the ocean as well. The most important form of nitrogen in the sea is **nitrate** (NO_3^{-1}). Phosphate is the main source of phosphorus for marine plants.

Primary production requires nutrients, as well as light. Nitrogen and phosphorus are the most important nutrients for plant growth in the ocean.

LIVING MACHINERY

The chemical reactions of metabolism make life possible, but chemical reactions are not living things and organisms are not just bags of chemical soup in which reactions take place. If you ran a fish, say, through a blender, you would still have all the same molecules but the fish would definitely not be alive. The molecules in living things are *organized* into structural and functional units that work in a coordinated way. It is this "living machinery" that is responsible for many of the properties of life.

Cells and Organelles

The basic structural unit of life is the **cell** (Figure 4-11). All organisms are made of one or more cells. Cells contain all the molecules needed for life

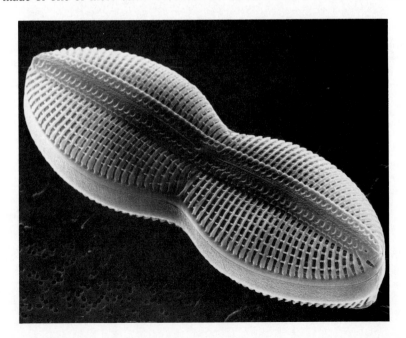

Figure 4-11   This diatom (*Diploneis*) consists of a single cell that lies inside an intricately sculptured "shell."

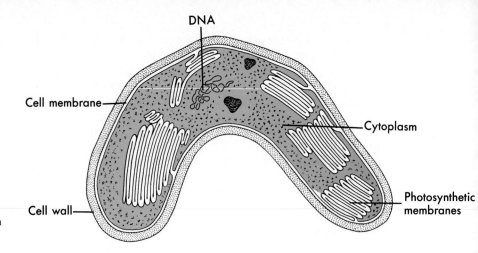

Figure 4-12 Prokaryotic cells like this bacterium are fairly simple. The cell is surrounded by a plasma membrane. Outside the membrane lies a cell wall. In this bacterium, which performs photosynthesis, the cell membrane is folded into the cell's interior. The membrane folds are where photosynthesis takes place. There is also a single circular molecule of DNA. Otherwise, there is little structure within the cytoplasm. Compare this photosynthetic bacterium with the heterotrophic one shown in Figure 5-2.

packaged in a living wrapper called the **cell membrane or plasma membrane**. This membrane isolates the gelatinous contents of the cell, or **cytoplasm,** from the outside world. The cell membrane allows some substances to pass in or out but prevents others from doing so.

There are a number of structures within the cytoplasm. Among the most important are membranes similar to the outer cell membrane. These interior membranes act as sites for such chemical processes as photosynthesis and respiration. Membranes may also divide the cell into compartments that enclose still more complex structures specialized for particular tasks.

Figure 4-13 Typical animal (**A**) and plant (**B**) cells and some of the structures they contain. Both are eukaryotic and thus have a nucleus, mitochondria, endoplasmic reticulum, and Golgi complex. In addition, plants have a cell wall and chloroplasts.

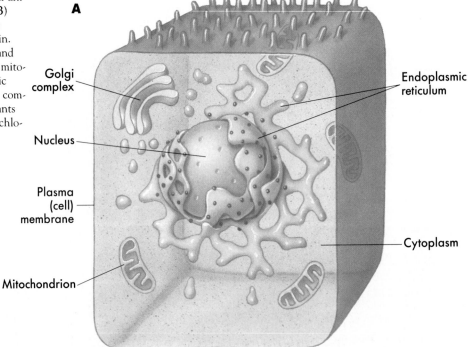

LIFE IN THE MARINE ENVIRONMENT

These membrane-bound structures are known as **organelles.** It is the presence of organelles that separates the two major types of cells.

Primitive Cells: Prokaryotes. The smallest, simplest, and most ancient type of cell is the **prokaryotic** cell (Figure 4-12). The distinguishing feature of prokaryotic cells is that they lack membrane-bound organelles. Only a few groups of organisms are prokaryotic. Microscopic organisms called **bacteria** have prokaryotic cells and are therefore known as **prokaryotes.**

Outside the cell membrane of bacteria lies a supportive **cell wall.** Inside, the cell has a single, circular molecule of DNA. In photosynthetic bacteria the cell membrane folds into the cell's interior. The photosynthetic pigments lie in these folds. Some bacteria possess one or more whiplike extensions known as **flagella** (singular, **flagellum**). These bacteria move by rotating their flagella like tiny propellers.

The most primitive cells are prokaryotic. They have little internal structure and lack membrane-bound organelles.

Advanced Cells: Eukaryotes. Most organisms are **eukaryotes,** whose cells are much more organized and complex than those of prokaryotes (Figure 4-13). The interior of eukaryotic cells is packed with a system of folded membranes called the **endoplasmic reticulum.** Together with the **Golgi complex**—another system of membranes—the endoplasmic reticulum is involved in making, packaging, and transporting many of the organic molecules used by the cell.

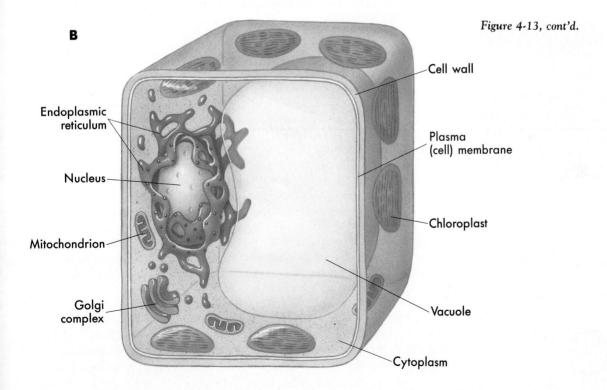

B

Figure 4-13, cont'd.

Cell wall

Endoplasmic reticulum

Plasma (cell) membrane

Nucleus

Mitochondrion

Chloroplast

Golgi complex

Vacuole

Cytoplasm

Eukaryotic cells contain other membrane-bound organelles. One of these is the **nucleus.** Within the nucleus lie structures called **chromosomes** that contain most of the cell's DNA. The nucleus thus contains the cell's genetic information and directs most of its activities. The nucleus can be thought of as the cell's headquarters.

Instead of occurring throughout the cytoplasm as it does in bacteria, respiration in eukaryotes takes place in special organelles called **mitochondria** (singular, **mitochondrion**). Mitochondria are the cell's power plants, breaking down organic molecules to provide energy. The inside of a mitochondrion contains yet another set of folded membranes. Many respiratory enzymes sit on these membranes.

Plants have organelles known as **chloroplasts.** The inside of a chloroplast is packed with stacks of folded membranes. These membranes contain the photosynthetic pigments and are the site of photosynthesis. Plants also have a cell wall, which is made of cellulose.

Like bacteria, many eukaryotic cells have flagella. When they are short and numerous the flagella are called **cilia,** but the structure is exactly the same. Like bacteria, eukaryotes often use their flagella or cilia to swim. They are also used to push water or particles through or past the organism, rather than to push the organism through the water. Many marine animals, for example, use cilia to push food particles to the mouth. Cilia in your lungs carry away dust and other irritants.

> The cells of eukaryotes are much more structured than those of prokaryotes. They have membrane-bound organelles, and the DNA is carried in chromosomes.

Levels of Organization

A cell is a self-contained unit that can carry out all the functions necessary for life. Indeed, some organisms get by just fine with only a single cell. One-celled organisms are called **unicellular.** All bacteria and some eukaryotes are unicellular. Most eukaryotic organisms, however, are **multicellular;** that is, they have more than one cell. The human body, for instance, contains something like 100 trillion cells.

There is a division of labor among the cells of multicellular organisms, with different cells becoming specialized for particular jobs. Groups of specialized cells that perform the same task may be organized into more complex structures. The extent of this specialization and organization is referred to as the **level of organization** (Table 4-1).

Organization Within the Body. At the simplest, or **cellular level,** each cell is essentially an independent, self-sufficient unit. Each can perform all of the functions needed to sustain the organism, and there is little or no cooperation among the cells.

Only a few multicellular organisms remain at the cellular level (Figure 4-14). In most, certain groups of cells act together to do a particular job. These specialized, coordinated groups of cells are called **tissues.** Some cells, for example, are specialized to contract and do work. These cells are bound together to form muscle tissue. Nervous tissue—specialized to collect, pro-

Figure 4-14 Sponges are the only multicellular animals that are at the cellular level of organization (see the section in Chapter 6, "Sponges").

LIFE IN THE MARINE ENVIRONMENT

TABLE 4-1

Levels of organization of biologic systems

LEVEL	DESCRIPTION	EXAMPLES		
Ecosystem	A community or communities in a large area, together with their physical environment	Nearshore ecosystem		
Community	All the populations in a particular habitat	Rocky shore community	Organismal	
Population	Group of organisms of the same species that occur together	All the mussels on a stretch of rocky shore		
Individual	A single organism	One mussel		
Organ system	A group of organs that work in cooperation	Digestive system		
Organ	Tissues organized into structures	Stomach		
Tissue	Groups of cells specialized for the same function that are bound together	Muscle tissue	Suborganismal	
Cell	Independent cells, the fundamental unit of living things	Muscle cell, single-celled organisms		
Organelle	A complex structure within cells; bound by a membrane	Nucleus, mitochondrion		
Molecule	Combinations of atoms that are bound together	Water, proteins	Subcellular	
Atom	The fundamental units of all matter	Oxygen, hydrogen		

INCREASING COMPLEXITY →

cess, and transmit information—and bone—specialized for structural support—are other examples of tissues.

Most animals don't stop at the tissue level either. Their tissues are further organized into structures known as **organs** that carry out specific functions. The heart is an example of an organ. There are several different tissues that make up the heart. Muscle tissue, for example, allows the heart to contract and pump blood, and nervous tissue controls the muscle.

Different organs act together in **organ systems.** The digestive system, for instance, includes many organs: a mouth, stomach, intestine, and so on. Animals usually have a number of organ systems, including nervous, digestive, circulatory, and reproductive systems.

Interactions Among Individuals. Not only are there different levels of organization within individual organisms, but individuals interact with each other at still higher levels of complexity. A **population** is a group of organisms of the same kind, or **species,** that lives at the same place. The mussels on a stretch of rocky shore, for example, form a population (Figure 4-15).

Populations of different species that occur in the same place form **communities.** A community is often more than just a collection of organisms that happen to live in the same area. The characteristics of communities are determined in large part by the way the organisms *interact*—which organ-

Figure 4-15 These blue mussels (*Mytilus edulis*) growing on a rocky shore in New Zealand are part of a population.

Figure 4-16 The community that lives on this rocky shore is made up of populations of many different types of organisms, including the mussels shown in Figure 4-15.

isms eat each other, which organisms compete, and which depend on each other. Rocky shores, for example, are home to many organisms in addition to mussels: seaweeds, crabs, barnacles, snails, and starfishes, to name a few (Figure 4-16). It is the interactions among all these organisms that give the rocky shore community its own unique structure.

A community, or several communities—together with the physical, or non-living, environment—in turn makes up an **ecosystem.** The rocky shore community, for instance, is part of a larger ecosystem, along with other nearby communities and the tides, currents, nutrients dissolved in the water, and other physical and chemical aspects of the area.

CHALLENGES OF LIFE IN THE SEA

Every habitat has its own unique set of characteristics and presents special challenges to the organisms that live there. Thus marine organisms must cope with problems unlike those of land dwellers. Even within the marine environment there are different habitats, each of which poses special difficulties. For example, **planktonic** organisms—those which drift in the water—face much different conditions than **benthic** organisms—those which live on the bottom—or **nekton,** organisms that swim (see Figure 9-21). Organisms have evolved innumerable ways to adapt to the conditions of their habitats.

Many of the adaptations of marine organisms have to do with maintaining suitable conditions inside their bodies. The living machinery inside most organisms is rather sensitive and can only operate properly within a narrow range of conditions. Living things have therefore devised ways to keep their internal environments within this range no matter what external conditions are like.

Salinity

Many enzymes and other organic molecules are very sensitive to changes in the concentrations of the ions that are common in seawater. Marine organisms are immersed in a medium that has the potential to profoundly alter their metabolism. To fully understand the problems posed by salinity, we must know something about how dissolved ions and molecules behave.

Diffusion and Osmosis. When in solution, ions and molecules move around just like water molecules do. If the molecules are concentrated in one part of the solution, they tend to spread out because of this random movement until they are evenly distributed (Figure 4-17). The overall result is that the molecules tend to move from areas of high concentration to areas of low concentration. This process is called **diffusion.**

Whenever the internal composition of a cell differs from that on the outside, substances will tend to move in or out of the cell by diffusion. If the surrounding seawater contains more sodium, say, than the cell's interior, sodium will tend to diffuse into the cell. If the organism is sensitive to sodium, this presents a problem. Similarly, things that are concentrated inside cells tend to diffuse out. Many of the cell's precious molecules—like ATP, amino acids, and nutrients—are much more concentrated in the cell than in seawater. Therefore they tend to "leak" out of the cell.

ions Atoms or groups of atoms that have a positive or negative charge. The most common ions in seawater are chloride, sodium, and potassium
Chapter 3, p. 55.

Lump
of salt

One answer to this problem is to have some kind of barrier that prevents materials from diffusing in and out. The cell membrane is just such a barrier. It blocks the passage of the common ions in seawater, as well as many organic molecules. The membrane cannot be a complete barrier, however, because the cell needs to exchange many materials—such as oxygen and carbon dioxide—with its surroundings. Therefore the cell membrane is **selectively permeable;** that is, it allows some substances to enter and leave the cell but prevents others from doing so. Water and other small molecules pass readily through the cell membrane, for example, whereas many proteins do not.

The selective permeability of the cell membrane solves the problem of diffusion of ions and organic molecules, but it creates a new problem. Like any other molecule, water will diffuse from areas where it is concentrated to areas of lower concentration. If the total concentration of *solutes* inside the cell is higher than that outside, the concentration of *water* will be lower. Since water molecules are free to cross the cell membrane, they will stream into the cell, causing it to swell (Figure 4-18). On the other hand, the cell will tend to lose water and shrivel if the total salt concentration outside the cell is higher than that inside. This diffusion of water across a selectively permeable membrane is known as **osmosis.**

> Diffusion is the movement of ions and molecules from areas of high concentration to areas of low concentration. Osmosis is the diffusion of water across a selectively permeable membrane.

Regulation of Salt and Water Balance. Marine organisms have adapted to the problems of maintaining the proper balance of water and salts in various ways. Actually, some do not actively maintain salt and water balance at all: their internal concentrations change as the salinity of the water changes. Such organisms are called **osmoconformers.** Osmoconformers often have to stay where the salinity of the water matches that of their fluids. Outside of a narrow range of salinity, they experience osmotic problems. If placed in fresh water, they would swell up and burst because of the osmotic flow of water into their tissues. In areas such as the open ocean,

Figure 4-17 Ions and molecules that are dissolved in water tend to move around randomly just like water molecules do. When a crystal of salt dissolves in water, the ions start out concentrated in one part of the solution but eventually spread out because of their random movement. This process, diffusion, results in the flow of ions from areas of high concentration to areas of low concentration.

solute Ions, organic molecules, or anything else that is dissolved in a solution
Chapter 3, p. 57.

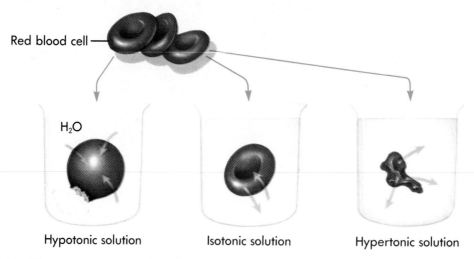

Red blood cell

H₂O

Hypotonic solution   Isotonic solution   Hypertonic solution

Figure 4-18   Water moves in or out of a cell by osmosis, depending on the concentration of solutes on the out-side. Water concentration is inversely proportional to the solute concentration. In a *hypotonic solution (left)* the solute concentration is low and water concentration is high, so that water moves into the cell from high to low concentration. The cell, in this case a red blood cell, swells and bursts. The reverse takes place if the cell is placed in a *hypertonic solution (right)*. The solute concentration is high and that of water is low, so water flows from inside the cell to the outside and the cell shrivels. No change takes place if the cell is in an *isotonic solution (center)*, one in which the solute concentration is identical to that of the cell. Osmotic problems can thus be avoided by match-ing the internal to the external concentration.

where salinity doesn't fluctuate much, osmoconformers have few difficulties.

Other marine organisms **osmoregulate,** or control their internal concentrations to avoid osmotic problems. One way they adapt to different salinities is by adjusting the concentration of solutes in their body fluids so that the overall concentration of their fluids matches that of the water outside. Some organisms use a particular chemical to osmoregulate, increasing or decreasing the amount of the chemical in response to salinity changes. Sharks, for example, do this with **urea** (see Figure 7-19), a waste product that is toxic to other animals. A single-celled marine plant called *Dunaliella* may be the champion osmoregulator (Figure 4-19). By altering the amount of a chemical called glycerol in its cells, *Dunaliella* can maintain osmotic balance in water ranging from nearly fresh to up to nine times saltier than normal seawater. At very high salinities, most of the volume of the cell is taken up by glycerol.

Osmoregulators may have blood concentrations of salts different from that of seawater. Most marine fishes have body fluids considerably more dilute than seawater (Figure 4-20, A) and tend to lose water by osmosis. They replace the lost water by drinking seawater. They also conserve water by producing only a very small amount of urine. Marine fishes must actively rid themselves of, or **excrete,** the excess salts that are taken in with the seawater they drink. Much of this excess is excreted with the urine, making the

LIFE IN THE MARINE ENVIRONMENT

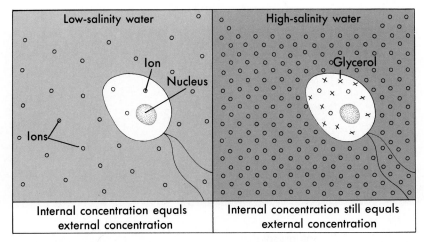

Figure 4-19 The microscopic plant *Dunaliella* uses glycerol to control osmosis. Most of the time the cell is in osmotic balance, with the internal concentration of solutes equal to the outside salinity. If the salinity of the water increases, the cell tends to lose water by osmosis and shrivel up. To prevent this, the cell produces glycerol until the concentration inside the cell once again matches the external concentration.

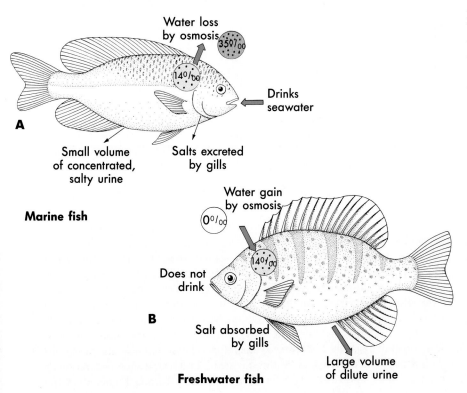

Figure 4-20 Water flows by osmosis from the relatively dilute body fluids of most marine fishes (**A**) to the more concentrated seawater. To compensate and avoid dehydration, the fish drinks seawater. It excretes as little urine as possible to conserve water, but the urine contains a lot of excess salt, which is taken in with the seawater. Excess ions are also excreted by the gills. Freshwater fishes (**B**) are in the reverse situation: the outside water has a very low salinity, so they tend to gain water by osmosis. To avoid swelling up, they refrain from drinking and produce large amounts of dilute urine. Salts are absorbed by the gills to replace those lost in the urine.

Figure 4-21 **(A)** Sea turtles have glands near the eyes that excrete excess ions, producing salty "tears." This is a green sea turtle *(Chelonia mydas).* **(B)** Mangrove trees like this black mangrove *(Avicennia germinans)* excrete excess salts through the leaves. The accumulation of salt on the outside surface of the leaves is visible in this photo.

urine very salty. Marine fishes also excrete excess salts through their gills.

Freshwater fishes have the reverse problem: their blood has a higher concentration of salt than the surrounding water, and they tend to take in water by osmosis. Their adaptations are opposite those of freshwater fishes (see Figure 4-20, B).

Marine birds and reptiles, and some marine plants, also have special cells or glands to rid themselves of excess salt (Figure 4-21). Marine plants have the added advantage of a fairly rigid cell wall that helps them resist the swelling caused by osmotic water gain.

Temperature

Organisms are greatly affected by temperature. Metabolic reactions proceed faster at high temperatures and slow down dramatically as it gets colder. A general "rule of thumb" is that most reactions occur about twice as fast with a 10° C (18° F) rise in temperature. At extreme temperatures, however, most enzymes cease to function properly.

Marine plants and most marine animals are **ectotherms,** or **poikilotherms,** or what is often referred to as "cold-blooded." As the temperature of the surrounding water rises and falls, so does their body temperature and, consequently, their metabolic rate. Many become quite sluggish in unusually cold water.

The "slowing down" caused by cold water is a disadvantage for active swimmers. Some large fishes, such as certain tunas and sharks, can maintain body temperatures that are considerably warmer than the surrounding water (see the section in Chapter 14, "Swimming: The Need for Speed"). They do this by retaining the heat produced by their large and active muscles. This allows them to remain active even in cold water. The temperature of their bodies, however, still rises and falls with that of their surroundings.

Mammals and birds are able to keep their body temperatures more or less constant regardless of water temperature. These animals are called **endotherms, homeotherms,** or "warm-blooded." Endotherms use respiration to produce heat, and they generally have body temperatures higher than the

LIFE IN THE MARINE ENVIRONMENT

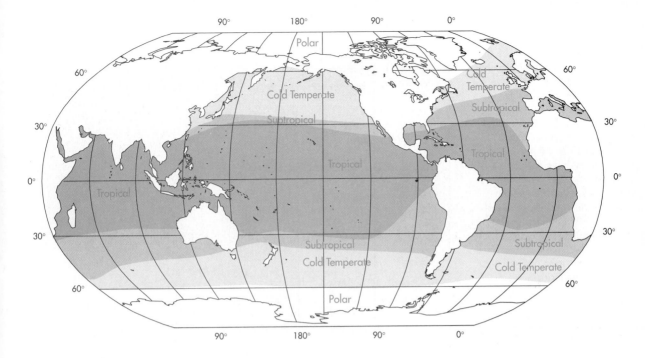

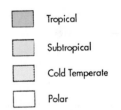

Tropical

Subtropical

Cold Temperate

Polar

Figure 4-22 Marine organisms are often restricted to specific regions that correspond to average water temperatures. The major divisions are the *polar, cold temperate, subtropical* (or *warm temperate*), and *tropical* regions. The boundaries of these regions are not absolute and change slightly with the seasons and shifts in current patterns.

surroundings. This allows them to remain highly active regardless of water temperature, but it also uses up a lot of energy. To reduce the energy cost of maintaining their body temperature, mammals and birds are often insulated with feathers, hair, or blubber.

Most marine organisms are adapted to live in particular temperature ranges. Many polar species, for instance, have enzymes that work best at low temperatures, and they cannot tolerate warm water. The reverse is true of many tropical species. As a result, temperature plays a major role in determining where different organisms are found in the ocean (Figure 4-22).

Surface-to-Volume Ratio

Adaptations to salinity and temperature are needed because salts and heat can flow into and out of organisms. Organisms also exchange nutrients, waste products, and gases. Such materials often move in and out across the surface of marine organisms, so the amount of surface area is very important. More precisely, it is the amount of surface area relative to the total volume of an organism—in other words, the **surface-to-volume ratio** or **S/V ratio**—that determines how rapidly heat and materials flow in and out.

One thing that determines the S/V ratio is the size of the organism. As organisms grow larger, their volume increases faster than their surface area. Small organisms have a larger S/V ratio than big ones (Figure 4-23). Small organisms, especially single-celled ones, can therefore rely on simple diffusion across their surfaces for the exchange of materials. Larger organisms must develop supplementary mechanisms, like respiratory and excretory systems.

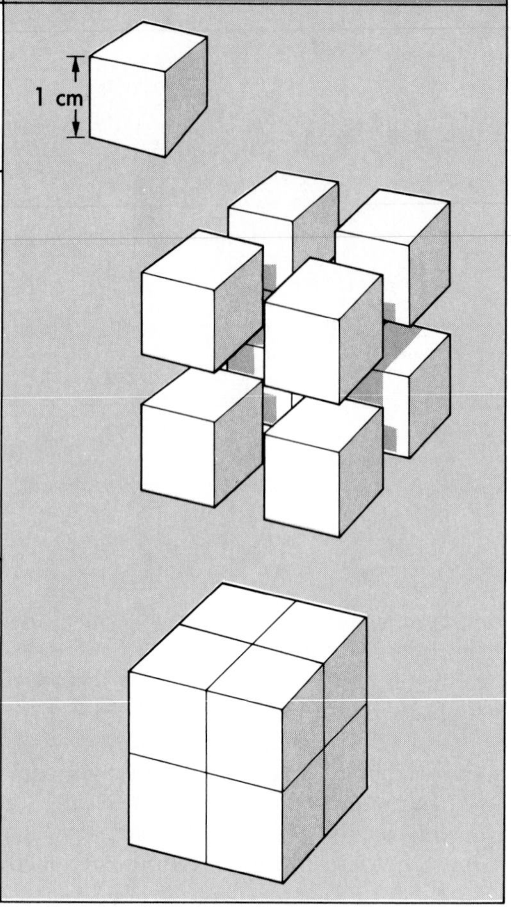

Figure 4-23 A cube that is 1 cm on a side (*top*) has a volume of 1 cm³ and, because it has 6 sides, a surface area of 6 cm². Eight such cubes (*middle*) will, of course, have 8 times the volume and 8 times the surface area. If the 8 cubes are combined into a single large one (*bottom*), half of the surfaces are hidden in the interior. The ratio of the surface area to the volume of the cube, or surface-to-volume ratio, is cut in half.

Volume	Surface area	
1 cm³	6 cm²	1 cm
8 cm³	48 cm³	
8 cm³	24 cm²	

PERPETUATING LIFE

A basic, perhaps the most basic, characteristic of living things is the ability to reproduce—to generate offspring similar to themselves. Any life form that failed to replace itself with new individuals of its own kind would soon vanish from the planet. It is only by reproducing that a species ensures its own survival.

Organisms must do two things when they reproduce. First, they must produce new individuals to perpetuate the species. Second, they must pass on to this new generation the characteristics of the species in the form of genetic information. The transfer of genetic information from one generation to the next is called **heredity.**

Given the amazing variety of organisms that inhabit the earth, it should not surprise you that organisms achieve the first objective, the production of offspring, in many different ways. What *is* a little surprising is that there are relatively few ways through which organisms pass on genetic information.

LIFE IN THE MARINE ENVIRONMENT

This similarity of hereditary mechanisms is evidence that living things are all fundamentally related.

Modes of Reproduction

The only way that individual cells reproduce is by dividing to form new **daughter cells.** In prokaryotes, where the DNA that carries the genetic information is not enclosed in a nucleus, cell division is relatively simple. Before dividing the cell copies, or **replicates,** its DNA molecule. Each daughter cell gets one of the copies, as well as the rest of the cellular machinery it needs to survive.

In eukaryotes the DNA lies on several different chromosomes. Before cell division, each of the chromosomes is replicated. The most common form of eukaryotic cell division, **mitosis,** is a complex process that takes place in the nucleus. Mitosis ensures that each daughter cell gets a copy of every chromosome. As in bacterial cell division, the end result is two daughter cells that are exact duplicates of the original, with the same genetic information.

Asexual Reproduction. Cell division is the primary way that single-celled organisms reproduce. Because a single individual can reproduce itself without the involvement of a partner, this form of reproduction is called **asexual reproduction.** All forms of asexual reproduction are similar in that the offspring inherit all the characteristics of the parent. They are, in fact, exact copies, or **clones.**

Many multicellular organisms also reproduce asexually (Figure 4-24). Some sea anemones, for example, simply split in half to create two smaller anemones. This process is known as **fission.** Another common form of asexual reproduction is **budding.** Instead of dividing into two new individuals of equal size, the parent develops small growths—or **buds**—that eventually break away and become separate individuals. Many plants reproduce asexually by sending out various kinds of runners that take root and then sever their connection to the parent. Because it is so common in plants, asexual reproduction is sometimes called **vegetative reproduction,** even when it occurs in animals.

Asexual, or vegetative, reproduction can be performed by single individuals and produces offspring that are genetically identical to the parent.

Figure 4-24 Seagrasses reproduce asexually by putting out "runners," or stolons, that take root and form new plants (**A**). The aggregating sea anemone (*Anthopleura elegantissima*) divides repeatedly by fission and forms dense groups called clones (**B**). Because it is low tide, the anemones in this group have contracted. They are covered by bits of shells. Some sponges form buds that grow into interconnected individuals like these (**C**).

A

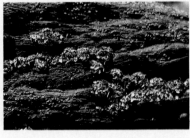

B

C

Sexual Reproduction. Most multicellular and some unicellular organisms reproduce part or all of the time by **sexual reproduction,** in which new offspring arise from the union of two separate cells called **gametes.** Usually each of the gametes comes from a different parent.

Organisms that reproduce sexually have a special kind of tissue called **germ tissue.** While all the other cells in the body divide only by mitosis, germ cells are capable of a second type of cell division, **meiosis.**

Most cells in the body have two similar sets of chromosomes. Such cells are said to be **diploid,** designated **2n.** Meiosis produces daughter cells that have copies of only half of the parent's chromosomes—one of each pair. Cells with half the normal number of chromosomes are called **haploid,** designated **n,** or **1n.** The haploid daughter cells produced in meiosis are the gametes. In some seaweeds and microscopic organisms, all the gametes are the same. Most organisms, however, have two types of gametes: female gametes called **eggs** and male gametes called **sperm,** or pollen in a few marine and most terrestrial plants. In animals the germ tissue is usually contained in **gonads,** the organs that produce the gametes. **Ovaries** are the female gonads, which of course produce eggs. Sperm are produced by the male gonads, the **testes.**

Eggs are large cells, usually an organism's largest (Figure 4-25). They contain all of the cellular machinery—organelles, cytoplasm, and so forth—characteristic of regular cells. The eggs of many species also contain large amounts of energy-rich **yolk.** Sperm, by contrast, are usually an organ-

Figure 4-25 The egg of the variegated sea urchin (*Lytechinus variegatus*) is many times larger than sperm from the same species. Though many sperm surround the egg, only one will fertilize it. The egg contains a large amount of cytoplasm and most of the organelles characteristic of other cells. It also contains yolk, which nourishes the developing embryo. This photograph was taken with an electron microscope, which is more powerful than ordinary light microscopes, and provides three-dimensional images at high magnifications.

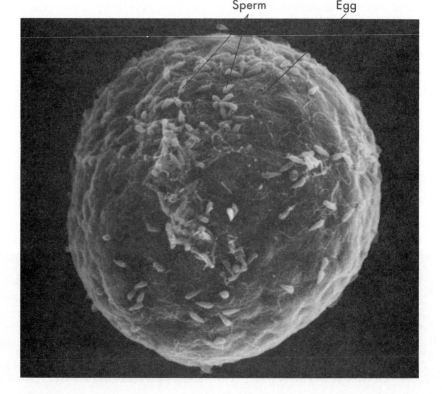

Sperm   Egg

LIFE IN THE MARINE ENVIRONMENT

ism's smallest cells, with almost no cytoplasm and few of the typical organelles (Figure 4-26). Sperm are little more than tiny packages of chromosomes equipped with flagella, powered by mitochondria, that enable the sperm to swim.

Mitosis, the most common form of cell division, produces two daughter cells that are identical to the original cell. Meiosis, on the other hand, produces daughter cells that have half the normal number of chromosomes, that is, are haploid. These daughter cells are called gametes. Eggs and sperm are the female and male gametes, respectively.

Sperm are attracted to eggs from the same species. When egg and sperm come into contact, they fuse; that is, **fertilization** occurs. The genetic material contained in the two gametes combines. Since both haploid gametes had 1n chromosomes—or half the normal number—the fertilized egg—or **zygote**— is back to the normal diploid or 2n number of chromosomes for the species. It has DNA from *both* parents, however, and is therefore different from either. This **recombination** of genetic information to produce offspring that differ slightly from their parents is probably the most important feature of sexual reproduction.

In sexual reproduction, eggs and sperm fuse to create a genetically distinct individual.

The fertilized egg begins to divide by ordinary mitosis. Now called an **embryo,** it is nourished by the yolk. Through an extraordinarily complex

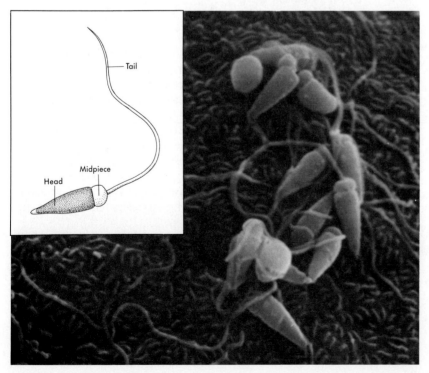

Figure 4-26 A sperm from the variegated sea urchin *(Lytechinus variegatus)* is shown here at a higher magnification than used in Figure 4-25. The sperm has three parts. The head is made up almost entirely of the cell nucleus and contains the genetic material. The midpiece contains mitochondria that power the tail or flagellum.

Figure 4-27 Marine organisms have many different reproductive strategies, a few of which are shown here. Corals and many other marine organisms simply release huge numbers of gametes into the water and have no further contact with their young (A). Some fish, like this jawfish (*Opistognathus macrognathus*), carry the eggs in their mouths until the larvae hatch (B). Seals feed and protect their young for a long time after they are born (C). The seeds of the coconut tree, which contain coconuts, can float for long periods in saltwater (D) and germinate when they wash up on shore.

process called **embryologic development,** the embryo eventually becomes a new adult of the species. Along the way, most marine organisms pass through immature, or **larval,** stages that often look completely different from the adults (see Figure 6-45).

Reproductive Strategies

The goal of reproduction for any organism is to pass on its hereditary characteristics to a new generation. There is a nearly endless variety of ways to achieve this (Figure 4-27). Some species release millions of eggs and sperm into the water and never see their offspring again. Others issue only a few offspring and invest a lot of time and energy in caring for them. Some species have many different larval stages, whereas others develop directly from egg to adult. Some reproduce asexually, some sexually, and some do both.

The particular combination of methods used by a given species is called its **reproductive strategy.** The strategy used by a species depends on its size, where and how it lives, what kind of organism it is, in fact, on just about everything about it. Learning how reproductive strategies relate to lifestyles is a common pursuit of marine biologists.

THE DIVERSITY OF LIFE IN THE SEA

There are so many different things that live in the sea that it almost boggles the mind. From microscopic bacteria to gigantic whales, marine organisms come in all shapes, sizes, and colors. The ways by which they live are just as varied. Making sense of all this **diversity** might seem a hopeless task. Fortunately, there is a unifying concept that helps make the bewildering diversity of life comprehensible. This concept is the theory of **evolution.** Remember that scientists do not use the term "theory" lightly. Evolution, the gradual alteration of a species' genetic makeup, is supported by a vast body of evidence. It is as well established as the theory of gravity. The *way* in which evolution occurs, on the other hand, never ceases to fascinate biologists.

scientific theory An idea about nature that is accepted as "true" for the time being because it has passed test after test and is supported by a large body of evidence
Chapter 1, p. 20

Natural Selection and Adaptation

Evolution occurs because individual organisms differ in their ability to find food and avoid being eaten, in their success at producing offspring, in metabolism, and in countless other attributes. The best **adapted** individuals are those who are most successful at meeting the challenges of the environment. They will tend to survive longer and produce more offspring than those who are not so well adapted. This process was called **natural selection** by Charles Darwin, the nineteenth-century English naturalist who along with another Englishman, Alfred Wallace, first proposed the theory of evolution.

Individuals who are favored by natural selection, those best adapted to their environment, will tend to pass their favorable characteristics on to their more numerous offspring. The favorable traits will grow more common, and the population as a whole will gradually become more similar to the best-adapted individuals. Thus the result of natural selection is that the population continually adapts to its environment; that is, it evolves.

> **Natural selection occurs when some members of a population survive and reproduce more successfully than others. Evolution is the genetic change in the population that results because these individuals pass their favorable characteristics to their young.**

Every population is constantly adapting to its environment. The world is an ever-changing place, however, and organisms are always being faced with new challenges. Populations either adapt to the changes in the environment or become extinct, making way for others. As a result, evolution is an endless process.

Classifying Living Things

Since the appearance of life on earth some 3.8 billion years ago (see Box 4-1), the adaptation of various populations to different environments has produced a fantastic variety of life forms. There are now perhaps 10 million different kinds of living organisms on earth. In order to study and discuss all these organisms, we must classify them, or sort them into groups. One of the goals of biologic classification is to give a universally accepted name to different types of organisms so that scientists from all over the world can call a species by the same name. Another goal is to group organisms according to their "relatedness." Most people instinctively understand that a seal is related to a sea lion, for example, or that oysters and clams are related. They might be hard-pressed, though, to explain exactly what they mean by "related." To a biologist, "related" refers to organisms that share a similar evolutionary history, or **phylogeny.**

BOX 4-1

How Did It All Begin?

Today the earth teems with living things, all of which arise by the reproduction of preexisting organisms: Life only comes from life. This process has been going on for billions of years. We know of organisms that look like simple bacteria from fossils that are 3.8 *billion* years old, nearly as old as the oceans themselves. As old as it is, though, life had to start somewhere. How did life first arise? After all, it couldn't have just started by itself, could it?

Some tantalizing evidence suggests that it could. Since all life depends on organic molecules, the first step in the origin of life must have been the formation of these molecules. Today, it is organisms that produce organic molecules, so where did they come from if there *were* no living things? In the 1920's A.I. Oparin and J.B.S. Haldane hypothesized that organic molecules could have arisen spontaneously in the conditions of the primitive earth.

The earth of 4 billion years ago was a very different place than it is today. The oceans had just begun to form and were quite shallow. The atmosphere was mostly carbon dioxide, nitrogen, and water vapor. It probably also con-

tained a noxious "soup" of methane, ammonia, hydrogen sulfide (the chemical that gives rotten eggs their smell), and other simple molecules. There was no oxygen in the primitive atmosphere, because there were no plants. Today the oxygen in our atmosphere, in the form of ozone (O_3), protects us from most of the harmful ultraviolet radiation of the sun. The early earth, because of the lack of oxygen in the atmosphere, was bombarded with huge amounts of ultraviolet radiation. Intense electrical storms were common, and volcanic eruptions wracked the earth. The Oparin-Haldane hypothesis holds that radiation, electrical discharges, and volcanic heat provided the energy for the formation of simple organic molecules in the primitive soup of the oceans.

In the 1950's, experiments by Stanley L. Miller and Harold C. Urey provided evidence to support the Oparin-Haldane hypothesis. Miller and Urey devised an apparatus that mimicked the conditions of the early earth. They created a mixture of gases thought to resemble the primitive atmosphere and subjected it to powerful electric sparks.

It is almost always difficult to determine the phylogenetic relationships among species. Very few groups have a good record of fossils that trace their evolutionary past (Figure 4-28). In most cases the fossil record is incomplete or even nonexistent. Biologists have to try to piece together an organism's evolutionary history from other evidence. Body structure, reproduction, embryologic and larval development, chemical composition, and behavior all provide clues to the past. The answers are not always certain, however, and the experts often argue about the phylogeny of their respective groups. Classification schemes change and new arguments arise as fresh information comes in.

The Species Concept. The fundamental unit of evolution, and therefore of biologic classification, is the species. We have loosely defined a species as a "kind" of organism, but what exactly does this mean? After all, fishes are all the same general kind of animal, but a sailfish is obviously

LIFE IN THE MARINE ENVIRONMENT

This procedure eventually resulted in the spontaneous formation of a whole variety of organic molecules, including amino acids and one of the constituents of nucleic acids. The experiment has since been repeated successfully with other hypothetic versions of the primitive atmoshpere. It is thus at least possible for organic molecules to be created without living things.

Some scientists believe that the original organic molecules necessary for life did not originate on earth. Organic molecules occur in space, and in interplanetary dust, meteorites, and comets. Halley's comet, for example, is thought to contain organic matter equivalent to about 10% of all the living matter on earth! In its early years the earth was struck by a barrage of comets and meteorites.

It is not difficult, then, to account for the presence of simple organic molecules. It is harder to explain how the molecules came to be organized into living systems. It has been found, however, that organic molecules will tend to cluster together, or aggregate, in much the same way that droplets of oil will clump together on the water's surface. Under the right conditions these organic aggregates resemble simple cells. In particular, if a mixture of lipids—the main component of the cell membrane—and DNA is repeatedly dried out and soaked again, the lipid tends to surround the DNA. The result is a bubble of water containing DNA surrounded by a membrane-like layer of lipid, in other words, a "protocell."

It has long been thought that life probably first arose in shallow pools at the edge of the ocean. Here, ultraviolet radiation would not be filtered out by the water so there would be a maximal amount of energy available for the formation of organic molecules. It has also been suggested that life arose not in shallow tide pools, but in warm pools on land left behind after the impact of comets, which are largely ice. Another hypothesis, though highly controversial, is that life arose at hydrothermal vents. The idea is that heat energy from vulcanism at mid-ocean ridges provided the energy for the formation of the first organic molecules. We may never have all the answers, but the possibilities are intriguing.

not the same species as a tuna (see Figures 1-30 and 1-32). No definition is perfect, but we will define species as populations of organisms that have common characteristics and can successfully breed with each other. Successful breeding means that the offspring produced are fertile and can themselves propagate the species. Dogs, to use a familiar example, are all the same species because all breeds of dog—no matter how different they may look—can interbreed. Similarly, all blue whales (*Balaenoptera musculus*) in the ocean are expected to be able to breed, as long as they get together and are of the opposite sex. Organisms that cannot breed with each other are not members of the same species no matter how similar they look. When two populations are unable to interbreed successfully, they are said to be **reproductively isolated.**

A species is a population of organisms that share common characteristics, can breed with each other, and are reproductively isolated from other populations.

Figure 4-28 A fossil trilobite. Trilobites dominated the oceans about 500 million years ago, but went extinct some 300 million years later.

Figure 4-29 *Panulirus interruptus,* the Pacific spiny lobster.

Biologic Nomenclature. Organisms are identified biologically by two names, their species name and the name of their **genus.** A genus is a group of very similar species. Dogs, for example, have the scientific name of *Canis familiaris.* They are closely related to other species in the genus *Canis,* like wolves, *Canis lupus,* and coyotes, *Canis latrans.* Similarly, the genus *Balaenoptera* contains several related species of whales, including the blue whale, *Balaenoptera musculus;* the fin whale, *Balaenoptera physalus;* and the minke whale, *Balaenoptera acutorostrata.* This two-name system is called **binomial nomenclature.** It was first introduced in the eighteenth century by the Swedish biologist Carl Linnaeus. At that time, Latin was the language of scholars, and Latin is still used for scientific names, along with Greek and other languages. By convention, binomial names are always underlined or italicized. The first letter of the generic name, but not the species name, is capitalized. When the name of the genus is obvious it may be abbreviated. Thus, the blue whale could be identified as *B. musculus.*

Students often ask why biologists have to use complicated, hard-to-pronounce Latin or Greek names rather than ordinary common names. The trouble with common names is that they are not very precise. The same common name may be applied to different species, and the same species may have different common names. The name "spiny lobster" (Figure 4-29), for example, is applied to many different species that are not even all in the same genus. Australians call spiny lobsters "crayfish." Americans reserve the name "crayfish" for freshwater relatives of lobsters. Things get even more confusing when people speak different languages. In Spanish, spiny lobsters are called *langostas,* which can also refer to grasshoppers. Another example is the name "dolphin," which is applied not only to lovable relatives of whales, but to a delicious game fish. Dolphin fish are called *dorado* in Latin America and *mahi-mahi* in Hawaii. At the local seafood grotto this confusion may only be a minor annoyance. To biologists, however, it is absolutely essential to identify precisely whatever species they are discussing. The use of scientific names that are accepted world-wide avoids any possible confusion.

Confusion over common names can have practical implications as well. After a dive on a coral reef near Australia, one of the authors mentioned to his partner that he had seen a grey reef shark. Grey reef sharks, although not so dangerous as to force you out of the water, can be aggressive and merit some attention and respect. The diving partner, who was familiar with the area, replied that the shark was not a grey reef, but a "graceful whaler." "Looks just like a grey reef shark to me," said the author but, reassured, he continued diving, blissfully ignoring the "graceful whalers" that swam by. Only later did he discover that "graceful whaler" was just a local name for the grey reef shark, *Carcharhinus amblyrhynchos!*

Higher Taxa. Just as similar species are grouped into a genus, genera with similar phylogenies are grouped together. This process can be continued, sorting organisms into progressively larger groups. All the members of a group, or **taxon** (plural, **taxa**), have certain common characteristics and are thought to share a common ancestry. Taxa are systematically arranged in a hierarchy that extends from the most general classification, the kingdom, down to the species (Table 4-2).

The Five Kingdoms of Life. The kingdom is the most general taxon in biologic classification. For many years, a simple two-kingdom system was

LIFE IN THE MARINE ENVIRONMENT

TABLE 4-2

Biologic classification of the sea lettuce, *Ulva lactuca*, and the American lobster, *Homarus americanus*.

TAXONOMIC LEVEL	EXAMPLE	
Kingdom	Plantae	
Division*	Chlorophyta	
Class	Chlorophyceae	
Order	Ulvales	
Family	Ulvaceae	
Genus	*Ulva*	Sea lettuce.
Species	*lactuca*	
Kingdom	Animalia	
Phylum*	Arthropoda	
Class	Malacostraca	
Order	Decapoda	
Family	Nephropidae	
Genus	*Homarus*	
Species	*americanus*	Lobster.

*"Division" is often used in place of "phylum" in the classification of plants.

satisfactory. All organisms could be classified as either plants (kingdom **Plantae**) or animals (kingdom **Animalia**). As biologists learned more about the living world, however, they began to discover many organisms that did not fit neatly into one of these two categories. The prokaryotic bacteria are clearly neither plant nor animal and are placed in a kingdom of their own, the **Monera.**

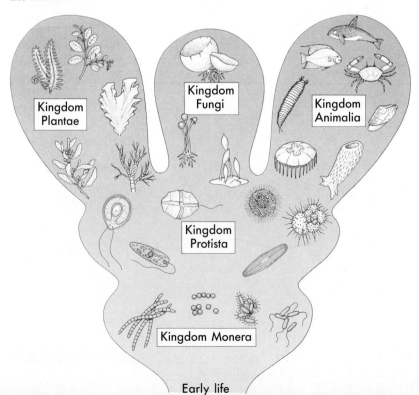

Figure 4-30 The "family tree" of the five kingdoms of life.

Biologists do not all agree on the kingdom to which some organisms belong. In fact, there is not even universal agreement about how many kingdoms there are. In this text we use a widely accepted classification that includes five kingdoms (Figure 4-30). The two kingdoms in addition to the Monera, Animalia, and Plantae are the **Protista** and the **Fungi.** Protists are an extremely varied group of mostly unicellular organisms, some plant-like, others animal-like. The fungi are multicellular plant-like organisms with unique characteristics that distinguish them from both plants and animals.

SUMMARY

1. There are four major groups of organic molecules: _____, _____, _____, and _____.
2. The energy in _____ is used by photosynthetic organisms to make _____. _____, _____ and water are used as raw materials, and _____ is released.
3. _____ is broken down by _____, and the energy that is released is stored in molecules of _____.
4. When primary production in the ocean is nutrient limited, the nutrient that is most often limiting is _____.
5. The major difference between prokaryotes and eukaryotes is the presence of _____-bound _____.
6. The movement of dissolved substances from areas of high concentration to areas of low concentration is called _____. _____ is the diffusion of water through a _____.
7. _____ is the control of the internal concentration of a solute.
8. The body temperatures of _____ go up and down as the water gets warmer or colder. _____ maintain a relatively constant body temperature.
9. Most cells divide by the process of _____, which produces two identical cells called _____.
10. Sexual reproduction involves the fusion of male and female _____, which are produced by a form of cell division called _____.
11. The fundamental unit of evolution, as well as of biologic classification, is the _____.
12. The five _____ of life are the Animalia, Plantae, _____, _____, and _____.

LIFE IN THE MARINE ENVIRONMENT

THOUGHT QUESTIONS

1. During the day, plants carry out both photosynthesis and respiration but at night, when there is no light, they perform only respiration. Small, isolated tide pools on rocky shores are often inhabited by thick growths of seaweeds, which are plants. Would you expect the amount of oxygen in the water to differ between night and day? How?
2. Some marine plants and animals are known to have high concentrations in their cells of ions found in minute amounts in seawater. Could these organisms accumulate the ions by diffusion? Formulate a hypothesis to explain how this accumulation is accomplished.

FOR FURTHER READING

Hall, S.S.: "James Watson and the Search for Biology's 'Holy Grail'." *Smithsonian*, vol. 20, no. 11, February 1990, pp. 40-49. *Scientists are embarking on what has been described as "biology's moon shot"—an attempt to identify every gene in human beings. Heading this search is James Watson, one of the scientists who orginally discovered the structure of DNA.*

Hendry, G.: "Making, Breaking, and Remaking Chlorophyll." *Natural History*, May 1990, pp. 36-41. *The continuing cycle in which chlorophyll—the pigment that allows plants to capture sunlight—is produced, broken down, and recycled within the plant is described.*

Horgan, J.: "In the Beginning . . ." *Scientific American*, vol. 264, no. 2, February 1991, pp. 116-125. *A review of current views on when, where, and how life first appeared on our planet.*

Trachtman, P.: "The Search for Life's Origins—and a First 'Synthetic Cell'." *Smithsonian*, vol. 15, no. 3, June 1984, pp. 42-51. *Various hypotheses about how life began are discussed.*

Marine Plants and Other Plantlike Organisms

5

As inhabitants of land, our perception of the world of plants is based mostly on things like trees, ferns, and mushrooms. The oceans are populated by a fascinating group of plants and plantlike organisms that are very different from the plants that surround us. The world of marine plants is full of surprising and unique forms of life.

This chapter, the first in our survey of life in the oceans, is devoted to the marine organisms traditionally grouped as "plants." Many of them, however, are not true plants at all. Nevertheless, many of these organisms are **primary producers** and thus capable of using light energy by means of photosynthesis. Still, this is not always the case! Many microscopic organisms normally thought of as plants do not photosynthesize. Some swim about in the water and may even "feed" like animals. Such organisms cannot be easily categorized as either plants or animals. As a result of all of these exceptions, the conventional, familiar concept of "plants" has been modified into the five-kingdom scheme outlined in Figure 4-30.

No matter how they are classified, marine primary producers have played a crucial role in the evolution and maintenance of life on our planet. Without them, the world we live in would be very different. The seaweeds illustrated in Figure 5-1 have transformed light energy from the sun into chemical energy, making it available to a long list of hungry creatures, which may include humans. Seaweeds may be inhabited by other organisms that live on, or even among, their tissues. The oxygen they produce is used by other organisms living on land, as well as in the ocean. Microscopic plantlike organisms will break down the seaweed after it dies, releasing nutrients that are needed for the growth of other plants.

Figure 5-1 Seaweeds, displaying many growth forms, textures, and colors, thrive in well-lit rocky bottoms along temperate coasts.

PROKARYOTES

The organisms known as **prokaryotes** appear to be among the simplest, most primitive forms of life. Some look very similar to the oldest fossils ever found, cells estimated to be at least 3.8 billion years old. Thousands of species of prokaryotes are still very much alive, and they play crucial roles in

prokaryotes Organisms with a simple cell structure without a nucleus or other organelles, such as bacteria and blue-green algae, or cyanobacteria
Chapter 4, p. 94; Figure 4-12

Figure 5-2 Photograph taken using a high-magnification electron microscope of *Cyclobacterium marinus*, a ring-forming marine bacterium. The bar represents 0.1 micron. Also see Figure 14-30.

Cell wall

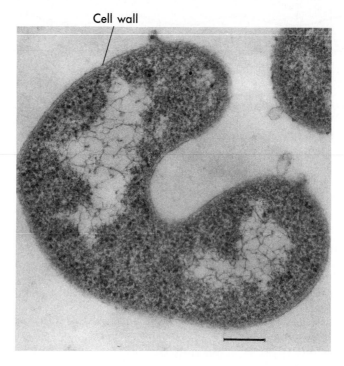

the ocean. Living prokaryotes are represented by two major groups, whose characteristic cell structure is enough to have them placed in their very own kingdom, **Monera** (see the general classification scheme in Figure 5-29).

Bacteria

Bacteria (division **Schizophyta** or **Eubacteria**) are microscopic, single celled, and simple in structure. A typical bacterial cell is rod shaped and bound by a rigid, protective **cell wall** (Figures 5-2 and 14-30). A gelatinous covering often surrounds the cell wall, perhaps as additional protection or to provide the bacterium with a means of attachment to its surroundings. A **plasma,** or **cell, membrane** is present immediately below the cell wall. Like all prokaryotes, the cell lacks microscopically visible organelles.

Bacteria represent the smallest living creatures on earth (see Appendix B for size scale). About 250,000 averaged-sized bacteria would fit on the dot at the end of this sentence. In large numbers, bacteria may sometimes be visible as iridescent or pink patches on the surface of stagnant pools in mud flats and salt marshes. Sometimes gigantic strands of cells are visible as whitish hairs on rotting seaweed. Under optimal conditions they reproduce rapidly by simple cell division, as many as two per hour, so extremely high numbers of them can be found in sediments or decomposing organic matter.

Though structurally primitive, bacteria are able to carry out nearly all of the basic chemical reactions found in more complex forms of life. In fact, most of these fundamental reactions probably first developed in bacteria. In

Five kingdom system of classification
Chapter 4, p. 113; Figure 4-30
  Monera   Prokaryotes: bacteria and blue-green algae
  Protista   Unicellular eukaryotes, many of which combine characteristics of both plants and animals: unicellular algae, protozoans
  Fungi   Nonphotosynthetic, plantlike eukaryotes: fungi and molds
  Plantae   Eukaryotic and mostly multicellular, photosynthetic: seaweeds, land plants
  Animalia   Eukaryotic and multicellular, nonphotosynthetic: animals

addition, bacteria can perform some chemical processes found in no other groups.

Heterotrophic Bacteria. Most bacteria, like animals, are **heterotrophs** and obtain their energy from other organisms. Most of these bacteria are **decomposers** and break down waste products and dead organic matter. These bacteria, the **decay bacteria,** are vital to life on earth because they ensure the recycling of essential nutrients. They are found everywhere in the marine environment—on almost all surfaces, as well as in the water. They are especially abundant in bottom sediments, which are usually rich in organic matter. All organic matter is sooner or later decomposed, though in very deep, cold water the process is slower than elsewhere (see the section in Chapter 15, "Bacteria in the Deep Sea").

Decay and other types of bacteria play a crucial role, because they are a major part of the organic matter that feeds countless bottom-dwelling animals. Even organic particles found free in the water are mostly made up of bacteria. Other marine bacteria are beneficial, because they are involved in degrading oil and other toxic pollutants that find their way into the environment. The same process of decomposition is unfortunately involved in the spoilage of valuable fish and shellfish catches.

Autotrophic Bacteria. **Autotrophic** bacteria make their own organic compounds. Some of them are **photosynthetic** (see Figure 4-12). They contain chlorophyll and, like green plants, derive energy from organic compounds they manufacture themselves by the use of light energy. Other bacterial autotrophs are called **chemosynthetic** (or **chemoautotrophic**) because they derive energy not from light, but by releasing energy stored in chemical compounds such as hydrogen sulfide (H_2S, responsible for the rotten-egg smell often detected in sewage or mud flats), other sulfur or iron compounds, or methane (CH_4).

> Bacteria are simple microorganisms that are especially significant as decomposers, breaking down organic compounds into nutrients that can be used by other organisms. They are also an important food source and help degrade pollutants.

Blue-Green Algae, or Cyanobacteria

Among the most primitive plantlike organisms, the **blue-green algae,** also known as **cyanobacteria,** (division **Cyanophyta** or **Cyanobacteria**) are prokaryotes specialized to perform photosynthesis. In addition to chlorophyll, most contain a bluish pigment called **phycocyanin** that is responsible for the common name of the group (Table 5-1). Most biologists consider them bacteria because they have a prokaryotic cellular organization, and they are technically called cyanobacteria. Photosynthesis takes place on folded membranes within the cell rather than in chloroplasts as in eukaryotic plants. There are, however, similarities between blue-green algae and more complex primary producers like the seaweeds in the way photosynthesis takes place: They share a type of chlorophyll, chlorophyll *a*, and liberate gaseous oxygen. Most marine blue-green algae are microscopic, although

heterotrophs Such organisms as animals that can obtain energy only by eating organic matter taken from other organisms
Chapter 4, p. 90

autotrophs Such organisms as plants and other producers that are able to manufacture their own organic matter by using energy from sunlight or other sources
Chapter 4, p. 90

Figure 5-3 Stromatolites, calcareous mounds deposited by blue-green algae, are frequently found as fossils. These, however, are living stromatolites recently discovered in shallow water in the Exuma Cays, Bahama Islands.

TABLE 5-1

Chemical compounds used in the characterization of marine plants and other primary producers

	PHOTOSYNTHETIC PIGMENTS	MAJOR STORAGE PRODUCTS	MAJOR CELL-WALL COMPONENTS
Blue-green algae (flexibacteria)	Chlorophyll *a* Phycobilins (phycocyanin, phycoerythrin, etc.) Carotenoids	Cyanophycean starch Cyanophycin (protein)	Chains of amino sugars and amino acids
Diatoms	Chlorophyll *a, c* Carotenoids	Chrysolaminarin Oil	Silica Pectin
Dinoflagellates	Chlorophyll *a, c* Carotenoids	Starch Oil	Cellulose
Green algae	Chlorophyll *a, b* Carotenoids	Starch	Cellulose Carbonates in coralline algae
Brown algae	Chlorophyll *a, c* Carotenoids (fucoxanthin, etc.)	Laminarin Oil	Cellulose Algin
Red algae	Chlorophyll *a, d* Phycobilins (phycocyanin, phycoerythrin) Carotenoids	Starch	Agar Carrageenan Cellulose Carbonates in coralline algae
Flowering plants	Chlorophyll *a, b* Carotenoids	Starch	Cellulose

some form long filaments, strands, or thick mats visible to the naked eye.

Blue-green algae are believed to have been among the first photosynthetic organisms on earth. They probably had an important role in the accumulation of oxygen in our atmosphere. Fossil **stromatolites,** massive **calcareous** mounds formed by blue-green algae, are known to date back some 3 billion years. Stromatolites are still being formed in tropical seas (Figure 5-3).

Many species of blue-green algae are marine. They can tolerate wide ranges of salinity and temperature and are therefore found practically everywhere, including the least expected places, including the hair of polar bears! Some, called **endolithic,** burrow into calcareous rocks and coral skeletons. Others form thick, dark crusts along the wave-splash zone of rocky coasts. Polluted, oxygen-poor sediments are exploited by some blue-greens. Planktonic species may rapidly multiply and change the color of the water. Even some of the so-called **red tides** (see Box 14-1) are caused by blue-green algae containing a red pigment. A few species are responsible for skin rashes on swimmers and divers.

Many blue-green algae are known to carry out **nitrogen fixation,** converting gaseous nitrogen (N_2) into precious nitrogen compounds that can be used by other primary producers (see the section in Chapter 9, "Cycles of

calcareous Made of calcium carbonate ($CaCO_3$)
Chapter 2, p. 40

planktonic organisms Those organisms found drifting in the water
Chapter 4, p. 98

Essential Nutrients"). Some blue-green algae live on the surface of seaweeds and seagrasses. Plants that live on other plants are called **epiphytes.** Still others are known to lose their ability to photosynthesize, thus becoming heterotrophs (Figure 5-4).

The blue-green algae, also known as cyanobacteria, are photosynthetic prokaryotes widely distributed in the marine environment.

UNICELLULAR ALGAE

"Algae" (singular "alga"), a widely used term already introduced in the previous section, actually has different meanings for different people. What are algae? It is useful to say they are "simple," mostly aquatic (that is, marine and fresh water), mostly photosynthetic organisms. Except for the blue-green algae, all are eukaryotic. Their cells contain organelles enclosed by membranes. Photosynthesis takes place in **chloroplasts**—green, brown, or red organelles with layers of internal membranes. The color of algae is a reflection of the pigments and their concentration. In contrast to the land plants with which we are familiar, algae lack flowers and have relatively simple reproductive structures. Their non-reproductive cells are also mostly simple and unspecialized. Algae lack true leaves, stems, and roots.

Most biologists refer to algae as plants. Many of the unicellular algae, however, show animal-like characteristics. Some swim by moving their **flagella.** Distinguishing these free-swimming algae from the simpler animals becomes rather difficult, especially when there are species that carry out photosynthesis like plants, while very similar ones move and eat food particles like animals. Some are claimed by both botanists (plant biologists) and zoologists (animal biologists)! A widely accepted compromise is to group all of these unicellular forms in a separate kingdom, **Protista** (see Figure 5-29). Protists that contain chlorophyll are discussed in this chapter with other marine primary producers.

Algae are simple, mostly aquatic organisms that lack the specialized tissues of higher plants. They range in size and complexity from single cells to multicellular seaweeds.

Diatoms

Diatoms (division **Chrysophyta,** class **Bacillariophyceae**) are unicellular, though many aggregate into chains or starlike groups. These microscopic organisms consist of cells characteristically enclosed by cell walls made largely of **silica (SiO_2),** a glasslike material (see Table 5-1). The glassy "shell," or **frustule,** consists of two tightly fitting halves often resembling a flat, round, or boat-shaped box (Figure 5-5). The frustule typically has intricate perforations and ornaments such as spines or ribs, making diatoms strikingly beautiful when seen under a microscope (see Figure 14-3). The frustule allows light to pass through, so that the conspicuous golden-brown chloroplasts can capture light energy for photosynthesis. The minute perforations allow dissolved gases and nutrients to enter and exit. Oil droplets and spines help

eukaryotes Organisms with cells containing a nucleus and other organelles that are enclosed by membranes, such as seaweeds, fungi, plants, and animals
Chapter 4, p. 95; Figure 4-13

Figure 5-4 An electron-microscope photograph of a sponge (*Hymenamphiastra cyanocrypta*) that reveals filaments of a blue-green alga, or cyanobacterium. Two large sponge cells are at the left; a cross-section of a round filament is at the top. The filaments lack chlorophyll. Is it correct to call this organism an "alga"?

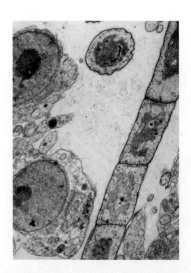

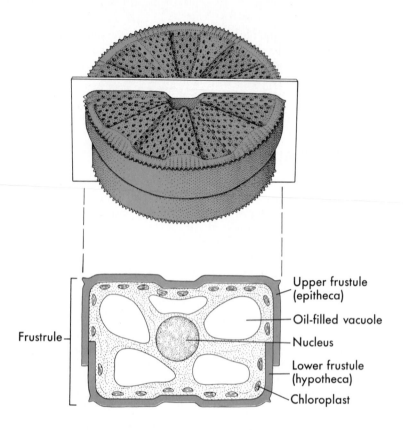

Figure 5-5 Diagrammatic representation of a diatom cell as seen with the cell cut across. Also see Figures 4-11 and 14-3.

Upper frustule (epitheca)

Oil-filled vacuole

Nucleus

Lower frustule (hypotheca)

Chloroplast

Frustrule

to keep planktonic cells from sinking below the well-lit surface water (see the section in Chapter 14, "Staying Afloat").

The characteristic color of diatoms is due to yellow and brown **carotenoid** pigments present in addition to two types of chlorophyll, *a* and *c* (see Table 5-1). Diatoms are efficient photosynthetic factories, producing much needed food (the food being the diatoms themselves), as well as oxygen for other forms of life. They are the most important open-water, nonattached primary producers in temperate and polar regions (see the section in Chapter 14, "The Nature of Epipelagic Food Webs"). In fact, billions of diatom cells in the ocean account for a hefty share of the food and oxygen produced on planet earth.

As many as half of the estimated 12,000 living species of diatoms are marine. Most are planktonic, but many produce a stalk for attachment to rocks, nets, buoys, and other surfaces. The brownish scum sometimes seen on mud flats or glass aquaria often consists of millions of diatom cells. Some, perhaps most, are able to slowly glide on surfaces. A few are colorless and, having no chlorophyll, live on the surfaces of seaweeds as heterotrophs.

Diatoms are unicellular organisms that live mostly as part of the plankton. A silica "shell" is their most distinctive feature. They are the most important open-water primary producers in cold waters.

LIFE IN THE MARINE ENVIRONMENT

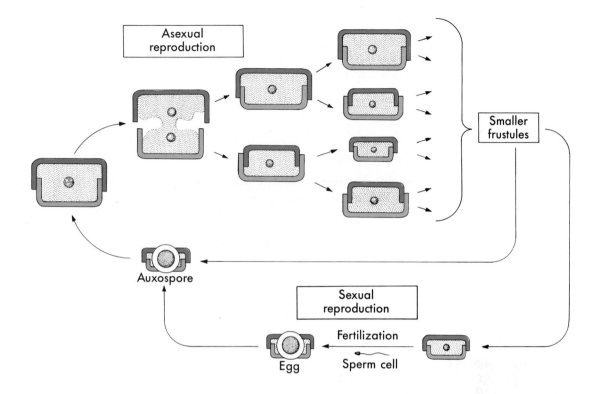

Figure 5-6 The usual method of reproduction in diatoms is by cell division. Most, but not all, frustules get smaller with time. Resistant auxospores are produced two ways: directly from the expansion of a smaller frustule or by sexual reproduction when an egg is fertilized by a sperm cell, both of which are liberated from separate frustules.

Diatoms reproduce mostly by cell division, a type of **asexual reproduction.** The overlapping halves of the frustule separate, and each secretes a new, smaller half (Figure 5-6). Optimal environmental conditions, such as the provision of adequate nutrients and temperature, trigger periods of rapid reproduction called **blooms.** This is a general phenomenon that may occur in other algae. During blooms, most diatoms get progressively smaller, partly because they use up the silica dissolved in the water. Maximal size may be regained by the development of resistant resting stages known as **auxospores.** The auxospores eventually give rise to cells that display the frustule characteristic of the species. Some auxospores are produced as a result of **sexual reproduction.** Some diatom cells develop eggs, others develop flagellated sperm (see Figure 5-6). Fertilization then results in the development of auxospores.

The glassy frustules of dead diatoms are fairly resistant to decomposition and eventually settle on the bottom of the sea floor. Here they may form

Types of reproduction:
asexual (or *vegetative*) the production of new individuals by simple division (or other means) without involving *gametes* so that the offspring are identical to the parents
Chapter 4, p. 105
sexual the production of new individuals by the formation of *gametes* (sperm and eggs) so that the offspring are different from the parents
Chapter 4, p. 106

thick deposits of siliceous material that cover large portions of the ocean floor. Such sediments are known as **diatomaceous ooze.** Huge fossil deposits of these sediments can now be found inland in various parts of the world. The siliceous material, or **diatomaceous earth,** is mined and used in products such as filters for swimming pools, for clarifying beer, as temperature and sound insulators, and as mild abrasives that may find their way into toothpaste.

Dinoflagellates

The **dinoglagellates** (division **Pyrrhophyta**) constitute another large group of planktonic, unicellular organisms important in the economy of the oceans. Their most outstanding characteristic is the possession of two unequal flagella, one wrapped around a groove along the middle of the cell and one trailing free (Figure 5-7). These flagella direct movement in practically any direction. Most have a cell wall that is externally armored with plates made of **cellulose,** the characteristic component of the cell walls of seaweeds and land plants (see Table 5-1). The plates may have spines, pores, or other ornaments.

Though most dinoflagellates have chlorophyll, many can also ingest food particles. A few have a light-sensitive pigment spot that acts as a crude "eye." For these reasons, dinoflagellates are sometimes classified as animals. It has been suggested that in the course of their evolution, dinoflagellates have gained the ability to function as primary producers by "capturing" and using chloroplasts from other algae. Almost all known dinoflagellates, around 1,200 living species, are marine. In cold waters they are usually second in importance to the diatoms among the planktonic primary producers. In the tropics, however, they are typically the dominant group.

Dinoflagellates reproduce almost exclusively by simple cell division. They may form blooms that often color the water red, reddish-brown, yellow, or other unusual shades. Such blooms are sometimes called red tides (see Box 14-1). Some of these dinoflagellates release substances poisonous to other forms of life, and seafood collected during red tide periods may be poisonous. Other dinoflagellates are noted for the production of light, a phenomenon referred to as **bioluminescence** (see Box 5-1). Though bioluminescence has also been observed in some bacteria and in many types of animals, dinoflagellates are generally responsible for the diffuse light sometimes observed in the open sea. This effect is of course seen only at night. It is especially bright if the water is disturbed by a boat or when a wave crashes on the shore.

A variety of marine animals contain in their tissues round, golden-brown photosynthetic cells known as **zooxanthellae.** These cells are actually dinoflagellates adapted to live in close association with an animal host (see Figure 13-9). Though hosts range from sponges and sea anemones to giant clams, it is perhaps in reef-building corals where zooxanthellae are most significant. They fix carbon dioxide by photosynthesis, release organic matter used by the coral, and also help in the formation of the coral skeleton (see the section in Chapter 13, "Coral Nutrition"). A few highly-modified di-

cellulose Complex carbohydrate characteristic of plants and other primary producers
Chapter 4, p. 87

Figure 5-7 Dinoflagellates such as *Gonyaulax (above)* and *Ceratium (below)* have a cell wall, or theca, that consists of cellulose plates. The theca is marked by grooves for the flagella. Species of *Gonyaulax* are responsible for red tides.

noflagellates are known to be parasites of seaweeds and some marine animals. Like zooxanthellae, these highly specialized forms reveal their true nature by having life cycles that include free-swimming stages resembling typical dinoflagellates.

> Dinoglagellates are unicellular organisms that have two unequal flagella. They are mostly marine and are more common in the tropics. Some are noted for their emission of light; others are closely associated with marine animals, especially reef corals.

Other Unicellular Algae

Three additional groups of primary producers may be abundant in some areas. **Silicoflagellates** (division **Chrysophyta**, class **Chrysophyceae**) are characterized by a star-shaped internal skeleton made of silica and a single flagellum (Figure 5-8). **Coccolithophorids** (division **Chrysophyta**, class **Prymnesiophyceae**) are flagellated, spheric cells covered with buttonlike structures called **coccoliths** that are made of calcium carbonate (see Figure 14-4). Coccoliths may be found in sediments, including those of fossils. **Cryptomonads** (division **Cryptophyta**) have two flagella and lack a skeleton. Members of these three groups are so small that hundreds may fit into a large diatom or dinoflagellate cell.

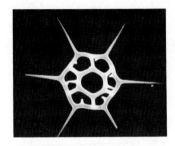

Figure 5-8 *Dictyocha speculum*, a silicoflagellate.

FUNGI

Fungi are eukaryotic organisms that are superficially plantlike but unable to photosynthesize because they lack chlorophyll. Most are multicellular. Fungi, including molds and yeasts, are most often grouped in the kingdom **fungi** (see Figure 4-30).

There are at least 500 species of marine fungi. Many, like bacteria, are decomposers of dead organic matter. Others are parasites that cause diseases of economically important seaweeds, sponges, shellfish, and fish. Still others live in close association with algae to form unique entities, the **lichens.** In lichens the long, filament-like growths of the fungi provide support, while the algae provide food manufactured by photosynthesis. Marine lichens can be found as thick, dark-brown or black patches on the wave-splashed zone of rocky shores, particularly on north Atlantic shores. By comparison with the multitude of lichen species on land, there are very few species of marine lichens.

MULTICELLULAR ALGAE: THE SEAWEEDS

Most species of marine algae are represented by the forms popularly known as **seaweeds.** This, however, is a rather unfortunate term. For one thing, the word "weeds" does not do justice to these conspicuous and often elegant inhabitants of rocky shores and other marine environments. Some biologists opt for the more formal name of **macrophytes.** On the other hand, the term "seaweeds" is useful in distinguishing them from the unicellular algae surveyed in the previous section. The structures of seaweeds are far more complex than those of unicellular algae, and reproduction is also more elaborate. Seaweeds are all eukaryotic. Most are multicellular, but some forms consisting of single cells or simple filaments are considered seaweeds. This is because the classification of seaweeds (not a simple matter, by the way!) is based not only on structure, but also on other features such as the types of pigments and food storage products (see Table 5-1).

Though more advanced than unicellular algae, seaweeds still lack the complex structures and reproductive mechanisms characteristic of the higher, mostly terrestrial plants. Nevertheless, most specialists prefer to include them, together with the higher plants, in the kingdom **Plantae** (see Figure 5-29). As usual, there are some who disagree and assign them instead to the kingdom Protista (together with the unicellular algae) because of their relatively simple structure.

The range of variation observed among the multicellular algae is spectacular. Those we see on rocky shores at low tide are usually small and sturdy as an adaptation to withstand waves (see Figure 5-20). Some small, delicate ones live as epiphytes or parasites on other seaweeds. Kelps found offshore in cold waters are true giants, known to form dense underwater forests (see Figure 5-19, A).

The multicellular condition of seaweeds allows many adaptations not available to unicellular forms. Seaweeds can grow tall and raise off the bottom. This provides new opportunities as well as challenges: wave action and turbulence, competition for space and light, and the problem of predatory sea urchins and fish.

General Structure

Seaweeds show a wide range of growth forms and complexity of structures. Nevertheless, several unifying features are worth mentioning. Seaweeds are algae, and as such they lack the true leaves, stems, and roots of the higher plants. The complete body is known as the **thallus,** whether it is a filament, a thin leafy sheet, a crusty cushion, or a giant kelp.

Leaflike, flattened portions of the thallus are known as **blades** (Figure 5-9, A). They increase the surface area and are the main photosynthetic regions. All portions of the thallus are able to photosynthesize in light as long as they have chlorophyll. Blades are technically not true leaves because the upper and lower surfaces are identical and there are no "veins." Blades are sometimes kept close to the sea surface by means of gas-filled bladders known as **pneumatocysts** (see Figure 5-9, A), thereby maximizing their exposure to the sunlight. The mixture of gases in the pneumatocysts of some seaweeds includes carbon monoxide, a gas toxic to humans!

Some seaweeds have a distinct stemlike structure to provide support, the **stipe,** from which blades originate. It is long and tough in the large kelps (see Figure 5-9, A). Structures that look like roots, the **holdfast** or simple **rhizoids,** anchor the thallus to the bottom. Holdfasts are particularly well developed in the kelps (Figure 5-9, B). They are not involved in any significant absorption of water and nutrients as are true roots. Water and nutrients, which bathe the entire thallus, are picked up directly across the sur-

BOX 5-1

The Bay of Fire

Imagine entering a quiet tropical bay at night. There are no lights on shore, and on a moonless night the spectacle is unforgettable. As the propeller of your boat churns the black water, an intense blue-green light is left behind, like an eerie trail of cool fire. Outside the bay a few sparks of light were observed as the boat approached, but once inside the bay things are very different. Long streaks of light shoot out like fireworks beneath the boat. They are created by fish fleeing from the path of the boat. Swimming in the bay is even more spectacular. Your dive into the water is accompanied by a blinding flood of light, and sparks scatter out as you wave your arms.

Such light effects result from an unusual concentration of bioluminescent dinoflagellates, a natural and permanent phenomenon in a few select locations such as *Bahía Fosforescente*, or Phosphorescent Bay, in southwestern Puerto Rico.

Bioluminescent Bay might be a more apt name, since the light is produced by living organisms. The bay's most important source of bioluminescence is *Pyrodinium bahamense*, a unicellular, photosynthetic dinoflagellate about 40 microns (0.004 cm, or 0.0015 inch) in diameter. As many as 720,000 individuals are found in a gallon of water, many more than outside the bay. There seem to be several reasons why the dinoflagellates are concentrated in this bay. It is small, about 90 acres, and fan-shaped, and it does not exceed 4.5 m (15 feet) in depth. It is connected to the open sea by a narrow and shallow channel. As water containing the dinoflagellates flows into the bay, evaporation in the shallow water causes the surface water to sink because of the increase in salinity and density. Evaporation is enhanced by the dry and sunny days. The *Pyrodinium* stays near the surface and is therefore not carried out as the denser water flows out along the bottom of the shallow entrance. The tidal range here is only about a foot at the most, so water exchange with the outside is limited. The bay thus acts as a trap that keeps the dinoflagellates from leaving.

Of all the planktonic organisms that enter the bay, why is *Pyrodinium* favored? A key factor seems to be the thick mangrove trees bordering the bay. They grow along the muddy shores, together with all kinds of organisms living attached to their roots. Mangrove leaves fall into the water, where intense bacterial decomposition increases the organic nutrient levels in the water. Some nutrients essential to the growth of *Pyrodinium*, perhaps a vitamin, may be released by bacteria or other organisms.

Bahía Fosforescente is protected to keep the critical natural balance intact. The same phenomenon in a bay in the Bahamas disappeared when its narrow entrance was dredged to allow larger boats, and therefore larger loads of tourists, to pass through.

Bahía Fosforescente in Puerto Rico.

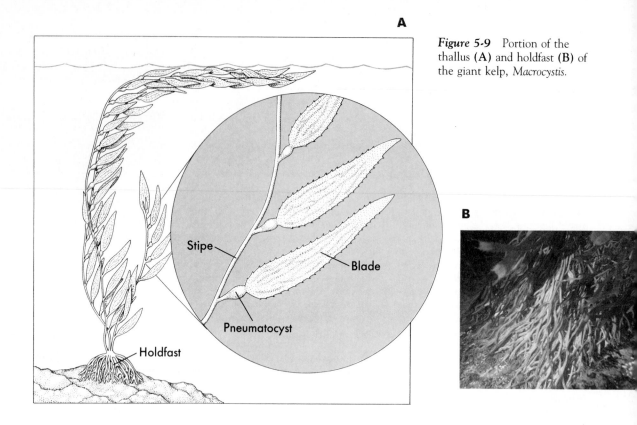

A

Figure 5-9 Portion of the thallus (**A**) and holdfast (**B**) of the giant kelp, *Macrocystis.*

Stipe

Blade

Pneumatocyst

Holdfast

B

face without the need of roots, as in land plants. Also in contrast to the higher plants, the stipe and holdfast usually lack tissues specialized in the transport of water and nutrients.

Seaweeds typically consist of a thallus, which is sometimes provided with leaflike blades, and a rootlike holdfast or rhizoid. They lack true leaves, stem, or roots.

Types of Seaweeds

Three types of seaweeds are recognized: the green, brown, and red algae. Though their pigment composition is used in their classification (see Table 5-1), their actual color may not mean much since it tends to vary a great deal.

Green Algae. The great majority of the **green algae** (division **Chlorophyta;** see Figure 5-29) are restricted to fresh water and terrestrial environments. Only around 10% of the estimated 6,000 to 7,000 species are marine. This, however, does not mean that green algae are uncommon in the sea. Some species are dominant in environments with wide variations in salinity such as bays and estuaries and in isolated tide pools on rocky coasts (Figure 5-10).

LIFE IN THE MARINE ENVIRONMENT

Few green algae are as complex as the other two groups of seaweeds in terms of the general structure of the thallus. Most are unicellular or filamentous; many are microscopic. Their pigments and food reserve, however, are the same as those in the higher plants (see Table 5-1), so it is thought that land plants evolved directly from green algae very similar to those we see today. Chlorophyll in both the green algae and the higher plants is not normally masked by any other pigments, so the thallus is typically bright, or "grass," green.

The green algae are largely unicellular and nonmarine. They are typically bright green, since chlorophyll is not masked by other pigments.

The simplest marine green algae are unicellular and planktonic and possess one to four flagella, which they use to swim. Some are responsible for blooms in salt marshes and tide pools, especially in the tropics. Some species are epiphytes on other seaweeds, and a few actually live within their tissues. Plants that live within the tissues of other plants are known as **endophytes.** Other species bore into coral skeletons or the shells of other animals. Filamentous forms grow on a wide variety of surfaces such as rocks in shallow water, other seaweeds, and rocky-shore tide pools (Figure 5-10). The filaments of these species may be branched or unbranched. Species of *Enteromorpha* are characterized by a thin thallus in the form of a hollow tube. They tend to flourish in areas disturbed by pollution. Sea lettuce (*Ulva*) forms paper-thin sheets whose shape varies depending on environmental factors (Figure 5-11). Different species of *Ulva* are widespread, from arctic to tropical waters. Some species are particularly common in brackish water. *Valonia,* another green alga, forms huge spheres or curious spheric clusters in the tropics and subtropics.

Several other green algae consist of a branched tube with many nuclei. Such is the case of *Caulerpa,* which is restricted to the tropics and subtropics. Its many species show a great variety of shapes. "Dead man's fingers" (*Codium*) is a green alga that extends into temperate waters, including both sides of North America (Figure 5-12). It consists of multinucleated filaments woven into a spongy, often branching thallus. *Halimeda* is characterized by a thallus consisting of numerous segments with deposits of calcium carbonate (Figure 5-13). Because calcium carbonate is the same material as

Figure 5-10 A few hardy green algae thrive in tide pools high on rocky shores. Some—like *Chaetomorpha*—are filamentous, whereas *Enteromorpha* grows as thin tubes.

Figure 5-11 *Ulva,* or sea lettuce, and *Enteromorpha* are green algae that are common on rocks where fresh water, often containing high amounts of nutrients, reaches the sea. Also see Figure 11-12.

Figure 5-13 *Halimeda* is a coralline green alga. Its calcareous accumulations discourage fish, sea urchins, and others from eating the thallus. Also see Figure 13-12.

Figure 5-12 The dead man's fingers seaweed (*Codium fragile*) forms branched clumps on rocky shores.

Figure 5-14 *Padina*, a brown alga, consists of clusters of flat blades that are rolled into circles. This species is from the Hawaiian Islands.

Figure 5-15 The spiral rockweed (*Fucus spiralis*) is common on rocky shores on the Atlantic coasts of temperate North America and Europe (see Figure 10-27). Its thallus lacks the air bladders that characterize a similar species, the bladder rockweed (*Fucus vesiculosus*).

Figure 5-16 The knotted rockweed (*Ascophyllum nodosum*) occurs on the North American and European coasts of the north Atlantic.

that which forms the skeleton of corals, *Halimeda* is known as a **coralline green alga.** The accumulation of its dead, calcified segments has an important role in the formation of coral reefs (see the section in Chapter 13, "Other Reef Builders").

Brown Algae. The characteristic color of the **brown algae** (division **Phaeophyta**), which actually varies from olive green to dark brown, is due to a preponderance of yellow pigments—particularly **fucoxanthin**—over chlorophyll (see Table 5-1). Almost all of the approximately 1,500 known species are marine. Brown algae are often the dominant primary producers on temperate and polar rocky coasts. They include the most complex and the largest of the seaweeds.

> The brown algae include seaweeds that are the largest and structurally most complex algae. Chlorophyll is associated with brown pigments.

The simplest brown algae have a finely filamentous thallus, as in the case of the widely distributed species of *Ectocarpus*. The thallus is flat and branched in *Dictyota* and fan-shaped in *Padina* (Figure 5-14). Both are found mostly in tropical and subtropical waters. The thallus of most species of *Desmarestia* is branched, though the branching pattern varies widely. *Desmarestia* is found in cold waters, ranging from that of the Antarctic, where it is one of the dominant species, to the deep water in the Caribbean and temperate shores elsewhere.

Some of the most conspicuous of all brown algae are those found exposed at low tides at the middle and upper levels of rocky shores. Their thick, leathery thalli can withstand exposure to air (see the section in Chapter 10, "Exposure at Low Tide"). Many species have gas-filled floats. Known locally as **rockweeds,** or **wracks,** *Fucus* (Figure 5-15) and *Pelvetia* are found on the Atlantic and Pacific coasts of North America and other temperate shores and *Ascophyllum* is found on the temperate Atlantic (Figure 5-16). In warm waters, including the gulfs of Mexico and California, these temperate species are replaced by sargasso weed (*Sargassum;* see Figure 14-16). The sargasso weed has spheric air bladders that keep the small, leaf-like blades afloat at the sea surface. Most species grow on rocks, but at least two float offshore in huge masses. They give the **Sargasso Sea,** an area in the Atlantic north of the West Indies (see Figure 7-26), its name. Sargasso weed drifts in other regions of the world as well. It is particularly common in the Gulf of Mexico.

The large group known as the **kelps** includes the most complex and largest of all brown algae. Most kelps are found in great abundance below the low-tide level in temperate latitudes and in the Arctic. In these environments, they are a most important element of the marine life, providing food and shelter for numerous other organisms (see the section in Chapter 12, "Kelp Communities"). Some kelps consists of a single large blade, as in the many species of *Laminaria* (see Figures 12-22, A and B and 12-27). Their large blades, up to 3 m (almost 10 feet) in length, are harvested for food in several parts of the world (see Box 5-2). In some species the blade is split or even branched; several blades may grow from a single holdfast. In *Agarum* and *Alaria* (see Figure 12-22, C), a conspicuous rib runs along the middle of

LIFE IN THE MARINE ENVIRONMENT

the single blade. The blade of *Alaria* can be as long as 25 m (82 feet). *Postelsia,* commonly known the sea palm because of its appearance (Figure 5-17), grows on rocks exposed to heavy waves at low tide. It occurs in thick clusters from central California to British Columbia. Two branched forms, the feather-boa kelp (*Egregia;* see Figure 12-27) and the southern sea palm (*Eisenia*), are also common on Pacific rocky shores.

In the Pacific the largest kelps are found in deeper water just below the lowest tide level. The bullwhip kelp, *Nereocystis,* consists of a long, whiplike stipe up to 30 m (almost 100 feet) in length with a large spheric pneumatocyst at the upper end (Figure 5-18; see also Figure 12-27). Another large kelp is *Pelagophycus* (see Figures 12-24 and 12-27). It is similar to the bullwhip kelp, but it has impressive antlerlike branches. The giant kelp, *Macrocystis,* is indeed the giant of them all. Its massive holdfast, which is attached to a hard bottom, may weigh several kilograms (see Figure 5-9, B). Several long stipes grow from the holdfast, from which elongate blades develop (Figure 5-19, A). At the base of each blade a pneumatocyst eventually develops, which is filled with gas, thereby helping to keep the blades close to the surface. This body plan rivals that of many land plants in its complexity. Individual plants as long as 100 m (330 feet) have been recorded! It has been estimated that such kelp plants can grow 50 cm (20 inches) or more per day for short periods. Many plants, each with many fast-growing and intertwined stipes, form dense and very productive **kelp beds,** or **forests** (Figure 5-19, B), in the colder waters of the north and south Pacific (see the map in Figure 12-23). Kelp beds are harvested very much like giant wheat fields for the extraction of several natural products (see the section "Economic Importance" on p. 135). The significance of kelp beds as among the richest, most productive environments in the marine realm will be discussed in the section "Kelp Communities" in Chapter 12.

Red Algae. There are more species of marine **red algae** (division **Rhodophyta**) than species of green and brown algae combined. Among

Figure 5-17 Strands of the sea palm *(Postelsia palmaeformis)* are common on rocky shores exposed to heavy wave action on the Pacific coast of North America. Also see Figure 10-21.

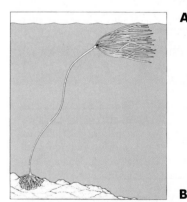

Figure 5-19 **(A)** Growing end of a stipe of the giant kelp, *Macrocystis pyrifera.* **(B)** A kelp forest in California. Also see Figures 12-25 and 12-27.

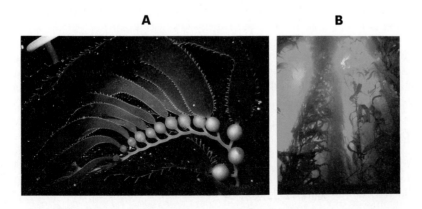

Figure 5-18 **(A)** Bullwhip kelp *(Nereocystis luetkeana)* consists of 30 to 100 blades, each up to 5 m (16 feet) long, that hang below the surface suspended by pneumatocyst. **(B)** Thallus of *Nereocystis* that drifted ashore.

Figure 5-20 *Endocladia muricata* is a red alga that is fairly resistant to exposure to air. It is found at the higher tide levels along the Pacific coast of North America.

Figure 5-21 Many species of *Porphyra* inhabit temperate, polar, and tropical rocky shores around the world. Some are of significant economic importance.

Figure 5-22 Irish moss (*Chondrus crispus*) is an edible red alga found on North Atlantic rocky shores.

other features, they are characterized by having red pigments called **phyco-bilins,** which mask chlorophyll (see Table 5-1). Most species are red, though some may show different colors depending on their daily exposure to light. The group is essentially marine; only a few of the approximately 4,000 species are restricted to fresh water or soil. Red algae are found in most marine environments. Many are of significant commercial importance and are harvested for food and for the extraction of many different products used by humans (see the section "Economic Importance" on p. 135).

The structure of the thallus of red algae does not show the wide variation in complexity and size that is observed in the brown algae. Some reds have become greatly simplified, at least in their structure, by becoming parasites of other seaweeds. A few have lost all trace of chlorophyll and depend entirely on their host for nutrition. Most red algae are filamentous, but the thickness, width, and arrangement of the filaments vary a great deal. Dense clumps are more common on the upper levels of rocky shores that are exposed at low tide; longer and flatter branches predominate in areas less exposed to air and in deeper water. These variations are observed, for example, among the numerous species of *Gelidium* and *Gracilaria* that are found worldwide. *Endocladia* forms wiry clumps on rocky shores from Alaska to southern California (Figure 5-20). The genus *Gigartina*, found on both sides of North America, includes many species characterized by large blades as long as 2 m (6 feet). These are among the most massive of the red algae. Numerous species of *Porphyra* are common on rocky shores and below the lowest tide marks from polar to tropical coasts (Figure 5-21). A thallus consisting of thin, large blades is the most common growth form. *Palmaria* is common in the north Atlantic. Its blades may reach 1.8 m (almost 6 feet) in length. Another north Atlantic red alga is the "Irish moss," *Chondrus* (Figure 5-22). It can tolerate wide ranges of temperature, salinity, and light, and its shape varies greatly in response to these physical factors.

The red algae constitute the largest group of algae. Chlorophyll is typically masked by a red pigment.

Red algae of significant importance in several marine environments include the **coralline red algae.** These are characterized by deposits of calcium carbonate around their cell walls. The calcified thallus takes a variety of shapes: thin discs growing over other seaweeds, branches with numerous joints (Figure 5-23), or smooth or rough **encrusting** growths on rocks. The color of these seaweeds varies from light to intense reddish-pink; dead calcified thalli are white. Warm-water coralline algae are actively involved in the formation and development of coral reefs (see the section in Chapter 13, "Other Reef Builders"). Other species thrive in temperate and polar waters, often attaining a large size.

Life History

Reproduction is a complex affair in the seaweeds. Asexual, or vegetative, reproduction is common. It is possibly more important than sexual reproduction in most species. Fragments of the thallus can often grow into new individuals, as occurs in the floating masses of *Sargassum* of Sargasso Sea fame. Some seaweeds produce a variety of asexual, single-celled stages. These stages, named **spores,** are cells specialized in dispersing to new locations. Some spores are protected by resistant cell walls; others are provided with flagella for movement and are known as **zoospores.**

Sexual reproduction, an uncommon phenomenon among the unicellular marine algae, is widespread in the seaweeds. The production of gametes is a key event in sexual reproduction. Gametes from two different individuals fuse, so that the new generation contains genetic information from both parents. Genetic variation, essential in the survival of a species, is thus ensured generation after generation. Gametes produced by all members of a seaweed species may be similar in appearance or may consist of larger, nonmotile eggs and smaller sperm that can swim by means of flagella. Male gametes are nonmotile in the red algae but may be released in strands of slime. Male and female gametes may be formed in the same thallus, but the chances are good that fusing gametes will be from separate thalli.

Body cells of seaweeds (and of us all—clam, fish, or human) divide and produce identical diploid cells by the process of **mitosis.** Seaweeds may also produce haploid cells by **meiosis.** The existence of diploid and haploid cells is fundamental in understanding the often complex life histories, or life cycles, of seaweeds. Their life histories can be divided into four basic types. The first type (Figure 5-24, A) is the most common among all three groups of seaweeds, and it involves two types of thalli. The first is a diploid (2n) **sporophyte** generation that divides by meiosis, producing not gametes but haploid (n, or 1n) spores. Except in the red algae, these spores are typically motile. They develop into the second kind of thallus, a haploid (n) **gametophyte** generation. The gametophyte is the one that produces haploid gametes. In some species there are separate male (sperm-producing) and female (egg-producing) thalli; in others, both types of gametes are produced by each thallus. The gametes are released and, after fertilization, develop into the diploid sporophyte. A life history with two generations of plants, a sporophyte and a gametophyte, is an example of the phenomenon of **alternation of generations.** In some algae such as the sea lettuce (*Ulva*) and the brown *Dictyota,* the sporophyte and gametophyte are structurally identical. On the other hand, in kelps (*Laminaria, Macrocystis,* and others) the large plant that we see is the sporophyte, whereas the gametophyte is minute and barely visible (see Figure 12-26). Alternation of generations also takes place in land plants as well. Ferns and other primitive land plants have a gametophyte as the more visible generation. The more advanced plants, like the flowering plants more familiar to us, are dominated by the sporophyte generation.

In the second type of life history (Figure 5-24, B), the adult thallus is haploid (n) and produces haploid gametes. On fertilization the gametes form a diploid (2n) zygote. It is in the zygote where meiosis takes place, resulting in haploid spores. Each of these spores develops into a haploid adult plant, the only kind of thallus in the cycle.

The third type of life history (Figure 5-24, C) is perhaps the easiest to understand since it is similar to that of animals, including humans. There is

Figure 5-23 *Corallina,* a coralline red alga.

Types of cell division:
mitosis   Cell division wherein the resulting cells are identical to the original cell, having their chromosomes in pairs (diploid cells, or 2n), as in the case of body cells
Chapter 4, p. 105
meiosis   Cell division wherein the resulting cells are haploid (n, or 1n), as in the case of gametes, since they contain only half the number of the parents' chromosomes
Chapter 4, p. 106

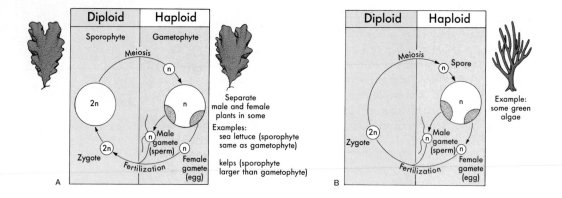

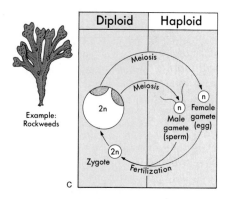

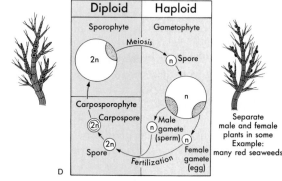

Figure 5-24 Basic patterns of sexual reproduction in the life histories of seaweeds. These patterns are not absolute, and many modifications and alternative patterns have been described.

only one thallus, and it is diploid (2n). The thallus then produces haploid (n) gametes by meiosis. After fertilization the resulting zygote develops into a new diploid thallus. This type of life history is observed in some brown algae such as *Fucus* and other rockweeds.

The life history of the red algae is more complex, involving three generations (Figure 5-24, *D*). It is similar to the life history illustrated in Figure 5-24, *A*, but a third generation, a diploid **carposporophyte,** results from the fertilization of gametes. **Carpospores,** diploid spores produced by the carposporophyte, develop into sporophytes.

> Reproduction in seaweeds is by asexual and sexual means. Sexual reproduction is usually complex, and it may involve an alternation of a haploid (or gametophyte) and a diploid (or sporophyte) generation.

There are many known exceptions to these basic schemes, and certainly there must be combinations and variants of life histories waiting to be discovered. Other aspects of the life history of seaweeds are also interesting. It has been shown, for example, that the development of gametes or spores in some seaweeds can be influenced by the amounts of nutrients in the water, by temperature, or even by day length. High levels of nitrogen nutrients in the water cause the development of asexual spores in the sea lettuce (*Ulva*), but low levels stimulate the development of gametes instead. The release of gametes and spores may be triggered by the splashing of water in an incoming tide (and hence by the cycles of the moon) or by chemical "messengers" received from cells of the opposite sex. It is also known that in some seaweeds the release of male and female gametes is timed to take place at about the same time. Furthermore, gametes must be able to recognize gametes of the opposite sex that belong to the same species. Such mechanisms help to ensure that gametes will not be wasted and that as many successful fertilizations as possible will take place.

Economic Importance

Humans have used seaweeds since time immemorial. Around the world, workers harvest seaweed to be used in many ways. The most obvious use is as a food source. People from different cultures have discovered that many seaweeds are edible, especially some of the red and brown algae. They are consumed in a variety of ways (see Box 5-2). The farming, or mariculture, of seaweed is big business in China, Japan, Korea, and other nations (see Table 16-3).

Seaweeds produce several types of starchlike chemicals called **phycocolloids** that are used in food processing and in the manufacture of different products. These phycocolloids are valuable because of their ability to form viscous suspensions or gels even at low concentrations.

One important phycocolloid, **algin** (which comprises alginic acid and its salts, the alginates), is used extensively as a stabilizer and emulsifier in the manufacture of dairy products such as ice cream, cheese, and toppings, which need to be smooth and not likely to separate. Algin is also used in the baking industry to prevent frostings and pies from becoming dry. Its ability as a thickener and emulsifier makes it suitable for use in the pharmaceutical and chemical industries, and in the manufacture of various products—from shampoo and shaving cream to plastics and pesticides. Algin also has uses in the making of rubber products, paper, paints, and cosmetics. One of its biggest applications is in the textile industry—algin thickens the printing paste and provides sharper prints. A major source of algin for commercial uses is the giant kelp, *Macrocystis*. The west coast of temperate North America, particularly California, is home to extensive kelp forests, making this an important algin-producing area. The forests are leased from the state of California, and large barges equipped with rotating blades cut and collect the stipes and fronds to a depth of 1 to 2 m (up to 6 feet) below the surface (Figure 5-25). The stipes quickly grow back toward the surface. An additional important source of algin is *Laminaria*, another brown alga that is harvested in the north Atlantic.

A second phycocolloid, **carrageenan,** is obtained from red algae such as "Irish moss" (*Chondrus*) in the north Atlantic (see Figure 5-22) and *Eucheuma* in the tropics. Several species of *Eucheuma* are farmed extensively in the Philippines. Carrageenan is especially valued as an emulsifier. It is used to give body to dairy products and an amazing variety of processed foods, including the quick-thickening "instant" puddings.

Figure 5-25 The *Kelstar*, a kelp harvester based in San Diego, California. It is 55 m (180 feet) long and it has a capacity of 600 tons.

BOX 5-2

Seaweeds for Gourmets

Drying seaweeds on a Japanese seaweed farm.

Seaweeds, raw, cooked, or dried, are used as food in many cultures. Seaweeds are good sources of some vitamins and minerals; some are said to contain substantial amounts of protein. Unfortunately, we cannot digest many of the complex carbohydrates in the plants, but this may be an advantage for those counting calories. Seaweeds can add variety and taste to bland foods and may be used to wrap such foods as rice. Cookbooks have been written to whet the appetites of the most demanding gourmets.

Ulva is not called "sea lettuce" for nothing: It can be eaten fresh in salads. Seaweeds (or *limu*), including *limu 'ele 'ele* (the green *Enteromorpha prolifera*) and *limu manauea* (the red *Gracilaria coronopifolia*), are beloved by the Hawaiians. Purple laver (a species of *Porphyra*, a red alga) prepared in various ways is still eaten in some parts of the British Isles. It is washed and boiled, then formed into flat cakes, rolled in oatmeal, and fried—it is then called laverbread. Purple laver is also eaten as a hot vegetable or fried with bacon. "Irish moss" (*Chondrus*, a red alga and a source of carrageenan) is dried and used in preparing *blancmange* and other desserts in eastern Canada, New England, and parts of northern Europe. *Palmaria*, another red alga, is dried and eaten, mostly by those living along the Atlantic coast of Canada and northern Europe. Called *dulse*, it is sometimes still used in making bread and several types of desserts. For those on a diet, dulse can also be chewed like tobacco—nicotine-free, of course!

It is in the Orient, however, that preparing seaweed for food has reached the level of an art. Several species are carefully cultivated, supporting multimillion-dollar operations. Seaweed culture is a very old tradition in Japan, and Japanese cuisine uses seaweed in many ways. Species of *Laminaria* and *Alaria*, a kelp called *kombu*, are dried and shredded, then eaten in many ways. They are even used to make tea and candy. *Undaria*, or *wakame*, is another edible kelp best when fresh or cooked for a very short time. *Porphyra*, a red alga, is used to make thin sheets of *nori*, widely used in soups, and for wrapping *sushi*, boiled rice stuffed with bits of raw fish, sea urchin roe, or other ingredients (see Figure 16-11).

The harvesting and marketing of edible seaweed is a growing business in the United States. There is no need for planting, fertilizing, weeding, or tilling! The ocean takes care of everything, though oil and sewage pollutants may spoil the best of harvests. Seaweeds are harvested by hand, rinsed in water, dried on lines, and sold at health food stores or through the mail. The number of seaweed cookbooks available is on the rise. Connoisseurs use seaweed in salads, soups, omelets, casseroles, and sandwiches. Sea palm (*Postelsia*), also known as "sea noodles," is a best-seller. They are reported to be excellent when sautéed in honey or in butter and garlic. Coastal Indians cooked it in ovens and made it into cakes. Pickled bullwhip kelp (*Nereocystis*) tastes even better than regular pickled cucumbers. Bladder rockweed (a type of *Fucus*) makes great tea. Are you ready for feather-boa burgers, French-fried sea palm, and *wakame* shakes?

Some additional food for thought: seaweeds at their best, Hawaiian style.

Limu Salad *

3 cups *limu*
⅓ cup vinegar
⅔ cups soy sauce
1 chopped onion
3 smashed cloves of garlic
1 small piece of minced ginger, preferably fresh
Wash the *limu* to remove any sand, and wilt with hot water. Drain after approximately 1 minute, rinse in cold water, and drain. Mix the remaining ingredients and add the *limu*. Marinate overnight before serving.

*Adapted from Major, A.: *The Book of Seaweed*, London, 1977, Gordon and Cremonesi.

LIFE IN THE MARINE ENVIRONMENT

Another phycocolloid extracted for its ability to form jellies is **agar.** Agar is employed to protect ham, fish, and meats during canning, in low-calorie foods (since it is not digestible by humans), and as a thickener. It is also used in laxatives and other pharmaceuticals (see the section in Chapter 16, "Drugs from the Sea") and in cosmetics. Biologists use agar as a medium in which to grow bacteria and molds; it is widely used in medicine for this purpose. Agar is obtained commercially from several red algae, especially from species of *Gelidium, Gelidiella,* and *Pterocladia.*

Seaweed may also be used as fertilizers, food additives in animal feeds, and wound dressings in hospitals. Coralline red algae are sometimes marketed in Europe to reduce the acidity of soils. The future may see new uses for seaweeds. Their fermentation to produce methane for fuel has already been proposed.

FLOWERING PLANTS

The 250,000 species of **flowering plants** or **angiosperms** (division **Anthophyta**) are the dominant plants on land. They have true leaves, stems, and roots, all three provided with specialized tissues to transport water, nutrients, and the food manufactured during photosynthesis. Reproduction involves a dominant sporophyte that features elaborate reproductive organs, the flowers. Few of these "higher" plants are successful in the oceans. Of all flowering plants, only the seagrasses are truly marine. They spend almost their entire lives submerged by seawater, rarely, if ever, exposed at low tide. Salt-marsh plants and mangroves inhabit estuaries and shores protected from wave action. They are not completely at home in the ocean, and only their roots are covered by water at high tide. There are also numerous flowering plants adapted to colonize coastal areas exposed to salt-laden winds and occasional sea spray, though they do not tolerate immersion in seawater. Such plants may be found colonizing sand dunes or living along the edges of salt marshes.

Seagrasses

Though superficially resembling grasses, **seagrasses** are not grasses at all. Like true grasses, seagrasses are flowering plants, but they are not closely related. The closest relatives of certain seagrasses appear to be members of the lily family.

Seagrasses have adapted to life in the marine environment. They have rootlike horizontal stems under the sediment that connect the erect shoots (Figure 5-26, A-E). Seagrass flowers are typically very small and inconspicuous, since there is no need to attract insects for pollination. **Pollen,** which contains the male gamete, is carried instead by water currents. For this reason it is often released in strands. In some seagrasses the pollen cells are long and threadlike, instead of tiny and round as in land plants. Tiny seeds, which in some species develop inside small fruits, are the result of successful fertilization. These seeds are dispersed by water currents and perhaps in the feces of the fish and other animals that browse on the plants.

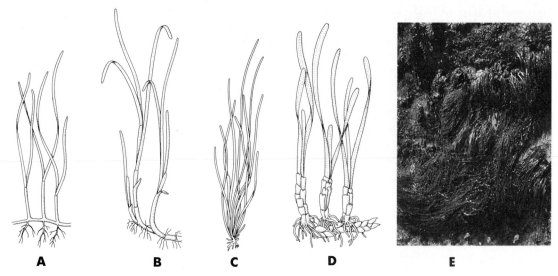

Figure 5-26 Seagrasses are flowering plants adapted to live under seawater. They most common are **(A)** manatee grass (*Syringodium*), **(B)** eelgrass (*Zostera*), **(C)** surf grass (*Phyllospadix*), **(D)** turtle grass (*Thalassia*), **(E)** surf grass (*Phyllospadix scouleri*).

Eelgrass (*Zostera*) is the most widely distributed of the roughly 50 species of seagrasses known. It is found along the Pacific and Atlantic coasts of temperate North America, as well as in other regions of the world, where it is an inhabitant of shallow, well-protected coastal waters such as bays and estuaries (see Figure 12-16). It has distinctively flat, ribbonlike leaves (Figures 5-26, *B* and 12-18). It is common in oxygen-poor sediments. Thick *Zostera* beds are highly productive and provide shelter and food to a wide variety of animals, some of considerable economic importance (see the section in Chapter 12, "Seagrass Beds"). Surf grass (*Phyllospadix*) (Figure 5-26, *C* and *E*) is an unusual seagrass, because it is an inhabitant of rocky coasts exposed to wave action, as its common name implies. Some species may become exposed at low tides. It is found on the Pacific coast of North America.

There are more species of seagrasses in the Caribbean Sea, the Florida Keys, and the Gulf of Mexico than anywhere else in the Western Hemisphere. The most common is turtle grass (*Thalassia*). It looks similar to eelgrass, but its leaves are broader and more straplike (Figure 5-26, *D*). Highly productive turtle grass beds can be found on muddy and sandy bottoms in water that is calm and of moderate depth, down to approximately 10 m (30 feet). The meadows are especially well developed in Caribbean coral reefs, where they play an important role in the stabilization of sediments on the less exposed side of the reef. Other western Atlantic seagrasses are manatee grass (*Syringodium*, Figure 5-26, A), *Halophila*, and—in estuaries—*Ruppia*.

Salt-Marsh Plants

The **cord grasses** (species of *Spartina*, see Figure 11-9) are true members of the grass family. They are not really marine species, but rather land plants

LIFE IN THE MARINE ENVIRONMENT

tolerant to salt. Unlike seagrasses, which are true marine species, cord grasses do not tolerate total submergence by seawater. They live in salt marshes and other soft-bottom coastal areas throughout temperate regions. Cord grasses inhabit the zone above mud flats that becomes submerged by seawater only at high tide, so their leaves are always exposed to air. Salt glands in the leaves excrete excess salt. Other salt-tolerant plants, or **halophytes,** such as pickle weed (*Salicornia*; see Figure 11-10) may be found at higher levels on the marsh. Salt-marsh plants and their adaptations to the estuarine environment along the mouth of rivers are discussed in the section "Salt Marshes" in Chapter 11.

Mangroves

Mangroves are shrubs and trees adapted in the most ingenious ways to live along tropical and subtropical shores around the world. They are land plants that can tolerate salt. Luxuriant and very productive mangrove forests flourish along muddy or sandy shores protected from waves (Figure 5-27). Only their roots are covered by seawater at high tide.

At least 80 mostly unrelated species of flowering plants receive the common name of mangrove. They are adapted in various ways to survive in an uninviting salty environment where water loss from leaves is high and where sediments are soft and poor in oxygen. Adaptations become more crucial in those species living right on the shore, such as the red mangrove (*Rhizophora*). Several species are found throughout the topics and subtropics. The extreme northern and southern limits of the red mangrove are those areas in which killing frosts begin. Salt marshes replace red mangrove forests in areas exposed to frosts.

The leaves of the red mangrove are thick, probably to reduce water loss. An adaptation shared with several other mangroves is the presence of seeds that germinate while still attached to the parent tree (Figure 5-28, A).

Figure 5-27 The red mangrove, *Rhizophora mangle*, forms lush forests along the shores in Florida, the Caribbean, the Gulf of California, and other tropical regions of the Western Hemisphere and West Africa. Notice the long roots extending into the mud, exposed here at low tide. Other species of mangroves can be found further inland. Also see Figures 11-1 and 11-21.

A   B

Figure 5-28 Seedlings of the red mangrove, *Rhizophora mangle,* as they appear in the trees **(A),** and one that has already taken root in the soft sediment **(B).**

They develop into elongated, pencil-shaped seedlings with a pointed end and growing as long as 30 cm (almost a foot), before they eventually fall off. Successful seedlings stick in the soft muddy sediment like a knife thrown into a lawn, or float in the water to be carried by currents to new locations (Figure 5-28, *B*). The types and distribution of mangroves and their significance in the marine environment are discussed in the section "Mangrove Forests" in Chapter 11.

Flowering plants, dominant on land, have few marine representatives. Seagrasses, which are true marine species, and salt-tolerant plants like salt-marsh plants and mangroves are exceptions. They have successfully adapted to soft-bottom coastal regions, developing as highly productive meadows and, in the case of mangroves, forests along the shore.

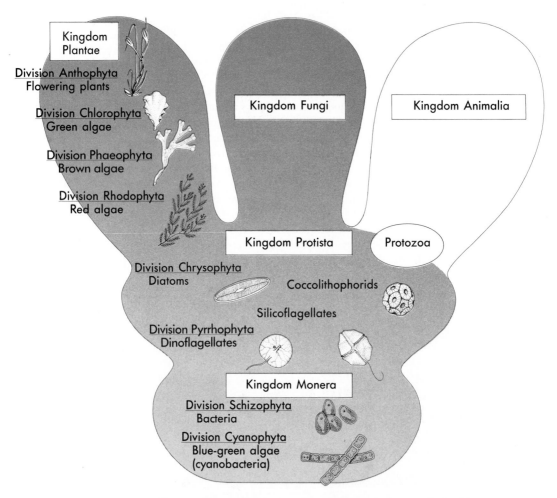

Figure 5-29 Classification scheme of marine plants and plantlike organisms. For convenience, these organisms are covered in this chapter, even though many of them are not true plants.

LIFE IN THE MARINE ENVIRONMENT

SUMMARY

1. _____ and _____ are prokaryotic organisms included in the kingdom Monera.
2. Heterotrophic bacteria involved in breaking down organic material into nutrients are referred to as _____.
3. Algae, found almost exclusively in marine and fresh-water environments, can be distinguished from the mostly terrestrial higher plants because they lack three types of specialized tissues: _____, _____, and _____.
4. The two most important planktonic, unicellular algae are the diatoms and the _____. Diatoms are especially important in _____, whereas the second group is more important in tropical waters.
5. Seaweeds are algae, but in contrast to the unicellular algae, they are _____.
6. Three groups of seaweeds are recognized. In the green algae, _____ is not masked by any other pigments so the thallus is bright green. The _____ algae include the largest and structurally most complex seaweeds, whereas the third group, the _____ algae includes the largest number of species.
7. Almost all flowering plants live on land except three major groups. The _____ are truly marine since they are rarely exposed to air. Two other groups, the salt-marsh plants and _____, are essentially land plants that are salt tolerant.

THOUGHT QUESTIONS

1. Some biologists place the seaweeds in the kingdom Protista, others in the kingdom Plantae. Assume that a better arrangement is to group the green, brown, and red algae in their own kingdom, which we will call Macrophyta. Characterize the new kingdom by first giving its unique characteristics and then differentiate it from the protists and the true plants. Be sure to be aware of major exceptions or overlaps.
2. Only very few flowering plants have been able to invade the oceans, but the few marine ones are very successful. What are some possible reasons for the small number of marine flowering plants? How do those that have taken the step manage to thrive in some environments?

FOR FURTHER READING

Borowitzka, M.A., and A.W.D. Larkun: "Reef Algae," *Oceanus*, vol. 29, no. 2, Summer 1986, pp. 49-54. *A survey of the algae inhabiting coral reefs, including planktonic algae and those living in association with animals.*

Hendry, G.: "Making, Breaking, and Remaking Chlorophyll," *Natural History*, May 1990, pp. 37-40. *Delineates the rather complex process by which primary producers manufacture but ultimately destroy chlorophyll, the world's most abundant pigment.*

Hoover, R.B.: "Those Marvelous, Myriad Diatoms," *National Geographic Magazine*, vol. 155, no. 6, June 1979, pp. 870-878. *Beautiful to look at under the microscope, diatoms are among the planet's most important organisms.*

Pettitt, J., S. Ducker, and B. Knox: "Submarine Pollination," *Scientific American*, vol. 244, no. 3, March 1981, pp. 134-143. *Reproduction in seagrasses includes underwater pollen and pollination, unique among flowering plants.*

Schafer, K.: "Mangroves," *Oceans*, vol. 21, no. 6, Nov.-Dec. 1988, pp. 44-49. *A beautifully illustrated account of the importance of mangroves in the marine environment.*

Marine Animals Without a Backbone

6

ost organisms inhabiting our planet are animals. In contrast to plants, we animals cannot manufacture our own food and must therefore obtain it from others. The need to eat has resulted in the evolution of ingenious and diverse ways of obtaining and processing food, as well as equally diverse ways of avoiding being eaten by others.

The colorful crab in Figure 6-1 is a good example. It inhabits reef-building corals, relying on them for food and shelter. The crabs feed on mucus, which the coral produces to keep its surface free from debris. The coral also is an animal, though it may not look like one. It gets some of its food from zooxanthellae that live in its tissues. The coral also eats small planktonic organisms that it captures by using special stinging cells in its tentacles. Though an absent-minded crab is occasionally captured by a fish or an octopus, the crabs are usually safe hiding among the coral branches. Crabs repay the favor by using their pincers to drive away still other animals that have a taste for coral tissue.

Our survey of the many kinds of marine animals follows the traditional way of classifying them into two major groups: the **vertebrates,** which have a **backbone** (a row of bones called **vertebrae**), and those without a backbone, the **invertebrates.** This is an artificial scheme that considers humans, who happen to be vertebrates, as something special. It is nevertheless a very convenient arrangement.

The most current estimate is that at least 97% of all species of animals are invertebrates. In fact, there are more invertebrates than all other kinds of organisms! All major groups of invertebrates have marine representatives, and many are exclusively marine. Only a few groups have successfully invaded dry land. Were it not for one of these groups, the insects, we could boast without hesitation that most species of invertebrates, and hence most animals, are marine.

PROTOZOANS: THE ANIMAL PROTISTS

The simplest animal-like creatures are the **protozoans.** A few form colonies, but most consist of a single cell and can be seen only under a microscope.

zooxanthellae Dino-flagellates, single-celled organisms, adapted for living within animals *Chapter 5, p. 124; Figure 13-9*

Figure 6-1 *Trapezia pays for the food and shelter it gets from its coral host by defending it against predators.*

Their minute size and apparent simplicity disguise a complex makeup. Each cell may be described as a "super cell," since it is able to perform many of the same functions carried out by the multitude of cells in "higher" animals.

Having a single cell is the only thing protozoans—which show an enormous diversity in structure, function, and lifestyle—have in common. It appears that the Protozoa, meaning the "first animals," is more than a single group. There is no agreement as to how to classify the estimated 50,000 species of protozoans. Some contain chlorophyll and function as plants. For this reason, most biologists group all organisms that share characteristics with true plants and animals in the **kingdom Protista.**

In our exploration of the living organisms of the sea, the photosynthetic, or **autotrophic,** plantlike protists (diatoms, dinoflagellates, and other unicellular organisms) were covered in Chapter 5 together with marine plants and other primary producers. The animal-like protists, the protozoans, are covered in this chapter dealing with the marine invertebrates, even if protozoans are not animals in the strict sense. Invertebrate and vertebrate animals, the true animals, are grouped in the **kingdom Animalia.** Animals and protozoans are eukaryotes and heterotrophs, but animals—in contrast to protozoans—consist of more than one cell.

> **Protozoans are the most animal-like of the protists. They are eukaryotic and unicellular. They are heterotrophic and ingest food like true animals.**

Protozoans inhabit water everywhere, not only marine and freshwater environments, but even inside other organisms. Many kinds of marine protozoans can be collected from sediments rich in organic debris, the surface of seaweeds, the guts of animals, and plankton samples.

Foraminiferans

The **foraminiferans,** better known as **forams,** are marine protozoans typically having a shell, or **test,** that usually consists of **calcium carbonate ($CaCO_3$).** The test is usually microscopic and may have several chambers that increase in size as the foram grows (Figure 6-2, A). **Pseudopodia**—extensions of the jellylike contents of the cell, or **cytoplasm**—are thin, long, and retractable in forams. The pseudopodia protrude through pores in the shell and form a network used to trap diatoms and other minute organisms suspended in the water. Food is then moved into the interior of the cell as if on a conveyor belt.

Most forams live on the bottom, either free or attached. Attached forams may develop into conspicuous growths as much as 5 cm (2 inches) in diameter (Figure 6-2, B). Each growth consists of a single cell that forms a shell. Some may be covered with sand grains or other materials and therefore lack a shell. The shells of bottom-inhabiting forams can be important contributors of calcareous material in coral reefs and sandy beaches. Only a few species are planktonic, but these can be very abundant. Their shells are smaller and thinner than those that live on the bottom and may have delicate spines that aid in flotation. The shells of planktonic forams eventually sink to the bottom in such high numbers that large stretches of the ocean floor are covered by **foraminiferan ooze,** a type of calcareous ooze consisting

---

kingdom Protista The single-celled and *eukaryotic* (made of a cell containing organelles) organisms that combine characteristics of the true plants (*Kingdom Plantae*) and animals (*Kingdom Animalia*)
Chapter 4, p. 114; Figure 4-30

plankton Organisms found drifting in the water
Chapter 4, p. 98

biogenous sediments
Made of the skeletons and shells of marine organism: *calcareous ooze* made of calcium carbonate and *siliceous ooze* of silica
Chapter 2, p. 40

LIFE IN THE MARINE ENVIRONMENT

A

B

Figure 6-2 (**A**) Foraminiferans typically have calcareous shells that consist of a spiral arrangement of chambers. (**B**) *Homotrema rubrum* is a foraminiferan that forms bright red, calcareous growths several millimeters in diameter at the base of corals in the tropics. It is so common in Bermuda that their skeletons are responsible for the island's famous "pink beaches."

mostly of foram shells. Many limestone and chalk beds around the world, like the white cliffs of Dover in England, are products of foram sediments that were uplifted from the ocean floor.

> **Forams are protozoans characterized by a shell usually made of calcium carbonate. Most live on the bottom, but shells of planktonic species are important components of marine sediments.**

Most species of forams are known only as microfossils. The distribution of these microfossils in sediments is very important to geologists. Shells of warm water species are slightly larger and more porous than those from colder water. Past water temperatures can be estimated from the distribution of certain "marker species." Their distribution is also valuable in the search for oil, since it used to determine the age of sediments.

Radiolarians

Radiolarians are planktonic marine protozoans that secrete elaborate and delicate shells made of glass (**silica, or SiO_2**) and other materials. Typical shells are spheric with radiating spines (Figure 6-3), though the structure varies. Thin, needlelike pseudopodia capture food, as in forams.

Most radiolarians are microscopic, but some form bizarre, sausage-shaped colonies known to reach 3 m (6 feet) in length, true giants among protozoans. Radiolarians inhabit open water throughout the ocean. When abundant, the remains of their shells settle to the bottom and form a siliceous ooze known as **radiolarian ooze.** This ooze is more extensive in deep water, since radiolarian shells are more resistant to the effect of pressure than those of forams.

> **The shells of radiolarians are made primarily of silica (glass). These shells form siliceous sediments that cover large areas of the ocean floor.**

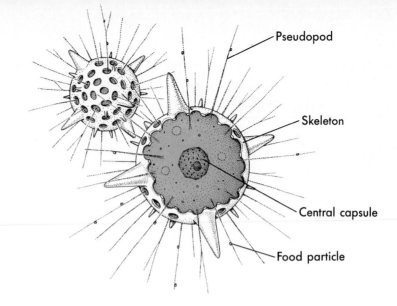

Figure 6-3 A typical radiolarian cell consists of a dense central portion surrounded by a less dense zone that is involved in the capture of food particles and in buoyancy.

Pseudopod

Skeleton

Central capsule

Food particle

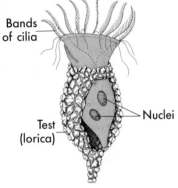

Bands of cilia

Test (lorica)

Nuclei

Figure 6-4 *Tintinnopsis* is a marine tintinnid that forms a vaselike lorica made of sand grains. Specialized cilia on one end are used in feeding.

tissues Specialized, coordinated groups of cells

organs Structures consisting of several types of tissues, grouped together to carry out particular functions.
Chapter 4, p. 96

Ciliates

Ciliates are protozoans that have many hairlike **cilia** that are used in locomotion and feeding. The most familiar ciliates are freshwater forms such as the paramecium. Marine ciliates are usually found creeping over seaweeds and bottom sediments. Some live on the gills of clams, in the intestine of sea urchins, on the skin of fish, and other unusual places. Other ciliates live attached to surfaces, even forming branched colonies of tiny individuals. **Tintinnids** are common ciliates that build their own quarters—vaselike cases, or **loricas**—that drift in the water (Figure 6-4). The cases may be transparent or made of bits of particles.

SPONGES

Sponges are animals that may be best described as complex aggregations of specialized cells. These cells are largely independent from each other and do not form true tissues and organs. They are among the simplest multicellular animals. Most sponges are marine. All are **sessile,** living attached to the bottom or a surface. They show an amazing variety of shapes, sizes, and colors but share a simple body plan. Numerous tiny pores on the surface allow water to enter and circulate through a series of canals where planktonic organisms and organic particles are filtered out and eaten (Figure 6-5, A). This network of canals and a relatively flexible skeletal framework give most sponges a characteristic spongy texture. Because of their unique body plan, sponges are classified as one of about 30 distinct and independent groups within the kingdom Animalia: the phylum **Porifera,** or the "pore bearers" (see classification scheme in Figure 6-57).

Sponges may not be too far from the first multicellular animals, which were probably simple colonies in which some of the cells became specialized for such functions as feeding and protection. Sponges have only a cellular level of organization, since the cells are not combined into different tissues. Sponge cells are very plastic and easily change from one type to another. If

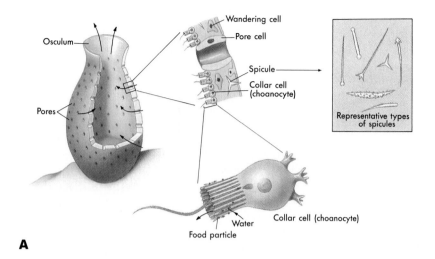

Representative types of spicules

A

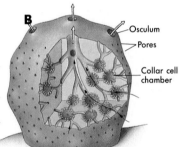

experimentally separated, the cells can even regroup and form a new sponge (Figure 6-6).

The architecture of sponges is best understood by examining the simplest kind of sponge (see Figure 6-5, A). The outer surface is covered with flat cells and occasional tubelike **pore cells** through which a microscopic canal allows water to enter. Water is pumped into a larger canal lined with **collar cells,** or **choanocytes.** These cells possess a flagellum that creates currents and a thin collar that traps food particles, which are then ingested by the body of the cell. Water then leaves through the **osculum,** a large opening on top of the sponge. Sponges are an example of **suspension feeders,** animals that eat food particles suspended in the water. Sponges actively filter the food particles and are thus known as **filter feeders,** a type of suspension feeder (see Figure 6-20).

Most marine sponges show a more complex arrangement in which the collar cells are restricted to pockets connected to the outer pores by a network of canals (Figure 6-5, B). Water leaves not through a single osculum, but through several ones that serve as the exit of a large number of canals.

Sponges are among the simplest multicellular animals, lacking true tissues and organs. They are mostly marine, living as attached filter feeders.

As sponges get larger, they need support. Many have **spicules,** transparent siliceous or calcareous supporting structures of varying shapes and sizes (see Figure 6-5, A). Most also have a skeleton of **spongin,** elastic but resistant fibers made of protein. Spongin may be the only means of support, or it may be found together with spicules. When present, spongin and spicules are mostly embedded in a gelatinous layer between the outer and inner layers of cells. Wandering cells in this layer are in charge of secreting the spicules and spongin (see Figure 6-5, A). Some of these wandering cells also

Figure 6-5 Sponges consist of complex aggregations of cells that carry out specific functions. In simple **(A)** as well as in more complex **(B)** sponges, collar cells are the cells in charge of trapping food particles.

gametes Specialized reproductive cells produced by organs called *gonads: sperm* (male gametes produced by the *testes*) and *eggs* (female gametes produced by the *ovaries*) Chapter 4, p. 106)

transport and store food particles. Some can even transform themselves into other types of cells, quickly repairing any damage to the sponge.

Many sponges reproduce **asexually** when branches or buds break off and grow into separate sponges identical to the original one. Sponges also reproduce **sexually** when collar cells or cells in the gelatinous layer develop into **gametes** like those of other animals: large, nutrient-rich eggs and smaller sperm cells provided with a flagellum (Figure 6-7). Most individual sponges can produce both male and female gametes, but in some species there are separate male and female individuals. Sperm are typically released from the sponge, and fertilization takes place internally after the sperm enter other egg-producing sponges. The release of gametes directly into the water like this is called **spawning.**

The early stages of development take place inside the sponge. Eventually a tiny sphere of cells surrounded by flagellated cells is released into the water (see Figure 6-7). This planktonic **larva** is carried by currents until it settles on the bottom and changes into a minute sponge. Most marine invertebrates have life histories involving characteristic larvae that eventually change into juvenile adults. This drastic change in the larva is referred to as **metamorphosis** (see Figure 12-5).

Almost all of the 5,000 to 10,000 species of sponges are marine. Sponges live from the poles to the tropics, but the largest number of species inhabits shallow tropical waters. Sponges (Figure 6-8, A-D) may grow into branching, tubular, round, or volcano-like masses that may reach a huge size. **Encrusting** sponges form thin, sometimes brightly colored growths on rocks or dead coral (Figure 6-8, C). **Glass sponges,** such as the Venus' flower basket sponge (*Euplectella*), live anchored in deep water sediments and are characterized by a lacelike skeleton of fused glass spicules. **Boring sponges** (*Cliona*) actively bore thin channels through calcium carbonate, such as coral skeletons and shells. In the **sclerosponges** or **coralline sponges** (*Ceratoporella*; Figure 6-8, D) a calcium carbonate skeleton forms beneath the body of the sponge, which contains siliceous spicules and spongin. Sclerosponges were first known as fossils, but living specimens were discovered in underwater caves and on steep coral reef slopes after the advent of scuba diving.

Some marine sponges are of commercial importance. **Bath sponges** (*Spongia*) are still harvested in a few locations in the Gulf of Mexico and the eastern Mediterranean in what remains of a once flourishing occupation. Bath sponges, not to be confused with synthetic "sponges," consist of the spongin fibers remaining after cells and debris are washed off. Some marine sponges produce chemicals that are potentially of commercial importance to humans (see the section in Chapter 16, "Drugs from the Sea").

Figure 6-6 Some sponges form new individuals after their cells are separated from one another. The separation of cells can be brought about by squeezing pieces through a very fine sieve **(A).** In a matter of hours, the cells begin to aggregate and reorganize **(B)** and eventually form new sponges **(C).**

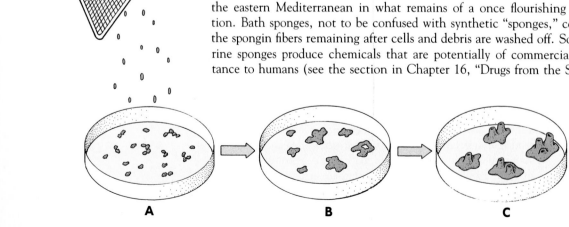

A   **B**   **C**

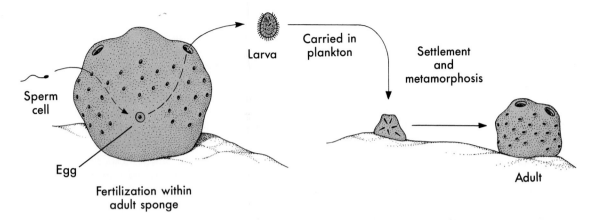

Figure 6-7 Sexual reproduction in many marine sponges involves fertilization and the release of a planktonic larva and its eventual settlement and metamorphosis on the bottom.

Figure 6-8 (A) *Verongia archeri*, a tubular sponge from the Caribbean; (B) a spherical sponge; (C) a thin, encrusting sponge from a tide pool in California; (D) *Ceratoporella nicholsoni*, a coralline sponge, or sclerosponge, photographed at a depth of 52 m (170 feet) in Puerto Rico. Also see Figures 4-14 and 4-24, C.

CNIDARIANS: A RADIALLY SYMMETRIC BODY PLAN

The next level of complexity after the sponges involves quite a big step. This step is the presence of tissues specialized to perform specific functions. Swimming is now possible and so is responding to external stimuli and engulfing prey. **Cnidarians,** sometimes called **coelenterates** (phylum **Cnidaria** or **Coelenterata**), comprise the sea anemones, jellyfishes, corals, and other familiar forms. These animals display the tissue level of organization. Another characteristic of cnidarians is **radial symmetry,** where similar parts of the body are arranged and repeated around a central axis (Figures 6-9 and 6-16). If a radially symmetric animal were cut like a pizza, all of the resulting slices would be similar. Animals with radial symmetry look the same from all sides and have neither a head nor a front or back.

Cnidarians can occur as one of two basic forms (Figure 6-9): a **polyp,** a saclike attached stage, or a **medusa,** or jellyfish, which is like an upside-down polyp adapted for swimming. The life history of some cnidarians includes both polyp and medusa stages. Others spend their entire lives as one of the two.

Figure 6-9   The flower-like appearance of many cnidarians is a consequence of their radial symmetry. In both the polyp and the medusa, tentacles are arranged and repeated around a central axis that runs along the mouth.

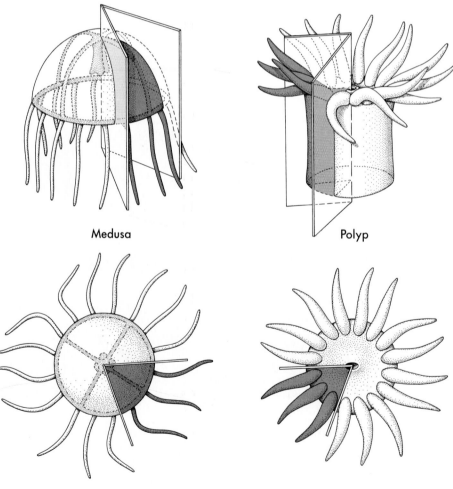

Medusa   Polyp

The polyp and the medusa share a similar body plan. Both have a centrally located mouth surrounded by **tentacles,** slender, fingerlike extensions used to capture and handle food. The mouth opens into a **gut** where food is digested. The cnidarians' gut is a blind cavity with only one opening, the mouth. Cnidarians capture small prey by discharging their **nematocysts,** unique stinging structures found on the tentacles (see Figure 6-11).

> The radially symmetric cnidarians exist as attached polyps and/or drifting medusae. Nematocysts, stinging cells unique to cnidarians, are present in tentacles used to capture prey.

Two layers of cells form the body wall of cnidarians. One of these is external, whereas the other lines the gut. There is also a thin middle layer that may contain cells. This middle layer is thick and gelatinous in medusae, hence their common name of "jellyfish." They are of course not related to fish at all.

Types of Cnidarians

The basic cnidarian body plan, though simple, has been very successful. More than 9,000 species are known, almost all of which are marine. The variety of shapes and colors of cnidarians contributes much to the beauty of the oceans.

Hydrozoans. The **hydrozoans** (class **Hydrozoa**) include cnidarians with a wide range of structures and life histories. Many consist of feathery or bushy colonies of tiny polyps. They attach to pilings, shells, seaweeds, and other surfaces (Figure 6-10). The polyps may be specialized for feeding, defense, or reproduction. Minute, transparent medusae are produced by the reproductive polyps. These medusae, usually planktonic, release gametes. Fertilized eggs develop into free-swimming larvae. The characteristic larva of most cnidarians is the **planula,** a cylindric, flagellated stage consisting of two layers of cells. After a time in the plankton, the planula settles on the bottom and metamorphoses into a polyp. This first polyp will divide and develop into a colony of many interconnected polyps. Some hydrozoans lack a polyp stage, and instead their planula develops into a medusa. Others lack a medusa stage and instead the polyp produces the gametes directly.

Siphonophores are hydrozoans that form drifting colonies. Polyps in a siphonophore colony may be specialized as floats, which may be gas-filled as in *Physalia,* or the Portuguese man-of-war (Figure 6-11). Other siphonophore polyps form long tentacles used to capture prey. Toxins from the nematocysts can produce very painful reactions in swimmers or divers (see Box 6-1).

Scyphozoans. The larger jellyfishes commonly seen in all oceans are quite different from the often invisible hydrozoan medusae. These medusae (Figure 6-12) are the dominant stage of the life cycle of **scyphozoans** (class **Scyphozoa**). Polyps, when they occur at all, are very small and release juvenile medusae. The rounded body, or **bell,** of some scyphozoan medusae

Figure 6-10 Colonial hydrozoans include the bushy clusters of *Tubularia crocea* from the Atlantic and Pacific coasts of North America **(A)** and the featherlike colonies of a hydroid living on the leaves of turtle grass, *Thallasia testudinum* **(B).**

A

B

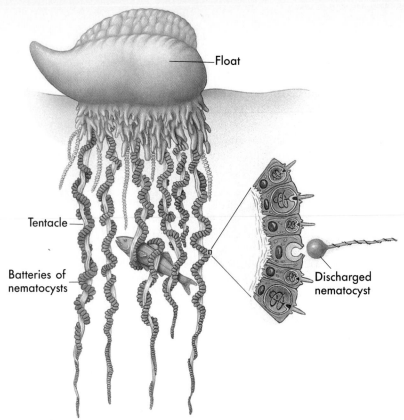

Figure 6-11 A diagrammatic representation of the Portuguese man-of-war (*Physalia physalis*). It consists of a colony of specialized polyps, one of which forms a gas-filled float that may reach 30 cm (12 inches) in length. Long tentacles, here contracted, are armed with nematocysts notorious for their ability to sting swimmers. Also see Figure 14-18.

Float

Tentacle

Batteries of nematocysts

Discharged nematocyst

A

B

Figure 6-12 Scyphozoan medusae are larger and more complex than hydrozoan medusae. Examples are *Phyllorhiza punctata* (**A**) and the sea nettle, *Chrysaora quinquecirrha* (**B**), which is found from Cape Cod to the Gulf of Mexico. It is especially common in Chesapeake Bay.

may reach a diameter of 2 m (6.6 feet) in some species. Their swimming ability, the result of the pulsating contraction of the bell, is limited, and scyphozoans are easily carried by currents. Some scyphozoan medusae (Figure 6-12) are among the most dangerous marine animals known to humans, notorious for their extremely painful and sometimes fatal stings (see Box 6-1).

Anthozoans. Most cnidarians are **anthozoans** (class **Anthozoa**), solitary or colonial polyps that lack a medusa stage. The anthozoan polyp is more complex than hydrozoan or scyphozoan polyps. The gut, for instance, contains several thin partitions or **septa** that provide additional surface area for the digestion of large prey (see Figure 13-2). **Sea anemones** are common and colorful anthozoans that often have large, muscular polyps (see Figure 10-32). Colonial anthozoans occur in an almost infinite variety of shapes and complexity. The mostly colonial **stony corals** have calcium carbonate skeletons that may form **coral reefs.** Though stony corals can be found in cold waters, it is in the tropics where they are active in reef-building (see the section in Chapter 13, "Reef Corals"). **Gorgonians,** such as sea fans (Figure 6-13), are colonial anthozoans secreting a tough branching skeleton made of protein. **Precious corals** are gorgonians with fused red or pink calcareous spicules in addition to a protein skeleton. **Black corals,** which are neither gorgonians nor stony corals, secrete a hard, black, protein skeleton (Figure 6-14). Both precious and black corals are carved into jewelry. Some anthozoans form fleshy colonies with large polyps and no hard skeletons. Examples of these are the soft corals (see the figure in Box 12-1), sea pens (see Figure 12-13), and sea pansies.

BOX 6-1

The Case of the Killer Cnidarians

The stings of most cnidarians are harmless to humans, but there are exceptions. Delicate and innocent-looking, some cnidarians are among the most dangerous marine animals. The sinister side of these creatures results from potent toxins released by their nematocysts.

The Portuguese man-of-war (*Physalia*), a siphonophore, is found in warm waters around the world. Though its blue, sail-like float can be seen fairly easily, its long tentacles are nearly invisible (see Figure 14-18). Armed with thick batteries of nematocysts, these tentacles reputedly may reach 50 m (164 feet) in length. Portuguese men-of-war may occur by the thousands, sometimes forcing the closure of beaches. Pieces of tentacle that wash ashore can be as nasty as the whole animal.

Physalia stings are very painful, similar to being repeatedly burned with a hot charcoal. The pain may last for hours, especially if sensitive areas of the body are affected. Red lines appear wherever tentacles have touched the skin, and welts usually follow. Both of the authors have had encounters with *Physalia*, painful but fortunately less severe than the experiences of others. One of us saw a man get stung on the hand. When the wave of intense pain reached his armpit, the man passed out. Even more severe reactions may occur. There can be nausea and difficulty in breathing. Contact of tentacles with the eye may result in damage to the cornea. Allergic reactions to the toxin may cause shock and even death, and swimmers may drown due to intense pain or shock.

If you are stung, the best thing to do is not to panic. Carefully wash the area with seawater, but don't rub the area or wash with fresh water because this stimulates firing of the nematocysts. Vinegar will also inactivate the nematocysts. Some recommend urine, useful if nothing else is available. The toxin is a protein, and some recommend papain, a protein-digesting enzyme found in meat tenderizer. The meat tenderizer is of little help, however, since the poison is injected into the skin while the meat tenderizer remains on the surface. Severe reactions should be treated in a hospital.

A group of scyphozoan medusae, the cubomedusae, release even more powerful toxins. The sea wasp, or box jellyfish, *Chironex fleckeri*, of northern and northeastern Australia, the Philippines, and the Indian Ocean, was responsible for more than 70 known deaths from 1900 to the early 1980's. Its stings cause immediate, extreme pain. Death due to heart failure may follow within minutes, especially in children. Skin that touches the tentacles swells up, followed by purple or dark brown lines that are slow to heal. Fortunately, an antivenom has been developed. Otherwise, first aid is the same as previously outlined. Other cubomedusae have been recorded as causing severe stings to humans in the tropics, particularly in Australia and the West Indies.

Cubomedusae are more common along the coast during summer. Their transparent, almost square bells are difficult to see in the water. Most are small, but in the sea wasp the bell may reach 25 cm (almost 10 inches) in diameter.

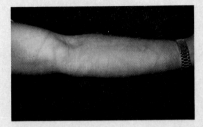

Sting of *Chiropsalmus*, a cubomedusa.

Figure 6-13 Sea fans are gorgonians with branches that grow in only one plane and have many cross-connections.

Figure 6-14 A treelike black coral (*Antipathes grandis*) from deep water off the island of Hawaii. Its black skeleton is much-valued for jewelry.

Figure 6-15 This comb jelly displays long tentacles used to capture food.

Biology of Cnidarians

The presence of tissues allows cnidarians to perform more complex functions than sponges can. In particular, cnidarians display advances in feeding and in the ability to sense and respond to their environment.

Digestion. Practically all cnidarians are **carnivores,** animals that prey on other animals, and must capture and digest prey much bigger than that of filter feeders such as sponges. Nematocysts are a clever way to capture prey. They consist of a fluid-filled capsule containing a thread able to be quickly ejected (see Figure 6-11). The thread can be sticky or armed with spines or consist of a long tube that wraps around parts of the prey. Some nematocysts contain toxins.

After ingestion, food passes into the gut where it is digested. The initial phase of digestion is said to be **extracellular,** since it takes place outside cells. **Intracellular digestion** within cells lining the gut completes the breakdown of food.

Behavior. Though cnidarians lack a brain or true nerves, they do have specialized **nerve cells.** These cells interconnect to form a **nerve net** that transmits impulses in all directions. This simple nervous system can produce some relatively sophisticated behaviors. Some anemones can tell whether other members of the same species are also members of the same group, or clone. They are known to attack, and sometimes kill anemones from other clones! Some medusae have primitive eyes. Medusae have **statocysts,** a small calcareous body in a fluid-filled chamber surrounded by sensitive hairs. Statocysts thus give medusae a sense of "balance."

COMB JELLIES: RADIAL SYMMETRY ONE MORE TIME

The **comb jellies,** or ctenophores (phylum **Ctenophora**), comprise an exclusively marine group of about 90 species. Their radially symmetric and gelatinous body resembles that of a medusa, but a closer look reveals some unique traits. Eight rows of **ciliary combs,** long cilia fused at the base like combs, beat in waves (Figure 6-15). The continuous beating of the ciliary combs refracts light, creating a prismlike multicolor effect. Length varies from a few mm in the sea gooseberry (*Pleurobrachia*) to 2 m (6.6 feet) in the elongated Venus's girdle (*Cestum*; see Figure 14-13, C).

> Comb jellies, or ctenophores, are radially symmetric invertebrates similar in appearance to cnidarians, but possessing eight rows of ciliary combs.

Comb jellies are common in both warm and cold waters. They are carnivores with a voracious appetite. Swarms of comb jellies may consume large numbers of fish larvae. Many capture their prey using two long tentacles armed with sticky cells named **colloblasts.** Unlike cnidarians, comb jellies lack nematocysts.

BILATERALLY SYMMETRIC WORMS

Radial symmetry works fairly well in animals that attach to surfaces or drift in currents, but active animals have different needs. Most animals show **bilateral symmetry,** the arrangement of body parts in such a way that there is only one way to cut the body and get two identical halves (Figure 6-16).

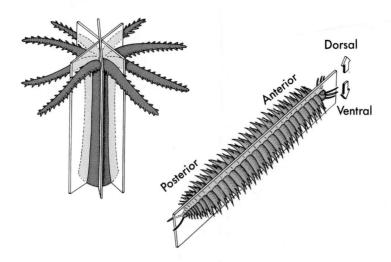

Figure 6-16 The radial symmetry of a soft coral's polyp in contrast with the bilateral symmetry of a worm. Notice that bilateral symmetry implies the development of an anterior end with a head, brain, eyes, and all the other features demanded by more complex behaviors.

Bilaterally symmetric animals, including ourselves, have a front—or **anterior end**—and a rear—or **posterior**—end. At the anterior end is a head with a brain, or at least an accumulation of nerve cells, and sensory organs such as eyes. Similarly, bilaterally symmetric animals have an upper, or **dorsal,** surface that is different from the lower, or **ventral,** surface. Bilateral symmetry provides new opportunities for animals to be more active in the pursuit of prey and to develop more sophisticated behaviors than those of radially symmetric animals.

Flatworms

The simplest bilaterally symmetric body plan is easily seen in elongate, creeping worms. In the marine world, worms come in all shapes and sizes. Among the simplest are the **flatworms** (phylum **Platyhelminthes**), so called because they are dorsoventrally flattened; that is, they have flat backs and bellies. Flatworms also are the simplest animals in whom tissues are organized into real organs and organ systems.

The presence of a **central nervous system** in which information is stored and processed is of special significance. In flatworms it typically consists of a very simple **brain,** which is actually an aggregation of nerve cells located in the head, and several nerve cords running from the brain through the length of the worm. The nervous system coordinates the movements of a well-developed muscular system. The gut is similar to those of cnidarians and ctenophores in having only one opening to the outside, the mouth. The space between the outer and inner tissue layers, however, is no longer thin or gelatinous as in cnidarians and ctenophores, but is filled with tissue. This middle layer of tissue gives rise to muscles, the reproductive system, and other organs—not only in flatworms, but also in higher animals.

> **Flatworms are bilaterally symmetric invertebrates typically flattened in appearance. They have true organs and organ systems, including a central nervous system.**

There are at least 15,000 species of flatworms. The most commonly seen marine flatworms are the **turbellarians,** a group consisting mostly of free-

Figure 6-17 A turbellarian flatworm from a coral reef in Truk Island, central Pacific. Cilia on the underside and sticky mucus help in their movements.

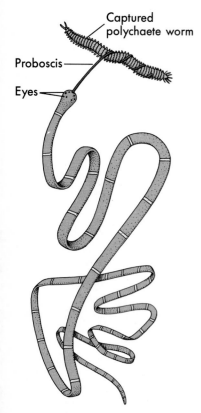

Figure 6-18 Many ribbon, or nemertean, worms are brightly colored and conspicuous but contain toxic substances that potential predators soon learn to avoid.

living carnivores (Figure 6-17). Most are small, but some are very obvious because of their striking color patterns. Some turbellarians live inside or on the surface of crabs, sea urchins, and other invertebrates.

The largest group of flatworms, more than 6,000 species, is the **flukes,** or **trematodes.** All flukes are **parasites,** which live in close association with other animals and feed on their tissues, blood, or intestinal contents. Like most parasites, flukes have complex life histories. Adult flukes always live in a vertebrate. Larvae may inhabit invertebrates like snails or clams or vertebrates like fish. The larva must then be eaten by the vertebrate destined to harbor the adult. Flukes are common in fishes, sea birds, and whales.

Tapeworms, or **cestodes,** are parasitic flatworms that, with a few exceptions, have a long body divided into repeated units. These unique worms hang inside the intestine of most species of vertebrates, including marine ones. The head of the worm attaches to the walls of the gut by means of suckers or other structures. Tapeworms lack a gut or mouth and absorb nutrients from their host's intestinal contents. Their larvae are found in invertebrates or vertebrates. Tapeworms may reach a prodigious size. The record appears to be a species found in sperm whales which is 15 m (50 feet) long!

Ribbon Worms

Though they look like long flatworms, **ribbon,** or **nemertean, worms** (phylum **Nemertea**) show several features that indicate a more complex degree of organization. Their digestive tract is complete, with a gut that includes a mouth and an anus to get rid of undigested material. They also have a **circulatory system,** by which blood transports nutrients and oxygen to tissues. The most distinctive feature of ribbon worms, however, is a **proboscis** (Figure 6-18), a long and fleshy tube used to entangle small prey. It is everted from a cavity above the mouth like the finger of a glove. The proboscis may be armed at the tip with a spine, or it may secrete toxins. Once the prey is captured, the proboscis is pulled back in and it's dinner time.

There are approximately 900 species of ribbon worms, most of which are marine. They are found throughout all oceans but are more common in shallow temperate waters. Some are nocturnal and not easily seen; others are brightly colored and may be found under rocks at low tide. Ribbon worms are incredibly elastic, and the proboscis may extend up to a meter or more beyond the body.

Nematodes

Nematodes (phylum **Nematoda**), some of which are known as **roundworms,** comprise one of the most underestimated groups of marine animals. They are hardly ever seen, but their numbers in sediments, particularly those rich in organic matter, can be staggering. Many species are parasites of marine animals and plants. Nematodes are perfectly adapted to live in sediments or the tissues of other organisms. They are mostly small, and their slender and cylindric body is typically pointed at both ends (see the figure in Box 11-1). Nematodes that inhabit sediments feed mostly on bacteria and organic matter. The gut, which ends in an anus, lies within a body cavity that is filled with fluid. Nutrients are transported by this fluid. A layer of muscles in the tough, but flexible, body wall pushes and squeezes against the fluid, which acts as a **hydrostatic skeleton,** a system used to provide support and aid in locomotion.

> **Nematodes are very common inhabitants of marine sediments and are widespread parasites of most groups of marine animals.**

The actual number of species of nematodes is debatable. Estimates vary between 10,000 and 15,000 species, but some biologists believe there may be as many as half a million species, most of which remain to be discovered.

The adults of *Anisakis* and a few related nematodes inhabit the intestine of seals and dolphins. Their larvae, however, are found in the flesh of many types of fish and may infect humans when raw or poorly cooked fish is eaten. Often the larvae are vomited or coughed up without further complications. Sometimes, however, larvae penetrate into the walls of the stomach or intestine, causing pain and symptoms similar to those of ulcers. It is a risk that lovers of raw fish dishes such as *sashimi* and *ceviche* take.

In addition to nematodes, other phyla of microscopic invertebrates inhabit marine sediments. They are discussed in Box 11-1.

Segmented Worms

A large group of perhaps as many as 13,000 species, the **segmented worms,** or **annelids** (phylum **Annelida**), includes earthworms and many marine worms. Their body plan includes innovations that have been incorporated in some of the more complex groups of animals. The body consists of a series of similar compartments or **segments,** a condition known as **segmentation.** Segmentation can be clearly seen in the familiar earthworm in the form of numerous rings. The gut, a complete tube as in nematodes and ribbon worms, lies in a cavity known as a **coelom.** The coelom is filled with fluid and divided with partitions corresponding to the external segments. The segments act as a hydrostatic skeleton and can be contracted in sequence by means of muscles in the body wall. These movements, plus the flexibility given by segmentation, make annelids efficient crawlers and burrowers.

Polychaetes. Almost all marine annelids are **polychaetes** (class **Polychaeta**), which are very common and important in many environments. Each of their body segments sports a pair of flattened extensions, or **parapodia,** which are provided with stiff and sometimes sharp bristles, or **setae** (Figure 6-19).

> **A body consisting of similar segments and the appearance of a coelom are features of the segmented worms. Most marine annelids are polychaetes, segmented worms that have parapodia.**

Like all annelids, polychaetes have a distinct circulatory system that transports nutrients, oxygen, and carbon dioxide. Blood circulates but always remains in blood vessels, making it a **closed circulatory system.** Muscular contraction of vessels helps in the circulation of blood.

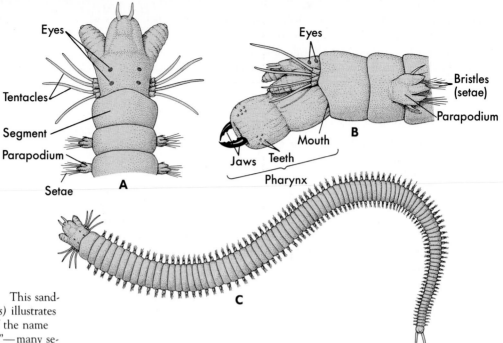

Eyes

Tentacles

Segment

Parapodium

Setae

A

Eyes

Bristles (setae)

Parapodium

B

Jaws   Teeth   Mouth

Pharynx

C

Figure 6-19 This sand-worm *(Nereis)* illustrates the origin of the name "polychaetes"—many se-tae, or bristles. **(A)** Dorsal view of the head, with the pharynx retracted, showing the sensory tentacles and eyes. **(B)** Side view of the head, showing the large pharynx in an extended position. **(C)** Dorsal view of the worm.

respiration   Organic mat-ter + $O_2 \longrightarrow CO_2 + H_2O$ + ENERGY
Chapter 4, p. 90

In small animals, oxygen—which is essential in the release of energy through **respiration**—can easily move from the water to reach all the tissues. In the larger and relatively more active polychaetes, however, obtaining enough oxygen from the water is a potential problem. Polychaetes have solved this problem by developing **gills** in the parapodia or elsewhere. Gills are extensions of the body wall that have many thin-walled blood vessels called **capillaries,** which allow for the easy absorption of oxygen. This absorption of oxygen, along with the elimination of carbon dioxide, is referred to as **gas,** or **respiratory, exchange.**

The life history of many polychaetes involves a planktonic larval stage known as the **trochophore,** which has a band of cilia around the body (see Figure 14-11, *D*). The trochophore is of considerable interest, because it is also present in other groups of invertebrates.

The 6,000 species of polychaetes are almost entirely marine. Length varies a great deal but is typically around 5 to 10 cm (2 to 4 inches). Many polychaetes crawl on the bottom, hiding under rocks or coral. These crawling worms, such as most sandworms *(Nereis, Neanthes)*, are mostly carnivores. They feature heads provided with several pairs of eyes and other sense organs (see Figures 6-19 and 6-21, *C*) used to search for small invertebrates. A proboscis, often armed with jaws, is used to capture prey. The parapodia are well developed and are used in locomotion.

Other polychaetes burrow in mud or sand (see Figure 10-37). Many, like bloodworms *(Glycera)*, capture small prey. Others, like lugworms *(Arenicola)*, feed on organic particles that settle *on the bottom*. This feeding technique is known as **deposit feeding,** in contrast with suspension feeding,

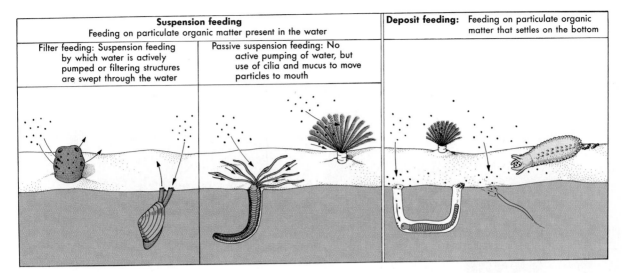

| Suspension feeding | Deposit feeding: |
| Feeding on particulate organic matter present in the water | Feeding on particulate organic matter that settles on the bottom |

| Filter feeding: Suspension feeding by which water is actively pumped or filtering structures are swept through the water | Passive suspension feeding: No active pumping of water, but use of cilia and mucus to move particles to mouth | |

Figure 6-20 Feeding on particulate organic matter can be classified into suspension and deposit feeding. The difference between these two types of feeding is not always well defined. Fanworms, for instance, are tube-dwelling polychaetes that are able to switch back and forth between suspension and deposit feeding, depending on the strength of the water current.

which involves feeding on organic particles and plankton present *in the water* (Figure 6-20).

Many polychaetes build and live in a wide variety of tubes, either temporary or permanent, single or in aggregations (see Figures 11-13, *H*, and 12-6, *I*). The tubes may be made of mucus, protein, bits of seaweed, cemented mud particles, sand grains, or tiny fragments of shells. Tube-dwelling polychaetes usually have reduced parapodia. Some, such as *Amphitrite* and related forms (Figure 6-21, *A*), are suspension feeders that extend tentacles provided with beating cilia and mucus to gather organic particles from the water and move them to the mouth (see Figure 6-20). Fanworms, or feather-duster worms (*Sabella* and others; Figure 6-21, *B*), use feathery tentacles covered with cilia to capture, sort, and transport particles. Serpulids (*Serpula*) and sprirorbids (*Spirorbis, Spirobranchus*), also suspension feeders, extend featherlike tentacles from calcium carbonate tubes built on rocks and other surfaces (see Figures 12-18, *D*).

Other lifestyles are successfully exploited by polychaetes. In the tropical Pacific the Palolo worm (*Eunice*) periodically swims up to the surface to spawn. This behavior, known as **swarming,** is timed in some areas with the phases of the moon, reaching its peak just after full moon. Species of *Tomopteris* are planktonic throughout life. Their parapodia are flat and expanded to help in swimming (see Figure 14-13, *D*). Some polychaetes may live on the external surface of such invertebrates as sea stars and sea urchins. Several species live in the burrows of other invertebrates or inhabit shells occupied by hermit crabs.

Leeches. Bloodsucking **leeches** (class **Hirudinea**) live mostly in fresh water but can be found attached to marine fishes and invertebrates. Leeches are highly specialized annelids distinguished by a sucker at each end and no parapodia.

Figure 6-21 Polychaetes are common inhabitants of most marine bottoms. **(A)** A terebellid worm (*Terebella*) showing the extended tentacles during suspension feeding. **(B)** *Sabella melanostigma,* a fanworm, inhabits leathery tubes. **(C)** The anterior end of a free-living polychaete that feeds on corals, *Hermodice carunculata.*

Odds and Ends in the World of Worms

A few unique groups, minor but hardy branches of the animal family tree, are variants of the worm body plan. Some show similarities to the segmented worms, but not enough to warrant their inclusion in the phylum Annelida.

Peanut Worms. Often called **peanut worms,** the **sipunculans** (phylum **Sipunculida**) bury their soft, unsegmented bodies in muddy bottoms, hide in empty shells, or burrow in rocks and corals. The long anterior portion contains a mouth and a set of small lobes or branching tentacles. These can be pulled into the remaining portion of the body, and the worm then becomes a compact bundle that looks like a large peanut (Figure 6-22, A). Most peanut worms are between 1 to 35 cm (0.4 to 14 inches) long. More than 300 species are known, all deposit feeders. Most live in shallow water.

Echiurans. All of the over 100 species of **echiurans** (phylum **Echiura**) are marine. They look like soft, unsegmented sausages buried in the mud.

Figure 6-22 **(A)** *Phascolosoma antillarum,* a peanut worm (or sipunculid), burrows in coral and rocks in the Caribbean. Also see Figure 12-10. **(B)** *Thalassema mellita,* an echiuran from soft bottoms in the Gulf of Mexico and the Atlantic coast of North America.

They are similar to peanut worms in shape and size except for a nonretractable and spoonlike or forked proboscis (Figure 6-22, B). Echiurans are deposit feeders that use the proboscis to gather organic matter. The fat innkeeper (*Urechis caupo*) of the western coast of North America lives in U-shaped tubes in mud (see Figure 11-13, E).

Beard Worms. **Beard worms,** or **pogonophorans** (phylum **Pogonophora**), are unique in several ways. These long and thin worms lack a digestive system—mouth and gut included. Except for sponges and tapeworms, this is an uncommon phenomenon in animals. A tuft of one to many thousand long tentacles (Figure 6-23), responsible for the group's common name, appears to be involved in absorbing nutrients dissolved in the water. Bacteria living in beard worms use dissolved nutrients to manufacture food that is in turn used by the worms.

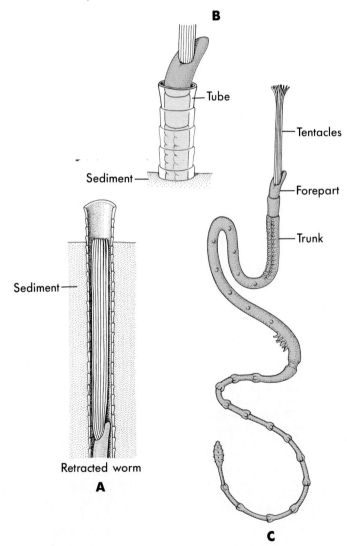

Figure 6-23 Diagrammatic representation of a beard worm, or pogonophoran. Most secrete and live in tubes buried in the soft sediment (**A**). Only the upper end of the tube protrudes out, with the tentacle or tentacles extending from it (**B**). Worm removed from its tube (**C**).

BOX 6-2

How to Discover a New Phylum

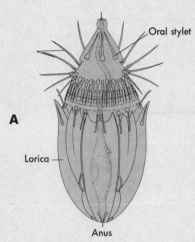

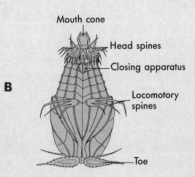

(A) Loriciferan and (B) Higgins larva.

It is not very difficult to discover a new marine invertebrate species. Small animals living in sediments, among rocky shore seaweeds, or in deep water are good candidates. Discovering a new phylum, however, is a different story.

The founding species of only three phyla escaped description until this century. All three are exclusively marine. The first new phylum of the century was created for the beard worms, or pogonophorans (see Figure 6-23). The first beard worm was dredged in Indonesia in 1900, but it was not formally described until 1914. The second species, from the Sea of Okhotsk in eastern Siberia, was described in 1933. More new species were and still are being described, mostly from deep water. A new phylum was officially created for them in 1955.

The first species of what eventually became the new phylum **Gnathostomulida** was not officially described until 1956. Gnathostomulids are a group of about 100 species of minute worms living among sediment particles around the world (see the Figure in Box 11-1). They are similar to the flatworms but possess unique features, including a set of toothed jaws to scrape bacteria, diatoms, and other organisms from sand grains.

The latest of the new phyla has a short but turbulent history. In 1961 Robert Higgins, of the National Museum of Natural History in Washington, D.C., predicted the existence of a group that lived in the spaces between clean, coarse sediments in deep water. He actually found a specimen in 1974 but unfortunately did not realize it was something new.

One year later, in 1975, Reinhardt Kristensen of the University of Copenhagen, Denmark, collected a specimen, but it was destroyed while being prepared for microscopic examination. Kristensen later found larvae of the elusive animal in coarse sediments from western Greenland and the Coral Sea. In 1982 he was working with a large sample taken off the coast of Brittany in France. It was his last day at the Roscoff Biological Station, and to save time he washed the sample with fresh water, instead of following the standard but more time-consuming method. It happened to loosen the grip of the animals on the sediment particles, and Kristensen got a complete series of larval and adult specimens!

The microscopic animals Kristensen found show similarities to rotifers and kinorhynchs (see the figure in Box 11-1). The body is encased by six plates. The head, which is protrusile, bears a set of spines and a mouth at the end of a cone. Kristensen got together with Higgins, and they concluded that the specimens Higgins examined in 1974 and those subsequently found by Kristensen were members of a new phylum. They found additional adults in eastern Florida, which further confirmed the new status of the group.

The new phylum **Loricifera** ("corset bearer") was officially born in 1983 when Kristensen published a paper in a German scientific journal. The first species was named *Nanaloricus mysticus* ("mystic, or enigmatic, dwarf armor") and the larva was baptized the Higgins larva in honor of Higgins, a nice consolation prize indeed!

Approximately 120 species of beard worms are known. They are mostly restricted to deep water, which helps explain why they remained unknown until 1900 (see Box 6-2). The total length of the worms ranges from 10 cm to 2 m (4 inches to 7 feet). **Vestimentiferans,** which some believe are polychaete worms and not beard worms, are even longer. Large numbers of these worms have been found at hydrothermal vents (see Figures 15-38 and 15-39).

Arrow Worms. In terms of number of species the **arrow worms,** or **chaetognaths** (phylum **Chaetognatha**), rank among the smallest animal phyla. Only about 60 species, all marine, are known. They are nevertheless one of the most common and important members of the plankton. Their body is much different from that of a bottom-dwelling worm. It is almost transparent, streamlined, with fishlike fins and tail (Figure 6-24). The head has eyes, grasping spines, and teeth. Total length ranges from a few millimeters to 10 cm (up to 4 inches).

Arrow worms are voracious carnivores. They prey on small crustaceans, eggs and larvae of other animals, other arrow worms, and practically anything else that is small. They spend most of their time motionless in the water but will swim in rapid, dartlike movements to grab their prey.

LOPHOPHORATES

Three groups of marine invertebrates are linked by the presence of a unique feeding structure, the **lophophore.** It consists of a set of ciliated tentacles arranged in a horseshoe-shaped, circular, or coiled fashion. Lophophorates are suspension feeders, using their cilia to create feeding currents. They share other important traits: lack of segmentation, bilateral symmetry, a coelomic cavity, and a U-shaped gut.

Bryozoans

Bryozoans, meaning "moss animals," are invertebrates that form delicate colonies on seaweed, rocks, and other surfaces. Approximately 4,000 species, almost all marine, are grouped in the phylum **Bryozoa,** also known as **Ectoprocta.** Bryozoan colonies consist of minute interconnected individuals called **zooids** that secrete skeletons of a variety of shapes (Figure 6-25). The colonies may be encrusting or take the form of tufts of crusty lace. A close look often reveals the rectangular, round, or vaselike compartments occupied by zooids (see Figure 12-18, *E*). The lophophore is retractable. The U-shaped gut ends in an anus outside the edge of the lophophore.

Most specialists separate bryozoans from a similar group of about 60 species, the **entoprocts** (phylum **Entoprocta**). Their minute colonies often escape notice but can be found on seaweeds, hydroids, and shells. Zooids possess a crown of tentacles that is not a true lophophore. Entoprocts also differ from bryozoans by lacking a coelom and by an anus that opens inside the crown of tentacles.

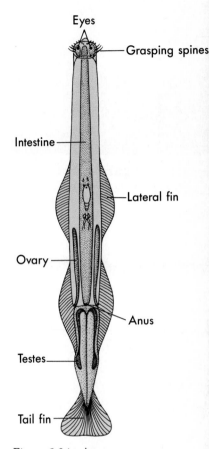

Figure 6-24 Arrow worms, or chaetognaths, are characterized by a transparent, fishlike body, an adaptation to live as planktonic predators. *Sagitta elegans* ("elegant arrow"), illustrated here, is widely distributed through all oceans. Notice that individuals have both testes and ovaries. A partially-digested copepod, a favorite food, is inside the transparent intestine. Also see Figure 15-6.

Phoronids

At first sight, **phoronids** (phylum **Phoronida**) may be easily confused with polychaete worms. They are wormlike and build tubes made in part of sand grains. They have a horseshoe-shaped or circular lophophore, however, and their gut is U-shaped in contrast to the straight gut of polychaetes. All 15 species are marine, burrowing in sand or attaching their tube to rocks and other hard surfaces in shallow water. Though total length may reach 25 cm (10 inches) in one California species, most phoronids are only a few centimeters long.

Figure 6-25 *Sertella* and other bryozoans form crusty, lacelike colonies.

Lamp Shells

There are close to 300 living species of **lamp shells** or **brachiopods** (phylum **Brachiopoda**). Thousands of other species are known only as fossils. Lamp shells have a shell with two parts, or **valves,** like the unrelated clams. It is only when the hard shell is opened that the major difference between lamp shells and clams is apparent. Lamp shells have a conspicuous lophophore consisting of two coiled and ciliated arms that occupies most of the space between the valves. Most brachiopods are found attached to rocks or burrowing in soft sediments in shallow water.

MOLLUSCS: THE SUCCESSFUL SOFT BODY

Snails, clams, octopuses, and other familiar forms are grouped as the phylum **Mollusca. Molluscs** have been very successful—there are more species of molluscs in the ocean than of any other group. Even if terrestrial and

Figure 6-26 The general body plan of a snail, indicating the most important internal structures. In many species the head and foot can be retracted into the shell, leaving a tough operculum blocking the shell opening.

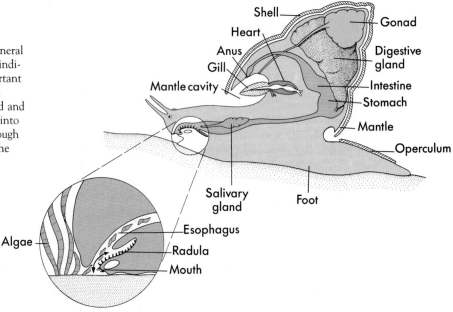

LIFE IN THE MARINE ENVIRONMENT

freshwater forms are included, the 110,000 species of molluscs are surpassed only by the arthropods as the largest phylum of animals.

Molluscs highlight a soft body protected by a calcium carbonate shell. The body is covered by a **mantle,** a thin layer of tissue that produces the shell (Figure 6-26). The unsegmented body is bilaterally symmetric. There is a ventral, muscular **foot,** typically used in locomotion. Most molluscs have a well-developed head that invariably includes eyes and other sensory organs. A unique feature is the **radula,** a ribbon of small teeth used in rasping food from surfaces (Figure 6-27). It is made largely of **chitin,** a highly-resistant material also found in other invertebrates. Gas exchange is through paired gills.

The molluscan body plan can be easily identified, but it may be greatly modified among members of the phylum. The shell, for example, is internal in squids and absent in sea slugs and octopuses. In snails, portions of the body are coiled and asymmetric. In some molluscs the radula is modified or even absent.

> The molluscs constitute the largest group of marine animals. Their body is soft with a muscular foot. They usually have a shell and most have a radula, a rasping "tongue" unique to the group.

Types of Molluscs

Molluscs exhibit an immense diversity of structures and habits. They occupy all marine environments from the wave-splashed zone of rocky shores to hydrothermal vents in the deep sea. They thrive on practically every conceivable type of diet. For all their diversity, most molluscs belong to one of three major groups.

Gastropods. The **gastropods** (class **Gastropoda**), best known as snails, are the largest, most common, and varied group of molluscs. There are perhaps 90,000 species, mostly marine. A typical gastropod (meaning "stomach footed") can best be described as a coiled mass of vital organs enclosed by a dorsal shell (see Figure 6-26). The shell rests on a ventral creeping foot and is usually coiled.

Most snails use their radula to scrape algae from rocks, as in the case of periwinkles (*Littorina;* see Figure 10-3), limpets (*Fissurella, Lottia;* Figure 6-28, A), and abalones (*Haliotis;* Figure 6-28, B). Some, like mud snails (*Hydrobia*), are deposit feeders on soft bottoms. Whelks (*Buccinum, Busycon*), oyster drills (*Murex, Urosalpinx*), and cone shells (*Conus;* Figure 6-28, C), are carnivores. These snails specialize in prey like clams, oysters, or even small fishes. The violet snail, *Janthina* (see Figure 14-19), has a thin shell and floats on the surface looking for siphonophores, its prey. Sea hares (*Aplysia*), which graze on seaweeds, have small and thin shells buried in tissue. **Nudibranchs,** or **sea slugs,** are snails that have lost the shell altogether. Colorful branches of the gut or exposed gills make sea slugs among the most beautiful of all marine animals (Figure 6-28, E). They prey on sponges, hydroids, and other invertebrates. As a defensive mechanism, sea slugs often retain noxious chemicals or nematocysts found in their prey (see Figure 9-9).

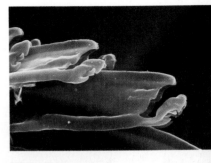

Figure 6-27 A high-magnification photo of the teeth of the radula of the Cooper's nutmeg snail (*Cancellaria cooperi*). The radula is adapted, of all things, for blood sucking. The snail actively seeks electric rays that rest partially buried in sand. It then uses its radula to cut through the tough ray's skin and sucks blood (see Figure 9-11).

A

B

C

D

E

Figure 6-28 Gastropods, or snails, come in all shapes, colors, and habits. **(A)** The giant keyhole limpet *(Megathura crenulata)*, from the Pacific coast of North America, photographed on a bottom covered by encrusting red coralline algae. **(B)** The red abalone *(Haliotis rufescens)* is much sought for food by humans. **(C)** Cone shells, such as *Conus geographus*, are carnivorous snails that bury in sand, waiting for prey such as small fishes. Their radula is modified into a dartlike tooth that is shot—together with a poison—into the unsuspected prey, which is eaten whole very much as in snakes. **(D)** *Simnia acicularis* lives on goronians in Florida and the Caribbean. **(E)** A flashy nudibranch, *Hermissenda crassicornis*.

Bivalves. Bivalves (class **Bivalvia**) are the clams, mussels, oysters, and similar molluscs. Bivalves retain the basic molluscan body plan, though modified (Figure 6-29). The body is laterally compressed and enclosed in a two-valved shell. The head is very small, and there is no radula. The gills, expanded and folded, are used not only to obtain oxygen, but also to filter and sort small food particles from the water. The inner surface of the shell is lined by the mantle, so that the whole body lies in the **mantle cavity,** a large space between the two halves of the mantle. Strong muscles are used to close the valves.

Clams *(Macoma, Mercenaria,* and others) use their shovel-shaped foot to burrow in sand or mud (see Figure 10-36). When the clam is buried, water is drawn in and out of the mantle cavity through a **siphon** formed by the fusion of the edge of the mantle (Figure 6-29, *A* and *D*). This allows clams to feed and obtain oxygen while buried in sediment.

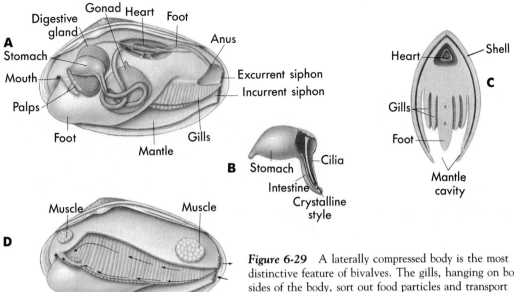

Figure 6-29 A laterally compressed body is the most distinctive feature of bivalves. The gills, hanging on both sides of the body, sort out food particles and transport them to the mouth with the help of cilia and mucus. The path of the particles from the incurrent siphon to the mouth is indicated by arrows *(D)*.

Not all bivalves are burrowers. Mussels (*Mytilus;* see Figure 4-15), for instance, secrete strong **byssal threads** that attach them to rocks and other surfaces. Oysters (*Ostraea, Crassostrea;* Figure 6-30, A) cement their left shell to a hard surface, often the shell of another oyster. Aphrodisiac or not, they have been lustfully swallowed by lovers of good food for thousands of years. Pearl oysters (*Pinctada*) are the source of most commercially valuable pearls. Pearls are formed when the oyster secretes shiny layers of calcium carbonate to coat irritating particles or parasites lodged between the mantle and the iridescent inner surface of the shell, which is called mother-of-pearl. Cultured pearls are obtained by carefully inserting a tiny bit of shell in the mantle. Scallops (*Pecten;* Figure 6-30, C) live unattached and can swim for short distances by rapidly ejecting water from the mantle cavity and clapping the valves. The largest bivalve is the giant clam (*Tridacna;* see Figure 13-42), which grows to 1 m (3 feet) in length.

Many bivalves bore in coral, rock, or wood. The "shipworm" (*Teredo*) bores in mangrove roots, driftwood, and wooden structures such as boats and pilings. Wood is excavated for food by the small valves, and the elongate body forms a tunnel lined with calcium carbonate. The rasping valves are at the inner end of the tunnel, while a tiny siphon protrudes from the entrance. Shipworms are an example of a **fouling organism,** one that settles on the bottom of boats, pilings, and other submerged structures.

Cephalopods. The **cephalopods** (class **Cephalopoda**), which are specialists in the art of locating and capturing prey, include the octopuses, squids, and other fascinating creatures. Cephalopods adapt the molluscan

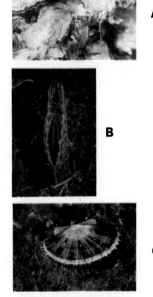

Figure 6-30 Oysters, such as *Crassostrea virginica* **(A),** are harvested commercially around the world. Other bivalves bury in sand or mud. An example is the pen shell (*Pinna carnea*) **(B).** Scallops, such as *Pecten ziczac* **(C),** live free on the bottom.

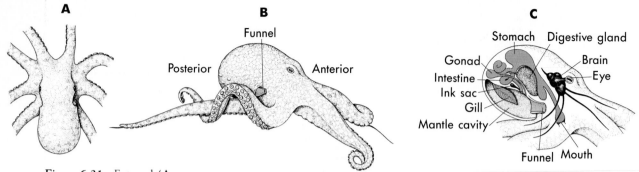

A

B

Funnel

Posterior Anterior

C

Stomach Digestive gland

Gonad Brain

Intestine Eye

Ink sac

Gill

Mantle cavity

Funnel Mouth

Figure 6-31 External (**A** and **B**) and internal (**C**) structure of the octopus. In the male the tip of the third right arm is modified to transfer packets of sperm from his siphon into the mantle cavity of the female. Copulation is preceded by courtship behavior that includes intricate color changes.

body plan to an active way of life. Most are agile swimmers displaying a complex nervous system and a reduction or loss of the shell. All 650 living species are marine. A cephalopod (meaning "head footed") is like a gastropod with its head pushed down toward the foot. The head is then below the rest of the body (Figure 6-31). The head has arms equipped with suckers to capture prey. The large eyes, usually set in the sides of the head, are remarkably like ours. The body, rounded in octopuses and elongate in squids, is protected by a thick and muscular mantle. The mantle forms a mantle cavity behind the head that encloses two or four gills. Water enters through the free edge of the mantle and leaves through the siphon, or **funnel,** a muscular tube that is formed by what remains of the foot and projects under the head. Cephalopods swim by forcing water out of the mantle cavity through the funnel. The flexible funnel can be moved around, allowing the animal to move in practically every direction. It is nature's own jet propulsion.

Octopuses *(Octopus),* not "octopi," are cephalopods that lack a shell and have eight long arms (Figures 6-31 and 6-32, A). They are common bottom-dwellers. Including arms, the size varies from 5 cm (2 inches) in the dwarf octopus to a record of 9 m (30 feet) in the Pacific giant octopus.

Octopuses are efficient hunters, with crabs, lobsters, and shrimps among their favorite dishes. The prey is bitten with a pair of beaklike jaws. The radula may help in rasping away flesh. They also secrete a paralyzing substance, and some have a highly toxic bite. Most, however, are harmless to humans. They use crevices in rocks, and even discarded bottles and cans, as homes. Their shelters are given away by the presence of rocks, which they move around, and by the remains of crabs. Like most other cephalopods, they can distract potential predators by emitting a dark fluid produced by the **ink sac** (Figure 6-31, C).

Squids *(Loligo;* Figure 6-32, B and C) are better adapted for swimming than are octopuses. The body is elongate, and the mantle forms two triangular fins. Squids can remain motionless in one place or move backward or forward just by changing the direction of the funnel. Ten arms with suckers circle the mouth. Two of them, the tentacles, are long and retractable and have suckers only at the broadened tips. These tentacles can be swiftly shot out to catch prey. A much reduced chitinous shell, or **pen,** is embedded in the upper surface of the mantle. Adult size varies from tiny individuals of a few centimeters in length to 20 m (66 feet) in the giant squid *(Architeuthis),* the largest living invertebrate. The giant squid is a deep-water species known mostly from specimens that have been washed ashore or found in the stomachs of sperm whales. Various aspects of the biology of deep-water squids are discussed in Chapter 15, "The Ocean Depths."

Cuttlefishes (*Sepia*) resemble squids in having eight arms and two tentacles, but the body is flattened and has a fin running along the sides. Cuttlefishes, which are not fish at all, have a calcified internal shell that aids in buoyancy. The shell is the "cuttlebone" sold as a source of calcium for cage birds.

An unusual external shell characterizes the chambered nautilus (*Nautilus*; see Box 15-1). The impressive shell is smooth, coiled, and up to 25 cm (10 inches) in diameter. The whorls of the shell contain a series of chambers, filled with gas and liquid, which serves as a buoyancy organ. The body—which occupies the outer, largest chamber—has 60 to 90 short and suckerless arms used to capture crabs and fish.

Other Molluscs. About 650 species of **chitons** (class **Polyplacophora**) are known, all marine. They can be readily identified by the eight overlapping shell plates that cover their slightly arched dorsal surface (Figure 6-33). Their internal organs are not coiled as in snails.

Almost all chitons are restricted to rocky shores, and for the most part they are not very active. Many species are known to return to a "home" site after feeding. The radula is used to graze on the algae that grows on rocks.

The 350 or so species of **tusk shells**, or **scaphopods**, (class **Scaphopoda**), have an elongate shell, open at both ends and tapered like an elephant tusk. They live in sandy or muddy bottoms. The narrow end of the shell protrudes from the bottom, whereas the foot projects from the wide end. Many species have thin tentacles with adhesive tips. They are used to capture forams, young bivalves, and other small organisms from the sediment. Tusk shells are most common in deep water, but empty shells are sometimes washed ashore.

The **monoplacophorans** (class **Monoplacophora**) are represented by only a handful of limpetlike molluscs known only as fossils until their discovery as "living fossils" in 1952. They have now been collected, mostly from deep water, in scattered locations around the world. Some biologists consider monoplacophorans a "missing link" with other groups of invertebrates. Others believe their particular characteristics are adaptations for survival in deep water.

Biology of Molluscs

The business of life is far more complex in molluscs that in cnidarians and worms. Does this complexity result from their abundance and variety, or is the variety and abundance of molluscs due to their complexity? The answer probably lies somewhere in the middle.

Digestion. The molluscan gut has a separate mouth and anus. Digestion involves **salivary** and **digestive glands** (Figure 6-26) that release digestive enzymes in charge of breaking down food into simpler molecules. Other aspects of the digestive system differ among groups and according to diet.

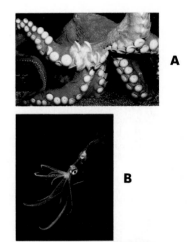

Figure 6-32 (**A**) A Pacific giant octopus (*Octopus dofleini*). (**B**) A deep-water squid. Other deep-water cephalopods are illustrated in Figures 15-7, 15-8, and 15-19. (**C**) Mating squids (*Loligo opalescens*). Notice the masses of white, gelatinous eggs on the bottom.

enzymes Substances that speed up specific chemical reactions
Chapter 4, p.88

chloroplasts Cell organelles in plants and other producers in which the process of photosynthesis takes place
Chapter 4, p. 96; Figure 4-13

Chitons and many snails are grazers. They have a rasping radula that removes minute algae from surfaces or cuts through large seaweeds. Their simple digestive system can efficiently process large amounts of hard-to-digest plant material. Digestion is partly extracellular in the gut cavity and partly intracellular in the digestive glands. Some shell-less snails that feed on seaweeds are able to keep the seaweed's chloroplasts intact. The chloroplasts are kept in the digestive gland and continue to carry out photosynthesis, providing nourishment for the snail.

Carnivorous snails have a radula modified to drill, cut, or even capture prey (see Figure 6-27). The radula and mouth are contained in a proboscis that can be protruded to strike the prey. Jaws may even be present. In these snails, digestion is entirely extracellular and takes place in the stomach.

Bivalves ingest food particles that are filtered and sorted out by the ciliated surface of the gills. The radula is absent, and food enters the mouth trapped in long strings of mucus. An enzyme-secreting rod in the stomach, the **crystalline style** (see Figure 6-29, B), continually rotates the food to help in its digestion. The stomach's contents eventually pass into a large digestive gland for intracellular digestion. The giant clam not only filters food, but obtains nutrients from zooxanthellae that live in its expanded mantle (see Figure 13-42). This extra nourishment may allow them to attain their giant size.

All cephalopods are carnivores that have to digest large prey. The stomach is sometimes connected to a sac in which digestion is rapidly and efficiently completed. It is entirely extracellular.

Molluscs have a circulatory system that transports nutrients, as well as oxygen. A muscular **heart** pumps blood to all tissues. Most molluscs have an **open circulatory system,** since blood flows out of vessels into open blood spaces. Cephalopods, on the other hand, have a **closed circulatory system** where the blood always remains in vessels and can be more effectively directed to oxygen-demanding organs such as the brain.

Nervous System and Behavior. The nervous system of molluscs shows a wide range of complexity. The simple behaviors of snails, bivalves, and chitons do not call for a well-developed brain. Chitons have a nervous system much like that of flatworms. Rather than a single brain, snails and bivalves have a set of "local brains," clusters of nerve cells located in several parts of the body.

In cephalopods, complexity of the nervous system reaches its highest point, not just in molluscs but in all invertebrates. The separate "local brains" of other molluscs are fused into a single large brain that coordinates and stores information received from the environment. Different functions and behaviors of cephalopods are known to be controlled by particular regions of the brain, as in humans. Giant nerve fibers rapidly conduct impulses, allowing cephalopods to capture prey or escape at amazing speeds. The strikingly complex eyes of cephalopods reflect the development of their nervous system. Octopuses and cuttlefishes have considerable intelligence and a remarkable capacity for learning. Most cephalopods, especially cuttlefishes, display color changes correlated with particular behaviors and moods, from intricate sexual displays to camouflage. Some cuttlefishes flash two

Figure 6-33 Chitons such as *Nuttallina californica* use their strong foot and the flexibility given by the eight articulated shells to fit tightly to the irregular surface of rocky shores.

LIFE IN THE MARINE ENVIRONMENT

large black spots resembling eyes, as if to petrify potential prey.

Reproduction and Life History. Molluscs usually have separate sexes, but some species are **hermaphrodites;** that is, all individuals have both male and female gonads. In bivalves, chitons, tusk shells, and some snails, sperm and eggs are released into the water and fertilization is external. Fertilization is internal in cephalopods and most snails. When cephalopods mate the male uses a modified arm to transfer a **spermatophore,** an elongate packet of sperm, to the female. Snails that copulate have a long, flexible penis.

Many molluscs have a trochophore larva like polychaetes, a characteristic often used as evidence for close affinities between molluscs, the segmented worms, and other groups. In snails and bivalves the trochophore usually develops into a **veliger,** a planktonic larva characterized by a tiny shell (see Figure 14-11, A). In many snails, part or all of development takes place within strings or capsules of eggs. Cephalopods lack larvae, and young develop in large yolk-filled eggs. Female octopuses protect their eggs until they hatch. The female usually dies afterward, since she eats little or nothing while guarding the eggs.

ARTHROPODS: THE ARMORED ACHIEVERS

The achievements of **arthropods** (phylum **Arthropoda**) are most impressive. Arthropods make up the largest phylum of animals, with almost one million known species and many remaining undiscovered. Of all the animals on earth, three out of four are arthropods. They have invaded all types of environments, including, of course, the oceans. Marine arthropods encompass a huge variety of animals such as barnacles, shrimps, lobsters, and crabs, to name a few.

The arthropod body is segmented and bilaterally symmetric. To the flexible, segmented body, arthropods have added the benefit of jointed appendages such as legs and mouthparts. Another characteristic of arthropods is a chitinous external skeleton, or **exoskeleton,** secreted by the underlying layer of tissue. The exoskeleton is tough and nonliving. The body and appendages, all covered by exoskeleton, are moved by sets of attached muscles. The exoskeleton works faster and is more efficient and reliable than the hydrostatic skeleton of worms.

The rigid exoskeleton, however, imposes limitations in terms of size and growth. We will never see arthropods as big as giant squids or whales. To grow, arthropods must **molt,** or shed their exoskeleton (Figure 6-34). The rigid old skeleton is discarded, and a new one develops after the animal takes in water to expand itself. Most arthropods are small, but they lead active lives. The exoskeleton and jointed appendages form a winning combination providing protection, support, and flexibility.

More species belong to the arthropods than to any other animal group. Arthropods have a segmented and bilaterally symmetric body. Their success in adapting to all types of environments is due in part to a resistant exoskeleton and jointed appendages.

Figure 6-34 This is not really a live Galápagos shore crab (*Grapsus grapsus*), but its exoskeleton—or molt. The old exoskeleton covered the entire external surface of the crab, even its mouthparts and eyes. Molting, which is regulated by hormones, results in a soft, helpless crab that must find shelter for a few days until its new, larger skeleton hardens.

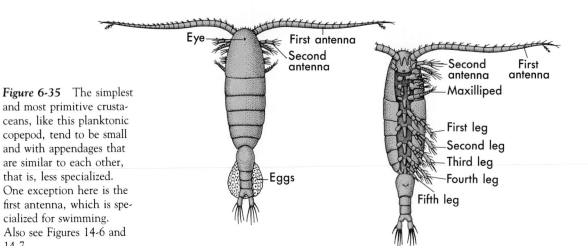

Figure 6-35 Eye — First antenna / Second antenna / Eggs

Second antenna / First antenna / Maxilliped / First leg / Second leg / Third leg / Fourth leg / Fifth leg

Figure 6-35 The simplest and most primitive crustaceans, like this planktonic copepod, tend to be small and with appendages that are similar to each other, that is, less specialized. One exception here is the first antenna, which is specialized for swimming. Also see Figures 14-6 and 14-7.

Crustaceans: Insects of the Sea

The overwhelming majority of marine arthropods are **crustaceans** (subphylum **Crustacea**), a large and extremely diverse group that includes shrimps, crabs, lobsters, and many less familiar animals. Crustaceans are specialized for life in water and possess gills to obtain oxygen. Their chitinous skeleton is hardened by calcium carbonate. The appendages are specialized for swimming, crawling, attaching to other animals, and feeding. Crustaceans possess two pairs of **antennae** (Figures 6-35 and 6-40) that are usually involved in sensing the surroundings.

There may be many as 35,000 species of crustaceans. Most are marine. While insects are by far the dominant arthropods on land, crustaceans are their counterparts at sea. The crustacean body plan is repeated in myriad forms, from the familiar to the not so familiar.

> **Crustaceans are arthropods adapted to live in water. They share the presence of two pairs of antennae, gills, and a calcified exoskeleton.**

The Small Crustaceans. Small crustaceans are everywhere: in the plankton, on the bottom, among sediments, on and in other animals, crawling among seaweeds. They are the bugs, flies, and mosquitoes of the sea.

Copepods are extremely abundant and important in the plankton (see the section in Chapter 14, "Copepods"). They use their mouthparts to filter out or capture food. Some species are so abundant that they are among the most common animals on earth. Many planktonic species keep from sinking by using their enlarged first pair of antennae (see Figure 6-35) to swim. Numerous species are parasitic, some becoming so simplified that they look like small bags of tissue.

Barnacles are filter feeders that usually live attached to surfaces, including living surfaces like whales and crabs. Many are very particular about the type of surface on which they live. Some are among the most important types of fouling organisms.

LIFE IN THE MARINE ENVIRONMENT

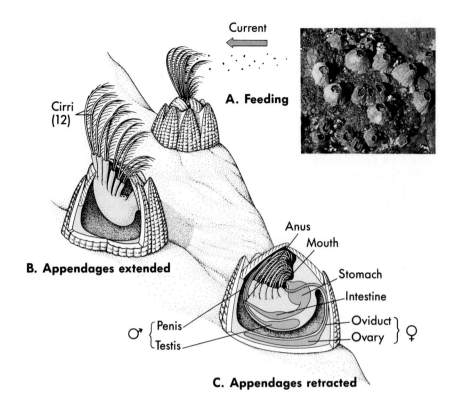

Current

Cirri (12)

A. Feeding

B. Appendages extended

Anus
Mouth
Stomach
Intestine
Oviduct
Ovary
♂ { Penis
Testis } ♀

C. Appendages retracted

Figure 6-36 Barnacles conceal a crustacean body beneath thick plates. They lie on their backs, using their legs to filter-feed. Notice that barnacles are hermaphrodites. Individuals, however, mate with each other, alternatively taking different sex roles. One individual acts as "male" by inserting "his" extended penis into a nearby "female." "He" may then turn into a "she" by welcoming the penis of a neighbor, perhaps a former "she." The photo inset shows the thatched barnacle (*Tetraclita squamosa*) from the Pacific coast of North America.

The common barnacles look almost like molluscs, since their bodies are enclosed by heavy calcareous plates (Figure 6-36). The plates on the upper surface open like windows to allow the feathery filtering appendages, actually the legs, to sweep the water. Some barnacles have become highly modified parasites and lack plates. All barnacles, however, have larvae typical of crustaceans, which swim and attach to surfaces before metamorphosis into adults (see Figure 6-45).

Beach hoppers and other **amphipods** are small crustaceans with a body distinctively compressed from side to side (Figure 6-37). Most amphipods are under 2 cm (¾ inch). The head and tail typically curve downward, and the appendages are specialized according to function. Beach hoppers, common among debris washed ashore, are strong jumpers that spring about by briskly stretching their curved body. Other amphipods crawl among seaweeds. Burrowing in the skin of whales (as "whale lice;" see Figure 9-10) and living as part of the plankton are some other lifestyles in this large, mostly marine group of over 5,000 species.

Isopods are found in many of the same environments exploited by amphipods. They are about the same size as amphipods, but isopods are easily identifiable because they are flat from top to bottom and have legs that are similar to each other (see Figure 10-24). Pill bugs, or roly-polies, are common land isopods that are similar to many marine species. **Fish lice** and other isopods are parasites of fishes (Figure 6-38) or other crustaceans.

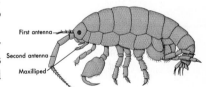

First antenna
Second antenna
Maxilliped

Figure 6-37 Most amphipods, like this sand hopper (*Orchestoidea*), can be recognized by their laterally compressed, curved body. The "skeleton shrimps" common among seaweeds and hydroids, however, are amphipods with bizarre skinny bodies. A giant deep-sea amphipod is illustrated in Figure 15-28.

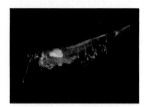

Figure 6-38 Many species of isopods suck blood from the gills, skin, and mouth of marine fishes. The photograph shows an isopod *(Anilocra)* below the eye of the coney *(Cephalopholis fulva).*

Figure 6-39 "Krill"—which comes from a Norwegian word for small, creeping animals—is also applied to the planktonic food of whales and to this group of filter-feeding crustaceans: the euphausids. *Euphausia pacifica,* shown here, has small light-producing organs that are characteristic of many species of krill.

Krill are planktonic, shrimplike crustaceans of moderate size, up to 6 cm (2.5 inches). Their head is fused with some of the body segments to form a distinctive **carapace** that covers the anterior half of the body like armor (Figures 6-39 and 15-4, A). Most krill are filter feeders that feed on diatoms and other plankton. They are extremely common in polar waters, aggregating in gigantic schools of billions of individuals. They are an almost exclusive food source for whales, penguins, and many types of fish (see Figure 9-14). A krill fishery has been developed, especially in Antarctica. There are concerns of the possible impact of taking large numbers of such an important resource (see the section in Chapter 16, "New Fisheries").

Shrimps, Lobsters, and Crabs. With around 10,000 species, the **decapods** are the largest group of crustaceans. They include the shrimps, lobsters, and crabs. Decapods are also the largest crustaceans in size. Many are prized as food and are of great commercial importance.

Decapods feature five pairs of walking legs, the first of which is heavier and usually provided with pincers used in obtaining food and in defense (Figure 6-40). The carapace is well developed and encloses the part of the body known as the **cephalothorax.** The rest of the body is called the **abdomen.**

Shrimps and **lobsters** are adapted for swimming. The body is laterally compressed, and there is a distinct and elongate abdomen, the "tail" we like to eat so much. Shrimps are typically **scavengers,** specialists in feeding on bits of dead plant and animal material on the bottom. They also exploit other lifestyles. Many colorful shrimps, particularly in the tropics, live on the surface of other invertebrates (Figure 6-41) or remove parasites from the skin of fish (see Box 12-2). Others live in deep water (see Figure 15-4, C and D). Ghost and mud shrimps burrow in muddy bottoms (see Figure 11-13, D). Lobsters, such as the American, or Maine, lobster *(Homarus;* see Figure 6-40) and the clawless spiny lobster *(Panulirus;* see Figure 4-29) are

Figure 6-40 The American lobster *(Homarus americanus)* illustrates the basic body plan of decapod crustaceans. The ducts from the gonads open at the base of the last pair of walking legs in males, but at the base of the second pair in females.

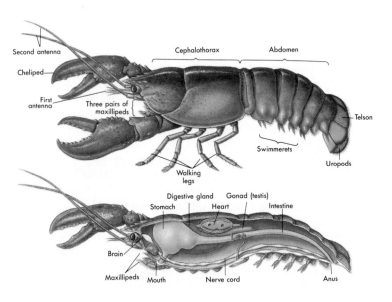

LIFE IN THE MARINE ENVIRONMENT

mostly nocturnal and hide during the day in rock or coral crevices. Their feeding habits are similar to those of shrimps, but they also are known to catch live prey. **Hermit crabs,** which are not "true" crabs, are also scavengers. They hide a long, soft abdomen in empty shells. Some hermit crabs cover their shells with sea anemones or sponges for added protection and camouflage. One type, however, doesn't hide the abdomen (Figure 6-42).

In the **crabs** the abdomen is small and tucked under the compact and typically broad cephalothorax. The abdomen is visible as a flat, V-shaped plate in males; in females it is expanded and U-shaped for carrying eggs (Figure 6-43). Crabs may be highly mobile, and most can easily move sideways when in a hurry. They comprise the largest and most diverse group of decapods, more than 4,500 species strong. Most are scavengers, but some have specialized diets such as seaweeds, organic matter in mud, molluscs, or even coral mucus (see Figure 6-1). Many crabs live along rocky shores or sandy beaches, exposed to air much of the time. Land crabs actually spend most of their lives on land, returning to the sea only to release their eggs.

Biology of Crustaceans

The diversity of forms among crustaceans is paralleled by equally diverse functional features.

Digestion. Filter feeding is a very common feeding habit among copepods and many of the small planktonic crustaceans (see Figure 14-7). Stiff, hairlike bristles on some of the appendages are used to catch food particles in the water. Particles are carried to the bristles by currents induced by the beating of other appendages. Still other specialized appendages move food from the bristles to the mouth. Parasitic crustaceans have appendages adapted for piercing and sucking.

Bottom crustaceans like shrimps have most of their body appendages specialized as walking legs. The appendages closest to the mouth, such as the **maxillipeds,** have turned forward and are specialized to sort out food and push it toward the mouth. Decapods have three pairs of maxillipeds (see Figure 6-40). They are used as filtering devices in those decapods that eat small food particles in water or mud.

Food passes to a stomach that typically has chitinous teeth or ridges for grinding and bristles for sifting. The stomach, two-chambered in decapods (see Figure 6-40), is connected to digestive glands that secrete digestive enzymes and absorb nutrients. Digestion is essentially extracellular. The intestine ends in an anus.

As in molluscs, absorbed nutrients are distributed by an open circulatory system. Gas exchange is carried out by gills attached to some of the appendages. In decapods the gills lie in a chamber under the carapace, where they are constantly bathed by water. In land crabs, however, the gill chamber, though moistened, is filled with air and acts almost like our lungs.

Nervous System and Behavior. The nervous system of the simplest crustaceans is ladderlike, but it is more centralized in decapod crustaceans. Crustaceans have a small, relatively simple brain (see Figure 6-40).

The sensory organs of crustaceans are well developed. Most have com-

Figure 6-41 This shrimp is beautifully camouflaged to live on the branching arms of crinoids (see Figure 6-53) in the tropical Pacific. Can you find it?

A

B

Figure 6-42 The coconut crab (*Birgus latro*) is a large land-dwelling hermit crab that does not use a shell as an adult **(A).** Females return to seawater only to release their eggs. After a planktonic existence, the young settle on the bottom and use shells for a home **(B)** when they crawl out of the sea to begin life on land. Coconut crabs, so called because they sometimes eat coconuts, are found in the tropical Pacific and Indian oceans.

Figure 6-43 The abdomen is V-shaped in male crabs (*below*) and larger and U-shaped in females. This is the mangrove crab (*Scylla serrata*).

Figure 6-44 Courtship behavior can be especially elaborate in decapod crustaceans. Male fiddler crabs (*Uca*), for instance, have one greatly enlarged claw that they wave in the air while performing a "dance" to attract females (see Box 11-2). Crabs whose claw is too small or who do not have the right moves can get pretty lonely!

pound eyes, which consist of a bundle of up to 14,000 light-sensitive units grouped in a mosaic. In decapods the compound eyes are at the end of movable stalks and can be used like periscopes. Crustaceans have a keen sense of "smell;" that is, they are acutely sensitive to chemicals in the water. Many crustaceans have a pair of statocysts for balance.

Crustaceans are among the most behaviorally complex invertebrates. They use a variety of signals to communicate with each other. Many of these signals involve special body postures or movements of the legs and antennae. This type of communication has been shown to be very important in settling disputes between neighbors and in courtship. Courtship behavior can be especially elaborate (Figure 6-44).

Reproduction and Life History. The sexes are separate in most crustaceans. Gametes are rarely shed into the water. Instead, males use specialized appendages to transfer sperm directly to the female. Even hermaphroditic species transfer sperm between individuals. Barnacles, for instance, have a penis that can stretch to reach other barnacles in the neighborhood. Mating in decapods usually takes place immediately after the female molts and the exoskeleton is still soft. Females of many species can store sperm for long periods and use it to fertilize separate batches of eggs. In amphipods, isopods, decapods, and other groups, females carry their eggs using specialized appendages beneath the body.

Most crustaceans have planktonic larvae that look nothing like the adult. Probably the most characteristic crustacean larva is the **nauplius,** but the type and number of larval stages vary widely from group to group (Figure 6-45).

Other Marine Arthropods

Very few arthropods other than crustaceans are common in the sea. Most belong to two small and entirely marine groups. A third group, huge and mostly terrestrial, includes a few shy invaders of the sea.

Horseshoe Crabs. The **horseshoe crabs** are the only surviving members of a group (class **Merostomata**) that is widely represented in the fossil record. The four living species of horseshoe crabs (*Limulus*) are not true crabs but "living fossils," not unlike forms that became extinct long ago. Horseshoe crabs live on soft bottoms in shallow water in the Atlantic and Gulf coasts of the United States. Their most distinctive feature is a horseshoe-shaped carapace that encloses a body provided with five pairs of legs (Figure 6-46).

Sea Spiders. Sea spiders (class **Pycnogonida**) only superficially resemble real spiders. Sea spiders are not to be forgotten. At least four pairs of segmented legs stretch from a body so small that it looks like it doesn't exist at all (Figure 6-47). A large proboscis with the mouth at the tip is used to feed on soft invertebrates. Sea spiders are most common in cold waters, but they occur throughout all oceans.

Insects. Insects (class **Insecta**) are distinguished from other arthropods by having only three pairs of legs as adults. They are the largest and

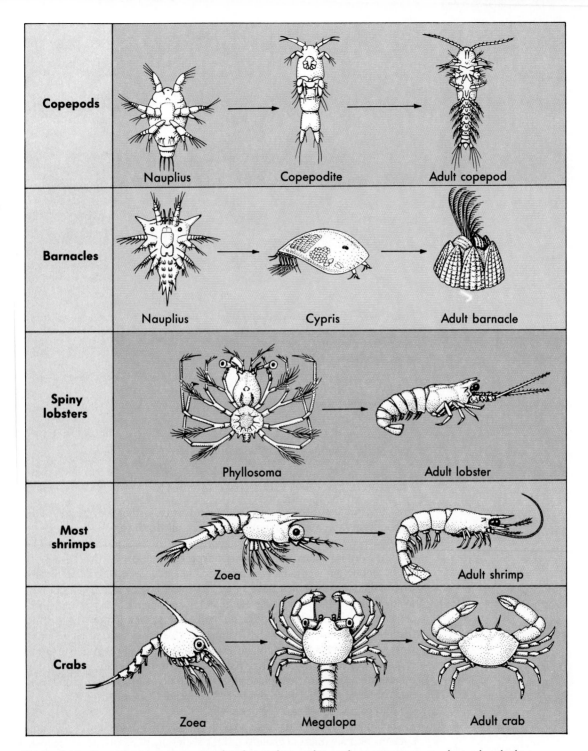

Figure 6-45 Eggs of marine crustaceans hatch into larvae that undergo consecutive molts in the plankton. Each molt adds a new pair of appendages, with the ones already present becoming more specialized. The last larval stage eventually metamorphoses into a juvenile (see Figure 12-5). The larvae and adults indicated here were not drawn to scale. Each arrow represents several molts.

Figure 6-46 Horseshoe crabs *(Limulus polyphemus)* grow as long as 60 cm (2 feet), including the tail. They feed on clams and other small soft-bottom invertebrates. A trilobite, member of a large fossil group related to the horseshoe crab, is shown in Figure 4-28.

Figure 6-47 Sea spiders, or pyconogonids, are often found on soft-bodied invertebrates such as sponges, sea anemones, jellyfishes, and bryozoans. Most shallow-water species are small, with body length rarely over 1 cm (about half an inch). Some deep-sea sea spiders are bigger (see Figure 15-32).

most diverse group of animals on earth but are relatively rare in the sea. Most marine insects live at the water's edge, where they scavenge among seaweeds, barnacles, and rocks. Many inhabit the decaying seaweed that accumulates at the high tide mark. One marine insect that *is* found in the open far from shore is the **water strider** (see Figure 14-21).

ECHINODERMS: THE SPINY MISFITS

Sea stars, sea urchins, sea cucumbers, and several other forms make up the **echinoderms** (phylum **Echinodermata**). Of all the invertebrate groups, the echinoderms stand alone. Their relationship with other animals is not clear, and they display many unique and unexpected traits.

Echinoderms are radially symmetric, like cnidarians and comb jellies. The radial symmetry of echinoderms, however, is only a *secondary* development. Their planktonic larvae are bilaterally symmetric, and only the adults develop radial symmetry. Unlike cnidarians and comb jellies, most echinoderms have **pentamerous** radial symmetry, that is, symmetry based on five parts (Figure 6-48). As might be expected in a radially symmetric animal, echinoderms lack a head. They have no anterior or posterior end or dorsal or ventral side. It is useful instead to refer to one surface of echinoderms as **oral,** because that is where the mouth is located (see Figure 6-48, B). The opposite surface is **aboral** (Figure 6-48, A).

Echinoderms typically have a complete digestive tract, a well-developed coelom, and an internal skeleton. This skeleton, like ours, is an **endoskeleton.** It is secreted within the tissues, rather than externally like the exoskeleton of arthropods. Though sometimes it looks external, as in the spines of sea urchins, the endoskeleton is covered by a thin layer of ciliated

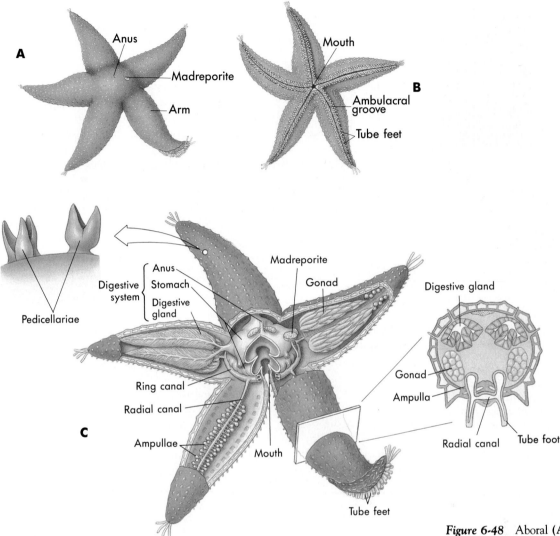

Labels in figure:

A — Anus, Madreporite, Arm

B — Mouth, Ambulacral groove, Tube feet

C — Pedicellariae; Digestive system: Anus, Stomach, Digestive gland; Madreporite; Gonad; Ring canal; Radial canal; Ampullae; Mouth; Tube feet; Digestive gland; Gonad; Ampulla; Radial canal; Tube foot

Figure 6-48 Aboral (**A**) and oral (**B**) surfaces of a sea star, *Asterias vulgaris*, common on the Atlantic and Gulf coasts of North America. Internal structure (**C**), with an arm cut across to show the relationship between tube feet, the internal sacs (ampullae), and the canals that make up the water vascular system. This and other carnivorous sea stars evert their thin-walled stomach and begin digesting prey without having to eat it (see Figure 13-38).

tissue. Spines and pointed bumps give many echinoderms a spiny appearance, and hence the name Echinodermata, or "spiny skinned."

Unique to echinoderms is the **water vascular system,** a network of water-filled canals (Figure 6-48, C). **Tube feet** are muscular extensions of these canals. They are extended when filled with water, sometimes by the action of muscular sacs—the **ampullae**—that extend inside the body opposite the tube feet. Tube feet often end in a sucker and are used for attachment and locomotion. In many species the system connects to the outside through the **madreporite,** a porous plate on the aboral surface.

Echinoderms are radially symmetric as adults. They are characterized by an endoskeleton and a unique water vascular system.

A

B

Figure 6-49 **(A)** The ochre sea star (*Pisaster ochraceus*—the one on the lower right in feeding position) and the smaller bat sea star (*Patiria miniata*) from the Pacific coast of North America. **(B)** *Linckia guildingi,* from the Caribbean, shows remarkable regenerating abilities: the larger arm regenerated four small ones to form a complete individual.

Types of Echinoderms

The echinoderms comprise a large group of about 6,000 species, all marine. They are important members of bottom communities from the poles to the tropics.

Sea Stars. Sea stars (class **Asteroidea**), sometimes called starfishes, most clearly display the distinctive echinoderm body plan (Figure 6-48). Most species have five arms that radiate from a central disk, though some have more than five, sometimes close to 50. Hundreds of tube feet protrude from the oral surface along radiating channels on each arm called **ambulacral grooves** (Figure 6-48, B). Sea stars can move in any direction, though slowly, by reaching out their tube feet and pulling themselves along.

The endoskeleton of sea stars consists of separate plates that form a flexible framework. This allows their arms to be somewhat flexible. The aboral surface of many sea stars is often covered with spines modified into minute pincerlike organs called **pedicellariae** (Figure 6-48, C). They help keep the surface clean.

Most sea stars are predators of bivalves, snails, barnacles, and other attached or slow-moving animals. Typical examples include *Asterias* (Figure 6-48), common on rocky shores from the north Atlantic to the Gulf of Mexico, and *Pisaster* (Figure 6-49, B), its counterpart on the west coast of North America.

Brittle Stars. The star-shaped body architecture is repeated in the **brittle stars** (class **Ophiuroidea**). The five arms, however, are long, very flexible, and sharply demarcated from the central disk (Figure 6-50). The swift, snakelike movements of the arms are used in locomotion. The tube feet, which lack suckers, are used in feeding.

Most brittle stars eat particulate organic matter and small animals they pick up from the bottom. Particles are collected by tube feet and passed from foot to foot to the mouth. There are more species of brittle stars, around 2,000, than of any other group of echinoderms. They are widely distributed but not always visible, often hiding under rocks and corals or covering themselves with mud or sand.

Figure 6-50 Brittle stars (*Ophiothrix suensonii*) on a sponge inhabited by a cnidarian (*Parazoanthus parasiticus*) in Jamaica **(A)** and another species (*Ophiothrix spiculata*) typically found under rocks and on holdfasts of seaweeds in California **(B)**.

A

B

LIFE IN THE MARINE ENVIRONMENT

Sea Urchins.   In the **sea urchins** (class **Echinoidea**) the endoskeleton forms a round, rigid test with movable spines and pedicellariae (Figure 6-51). Locomotion is achieved by the movable spines, jointed to sockets in the test, and the suckertipped tube feet. The flat and radiating body plan of sea stars can be transformed into a sea urchin by moving the arms upward and joining them at the tips. The five rows of ambulacral grooves with their tube feet now extend from pole to pole. The mouth is on the bottom, the "South Pole," and the anus on top, the "North Pole." The plates that make up the test can be seen in a sea urchin cleaned of spines and tissue. Bands of pores along the ambulacral grooves correspond to the bands of tube feet.

Sea urchins graze on attached or drifting plant material. In the process, they also ingest encrusting animals, like sponges and bryozoans, and dead organic matter. The mouth, directed downward, has an intricate system of jaws and muscles called **Aristotle's lantern,** and it is used to bite off algae and other bits of food from the bottom. Sea urchins are a common sight on rocky shores throughout the world. Examples are species of *Arbacia* from the Atlantic Ocean and Gulf of Mexico and *Strongylocentrotus* from most North American polar and temperate coasts (Figures 6-51, A; 12-21; and 12-29). Those in the tropics show an even richer variety of shapes and sizes, particularly in coral reefs (Figures 6-51, B and C).

Not all of the approximately 900 species in the class Echinoidea have round tests with prominent spines. **Heart urchins** (see Figure 12-8) and

A

B

C

Figure 6-51   **(A)** The green sea urchin (*Strongylocentrotus droebachiensis*) is found on rocky shores and kelp forests on the Atlantic, Artic, and Pacific coasts of North America. **(B)** The slate-pencil sea urchin (*Heterocentrotus mammilatus*) from the tropical Indian and Pacific oceans. **(C)** *Mellita sexiesperforata,* the six-hole sand dollar from sandy bottoms along the southern United States and the Caribbean.

Figure 6-52 A sea cucumber (*Thelenota rubralineata*) from Papua New Guinea. Sea cucumbers, among the largest echinoderms, are known to reach 2 m (6.6 feet) in length.

sand dollars (Figures 6-51, C and 12-9) are echinoids adapted to live in soft bottoms by having flattened bodies and short spines. They are deposit feeders that use their tube feet and sometimes strands of mucus to pick up organic particles.

Sea Cucumbers. In yet another modification of the echinoderm body plan, **sea cucumbers** (class **Holothuroidea**) are wormlike. They do not have spines and lack an obvious radial symmetry. The basic body plan of sea urchins appears to have been elongated along a "North to South poles" axis, as if they were pulled from the mouth and anus and stretched. The animal lies on one side, and the oral and aboral surfaces are at the ends (Figure 6-52). The endoskeleton consists of microscopic, calcareous spicules scattered through the warty skin. Like sea urchins, most species have five rows of tube feet extending from mouth to anus.

Many sea cucumbers are deposit feeders. The tube feet around the mouth are modified into branched tentacles that are used to pick up organic matter from the bottom or scoop sediment into the mouth (see Figures 6-20 and 10-38). Some sea cucumbers burrow or hide and extend their tentacles to obtain food directly from the water.

Many sea cucumbers have evolved novel defensive mechanisms that compensate for the lack of a test and spines. Some secrete toxic substances. When disturbed, some species discharge sticky filaments through the anus to discourage potential predators. Others resort to a startling response, the sudden expulsion of the gut and other internal organs through the mouth or anus, a response referred to as **evisceration.** It is assumed that evisceration distracts the offender while the sea cucumber, who will eventually grow back the lost organs, escapes. Messy, perhaps, but effective!

Crinoids. Crinoids (class **Crinoidea**) are suspension feeders that use outstretched, feathery arms to obtain food from the water. Crinoids are represented by close to 600 species of **feather stars** (Figure 6-53) and **sea lilies.** Sea lilies are restricted to deep water and attach to the bottom. Feather stars, on the other hand, crawl on hard bottoms in shallow to deep waters, especially in the tropical Pacific and Indian oceans.

The body plan of crinoids is best described as an upside-down brittle star with the ambulacral grooves and mouth directed upward. The mouth and larger organs are restricted to a small cup-shaped body from which the arms radiate. Some crinoids have only five arms but most have many—up to 200—due to branching of the initial five. The arms also have small side branches (see Figure 6-41). Tiny tube feet along these side branches secrete mucus, which aids in catching food particles. Food makes its way into the mouth by way of ciliated ambulacral grooves. Feather stars perch on hard surfaces using clawlike appendages. These appendages tilt the body so that the extended arms point downstream of the current for efficient feeding.

Biology of Echinoderms

Radial symmetry, so effective for suspension feeding, also has imposed limitations on echinoderm lifestyles. With the exception of limited swimming in feather stars and some deep water sea cucumbers, adult echinoderms are

LIFE IN THE MARINE ENVIRONMENT

slow bottom crawlers. But do they need to be fast to be successful? Certainly not. More than anything, echinoderms are fascinating because of the extraordinary ways in which they handle day-to-day life.

Digestion. The digestive system of echinoderms is relatively simple. The mostly carnivorous sea stars feed by extending, or everting, part of their stomach inside-out through the mouth to envelop the food (see Figure 13-38). The stomach then secretes digestive enzymes produced by large digestive glands that extend into the arms (see Figure 6-48, C). The digested food is carried into the glands for absorption and the stomach pulled back inside the body. The intestine is short or missing. Brittle stars and crinoids also have simple, short guts.

The gut of sea urchins and sea cucumbers is long and coiled. In sea urchins, this is an adaptation for the lengthier digestion needed for the breakdown of plant material. A long gut is advantageous in sea cucumbers, because they need to remove organic matter from the large amounts of sediment usually ingested.

In all echinoderms, nutrients are transported in the fluid that fills the extensive body cavity. The fluid is called **coelomic fluid,** since the body cavity of echinoderms is a coelom.

The coelomic fluid also transports oxygen, since echinoderms lack a distinct circulatory system. In sea stars and sea urchins, gas exchange takes place across small, branched projections of the body wall connected at the base to the coelomic cavity. In sea cucumbers, water is drawn in through the anus into a pair of thin, branched tubes called **respiratory trees.** The respiratory trees are extensions of the gut and are suspended in the body cavity, surrounded by coelomic fluid. They provide a large surface area in close proximity to the coelomic fluid, thus allowing considerable gas exchange.

Nervous System and Behavior. Our knowledge of the nervous system of echinoderms is rather limited. The presence of a nerve net is reminiscent of cnidarians. The nervous system coordinates movements of tube feet and spines, even in the absence of a brain. Nevertheless, more complex behaviors, such as the righting of the body after being turned over and camouflaging with bits of debris in sea urchins, are evidence that the nervous system may not be as simple as it looks at first sight.

Reproduction and Life History. Sexes are separate in most echinoderms. In most groups, five, ten, or more gonads shed sperm or eggs directly into the water. Gonads are usually located in the body cavity and open to the outside by way of a duct (see Figure 6-48, C). Gametes do not survive long in the water, so in many species individuals spawn all at once to ensure fertilization.

Development of the fertilized egg proceeds in the plankton and typically results in a ciliated larva characteristic of each group (see Figure 14-11, B and C). Echinoderm larvae are bilaterally symmetric, and it is not until metamorphosis that radial symmetry develops. Some echinoderms do not have plaktonic larvae and brood fertilized eggs in special pouches or under the body.

Asexual reproduction takes place regularly in some sea stars, brittle stars, and sea cucumbers by the separation of the central disk or body into

Figure 6-53 A feather star, or crinoid, photographed at night in Kwajalein atoll, Marshall Islands. Feather stars are among the most spectacular of all marine animals. The outstretched arms of feeding individuals, which can reach a diameter of 70 cm (more than 2 feet), are a truly impressive sight. See Figure 6-41 for a close-up of an arm.

two pieces. The resulting halves then proceed to grow into complete individuals. **Regeneration,** the ability to grow lost or damaged body parts, is highly developed in echinoderms. Sea stars, brittle stars, and crinoids regenerate lost arms. In some sea stars a severed arm can grow into a new individual (see Figure 6-49, *B*). In most sea stars, however, only arms that include portions of the central disk can regenerate.

HEMICHORDATES: A MISSING LINK?

The search for links between the chordates, our own phylum, and other groups of animals has been a most provocative challenge. As strange as it may seem, both echinoderms and chordates share a number of features related to the development of the embryos. The wide gap between echinoderms and chordates, however, may be filled by a small and infrequently seen group of worms, the **hemichordates** (phylum **Hemichordata**). Hemichordates share the same basic developmental characteristics of chordates and echinoderms. Some hemichordates also have a larva similar to that of some echinoderms. More puzzling is that hemichordates share with us chordates some of the features used to define our phylum. These characteristics, a dorsal nerve cord and openings along the anterior part of the gut, will be discussed below.

The hemichordates comprise approximately 100 species. Most of these are **acorn worms,** or **enteropneusts,** wormlike deposit feeders living free or in U-shaped tubes. Some acorn worms have been found around hydrothermal vents, often occurring in large numbers (see the section in Chapter 15, "Deep-Sea Hot Springs"). They generally range in length from about 8 to 45 cm (3 to 18 inches) but some reach 2.5 m (more than 8 feet). Like sea cucumbers they ingest sediment, but they use a thick, mucus-secreting proboscis to collect organic material that is then swept toward the mouth.

CHORDATES WITHOUT A BACKBONE

We share our phylum **Chordata** with many unexpected distant cousins. Some of these **chordates** are in fact invertebrates similar in appearance and lifestyles to some of the groups we have reviewed. The estimated 44,000 living species of chordates share many characteristics, but three stand out. At least during part of our development, chordates have (1) a single, hollow **nerve cord** that runs along the dorsal length of the animal, (2) **gill** (or **pharyngeal) slits,** small openings along the anterior part of the gut (or **pharynx**), and (3) a **notochord,** a flexible rod for support that lies between the nerve cord and the gut (see Figures 6-54, *B* and 6-56). In most chordates the notochord is surrounded or replaced by a series of articulating bones, the **backbone,** or vertebral column. Recall that this is the characteristic that defines and separates vertebrates from the invertebrates. The realm of invertebrates extends well into the chordates since the simplest, most primitive chordates lack a backbone.

> **All chordates possess—at least during part of their lives—a dorsal nerve cord, gill slits, and a notochord. Vertebrate chordates also have a backbone.**

LIFE IN THE MARINE ENVIRONMENT

The phylum Chordata is divided into three major groups, or subphyla. Two of these lack a backbone and, for this reason, are discussed here with the invertebrates. These invertebrate chordates are collectively called **protochordates.** The third and by far the largest chordate subphylum comprises the vertebrates, the subject of the next two chapters.

Tunicates

The largest group of protochordates are the **tunicates** (subphylum **Urochordata**). All 1,300 species are marine. Those we are most apt to see are the **sea squirts,** or **ascidians** (class **Ascidiacea**). Their saclike bodies are attached, often as fouling organisms, or anchored in soft sediments. They are the only sessile, or attached, chordates. To the inexperienced eye, some sea squirts may be confused with sponges because of their general appearance. The body of sea squirts, however, is protected by a **tunic,** a leathery or gelatinous outer covering that has a different texture than the mushy sponges.

Sea squirts are filter feeders. Water typically flows through the mouth, or **incurrent siphon,** and is filtered by a ciliated, sievelike sac. This sac represents the pharynx, and the openings are derived from the gill slits. Food is filtered from the water and passed into a U-shaped gut. Filtered water is expelled through a second opening, the **excurrent siphon.** When disturbed or expelling debris, ascidians force a jet of water out of both siphons, and hence their common name of "sea squirts." Some sea squirts are colonial. In some species the colorful colonies are just clumps of individuals (Figures 6-54, A) others consist of a circular, flowerlike arrangement of individuals sharing a common tunic and excurrent siphon (see Figure 12-18, C).

Were it not for their planktonic larvae, the headless, attached sea squirts could easily pass as anything but chordates. They possess neither notochord nor a dorsal nerve cord. Ascidian larvae are known as **tadpole larvae** because of their superficial resemblance to the tadpole of frogs (Figure 6-54, B). Tadpole larvae clearly display the fundamental chordate traits. In addition to gill slits, they have a dorsal nerve cord and a notochord. They also have an eye and a well-developed tail. Tadpole larvae do not feed, and their only purpose is to find a suitable surface on which to settle. The metamorphosis of a tadpole larva into a juvenile ascidian is nothing short of spectacular. The notochord and tail are absorbed, the filtering sac and siphons develop, and free existence is no more.

A

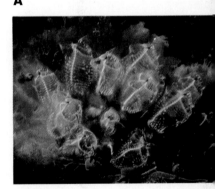

Figure 6-54 **(A)** *Clavelina lepadiformis* is a colonial sea squirt, or ascidian. Its outer covering—the tunic—contains cellulose, a substance typical of plants. **(B)** The tadpole larva is peculiar to sea squirts. It exhibits the distinguishing characteristics of chordates, some of which are missing in the adults.

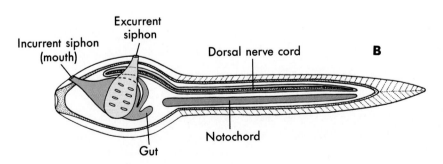

B

Salp

Some tunicates lead a planktonic existence throughout their lives. These are among the most remarkable examples of life in the sea. **Salps** (class **Thaliacea**) have a transparent and barrel-shaped body provided with muscle bands for locomotion. Water enters through the anterior mouth, or incurrent siphon, and it is forced through the excurrent siphon on the posterior end like a jet engine. Salps can be extremely abundant, particularly in warm water. Some are colonial, much like floating colonies of sea squirts, and may reach several meters in length (Figure 6-55). **Larvaceans,** or **appendicularians,** (class **Larvacea**) retain the body of a tadpole larva through life. Each tiny individual secretes a complex but delicate gelatinous house for protection and to filter water for food (see Figure 14-8).

Figure 6-55 This salp is made of a string of individuals that form a colony of several meters in length.

Lancelets

The second group of invertebrate chordates consists of about 25 species of **lancelets** (subphylum **Cephalochordata**). The body, up to 7 cm (close to 3 inches) long, is laterally compressed and elongate like that of a fish (Figure 6-56). The three basic chordate characteristics are well developed through life. Only the lack of a backbone separates lancelets from vertebrates. Lancelets are inhabitants of soft bottoms. They are filter feeders, using the gill slits to remove and concentrate organic particles.

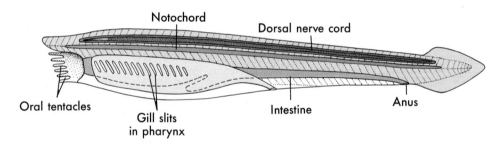

Notochord

Dorsal nerve cord

Oral tentacles

Gill slits in pharynx

Intestine

Anus

Figure 6-56 The lancelet, *Branchiostoma,* displays all major characteristics of chordates. Although it looks like a fish, it is an invertebrate since it lacks a backbone.

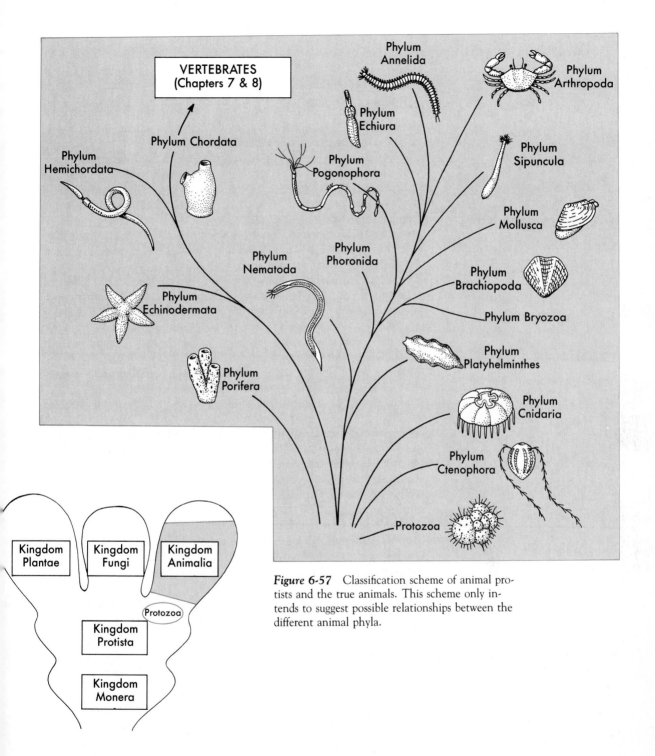

Figure 6-57 Classification scheme of animal protists and the true animals. This scheme only intends to suggest possible relationships between the different animal phyla.

SUMMARY

1. Protozoans such as forams and radiolarians are unicellular and are therefore included in the kingdom _____.

2. Sponges are considered simpler and more primitive than most other multicellular animals, because they lack true _____ and _____.

3. The presence of nematocysts and _____ symmetry are fundamental characteristics of _____.

4. In bilaterally symmetric animals, the body displays a _____ surface different from a ventral one and an _____ end, usually a head, differentiated from a posterior end.

5. Examples of bilaterally symmetric worms are the _____, which lack an anus as in cnidarians, and the _____, a large group common in marine sediments that may consist of as many as half a million species.

6. Annelids, the segmented worms, have a body cavity called _____, which acts as a hydrostatic skeleton.

7. Molluscs are characterized by a soft body and by two structures that may be absent in some: a calcareous, external or internal _____ and a _____, a chitinous structure in the mouth region used in feeding.

8. An exoskeleton and paired, _____ appendages characterize members of the phylum _____. The great majority of the marine representatives of this phylum belong to one group, the _____.

9. Echinoderms are different in several ways. Besides their _____ symmetry, locomotion and sometimes feeding are carried out by means of the _____.

10. Only relatively few chordates lack a backbone. Nevertheless, all chordates have during at least some stage of their lives small openings in the anterior part of the gut called _____ , a flexible _____ for support, and a single _____ nerve cord.

LIFE IN THE MARINE ENVIRONMENT

THOUGHT QUESTIONS

1. If bilateral symmetry was to evolve among the cnidarians, in which group or groups you would expect this to take place? Why?
2. Cephalopods, the squids, octopuses, and allies, show a much higher degree of complexity than the other groups of molluscs. What factors triggered the evolution of these changes? A rich fossil record among cephalopods shows that at one time they were very common and even dominant in some marine environments. Now there are only about 650 living species of cephalopods, a lot less than gastropods. In the long run, were cephalopods successful? What do you think happened along the way?
3. A new class of echinoderms, the sea daisies or concentricycloids, was discovered in 1986. They are deep-water animals living on sunken wood. They are flat and round, looking very much like a small sea star without arms. They also lack a gut. Without ever having seen them, why do you think they were classified as echinoderms, not as members of a new phylum? Any hypotheses as to how they feed or move around?

FOR ADDITIONAL READING

Bavendam, F.: "Sea Stars Deploy a Bag of Tricks in Marine Wars," *Smithsonian*, vol. 16, no. 8, November 1985, pp. 104-109. *Some sea stars ward off attackers, whereas others are adapted to attack their prey.*

Bavendam, F.: "Even for Ethereal Phantasms, It's a Dog-Eat-Dog World," *Smithsonian*, vol. 20, no. 5, August 1989, pp. 94-101. *A superbly photographed report of feeding habits of nudibranchs and how they deter potential predators.*

Brownlee, S.: "Jellyfish Aren't Out to Get Us," *Discover*, August 1987, pp. 42-54. *A review of the jellyfishes that are potentially dangerous to humans.*

Cameron, J.N.: "Molting in the Blue Crab," *Scientific American*, vol. 252, no. 5, May 1985, pp. 102-109. *Molting in crabs and other crustaceans is not a simple affair. The article reveals the chemical mechanisms related to molting.*

Gosline, J.M. and M.E. De Mont: "Jet-Propelled Swimming in Squids," *Scientific American*, vol. 252, no. 1, January 1985, pp. 96-103. *The mechanics of swimming in squids and other cephalopods.*

Grober, M.S.: "Starlight on the Reef," *Natural History*, October 1989, pp. 72-80. *An account, with excellent photographs, of brittle stars that emit light in Caribbean coral reefs.*

McMenamin, M.A.S.: "The Emergence of Animals," *Scientific American*, vol. 256, April 1987, pp. 94-102. *A technical but uncomplicated account of the appearance of the earliest marine invertebrates.*

Rudloe, J. and A. Rudloe: "Jellyfishes Do More With Less Than Almost Anything Else," *Smithsonian*, vol. 21, no. 11, February 1991, pp. 100-111. *Jellyfishes can be deadly to humans and other animals, but they are among the most beautiful of all marine animals.*

7 Marine Fishes

The fishes were the first vertebrates, appearing more than 500 million years ago. The first fishes probably evolved from an invertebrate chordate not much different from the lancelets or the tadpole larvae of sea squirts that still inhabit the oceans.

Fishes soon made their presence felt and have had a tremendous impact on the marine environment (Figure 7-1). They feed on nearly all types of marine plants and animals. Some of the organisms discussed in the previous two chapters, from bacteria to crustaceans to a unique sea cucumber, use fishes as their home. Many other animals eat them.

Fishes are the most economically important marine organisms, and marine fishes are a vital source of protein for millions of people. Some are ground up as fertilizer or chicken feed. Leather, glue, vitamins, and other products are obtained from them. Many marine fishes are chased by sport fishing enthusiasts. Others are kept as pets and have brought the wonders of ocean life into many homes.

Figure 7-1 A school of squirrel fish (*Holocentrus*).

The three basic characteristics of chordates
Chapter 6, p. 184:
1. a single dorsal nerve cord
2. gill, or pharyngeal, slits
3. a notochord

THE VERTEBRATES: AN INTRODUCTION

Vertebrates (subphylum **Vertebrata**) share the three fundamental characteristics of the phylum Chordata with invertebrate chordates like lancelets and sea squirts. They differ from these simpler chordates in having a **backbone,** also called the **vertebral column** or **spine,** which is a dorsal row of hollow skeletal elements—usually bone—called **vertebrae.** The vertebrae enclose and protect the **nerve cord,** also called the **spinal cord,** which ends in a complex brain. Vertebrates also are characterized by a bilaterally symmetric body and the presence of an endoskeleton.

> Vertebrates are chordates with a backbone made of segments of skeletal elements that enclose a nerve cord, or spinal cord.

TYPES OF FISHES

Fishes are the simplest and oldest of all living vertebrates (Figure 7-2). They also are the most abundant vertebrates in terms of both species and individuals. The estimated 22,000 species of fishes make up about half of all species of vertebrates on earth. Most fishes, about 58%, are marine.

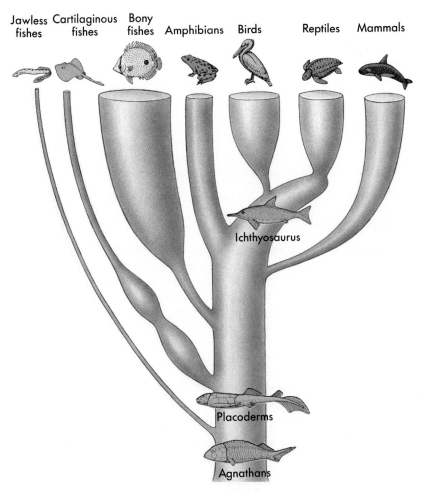

Jawless fishes | Cartilaginous fishes | Bony fishes | Amphibians | Birds | Reptiles | Mammals

Ichthyosaurus

Placoderms

Agnathans

Early vertebrates

Figure 7-2 Diagram of the evolutionary relationships of the major groups of vertebrates. The bony skeleton of most vertebrates is easily fossilized, allowing us to learn about forms now long gone.

Jawless Fishes

Fishes are divided into three major groups (see the classification scheme shown in Figure 7-34). The most primitive fishes living today are the **jawless fishes** (class **Agnatha**). Because they lack jaws, they feed by suction with the aid of a round, muscular mouth and rows of teeth. The body is cylindric and elongate like that of eels or snakes (Figure 7-3). They lack the paired fins and scales of most fishes.

Hagfishes, or **slime eels** (*Mixine, Eptatretus*) are jawless fishes that feed mostly on dead or dying fishes. They sometimes bore into their prey and eat

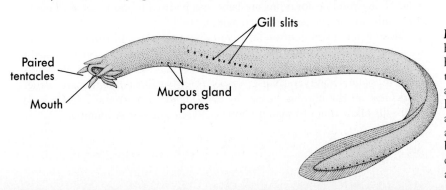

Gill slits

Paired tentacles

Mouth

Mucous gland pores

Figure 7-3 The Pacific hagfish (*Eptatretus stouti*) has 4 pairs of sensory tentacles around the mouth and 12 pairs of gill slits. Hagfishes are also known as slime eels because of the abundant mucus produced by glands found throughout the body. Notice the absence of paired fins.

Figure 7-4 Shark jaws display triangular teeth and rows of spare ones ready to take over. Each tooth has even smaller teeth along its edges. It has been estimated that a tiger shark may grow and discard 24,000 teeth in a 10-year period!

the inside out. Hagfishes live in tunnels they dig in muddy bottoms, mostly at moderate depths in cold waters. Only about 20 species are known. They reach a maximum length of approximately 80 cm (2.6 feet). Their skin is used in the manufacture of leather goods, but they are mostly known for attacking bait or fishes on fishing lines and in nets.

Lampreys (*Petromyzon*), found in most temperate regions, are primarily freshwater fishes. They breed in rivers and lakes, but some move to the sea as adults. They attach to other fishes and suck their blood or feed on bottom invertebrates. There are roughly 30 species of lampreys.

Hagfishes and lampreys lack jaws and are the most primitive living fishes.

Cartilaginous Fishes

More familiar are the **cartilaginous fishes** (class **Chondrichthyes**), a fascinating and ancient group that includes the sharks, rays, skates, and ratfishes. Cartilaginous fishes have a skeleton made of **cartilage,** a material that is lighter and more flexible than bone. Though the skeleton of jawless fishes is also cartilaginous, sharks and related fishes highlight some notable and significant advances. They possess movable jaws that are usually armed with well-developed teeth (Figure 7-4). The mouth is typically ventral, that is, underneath the head. In sharks it is invariably curved downward in a seemingly menacing grimace. Another important development is the presence of paired lateral fins for efficient swimming. Cartilaginous fishes have rough, sandpaper-like skin because of the presence of tiny **scales** (see Figure 7-9, A), which have the same origin and composition as teeth.

Sharks, rays, skates and related fishes are characterized by a cartilaginous skeleton and a rough skin covered by minute scales. They also have movable jaws and paired fins.

Sharks. **Sharks** are cartilaginous fishes magnificently adapted for fast swimming and predatory feeding. They are often described as "mysterious," "evil," and "formidable"—evidence of our fascination with sharks.

Sharks are sometimes referred to as "living fossils" because many of them are similar to species that swam the seas over 100 million years ago. They have changed little through time because they are very successful just the way they are. Their spindle-shaped bodies, tapering from the rounded middle toward each end, slips easily through the water. The tail, or **caudal fin,** is well-developed and muscular. The upper lobe of the tail is usually longer than the lower lobe (see Figure 7-9, A). The upper surface of the body features two **dorsal fins,** the first of which is typically larger and nearly triangular. The paired **pectoral fins** are large and pointed in most species. Five to seven gill slits are present on each side of the body.

Most sharks swim continuously, forcing water through the mouth, over the gills, and out through the gill slits (see Figure 7-17, A). Sharks caught in fishing nets cannot force in water and therefore "drown." Not all sharks need to swim, however. Nurse sharks (*Ginglymostoma*) and many other sharks rest on the bottom during the day (Figures 7-5, A and 7-6, C), and their gills allow them to obtain enough oxygen without swimming.

A

B

C

Figure 7-5 **(A)** The Atlantic nurse shark *(Ginglymostoma cirratum)*, rests on
the bottom during the day
and feeds at night. **(B)**
The tiger shark *(Galeocerdo
cuvier)* is an open-water
species that is potentially
dangerous to humans. **(C)**
The goblin shark *(Mitsukurina owstoni)*, a bizarre
deep-water shark.

Figure 7-6 Sharks are found practically everywhere in the ocean. Some of those mentioned in the text include
(A) hammerhead shark *(Sphyrna zygaena)*, **(B)** angel shark *(Squatina)*, **(C)** nurse shark *(Ginglymostoma cirratum)*,
(D) bull shark *(Carcharhinus leucas)*, **(E)** sawshark *(Pristiophorus)*, **(F)** short-fin mako shark *(Isurus oxyrinchus)*, **(G)**
great white shark *(Carcharodon carcharias)*, **(H)** basking shark *(Cetorhinus maximus)*, **(I)** whale shark *(Rhincodon
typus)*, **(J)** tiger shark *(Galeocerdo cuvier)*, **(K)** thresher shark *(Alopias vulpinus)*, **(L)** spiny pygmy shark *(Squaliolus
laticaudus)*, and **(M)** megamouth shark *(Megachasma pelagios)*.

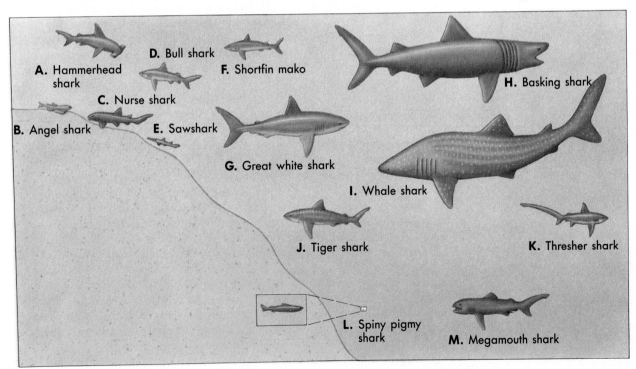

A. Hammerhead
shark

D. Bull shark

F. Shortfin mako

C. Nurse shark

B. Angel shark

E. Sawshark

G. Great white shark

H. Basking shark

I. Whale shark

J. Tiger shark

K. Thresher shark

L. Spiny pigmy
shark

M. Megamouth shark

The powerful jaws of sharks have rows of numerous sharp, often triangular teeth (see Figure 7-4). The teeth are embedded in a tough, fibrous membrane that covers the jaws. A lost or broken tooth is replaced by another which slowly shifts forward from the row behind it as if on a conveyor belt.

Not all of the nearly 250 living species of sharks conform to these specifications. Hammerhead sharks (*Sphyrna*), for example, have flattened heads with eyes at the tip of bizarre lateral extensions (Figure 7-6, A). The head serves as a sort of rudder. The wide head also separates the eyes and nostrils, which improves the shark's sensory perception. The head of sawsharks (*Pristiophorus*) extends in a long, flat blade armed with teeth along the edges (Figure 7-6, E). The upper lobe of the tail is very long in the thresher sharks (*Alopias*; Figure 7-6, K). They use the tail to herd and stun schooling fishes, which they then eat.

The size of fully grown adults also varies. The spiny pygmy shark (*Squaliolus laticaudus*; Figure 7-6, L) grows to no longer than 25 cm (almost 10 inches). At the other extreme, the whale shark (*Rhincodon typus*; Figure 7-6, I) is the largest of all fishes. These huge animals, found in tropical waters around the world, may be as long as 18 m (60 feet), though specimens longer than 12 m (40 feet) are rare. Whale sharks pose no danger to swimmers; they are **filter feeders** that feed on plankton. Another giant, second in size only to the whale shark, is the basking shark (*Cetorhinus maximus*; Figure 7-6, H), also a plankton eater. There are reports of basking sharks 15 m (50 feet) long, but most do not exceed 10 m (33 feet). Another very large shark is the most dangerous shark of all—the great white shark (*Carcharodon carcharias*; Figure 7-6, G). Individuals may exceed 6 m (20 feet) in length.

Sharks are found throughout the oceans at practically all depths. They are more prevalent, however, in coastal tropical waters. Sharks are primarily marine, but a few species travel far up rivers. Some believe that the bull shark (*Carcharhinus leucas*; Figure 7-6, D) has become permanently established in some rivers and lakes in the tropics. Several sharks, including some rare species, are restricted to deep water (see Figure 7-5, C).

Shark meat is eaten around the world and is delicious if prepared properly. Many people have tried shark without knowing it; it is often illegally sold as "regular" fish or scallops. Sharks are still fished for their oil, once used extensively in all kinds of products, and for their skin, popularly known as shagreen.

Rays and Skates. The appoximately 300 species of **rays** and **skates** have flattened bodies and for the most part live on the bottom (Figure 7-7). Fishes that live on the bottom are frequently referred to as **demersal**. Some sharks, like angel sharks (*Squatina*) and sawsharks, also have a flattened body (Figures 7-6, B and E; 7-7, A). To complicate things, some true rays, like the guitarfishes *Rhinobatos*, have a body that looks very much like a shark. Only rays, skates, and related fishes, however, have their gill slits (always five pairs) on the underside of the body—that is, located ventrally (see Figure 7-7, E), rather than on the sides. The pectoral fins are flat and greatly expanded, looking almost like wings. They are typically fused with the head. The eyes are usually on top of the head.

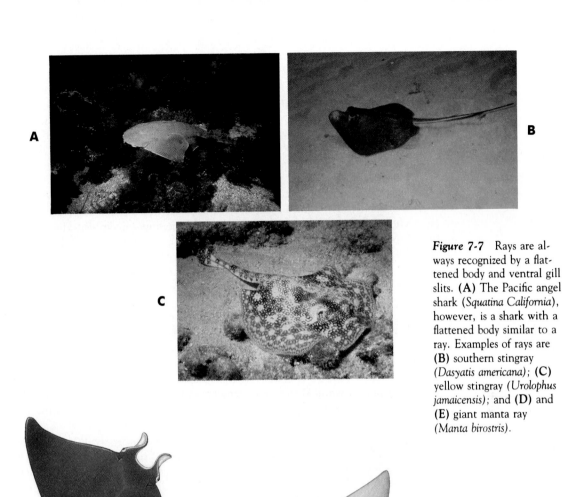

Figure 7-7 Rays are always recognized by a flattened body and ventral gill slits. **(A)** The Pacific angel shark (*Squatina California*), however, is a shark with a flattened body similar to a ray. Examples of rays are **(B)** southern stingray *(Dasyatis americana)*; **(C)** yellow stingray *(Urolophus jamaicensis)*; and **(D)** and **(E)** giant manta ray *(Manta birostris)*.

The tropical sawfishes *(Pristis)* look very much like sawsharks but have ventral gill slits. They feed by swimming through schools of fish and swinging their blade back and forth to disable their prey. They are known to grow up to 11 m (36 feet) long.

Numerous species of **stingrays** (Figures 7-7, B and C) and their relatives—the eagle, bat, and cow-nosed rays—have a whiplike tail usually equipped with stinging spines for defense. Poison glands produce venom that can cause serious wounds to anyone who steps or falls on them. Abdominal wounds caused by rays caught in nets may result in death. Many stingrays cover themselves with sand, becoming nearly invisible. They feed on clams, crabs, small fishes, and other small animals that live in sediment; stingrays have been known to damage valuable shellfish beds. They expose their food by excavating sediment with their pectoral fins. Their teeth are modified into grinding plates that crush the prey.

Electric rays (*Torpedo*) are rays that have special organs on each side of the head that produce electricity. They can deliver shocks of up to 200 volts that can stun fishes for food and discourage predators. The ancient Greeks and Romans used the shocks of electric rays to cure headaches and other ailments—the original shock treatment.

Not all rays spend their lives on the bottom. Eagle rays (*Aetobatus*) and the spectacular manta, or devil, rays (*Manta*) "fly" through the water, using their pectoral fins like wings. Eagle rays return to the bottom to feed. Mantas feed in midwater on plankton. They have at times been observed to leap out of the water. The giant manta ray (*Manta birostris*; Figures 7-7, *D* and *E*) may grow into a majestic giant. One individual was found to be almost 7 m (23 feet) wide.

Skates (*Raja*) are similar to rays in appearance and feeding habits, but they lack a whiplike tail and stinging spines. Some have electric organs. They can be extremely abundant, and the larger species are fished for food in some parts of the world.

Ratfishes.   About 25 species of strange-looking, mostly deep-water cartilaginous fishes are grouped separately because of their unique features. The **ratfishes** or **chimaeras** (Figure 7-8), for instance, have gill slits covered by a flap of skin. Some have a long tail, hence the name of "ratfish." They feed on bottom-dwelling crustaceans and molluscs.

Bony Fishes

The great majority of fishes are **bony fishes** (class **Osteichthyes**). As the name implies, they have a skeleton made at least partially of bone. There are roughly 21,500 species of bony fishes—about 98% of all fishes and almost half of all vertebrates. Between 75 and 100 new species are described

Figure 7-8 The elephant fish (*Callorhinchus*), an example of a ratfish, or chimaera. The elephant fish, which is caught commercially for food in the Southern Hemisphere, receives its name because of a snout that hangs down like an elephant's trunk.

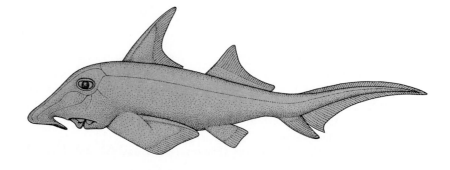

LIFE IN THE MARINE ENVIRONMENT

every year. A little more than half of all bony fishes live in the ocean, where they are by far the dominant vertebrates.

The composition of the skeleton is not the only distinguishing feature of bony fishes. In contrast to the tiny pointed scales of cartilaginous fishes, bony fishes usually have thin, flexible, overlapping scales that develop from bone (Figure 7-9, B). The scales are covered by a thin layer of tissue (see Figure 7-20), as well as protective mucus. A flap of bony plates and tissue known as the **gill cover,** or **operculum,** protects the gills.

The upper and lower lobes of the tail, the caudal fin, are almost always the same size (see Figure 7-9, B). The fins of bony fishes generally consist of thin membranes that are supported by bony spines, or **fin rays.** This arrangement gives the fins great flexibility and dexterity, in contrast to the stiff, fleshy fins of cartilaginous fishes.

While cartilaginous fishes have a ventral mouth, the mouth of most bony fishes is directed forward. Bony fishes have jaws with much more freedom of movement than those of sharks. The teeth are generally fused to the

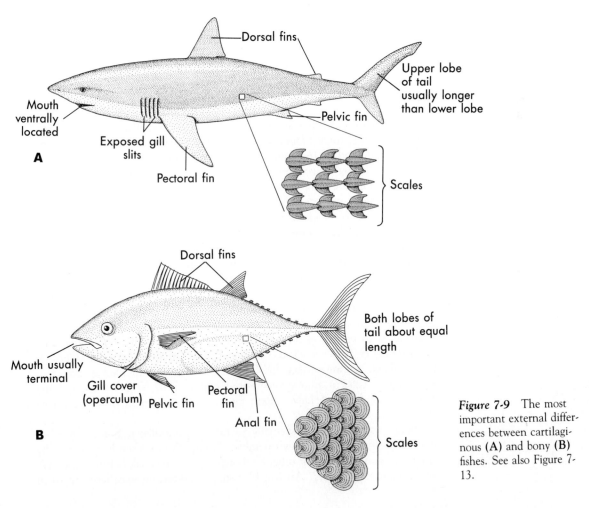

Figure 7-9 The most important external differences between cartilaginous (**A**) and bony (**B**) fishes. See also Figure 7-13.

jawbones. Though they are usually replaced, the new teeth do not move forward in rows as those in sharks.

Another important characteristic is the presence of a **swim bladder,** a gas-filled sac just above the stomach and intestine (see Figure 7-13). It allows the fish to adjust its buoyancy to keep from sinking or rising (see the section in Chapter 14, "Increased Buoyancy"). This is a significant development that compensates for the relatively heavy bony skeleton.

Bony fishes are the largest group of living vertebrates. In addition to their bony skeleton, they have gills covered by an operculum, highly maneuverable fins, protrusible jaws, and a swim bladder.

As we shall see in the next section, bony fishes are extraordinarily diverse in shape, size, color, feeding habits, reproductive patterns, and behavior. They have adapted to nearly every type of marine environment. All land vertebrates evolved from bony fishes.

BIOLOGY OF FISHES

Discovering how fishes, cartilaginous and bony alike, have conquered the demands of the aquatic environment so successfully has been a challenge to many investigators. The scientific study of fishes is called **ichthyology.**

Body Shape

The body shapes of fishes are directly related to their lifestyles. Fast swimmers like sharks, tunas (*Thunnus, Euthynnus*), mackerels (*Scomber, Scomberomorus*), and marlins (*Makaira*) have a streamlined body shape that helps them move through the water (Figures 7-9 and 7-10, A; see also Box 14-2). Laterally compressed bodies are good for leisurely swimming around coral reefs, kelp beds, or rocky reefs, but are still efficient enough to allow for bursts of speed to escape from enemies or capture food. This body form is seen in many shallow-water inshore fishes like snappers (*Lutjanus*), wrasses (*Labroides, Thalassoma*), damselfishes (*Amphiprion, Pomacentrus*), and butterflyfishes (*Chaetodon*; see Figure 7-30). Many demersal fishes, like rays, skates, and sea moths (*Pegasus*; Figure 7-10, B), are flattened from top to bottom. Flatfishes such as flounders (*Platichthys*), soles (*Solea*), and halibuts (*Hippoglossus*) are flat and beautifully adapted to live on the bottom, but their bodies are actually *laterally* compressed (Figure 7-10, C). They lie on one side, with both eyes on top. Distinctly elongate bodies are characteristic of moray eels (*Gymnothorax*), trumpetfishes (*Aulostomus*), and pipefishes (*Syngnathus*), among others. Eel-like fishes often live in narrow spaces in rocks, coral reefs, or among vegetation (Figures 7-10, D to F). Many bony fishes, like seahorses (*Hippocampus*; Figure 7-10, G) depart from these generalized shapes.

Body shapes may be especially useful for camouflage. Some pipefishes live among the eel grass they resemble. Trumpetfishes often hang vertically among gorgonian corals or tubelike sponges, or even sneak behind other fish when approaching prey. An irregular shape is often an excellent means of

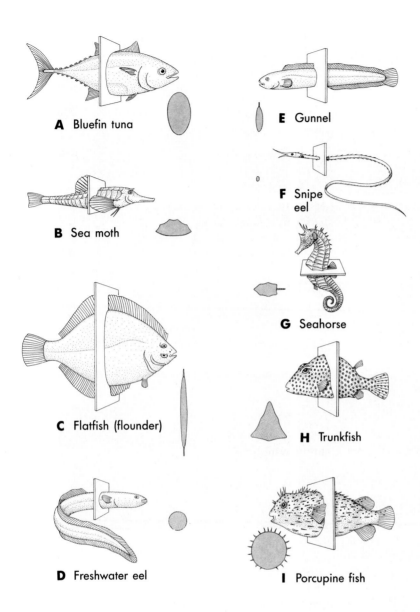

Figure 7-10 Body shape among bony fishes varies as an adaptation to habits. It is streamlined (**A**) for fast swimming as in tunas (*Thunnus*), flattened from top to bottom (**B**) in bottom-dwellers such as the sea moth (*Pegasus*), or flattened sideways (**C**) in bottom-dwellers such as the flounder (*Platichthys*). Fishes living among vegetation or rocks have eel-like bodies (**D**) as in the freshwater eel (*Anguilla*), ribbon-shaped (**E**) as in the gunnel (*Pholis*), or threadlike (**F**) as in the snipe eel (*Nemichthys*). Slow swimmers feature bodies that are elongate on a vertical plane (**G**) as in the seahorse (*Hippocampus*), triangular (**H**) as in the trunkfish (*Ostracion*) or round (**I**) as in the porcupine fish (*Diodon*).

concealment. Slow-moving bottom fishes such as blennies (*Blennius*) and sculpins (*Oligocottus*) have their outline broken up with irregular growths, particularly on the head, that resemble seaweeds. The body of the stonefish (*Synanceia verrucosa*; Figure 7-11, A) resembles a rock so closely that it is almost invisible to both potential prey and humans. Unfortunately for humans, this shallow-water fish from the tropical Pacific and Indian oceans possesses the most potent venom known in fishes. Stepping on a stonefish has been known to cause death.

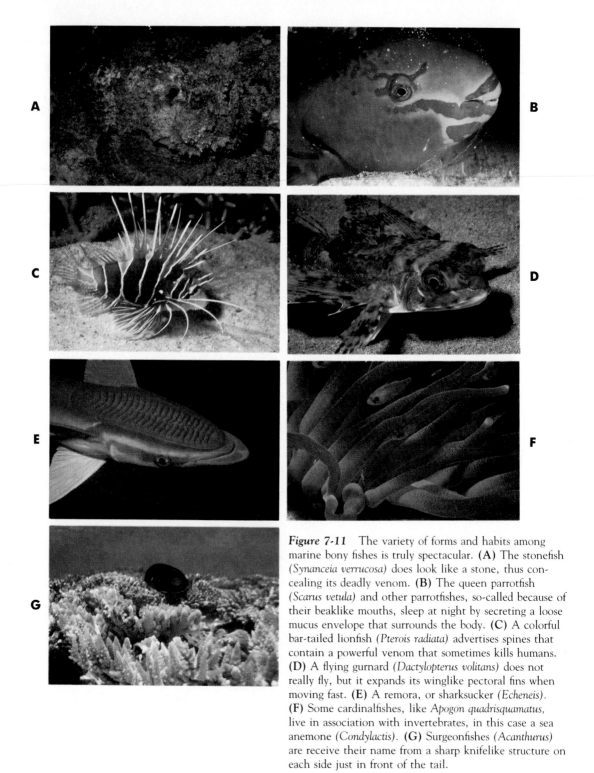

Figure 7-11 The variety of forms and habits among marine bony fishes is truly spectacular. **(A)** The stonefish (*Synanceia verrucosa*) does look like a stone, thus concealing its deadly venom. **(B)** The queen parrotfish (*Scarus vetula*) and other parrotfishes, so-called because of their beaklike mouths, sleep at night by secreting a loose mucus envelope that surrounds the body. **(C)** A colorful bar-tailed lionfish (*Pterois radiata*) advertises spines that contain a powerful venom that sometimes kills humans. **(D)** A flying gurnard (*Dactylopterus volitans*) does not really fly, but it expands its winglike pectoral fins when moving fast. **(E)** A remora, or sharksucker (*Echeneis*). **(F)** Some cardinalfishes, like *Apogon quadrisquamatus*, live in association with invertebrates, in this case a sea anemone (*Condylactis*). **(G)** Surgeonfishes (*Acanthurus*) are receive their name from a sharp knifelike structure on each side just in front of the tail.

Coloration

Some bony fishes use color for camouflage, but others, particularly those living in the tropics, are among the most brightly colored animals in the sea. The colored pigments in bony fishes are mostly found in special cells in the skin called **chromatophores.** These cells are irregular in shape and have branches radiating from the center. Flatfishes and others can rapidly change color by contracting and expanding the pigment in the chromatophores. The amazing variety of colors and hues observed among marine fishes results from combinations of chromatophores with varying amounts of different pigments. Fishes may also have **structural colors** that result when a special surface reflects only certain colors of light. Most structural colors in fishes are the consequence of crystals that act like tiny mirrors. The crystals are contained in special chromatophores called **iridophores.** The iridescent, shiny quality of many fishes is produced by structural colors in combination with pigments (Figure 7-11, B).

Colors can tell us a lot about fishes. Some change color with their mood or reproductive condition. They may also use color to advertise themselves, known as **warning coloration** (see Figure 7-11, C). **Cryptic coloration,** blending with the environment to deceive predators or prey, is a common adaptation (see Figure 7-11, A). Not only flatfishes, but also some blennies, sculpins, rockfishes *(Sebastes),* and others can change color to match their surroundings. Another use of color is **disruptive coloration,** the presence of color stripes, bars, or spots that help break up the outline of a fish (see Figures 7-30 and 14-24). These and other ingenious uses of color are especially common among coral reef fishes.

Open-water fishes and many shallow-water predators, on the other hand, are rarely as colorful. Most of them have silver or white bellies in sharp contrast to dark backs (see Figure 7-9). This distinctive color pattern, known as **countershading,** is a form of disguise in open water (see the section in Chapter 14, "Coloration and Camouflage"). When viewed from below, the light belly blends with the bright light coming from the surface. The dark back blends into the ocean's color as seen from above. Deep-water fishes also use color for concealment. They tend to be black or red, which are hard to see in the ocean depths (see the section in Chapter 15, "Coloration and Body Shape").

Locomotion

Swimming is obviously a major part of the life of fishes. Fishes swim to obtain food, escape from predators, and find mates. Many cartilaginous fishes must also swim to flush their gills with water to obtain oxygen.

Most fishes swim with a rhythmic side-to-side motion of the body or tail. S-shaped waves of contractions moving from head to tail push against the water and force the body forward. Variations on this theme are illustrated in Figure 7-12.

The rhythmic contractions are produced by bands of muscle called **myomeres,** which run along the sides of the body (Figure 7-13). The distinc-

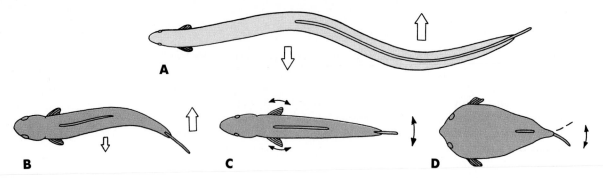

A

B   **C**   **D**

Figure 7-12 Locomotion in marine fishes involves many swimming styles. Eels and other elongate fishes, for instance, swim by flexing their body in lateral waves (**A**). Fast fishes with shorter bodies—like tunas, snappers, and jacks—swim by flexing the body in lateral waves that get longer toward the tail (**B**). Surgeonfishes, parrotfishes, and others swim by moving only the fins, either caudal (tail), pectoral, anal, and/or dorsal fins (**C**). Trunkfishes and porcupine fishes swim slowly by moving the base of the tail while the rest of the heavy body remains immobile.

tive bands of muscle are easily seen in fish fillets. Myomeres are attached to the backbone for support. Muscles make up a large percentage of the body weight of a fish—as much as 75% in tunas and other active swimmers.

Sharks tend to sink because they lack the buoyant swim bladder of bony fishes. To compensate, they have large, stiff pectoral fins that provide lift as do the wings of a plane (see Figure 7-13). The longer upper lobe of the tail tends to tilt the body upward, also generating some lift. The large amount of oil in the huge liver also provides buoyancy. In rays and skates, whose winglike pectoral fins are the main source of both thrust and lift, the tail is greatly reduced.

Since bony fishes have a swim bladder, they do not have to rely on their pectoral fins to provide lift. The pectorals are thus free to serve other purposes. This gives bony fishes great maneuverability. They can literally turn on a dime! Some can hover in the water, or even swim backward, things that sharks cannot do. The other fins of bony fishes also help provide maneuverability. The dorsal and **anal** fins (see Figure 7-9, B) are employed as rudders, at least part of the time, to "steer" and provide stability. The paired **pelvic** fins (see Figure 7-9, B) also help the fish turn, balance, and "brake."

The flexibility of their fins has allowed many fishes to depart from the standard undulating style of swimming. Some emphasize sheer speed (see Box 14-2). Many fishes—particularly those living around coral reefs, rocks, or kelp beds—swim mainly by moving their fins rather than their bodies, especially for the precise movements needed in feeding. Their tails are used almost solely as rudders. Wrasses, surgeonfishes (*Acanthurus;* see Figure 7-11, G), and parrotfishes (*Scarus, Sparisoma*)—which live on coral reefs—and the California sheephead (*Semicossyphus pulcher*)—common in kelp beds—swim mainly with their pectoral fins. Triggerfishes (*Balistes*) undulate their dorsal and anal fins to swim. This style is perfect for hovering over the bottom while hunting for crabs and sea urchins. Flying fishes (*Cypselurus*) have greatly expanded pectoral fins that they use to glide through the air (see Figure 14-26). A large variety of bottom fishes (gobies, sculpins, and many others) "crawl" or rest on the bottom on their pectoral and/or pelvic fins (Figure 7-11, D). Clingfishes (*Gobiesox*) are small fishes that have their pelvic fins modified into part of a sucker that allows them to attach to rocks (see Figure 10-13). Remoras, or sharksuckers (*Echeneis*), attach to sharks, whales, turtles, and many types of large fishes using a large sucker on top of

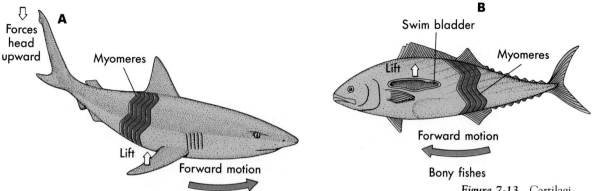

A. Cartilaginous fishes

Figure 7-13 Cartilaginous (**A**) and bony (**B**) fishes use different adaptations to maintain their position in the water. Sharks use their fins for lift. Bony fishes have evolved a gas-filled swim bladder to compensate for their heavier bony skeleton. This frees their fins for locomotion, resulting in a much greater diversity of swimming styles (see Figure 7-12).

their heads (see Figure 7-11, *E*). This sucker is derived from part of the dorsal fin.

> Locomotion of fishes usually involves sideways undulations of the body and tail. In contrast to those of sharks, the more flexible fins and tail of bony fishes do not have a role in buoyancy control and, as a result, have become highly maneuverable and important in locomotion.

Feeding

Sharks are carnivores, but in contrast to typical carnivores—who capture smaller prey—many sharks feed by taking bites from prey larger than themselves. For this they use their formidable jaws coupled with shaking of the head. Most sharks are not particular; almost anything can be found in a shark's stomach. A tiger shark (*Galeocerdo cuvier*; Figures 7-5, *B* and 7-6, *J*) that was caught off South Africa was found to contain the front half of a crocodile, the hind leg of a sheep, three gulls, and two cans of peas, among other goodies. Nurse sharks feed mostly on bottom invertebrates, including lobsters and sea urchins. Some deep-water sharks subsist mainly on squids. Cookie-cutter sharks (*Isistius*) are small, deep-water sharks that attack larger fishes and dolphins and cut out chunks of flesh with their razor-sharp teeth and sucking lips. Even the rubber sonardomes of nuclear submarines have not escaped their bites. It has been suggested that these sharks use bioluminescence to imitate squids in order to lure their prey.

Only three species of sharks are filter feeders: the whale shark, the basking shark, and the megamouth shark (*Megachasma pelagios*; Figure 7-6, *M*), a gigantic deep-water shark discovered off the Hawaiian Islands in 1976 and more recently in southern California and other parts of the Pacific and Indian oceans. Like other filter-feeding fishes, the whale and basking sharks filter the water with their **gill rakers,** slender projections on the inner surface of the gill (see Figure 7-18, *B*). The large mouths of these sharks have many small teeth and, excluding the megamouth, very long gill slits. The width of the spaces between the gill rakers determines the size of the food captured. Water is strained through the gill rakers, and the shark swallows the food that is left behind. Whale sharks feed in warm water on small

BOX 7-1

Shark!

Most sharks are harmless—at least to humans. Nevertheless, 25 species of sharks are known to have attacked humans, and at least 12 more are suspected of doing so. Three are particularly dangerous: the great white, tiger, and bull sharks (see Figures 7-5, B and 7-6). Even so, shark attacks are rare. The chances of being attacked by a shark are lower than those of being hit by lightning.

Many shark attacks, however, have been documented over the years: a U.S. naval officer killed by massive bites while swimming in the Virgin Islands in 1963, an abalone diver in southern California last seen protruding from the mouth of a shark, parts of arms and legs in the stomach of a tiger shark caught after a man had been mortally wounded in Australia in 1937. One of the authors knew a young man whose promising life ended horribly in the jaws of a tiger shark off Western Samoa.

World War II exposed the crews of torpedoed ships and downed airplanes to shark attacks. Many grim stories about bleeding bodies surrounded by sharks began to spread. They prompted research on the aggressive behavior of sharks and the circumstances leading to shark attacks.

Great white sharks typically inflict a massive wound on their prey (such as seals and sea lions) and then release it. The sharks wait until the bleeding prey is too weak to resist, then move in for the kill. White shark attacks on humans wearing wet suits may be cases of mistaken identity. Sometimes people are able to escape when sharks release them after the first bite. It has been discovered that before attacking, the small but dangerous gray reef shark performs a distinct aggressive display—a "warning" unrelated to feeding. Displaying sharks may attack if someone approaches.

So far there is no guaranteed repellent. Copper acetate was used during World War II as a shark repellent but was eventually found to be ineffective. A black chemical dye was used, but this helped only by obscuring the shark's vision. A repellent based on a poison obtained from a flatfish seems more promising. Chain mail suits offer effective protection from sharks but are too expensive and cumbersome for widespread use. For someone like a downed flyer or shipwrecked sailor, perhaps the best protection is a black plastic bag large enough to float inside.

How can you decrease the risk of an attack? First of all, do not swim, dive, or surf in an area known to be frequented by dangerous sharks. Sea lion colonies and coastal garbage dumps attract them. Blood, urine, and feces also attract sharks. Avoid murky water. Many sharks are more active at night, so avoid night swims. Sharks should not be provoked in any way. Even resting nurse sharks can turn and bite. Leave the water if fish suddenly appear in large numbers and behave erratically, which may be an indication that sharks are around. If you see a large shark, get out of the water with as little splashing as possible.

Actually, we threaten the survival of sharks more than they threaten us. They reproduce slowly, and their numbers are already being depleted by overfishing in many areas. This attitude toward sharks may be short sighted, because they play an important role in marine communities. When large sharks were netted off South Africa, for example, the number of species of small sharks, which are eaten by large sharks, increased. As an apparent result, the number of bluefish, a commercially important species, decreased. Some people catch shark only for the shark's dorsal fin. Others practice shark hunting for sport, leaving the meat to waste. The magnificent predator, the shark, may soon be exterminated by human beings, the bloodiest predators of them all.

schooling fishes, squids, and planktonic crustaceans. Basking sharks, which live in colder water, feed on plankton by opening their mouths and slowly swimming through the water (see Figure 7-6, *H*). The megamouth shark has bioluminescent organs in its mouth. The light is thought to attract planktonic crustaceans into the shark's mouth.

A Barracuda

Mantas are rays that feed on plankton and small fishes, filtering them from the water with their gill rakers. Two horn-shaped projections on the sides of the manta's mouth help channel food into the cavernous mouth (see Figure 7-7, *E*).

Bony fishes are very diverse in the ways they feed. Their protrusible jaws allow them much more flexibility in feeding habits than sharks and rays have. Most bony fishes are carnivores; almost no animal in the ocean is immune to being eaten by some bony fish. Bony fishes capture their prey from sediments, the water column, the surface of rocks, or from other organisms—including other fishes—or combinations of these. Some chase their prey; others sit and wait.

Carnivorous bony fishes typically have well-developed teeth for catching, grasping, and holding their prey (Figure 7-14, *A*), which is usually swallowed whole. The roof of the mouth, gill rakers, and pharynx may also have teeth to help hold the prey. Deep-water fishes often have huge mouths and teeth (see Figure 15-13), and a few are able to capture and swallow prey larger than themselves.

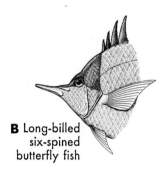

B Long-billed
six-spined
butterfly fish

Unusual food preferences have evolved in fishes. Some prefer sponges, infrequently touched by other carnivores. Others prefer sea urchins, sea squirts, or other seemingly unsavory or tough-to-eat food items. Reef corals, skeleton and all, are eaten by several types of fishes (see Figures 7-14, *B* and 7-30). Many species, however, are non-specialists and capture a wide variety of prey. Some feed on small invertebrates and dead animal material from the bottom. These bottom-feeders have a downward-oriented mouth adapted to suck food from the bottom. Frogfishes (*Antennarius*) use a modified spine on the head to lure small fishes.

C Slipmouth

Fishes that feed primarily on seaweeds and other plants are known as **grazers.** Parrotfishes, for example, graze on small algae growing on hard surfaces. Their front teeth are fused to form a beaklike structure (Figure 7-14, *D*). Some species use the beak to scrape off bits of live coral.

Fishes such as herrings (*Clupea*), sardines (*Sardinops*), anchovies (*Engraulis*), and menhaden (*Brevoortia*) filter plankton with their gill rakers. They typically strain their food by swimming with their large mouths open (Figure 7-14, *E*). These suspension feeders are small, in contrast to the huge plankton-feeding sharks. They usually occur in large, often immense schools (see Figure 16-7). Plankton feeders are the most abundant fishes in the ocean, and they are a most important food source for many types of carni-

D Parrotfish

Figure 7-14 The shape of the mouth of bony fishes may give us a hint of their diets. The barracuda (*Sphyraena*) uses its large mouth (**A**) to tear off chunks of prey that are too large to swallow, very much as sharks do. Most bony fishes, however, swallow their prey whole. Many butterflyfishes (*Chaetodon*) use a long snout and small mouth (**B**) to feed on very small prey. The extremely protrusible mouth (**C**) of the slipmouth (*Leiognathus*) is used for feeding on relatively small prey. Parrotfishes (*Scarus*) use their beaklike mouth (**D**) to graze on small algae (also see Figure 7-11, *B*). Herring (*Clupea*) and other filter feeders highlight a large mouth (**E**).

E Herring

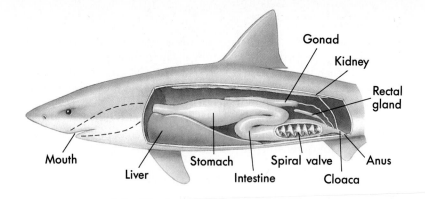

Figure 7-15 The digestive systems of cartilaginous (A) and bony (B) fishes display many of the basic features found in all vertebrates.

Gonad
Kidney
Rectal gland
Mouth
Liver
Stomach
Intestine
Spiral valve
Cloaca
Anus

A. Cartilaginous fishes

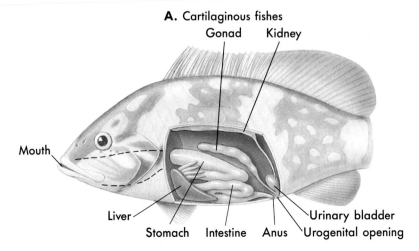

Gonad   Kidney
Mouth
Liver
Stomach   Intestine   Anus
Urinary bladder
Urogenital opening

B. Bony fishes

vores. The fishes previously listed account for a large share of the world's fish catch (see the section in Chapter 16, "Major Food Species").

Digestion

After being swallowed, food passes into the **stomach** through the pharynx and a short tube called the **esophagus** (Figure 7-15). The stomach is where the process of chemical digestion usually begins. It is typically J-curved or elongate but may be modified into a grinding structure or even lost altogether. The food passes from the stomach into the **intestine.** In most bony fishes the anterior portion of the intestine has many slender blind tubes, the **pyloric caecae,** which secrete digestive enzymes. Other digestive enzymes are secreted by the inner walls of the intestine and the **pancreas.** Another organ important in digestion is the **liver,** which secretes **bile** needed for the breakdown of fats. The liver is particularly large and oil-rich in sharks, sometimes constituting as much as 20% of their body weight.

A few fishes lack a stomach and tend to have a portion of the intestine expanded for the digestion of food. Carnivorous fishes have short, straight intestines. Fishes that eat hard-to-digest plant material, on the other hand,

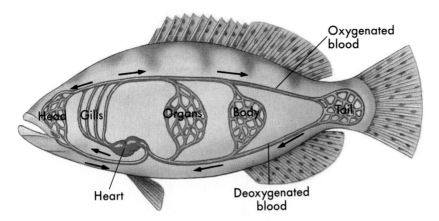

Figure 7-16 The circulatory system of cartilaginous and bony fishes consists of veins that carry deoxygenated blood *(in blue)* from the body, a two-chambered heart that pumps blood to the gills for oxygenation, and arteries that carry oxygenated blood *(in red)* to the rest of the body.

have coiled intestines, which may be much longer than the fish itself. The intestine of cartilaginous fishes and a few primitive bony fishes contains a spiraling portion called the **spiral valve,** which increases the internal surface area of the intestine (Figure 7-15, A). The intestine is responsible for absorbing the nutrients resulting from digestion. These nutrients pass into the circulatory system to be distributed through the body. Undigested material exits through the anus.

Circulatory System

All fishes have a two-chambered **heart** located below the gills (Figure 7-16). Deoxygenated blood comes into the heart from the body. The blood is pumped to the gills where gas exchange takes place. The oxygenated blood is then carried back to the body by blood vessels called **arteries.** The arteries branch out into thin-walled **capillaries** that allow oxygen and nutrients to reach every cell. The capillaries then collect into larger blood vessels, **veins,** that carry deoxygenated blood—along with carbon dioxide—back to the heart to complete the cycle.

gas exchange The absorption of oxygen used in respiration, the breakdown of glucose to release energy, and the elimination of carbon dioxide that results from the same process
Chapter 6, p. 158

Respiratory System

Fishes obtain oxygen dissolved in water and release carbon dioxide through paired gills. The gills lie in the pharynx, a chamber just behind the mouth that represents the front part of the gut.

   Irrigation of the Gills.   Fishes get the oxygen they need by extracting it from the water. To do this, they must make sure that water flows over the gills; that is, they must **irrigate** the gills.

   Cartilaginous fishes open and close the mouth to force water over the gills (Figure 7-17, A). Expansion and contraction of the walls of the pharynx and the gill slits assist in the process. Every gill lies in its own chamber, and each gill chamber opens to the outside by a separate gill slit. The first pair of gill slits of cartilaginous fishes is modified into **spiracles,** a pair of round openings just behind the eyes. The spiracles are located on the dorsal surface of rays and skates (see Figure 7-7, D). They allow these fishes, many of which live on the bottom, to take in water even when the ventral mouth

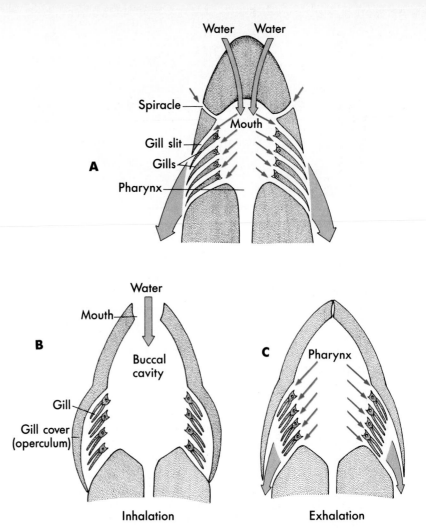

Figure 7-17 The mechanism behind the irrigation of the gills by water follows some simple principles. Cartilaginous fishes like sharks (**A**) allow water in (*path of water indicated by arrows*) by opening their mouth, which is then closed. The rising of the floor of the mouth causes water to be pumped through the gills and out by way of the separate gill slits. The spiracles also allow water to move in, but it is in rays and skates where they play a crucial role. Bony fishes are more efficient due to the precence of gill covers that close as the pharynx expands, sucking the water in (**B**). Closing of the mouth and opening of the gill covers forces water through the gills and out.

is buried in the sediment. Lampreys and other jawless fishes pump water directly in and out through their gill slits during feeding when the passage of water through the mouth is blocked.

Most bony fishes have a more efficient mechanism to bring in water to the gills. The gills on each side share a common gill chamber, which opens to the outside through a single opening. The opening is covered by an operculum (Figure 7-17, B). When the mouth opens, the opercula close and the pharynx expands, sucking water in. The fish does the reverse to pump water out: The mouth closes, the pharynx contracts, and the opercula open. Some fast swimmers simplify things by just opening their mouths to force water into the gills.

Structure of the Gills. Fish gills are supported by a cartilaginous or bony structure, the **gill arch** (Figure 7-18, B). Each gill arch bears two rows of slender fleshy projections called **gill filaments.** Gill rakers project along the inner surface of the gill arch. They prevent food particles from entering the gill slits or may be specialized for filtering the water in filter-feeding fishes.

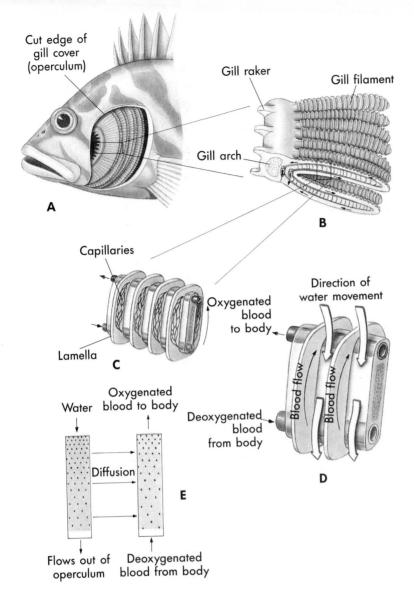

A

Cut edge of gill cover (operculum)

Gill raker

Gill filament

Gill arch

B

Capillaries

Oxygenated blood to body

Lamella

C

Water

Oxygenated blood to body

Diffusion

E

Flows out of operculum

Deoxygenated blood from body

Direction of water movement

Blood flow

Deoxygenated blood from body

D

Figure 7-18 The gills of fishes are very efficient structures for gas exchange. Bony fishes have four pairs of gills (**A**), each containing two rows of numerous gill filaments (**B**). Lamellae in the gill filaments (**C**) increase the surface area of the gill filaments. Diffusion of oxygen from seawater into the blood gets a boost because the water flows across the lamellae in the opposite direction to that of the blood (**D**). The concentration of oxygen (*indicated in* **E** *by dots*) is always higher in the water than in the blood. If circulation was not reversed, blood to the body would have less oxygen.

The gill filaments have a rich supply of capillaries (Figure 7-18, C), the blood of which gives them a bright red color. Each gill filament contains many rows of thin plates or discs called **lamellae** that greatly increase the surface area through which gas exchange can take place. The number of lamellae is higher in active swimmers, who need large supplies of oxygen.

Gas Exchange. Oxygen dissolved in the water diffuses through the capillary walls to oxygenate the blood. Diffusion will take place only if oxygen is more concentrated in the water than in the blood. This is usually true because the blood coming to the gills has already traveled through the rest of the body and is depleted of oxygen (see Figure 7-16). As oxygen diffuses from the water to the blood, the amount of oxygen in the water decreases and that in the blood increases. This could reduce the efficiency of gas exchange, which depends on the water having more oxygen than the

diffusion The movement of a substance from areas of high concentration to areas of low concentration
Chapter 4, p. 98

blood. Fishes have evolved a clever adaptation to ensure efficiency. The blood in the gills flows in the *opposite* direction to the water (Figure 7-18, *D*). When the water has passed over the gill and given up much of its oxygen, it meets blood that has just come from the body and is "hungry" for what oxygen remains in the water (Figure 7-18, *E*). By the time the blood has flowed most of the way through the gill, picking up oxygen, it encounters water that is just entering the gill chamber and is rich in oxygen. Thus the oxygen content of the water is always higher than that of the blood. This system makes the gills very efficient at extracting oxygen. If circulation was not reversed, blood returning to the body would have less oxygen.

Blood disposes of its carbon dioxide using the same mechanism. Blood flowing into the gills from the body has a high concentration of carbon dioxide, a product of respiration. It easily diffuses out into the water.

> **Gas exchange in the gills of fishes is highly efficient. The surface area of gills is greatly increased by lamellae, and the flow of water through them is in a direction opposite to that of blood.**

Once oxygen enters the blood it is carried through the body by **hemoglobin,** a red protein that gives blood its characteristic color. Hemoglobin is contained in specialized cells called **red blood cells.** The hemoglobin releases oxygen to the tissues as it is needed. After it gives off its oxygen, the hemoglobin picks up carbon dioxide from the body and carries it to the gills, where it diffuses into the water.

Muscles use a lot of oxygen during exertion. They have a protein called **myoglobin,** similar to hemoglobin, that can store oxygen. Hard-working muscles tend to have a lot of myoglobin, which makes them red. Strong swimmers, such as open-water sharks and tunas, have a high proportion of red, as opposed to white, muscle. Many other fishes have concentrations of red muscles at the base of heavily used fins (see the section in Chapter 14, "Swimming: The Need for Speed").

Regulation of the Internal Environment

In contrast to most marine organisms, the blood of marine fishes is less salty than seawater (see Figure 4-20). As a result they lose water by **osmosis.** Marine fishes therefore need to osmoregulate to prevent dehydration. To replace lost water, they swallow seawater (Figure 7-19, *B*). Seawater contains excess salts, which are excreted by the **kidneys**—the most important excretory organs of vertebrates—and the **chloride cells** of gills. The kidneys conserve water by producing only small amounts of concentrated urine.

Cartilaginous fishes are a special case (Figure 7-19, *A*). They reduce osmosis by increasing the amount of dissolved molecules—or solutes—in their blood, making the blood concentration closer to that of seawater. They do this by retaining a chemical called **urea,** a waste product that results from the breakdown of proteins. The amount of urea in the blood is controlled by the kidneys. As in all vertebrates, the kidneys remove wastes from the blood and eliminate them in the urine. In most animals, urea is toxic and is excreted, but sharks and other cartilaginous fishes excrete only

osmosis The diffusion of water across a selectively permeable membrane, such as a cell membrane
Chapter 4, p. 99

osmoregulation The active control by an organism of its internal solute concentration to avoid osmotic problems
Chapter 4, p. 100

LIFE IN THE MARINE ENVIRONMENT

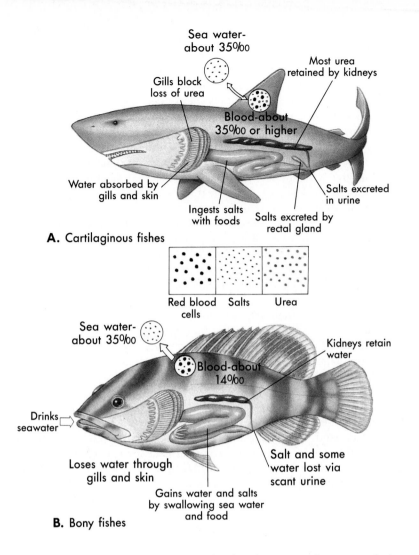

Sea water-
about 35‰

Most urea
retained by kidneys

Gills block
loss of urea

Blood-about
35‰ or higher

Water absorbed by
gills and skin

Salts excreted
in urine

Ingests salts
with foods

Salts excreted by
rectal gland

A. Cartilaginous fishes

Red blood   Salts   Urea
cells

Sea water-
about 35‰

Kidneys retain
water

Blood-about
14‰

Drinks
seawater

Salt and some
water lost via
scant urine

Loses water through
gills and skin

Gains water and salts
by swallowing sea water
and food

B. Bony fishes

Figure 7-19 Marine fishes live in water that is more salty than their body fluids. For this reason, water tends to diffuse out of the body. To prevent dehydration, sharks and other cartilaginous fishes (**A**) concentrate urea, absorb seawater through their gills and skin, and excrete excess salts by way of the urine and feces. A special gland, the rectal gland, also excretes excess salts. Their blood ends up with a solute concentration almost equal to or even higher than that of seawater. Marine bony fishes (**B**) do not concentrate urea, which is toxic to them. Instead, their kidneys conserve water, and they drink seawater, excreting excess salts as do cartilaginous fishes. Compare with Figure 4-20.

small amounts. Surprisingly, urea and related compounds are much less toxic to cartilaginous fishes. They retain most of the urea in their blood; their gills help in this process by blocking the loss of urea.

Cartilaginous fishes also absorb water to prevent dehydration, mostly through the gills (see Figure 7-19, A). Excess salts are excreted by the kidneys, intestine, and a special gland near the anus.

> **Marine fishes are able to keep a constant internal environment and check water loss through the osmoregulatory activities of the kidneys, gills, and other mechanisms.**

Nervous System and Sensory Organs

Vertebrates have the most complex and advanced nervous systems of any animal group. At the heart of the system is the **central nervous system,** consisting of the brain and spinal cord. The central nervous system coordinates and integrates all body activities and stores information. The brain is divided into several regions known to serve as centers for particular functions such as olfaction and vision. It is protected by a cartilaginous or bony skull. Nerves connect the central nervous system with various organs of the body and with sense organs that receive information from the surroundings. This information is sent to the brain in the form of nerve impulses.

Most fishes have a highly developed sense of smell, which they use to detect food, mates, and predators, and sometimes to find their way home. Fishes do this with special sensory cells located in **olfactory sacs** on both sides of the head. Each sac opens to the outside through one or two openings, the **nostrils.**

The sense of smell is particularly well developed in sharks. They are able to detect blood and other substances in concentrations as low as fractions of one part per million. Salmon (*Oncorhynchus*), which live as adults at sea but reproduce in fresh water, use olfaction to find the stream where they were born years earlier (see the section "Migrations" on p. 216). There is evidence that they accomplish this remarkable feat by "memorizing" the sequence of smells on their way out to sea.

Fishes detect other chemical stimuli with **taste buds** located in the mouth and on the lips, fins, and skin. Taste buds also are found on **barbels,** whiskerlike organs near the mouth of many bottom feeders such as goatfishes (*Pseudopeneus*). Fishes that have them use their barbels to detect food on the bottom.

Most bony fishes, unlike cartilaginous fishes, rely on vision more than any other sense. Fish eyes are not very different from those of vertebrates that live on land. One important difference, though, is the way they focus. Whereas the eyes of most land vertebrates focus by changing the shape of the lens, the round lens of the fish eye focuses by moving closer or farther away from the subject. This is partially why fish eyes tend to bulge. Many bony fishes—particularly shallow water species—have color vision, but most cartilaginous fishes have little or none. Some sharks have a distinct **nictitating membrane** that can be drawn across the eye to reduce brightness.

Fishes have a unique sense organ called the **lateral line** that enables them to detect vibrations in the water. The lateral line consists of a system of small canals that run along the body (Figure 7-20). The canals lie in the skin and in the bone or cartilage of the head. They are lined with sensory cells that are sensitive to vibration. The canals usually open to the surface through pores that are quite visible.

The lateral line system picks up vibrations resulting from the swimming of other animals, as well as water displacements caused by sound waves. It allows fishes to avoid obstacles, orient to currents, and keep their position in a school.

Cartilaginous fishes also have sense organs in the head called the **ampullae of Lorenzini** that can detect weak electrical fields. This system has been shown to help them locate prey—a kind of electrical sensing device. It may also assist in navigation as a sort of electromagnetic compass or perhaps a detector of currents.

Fishes can also perceive sound waves with their **inner ears,** paired hear-

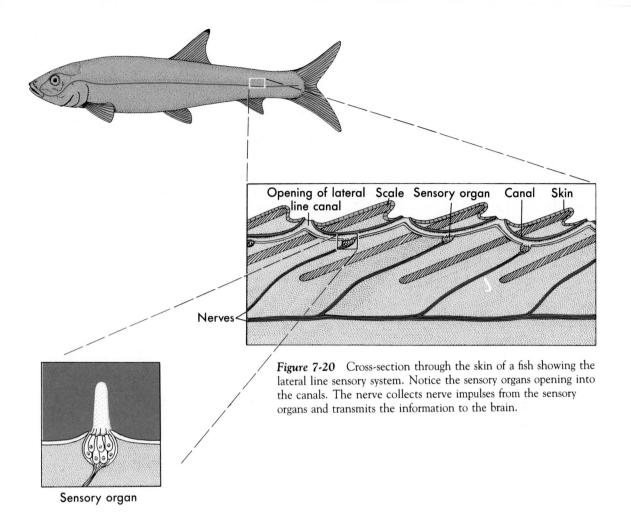

Opening of lateral line canal   Scale   Sensory organ   Canal   Skin

Nerves

Sensory organ

Figure 7-20 Cross-section through the skin of a fish showing the lateral line sensory system. Notice the sensory organs opening into the canals. The nerve collects nerve impulses from the sensory organs and transmits the information to the brain.

ing organs located to the sides of the brain just behind the eyes. The inner ears are a set of fluid-filled canals that contain sensory cells similar to those in the lateral line canals. In some fishes the swim bladder is known to amplify sound by vibrating and transmitting sound waves to the inner ear. The inner ear is also involved in equilibrium and balance. Changes in the position of many fishes are detected by shifts in the position of calcareous **ear stones,** or **otoliths,** that rest on hairs of sensory cells, a mechanism similar to the statocysts of invertebrates.

The sense organs of fishes include eyes, olfactory sacs, taste buds, and inner ears, as well as a lateral line and other specialized organs that pick up vibrations or electrical stimuli from the water.

statocysts Sense organs of many invertebrates consisting of one or more grains or hard bodies surrounded by sensitive hairs and used to orient the animal with gravity
Chapter 6, p. 154

Behavior

The well-developed nervous system of fishes allows them to respond to their environment in complex ways. Among the most important aspects of their response is their behavior, which in general is much more sophisticated than that of invertebrates. Complex behavior dominates nearly every phase in the lives of fishes. They use behavior to adapt to such physical factors as

BOX 7-2

A *Fish Called* Latimeria

The coelacanth *(Latimeria chalumnae)*, a living fossil.

In December of 1938 the skipper of a fishing trawler operating in deep water off the Chalumna River in South Africa found a very strange fish in his catch. He took it to Marjorie Courtnay-Latimer, a young curator at the local museum, who recognized the fish as something special. She sent a sketch of the 1.5-m (5-foot) specimen to Dr. J.L.B. Smith at nearby Rhodes University, and history was made.

The fish was a big catch indeed. It was a **coelacanth,** a type of fish thought to have become extinct 60 million years ago. Coelacanths were previously known only from fossils, some at least 400 million years old. They are crossopterygian fishes that gave rise to the first land vertebrates. About 350 million years ago a crossopterygian fish with paddlelike bony fins crawled out of the water and changed life on earth forever.

Dr. Smith officially described the fish and named it *Latimeria chalumnae* in honor of its discoverer and the river near which it was caught. Unfortunately, the internal organs of the fish had been discarded by the time Dr. Smith got to it, so nothing was known of its internal structure. A reward was offered for more specimens of this incredible living fossil. It was not until 1952 that a second specimen was caught near the Comoro Islands, between Madagascar and mainland Africa. Ironically, the fish is well known to the natives of the islands. They eat its oily flesh after drying and salting it, and they use its rough skin for sandpaper!

The fish is still very rare. None of the captured specimens has survived more than 20 hours; thus little is known about their habits. In 1987 a small submersible was used to film and observe live *Latimeria* in its natural environment for the first time. The fish was observed only at night, at depths of 117 to 200 m (386 to 660 feet).

Latimeria is a large fish, up to 2.7 m (8 feet) in length and weighing as much as 85 kg (187 pounds). The body is covered by large blue scales. It feeds on fish and squid. This living fossil is unique in many ways. It has heavy, stalked fins that have bones like land vertebrates. The fish appears to stand on the fins, but not crawl over the bottom with them as once thought. The pectoral fins can rotate nearly 180 degrees, allowing the fish to slowly swim over the bottom, sometimes standing on its head or with its belly up.

Much remains to be discovered about this fascinating creature. Jelly-filled organs on the head may be used to detect electrical fields and thus help in prey location. Little is known about its reproduction. Females bear live young, and the huge eggs (about 9 cm, or 3.5 inches, in diameter) develop in the reproductive tract.

Latimeria remains a priceless catch, and several aquaria around the world would like very much to capture live specimens. As its value in dollars soars, so do the concerns about the few hundred individuals that may still survive. International trade has now been officially outlawed.

Are there any new coelacanths waiting to be discovered somewhere else? Only time and alert fishermen will tell.

light and currents. Behavior is of key importance in finding food and shelter and avoiding enemies. Fishes also display a fascinating variety of behaviors related to courtship and reproduction. We can give only a glimpse of some of the important aspects of fish behavior. Reproductive behavior will be covered in the next section, and some other highlights will be discussed in later chapters.

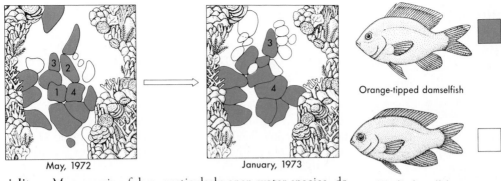

May, 1972   January, 1973

Orange-tipped damselfish

Ward's damselfish

Territoriality.   Many marine fishes, particularly open-water species, do not reside in any particular area. Many others, however, are known to establish **territories,** home areas that they defend against intruders. Some fishes defend territories only during reproduction. Many, however, have more-or-less permanent territories that they use for feeding and resting or as shelter. It is thought that fishes often guard territories to ensure that they have enough food and other resources. Thus territoriality is most common in crowded environments like kelp beds and coral reefs, where resources are most likely to be in short supply. Coral reef damselfishes are famous for fiercely defending their territories, often "attacking" fishes many times their size or even divers (Figure 7-21).

Fishes use a variety of **aggressive behaviors** to defend their territories. Actual fights are surprisingly rare. Instead, fishes usually prefer to avoid risking injury by bluffing. Many aggressive behaviors seem to have evolved to make the fish look as fierce as possible. Raised fins, an open mouth, and rapid darting about are examples of such threatening postures. Territorial defense may also involve sound production. Marine bony fishes may make sound by grinding their teeth or rubbing bones or fin spines on another bone. Some fishes "drum" by pulling muscles on the swim bladder, with sound amplified by the air-filled bladder.

Sometimes a solitary individual defends a territory. In other species, like some butterflyfishes, territories are established by a male-female pair. Territories may also be inhabited by groups that belong to the same species. This is the case in damselfishes, which inhabit spaces between the branches of corals (Figure 7-22), and anemonefishes, or clownfishes (*Amphiprion;* see Figure 13-43), and some cardinalfishes (*Apogon*), which live among the tentacles of large sea anemones (Figure 7-11, *F*). Members of such groups often divide the territory into sub-territories.

Schooling.   Many fishes form well-defined groups, or **schools.** Some—including herrings, sardines, mullets (*Mugil*), and some mackerels—school throughout their lives. Others are part-time schoolers, especially as juveniles or during feeding. Most cartilaginous fishes are solitary, but a few—such as hammerhead sharks, mantas, and other rays—sometimes travel in schools. It has been estimated that roughly 4,000 species, including both marine and freshwater species, school as adults. Schools can be huge, as large as 4,580 million cubic m (161,720 million cubic feet) in the Atlantic herring (*Clupea harengus*). The members of a school are typically all about the same size. The stationary schools that are common around coral reefs,

Figure 7-21   Map of the territories established by two species of damselfishes on Heron Island, Great Barrier Reef, Australia. The bottom was covered with fragments of dead coral, which offer sheltered spaces. The numbers refer to individuals recognized at the beginning of the experiment and 8 months afterward. Notice the tight, mosaic-like grouping of territories. There were never any overlaps between territories belonging to members of the same species. There were some overlaps, however, between those of different species. Some territories were relatively stable over the period of observation. Some fish disappeared, as is the case of 1 and 2 after May 1972.

Figure 7-22 Many damselfishes, such as species of *Chromis*, live among corals. They dash into the spaces between the coral branches whenever danger approaches. Also see Figure 13-31.

Figure 7-23 Fishes follow different recognizable patterns when schooling. Some common patterns include traveling **(A)**, feeding on plankton **(B)**, encirclement of a predator **(C)**, and streaming to avoid a predator **(D)**.

kelp beds, rocks, and shipwrecks, however, may include members of different sizes or even different species.

Schools function as well-coordinated units, though they appear to have no leaders (Figure 7-23). The individual fishes tend to keep a constant distance among themselves, turning, stopping, and starting in near-perfect unison. Vision has been found to play an important role in the orientation of individuals within a school. In some species, though, blinded fish can school in a coordinated way. These fish probably use the lateral line, olfaction, and sound they emit to keep track of each other. The tight coordination of schooling fishes may break down when they are feeding or attacked by a predator.

There is much debate about why fishes school. One explanation is that schooling offers protection against predation. Predators may be confused if, for instance, the school circles the predator (Figure 7-23, C) or splits into several groups. It is also difficult for a predator to aim for just one fish in a cloud of shifting, darting individuals (Figure 7-23, D). On the other hand, some predators, such as jacks (*Caranx*), are more efficient when they attack schools of prey rather than individuals. It also has been suggested that schooling increases the swimming efficiency of the fish because the fish in front form an eddy that reduces water resistance for those behind. There is experimental evidence, however, that this is not always the case and that fish do not always align in a hydrodynamically efficient way. In at least some fishes, schooling is advantageous in feeding or mating. There is probably no single reason that fishes school, and the reasons probably vary from species to species.

Migrations. Another fascinating aspect of the behavior of marine fishes is **migration,** regular mass movements from one place to another—once a day, once a year, or once in a lifetime. Schools of parrotfishes and other fishes migrate on and off shore to feed. Several open-water fishes migrate several thousand meters up and down the water column every day (see the section in Chapter 14, "Vertical Migration"). The most spectacular migrations, however, are the transoceanic journeys made by tunas, salmon, and other fishes. We still know little about why fishes migrate, but most migrations seem to be related to feeding and/or reproduction.

There is little doubt that feeding is the main reason behind the migration of open-water species like tunas. Tagged fish recaptured by fishermen

A   **B**   **C**   **D**

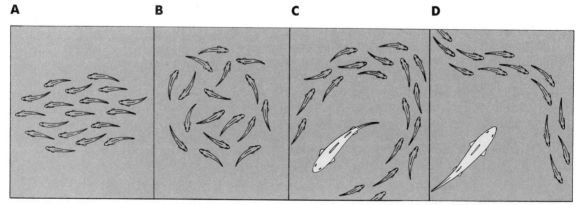

LIFE IN THE MARINE ENVIRONMENT

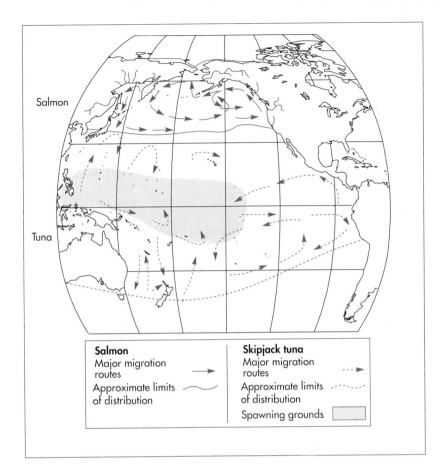

Figure 7-24 The skipjack tuna *Katsuwonus pelamis* the leading commercial catch among tunas, undertakes extensive migrations that every year takes it almost halfway across the globe. The Pacific salmon (*Oncorhynchus*) migrates to spawn only once in a lifetime. Approximate limit of distribution and migration routes indicated are for all seven species of salmon.

Salmon

Tuna

Salmon

Major migration routes

Approximate limits of distribution

Skipjack tuna

Major migration routes

Approximate limits of distribution

Spawning grounds

have provided much information on how far, how fast, and when tunas migrate (see the section in Chapter 14, "Swimming: The Need for Speed"). Though essentially tropical, some species migrate long distances to feed in temperate waters. Such is the case of the skipjack tuna (*Euthynnus pelamis*) and other tunas (Figure 7-24).

Even more amazing are the migrations between the sea and fresh water undertaken by some fishes that are dependent on fresh water for reproduction. **Anadromous** fishes spend most of their lives at sea but migrate to fresh water to breed. Sturgeons (*Acipenser*), whose eggs are eaten as caviar; some lampreys; and smelts (*Osmerus*) are examples. By far the best known anadromous fish, however, is the salmon.

There are seven species of salmon in the Pacific, all known by several common names. They spend their adult lives in the north Pacific, traveling thousands of miles in vast sweeps along the coast, the Aleutian Islands, and the open sea (see Figure 7-24). Some even venture into the Bering Sea and the Arctic Ocean. How they navigate at sea is not known. It has been hypothesized that they use land features at least part of the way. Currents, salinity, temperature, and other water characteristics might provide clues.

Figure 7-25 The journey of Pacific salmon, like these sockeye salmon (*Oncorhynchus nerka*), is an arduous one indeed. The fish have to swim far upstream, often leaping over rapids and waterfalls. Many fall prey to hungry predators or to the nets and lines of fishermen. Those that make it to their spawning grounds die soon after spawning.

Sargasso Sea   Section of the Atlantic Ocean north of the West Indies that is characterized by masses of drifting Sargasso weed, a brown seaweed
Chapter 5, p. 130

Other possibilities include orientation to polarized light, the sun, or the earth's magnetic field.

After several years at sea, a period characteristic of each species, salmon mature sexually and start migrating into rivers. They do not feed once they enter fresh water, living instead on stored fat. Their kidneys must adjust to the change from salt to fresh water. Eventually they reach the exact stream of their birth, sometimes far upstream (Figure 7-25). The king, or chinook (*Oncorhynchus tschawytscha*), and chum, or dog, (*O. keta*) salmon reach as far inland as Idaho and the headwaters of the Yukon River.

Salmon find their home stream with a remarkable accuracy, the result of a type of chemical memory. They have been found to recognize not only the "smell" of their own stream, but of others they pass along the way. There also is evidence that they respond to chemicals released by other members of their own species. The ability of an animal to find its way back to a home area is known as **homing behavior.**

Salmon spawn on beds of clean gravel in the shallows. The female digs out a shallow nest into which she deposits her eggs. The eggs are fertilized by the male and covered with gravel. After defending the nest for a while, the salmon die.

After hatching, the young salmon may return to sea immediately, as in the pink, or humpback, salmon (*O. gorbuscha*). The young of other species remain in fresh water for a time, as long as 5 years for the sockeye, or red, salmon (*O. nerka*). A race of this last species is landlocked and does not migrate to sea at all.

The Atlantic salmon (*Salmo salar*) breeds on both sides of the north Atlantic. It migrates across the ocean before returning to rivers from New England to Portugal. Atlantic salmon may survive after spawning; some females are known to have made as many as four round trips to their home.

Catadromous fishes feature a migratory pattern opposite that of salmon. They breed at sea and migrate into rivers to grow and mature. Several catadromous fishes are known, but the longest migration of any of them is that of freshwater eels (*Anguilla*). There are at least 16 species, of which the American (*A. rostrata*) and European (*A. anguilla*) eels are the best known.

Both American and European eels spawn in the Sargasso Sea at depths of at least 400 to 700 m (1,300 to 2,300 feet) and then die (Figure 7-26). The eggs hatch into tiny, transparent larvae that gradually develop into elongate, leaf-shaped **leptocephalus larvae.** The larvae of the American eel drift in the plankton for at least a year before metamorphosis. The juveniles then move into rivers along the Atlantic coast of North America. The leptocephalus larvae of the European eel are believed to spend at least an extra 2 or 3 years drifting in the Gulf Stream to reach rivers throughout western Europe. Adults of both species eventually grow to more than 1 m (40 inches) in length. Both juveniles and adults are highly valued as food, particularly in Europe. After 10 to 15 years in fresh water, adults turn silver and their eyes become larger. Soon after this, they head out to sea.

The migration of eels back to the Sargasso Sea is not completely understood. How do they navigate to reach such a distant spot? There is experimental evidence that they use the earth's magnetic field as a cue for navi-

218

Adult

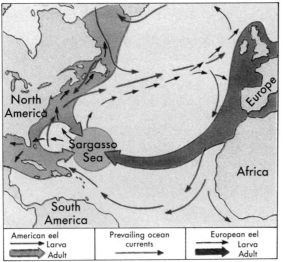

American eel	Prevailing ocean	European eel
→ Larva	currents	→ Larva
⇒ Adult	→	⇒ Adult

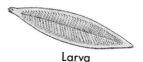

Larva

Figure 7-26 Two species of freshwater eels, the American (*Anguilla rostrata*) and European (*A. anguilla*) eels, breed in the Sargasso Sea and migrate to rivers in North America and Europe. The return trip of larvae is not precisely known, so arrows indicate only the most *likely* routes.

gation. It has been assumed that European eels follow favorable currents that take them first along the coast of northern Africa (see Figure 7-26), much like the Portuguese navigators of old (see Box 3-1).

Some biologists had suggested that the adults of the European eel did not return to the Sargasso Sea at all but died at sea. According to this view, American and European eels were really the same species. Larvae that remained too long at sea simply ended up drifting to Europe. We now know that this is not true. The American and European eels are distinct species. Both spawn at different, though overlapping, periods and locations in the Sargasso Sea.

Reproduction and Life History

Marine fishes have evolved an enormous variety of ways to produce offspring. Reproduction is a complex enterprise involving, in particular, adaptations of the reproductive system and the behavior that bring the two sexes together and ensure successful spawning.

Reproductive System. The reproductive system of fishes is relatively simple. The sexes are usually separate. Both sexes have paired gonads located in the body cavity (see Figure 7-15).

In cartilaginous fishes, ducts lead from the ovaries and testes into the **cloaca,** a common passage for the digestive, excretory, and reproductive systems (see Figure 7-15, A). The cloaca opens to the outside through the anus. Jawless and bony fishes, on the other hand, have a separate opening for urine and gametes, the **urogenital opening,** which is located just behind the anus (see Figure 7-15, B).

In most marine fishes the gonads produce gametes only during certain periods of time. The timing of gamete production is crucial. Both sexes must be ready to spawn at the same time. Spawning, as well as larval development, must take place during the period with the most favorable conditions. The exact timing of reproduction is especially critical for those fishes which make long migrations to breed.

hormones Molecules, pro-teins or lipids, that act as chemical messengers within the body
Chapter 4, p. 88

The timing of reproduction is controlled by **sex hormones.** Sex hormones are produced in the gonads and released in small amounts into the blood. They stimulate the maturation of gametes and may cause changes in color, shape, and behavior before breeding.

The release of sex hormones is triggered by environmental factors such as temperature, light, and the availability of food. Fishes can be artificially induced to spawn when these environmental factors are controlled or when hormones are injected. This discovery is being used by scientists interested in increasing the reproductive potential of fishes grown for food (see the section in Chapter 16, "Mariculture").

hermaphrodites Individu-als that have both male and female gonads
Chapter 6, p. 171

A few marine fishes are normally **hermaphrodites.** Though able to fertilize their own eggs, these fishes usually breed with one or more other individuals, ensuring fertilization between different individuals. Hermaphroditism also is found among several deep-water fishes, an adaptation to the dark depths of the ocean where it may not always be easy to find members of the opposite sex (see the section in Chapter 15, "Sex in the Deep Sea").

A variation of hermaphroditism among fishes is **sex reversal,** in which individuals begin life as males but eventually change into females, or females change into males. These changes are known to be controlled by sex hormones. Sex reversal occurs in several groups of marine fishes, but it is most prevalent among sea basses and groupers (*Serranus, Epinephelus*), parrotfishes, and wrasses. Some rather complicated reproductive strategies have been discovered among these fishes.

In at least some species of anemonefishes, all individuals begin as males. Each sea anemone is inhabited by a single large female who mates only with a large, dominant male. All the other fishes that live on the anemone are small nonbreeding males. If the female disappears or is experimentally removed, her mate changes into a female and the largest of the nonbreeding males becomes the new dominant male. The new female can start spawning as soon as 26 days after her sex change!

Males of some wrasses form "harems" of numerous females. If the male disappears, the largest, dominant female immediately begins to act like a male and within a relatively short period of time changes color and transforms into a male. This last case represents considerable savings in resources because fewer males are needed for reproduction since one can fertilize many females. Another possibility is that subordinate males would have no access to females, so they become females and produce eggs.

Reproductive Behavior. Potential mates must get together at the right time to breed. Many species migrate and congregate in specific breeding grounds, as in the salmon and freshwater eels previously discussed. Sharks are usually "loners" but may come together during the breeding season. Many of them appear to stop feeding at spawning time.

Many fishes change color to advertise their readiness to breed. Most salmon undergo dramatic changes. Both sexes of the sockeye salmon turn from silver to bright red, giving rise to another of its common names, the "red salmon." In male sockeye and pink—or humpback—salmon, the jaws grow into vicious-looking hooks. Males of the latter species also develop a large hump. Color changes also can be observed in tropical fishes. Many male

LIFE IN THE MARINE ENVIRONMENT

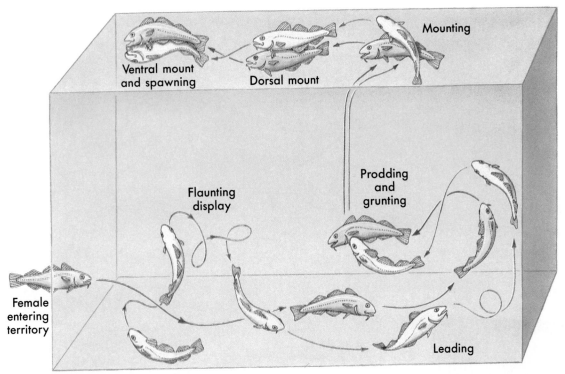

Labels in figure:
Ventral mount and spawning
Dorsal mount
Mounting
Flaunting display
Prodding and grunting
Female entering territory
Leading

wrasses, colorful all the time, appear even more spectacular before breeding.

The first step in reproduction is **courtship,** a series of behaviors that serves to attract mates. These behaviors involve an exchange of such active displays as "dances," special postures that display colors, swimming upside-down, and so forth (Figure 7-27). Each species has its own unique courtship behavior. This is thought to help keep fishes from mistakenly mating with members of the wrong species.

> Reproduction in fishes involves many adaptations that help individuals get together for mating. These include migrations, particular colors as sex signals, and courtship behavior.

Mating is the actual release of gametes leading to the fertilization of eggs by sperm. It can involve **copulation,** the direct transfer of sperm by males into the female, a case of **internal fertilization. External fertilization,** the release of gametes into the water, or **spawning,** is more common in marine fishes, however.

Internal fertilization occurs mainly in cartilaginous fishes. Unfortunately, not much is known about their sex life. Male sharks, rays, and skates have a pair of copulatory organs called **claspers** located along the inner edge of the pelvic fins (Figure 7-28). The typical approach of a romantic male shark consists of biting his potential mate on the back. The male copulates by inserting his claspers into the female's cloaca. He bites and hangs from her or partly coils around her middle. Some skates mate after the male bites the female's pectoral fins. He presses his ventral surface against hers and inserts the claspers.

Figure 7-27 The Atlantic cod (*Gadus morhua*), one of the most important food fishes in the world, spawns in large aggregations in the north Atlantic. The two sexes do not spawn indiscriminately but follow a strictly choreographed series of behaviors in a chain reaction. The male establishes a territory, and the action starts after an interested female enters his territory. Most of the display is made by the male, who uses fully spread fins, grunting sounds, and a series of swimming behaviors. If the female does not follow the male, he has to start all over again and try to attract another female. The spawning of gametes into the water thus climaxes a series of behaviors that includes visual, sound, and tactile signals.

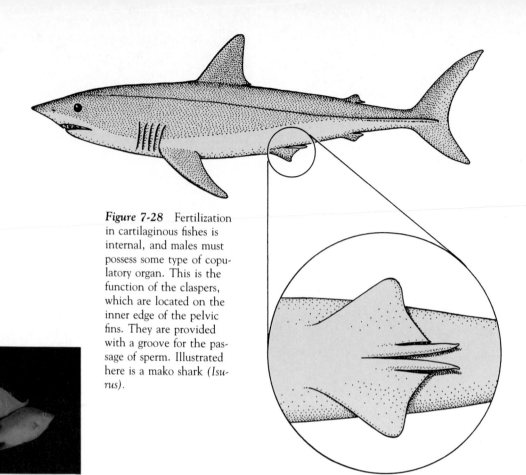

Figure 7-28 Fertilization in cartilaginous fishes is internal, and males must possess some type of copulatory organ. This is the function of the claspers, which are located on the inner edge of the pelvic fins. They are provided with a groove for the passage of sperm. Illustrated here is a mako shark (*Isurus*).

Figure 7-29 A pair of spawning hogfish (*Lachnolaimus maximus*), an Atlantic wrasse. The flashier and slightly larger male is above the female.

Bony fishes exhibit an almost infinite array of ways to spawn. Open-water fishes (sardines, tunas, jacks, and others) and those living around coral reefs and other inshore environments (such as surgeonfishes, parrotfishes, and wrasses) spawn directly into the water after courtship (Figure 7-29). Females typically release large numbers of eggs. In the Atlantic cod (*Gadus morhua*), for example, a female 1 m (40 inches) long can release up to 5 million eggs. The Atlantic tarpon (*Megalops atlanticus*) releases more than 100 million eggs every time it spawns!

Some fishes spawn in pairs, others in groups. Individual males may establish territories or aggregate into groups. Groups of males may be approached by single females or by females in groups. Usually males seek out the females and entice them to spawn via courtship. Two individuals may pair only during spawning time, or may establish long-lasting bonds (Figure 7-30).

Eggs fertilized in the water column drift in currents and develop as part of the plankton. Most eggs contain oil droplets and are buoyant. Other eggs sink to the bottom. Herring eggs stick to the surface of rocks and seaweeds. Lampreys and salmon bury their eggs after spawning. The California grun-

LIFE IN THE MARINE ENVIRONMENT

ion (*Leuresthes tenuis*) buries its eggs on sandy beaches during high tides; they will not hatch until the next high tide (Figure 7-31).

Most of the eggs that are released into the plankton will never survive. Fishes and other marine animals that spawn into the plankton are broadcasters that release as many eggs as possible to ensure that at least some hatch and make it to adulthood. Eggs require much energy to produce since they must contain food for early development.

Fishes that spawn fewer and larger eggs have evolved ways to take care of them. In many damselfishes, males establish and defend breeding sites or nests (Figure 7-32) in holes among rocks or coral, empty mollusc shells, and other shelters—even discarded tires! After spawning, the eggs are retained in the nest and guarded by the males; females leave after spawning. Males are promiscuous and will guard eggs they have fertilized from different females. Nests also are guarded by males in some gobies, blennies, and sculpins. In the Antarctic plunderfish (*Harpagifer bispinis*), the female prepares a breeding site and guards it for 4 to 5 months after spawning. If she disappears or is removed, her job is taken over by another plunderfish, usually a male.

Some fishes go even further and physically carry the eggs after they have been fertilized. Male pipefishes carry the eggs attached in neat rows to their bellies. A male seahorse literally becomes pregnant after his mate deposits eggs in a special pouch on his belly! In some of the cardinalfishes and marine catfishes (*Arius*), males brood the fertilized eggs in their mouths (see Figure 4-27, C).

Early Development. In egg-laying sharks, skates, and other cartilaginous fishes, the embryo is enclosed by a large, leathery egg case (the "mermaid's purses" of skates) that drops to the bottom after spawning (Figure 7-33). The egg cases are rather large (up to 30 cm, or almost 12 inches, in the whale shark) and often have thin extensions that attach them to surfaces. Only a few are spawned at a time. The eggs have a large amount of yolk in a **yolk sac** that is attached to the embryo's belly. Yolk provides energy for several months of development, a long time by fish standards. As a result, the pup is well developed when it finally hatches.

In many cartilaginous fishes the female retains the eggs inside her reproductive tract for additional protection. The eggs develop inside the female, who gives birth to live young. Such fishes are known as **ovoviviparous.** Most ovoviviparous fishes are cartilaginous. Some rockfishes are among the few marine bony fishes that can be classified as ovoviviparous. Most bony

A

B

Figure 7-30 Butterflyfishes, are among the most colorful of coral-reef fishes. Adults of many species occur in male and female pairs. In *Chaetodon trifasciatus*, pairs establish territories around coral colonies **(A)**, while *C. ornatissimus*, pairs patrol larger areas of the reef **(B)**.

Figure 7-31 The California grunion (*Leuresthes tenuis*) migrates to sandy beaches to spawn. It will do so only during the spring and summer high tides—on the second, third, and fourth nights after each full and new moon from March to August. The grunion "runs" begin when thousands of the small fish (length up to 20 cm, or 8 inches) move high up on the beach with the waves. Females burrow tail first in the wet sand and deposit their eggs while males curl around them to fertilize the eggs. They return back to sea with the waves, but the eggs remain buried by sand until hatching 2 weeks later when the next set of high tides wet them.

Figure 7-32 The golden damsel (*Amblyglyphidodon aureus*) and many other damselfishes establish and actively defend breeding sites where females deposit their eggs.

fishes spawn their eggs for external fertilization. These fishes are said to be **oviparous.**

In some ovoviviparous sharks, the embryos rely on other sources of nutrition once they have consumed their yolk. In the sand tiger sharks (*Odontaspis*) only two pups, which are large (up to 1 m or 3.3 feet) and active, are born. Each survived in one of the two branches of its mother's reproductive tract by eating its brothers and sisters. When that source of food is gone, they are known to consume other eggs produced by their mother's ovaries!

Some sharks and rays have embryos that actually absorb nutrients from the walls of the mother's reproductive tract. This is truly remarkable since it is very similar to the development of the embryo in the mammals, the higher vertebrates like us. These sharks are said to be **viviparous.** Not only are they live bearers, but nutrition for development before birth is provided by direct contact with the reproductive tract of the female. The surfperches (*Embiotoca*), which are bony fishes, are also viviparous. Their young have large fins that absorb nutrients from the walls of the mother's ovaries.

Development of the embryo proceeds rather quickly in most bony fishes. The transparent outer envelope of the eggs is thin, allowing oxygen to diffuse through. Eggs are almost always round.

The embryo is supplied with nutrient-rich yolk. After one or more days of development, eggs hatch into free-swimming larvae, or fry. When they first hatch, the larvae still carry the yolk in a yolk sac. The yolk is eventually consumed, and the larvae begin feeding in the plankton. Many larvae, like the leptocephalus larvae of eels (see Figures 7-26, 14-12, and 14-14, *D*), do not resemble their parents at all and undergo metamorphosis to a juvenile stage that resembles the adult. The larvae of flatfishes have eyes on both sides of the head, but during metamorphosis they migrate to one side.

Most marine fishes are oviparous and release eggs into the water. Internal fertilization leads to ovoviviparous or viviparous conditions in some, especially cartilaginous fishes.

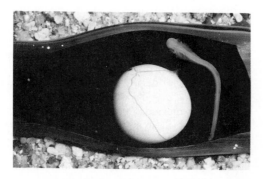

Figure 7-33 An egg of the swell shark (*Cephaloscyllium ventriosum*) containing a 1-month-old embryo.

LIFE IN THE MARINE ENVIRONMENT

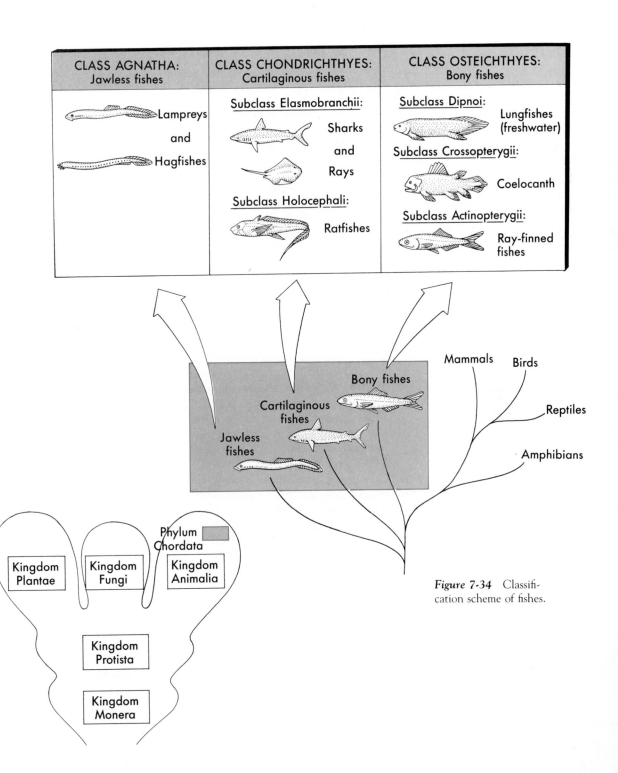

CLASS AGNATHA: Jawless fishes	CLASS CHONDRICHTHYES: Cartilaginous fishes	CLASS OSTEICHTHYES: Bony fishes
Lampreys and Hagfishes	Subclass Elasmobranchii: Sharks and Rays Subclass Holocephali: Ratfishes	Subclass Dipnoi: Lungfishes (freshwater) Subclass Crossopterygii: Coelocanth Subclass Actinopterygii: Ray-finned fishes

Mammals   Birds

Reptiles

Amphibians

Bony fishes

Cartilaginous fishes

Jawless fishes

Phylum Chordata

Kingdom Plantae   Kingdom Fungi   Kingdom Animalia

Kingdom Protista

Kingdom Monera

Figure 7-34 Classification scheme of fishes.

SUMMARY

1. Vertebrates are those members of phylum _____ possessing a backbone.

2. The most primitive of living fishes are the jawless fishes, represented by _____ and _____.

3. Cartilaginous fishes, sharks and related forms, are characterized by a cartilaginous skeleton and exposed _____ visible on the sides or, in skates and rays, on the _____ surface.

4. Most sharks are carnivorous. One exception is the whale shark, a _____ feeder.

5. Bony fishes, in contrast to cartilaginous fishes, have a bony skeleton, an operculum covering the _____, and a _____ that controls buoyancy control.

6. Locomotion in fishes is mostly the result of undulation of the _____ and tail. In bony fishes, the _____ are highly maneuverable, whereas in sharks they have a role in the control of _____.

7. Oxygen enters the blood in the gills of fishes by its _____ from water. The efficiency of oxygenation is increased by water flowing over the gills in the _____ direction to the blood.

8. Marine fishes live in water saltier than their blood and must therefore control the tendency for water _____ the body. Cartilaginous fishes have solved this problem in part by accumulating _____. Bony fishes swallow seawater, and their kidneys and _____ excrete excess salts.

9. Fishes feature a variety of sense organs. One, their _____, detects vibrations in the water.

10. Schooling is a common phenomenon among fishes. Two of the most widely given explanations for this behavior are _____ and _____.

11. Fishes migrate to feed and breed. Some, the _____ fishes, breed at sea but migrate to fresh water, whereas the _____ fishes migrate the opposite way.

12. The eggs of most marine fishes develop free in the water. These fishes are known as _____ since the eggs are spawned. Others, mostly cartilaginous fishes, are called _____ if the eggs develop in the female, or _____ if the developing embryo derives nutrition from the walls of the mother's reproductive tract.

THOUGHT QUESTIONS

1. Hagfishes and lampreys are the only living representatives of a very ancient group. Why do you suppose there are still some of these jawless fishes around?
2. A deep-water shark, new to science, is collected for the first time. The specimen is studied in detail, but its stomach is empty. How could you get a rough idea of its feeding habits? The specimen is a female, and its reproductive tract is found to contain 20 eggs. Can you tell the type of development characteristic of this species?
3. Individuals of some species of bony fishes change sex, some to maintain more males than females, others more females than males. What are the advantages and disadvantages of each situation? Are there any advantages and disadvantages in having an equal number of males and females?

FOR FURTHER READING

Brownlee, S.: "On the Track of the Real Shark," *Discover*, vol. 6, no. 6, July 1985, pp. 26-32. *The fascinating world of sharks and the many unanswered questions about their biology.*

Carter, J.: "Grouper Sex in Belize," *Natural History*, October 1989, pp. 60-69. *Courtship and spawning behavior in the Nassau grouper, a Caribbean fish whose reproductive behavior is timed with the lunar cycle.*

Doubilet, D.: "Scorpionfish: Danger in Disguise," *National Geographic Magazine*, vol. 172, no. 5, November 1987, pp. 634-643. *Superbly photographed survey of the poisonous scorpionfishes of the tropical Pacific, including the stonefish and lionfishes.*

Fricke, H.: "Coelacanths: The Fish that Time Forgot," *National Geographic Magazine*, vol. 173, no. 6, June 1988, pp. 825-838. *A report on an expedition to photograph live coelacanths using a submersible.*

Leggett, W.C., and K.T. Frank: "The Spawning of the Capelin," *Scientific American*, vol. 262, no. 5, May 1990, pp. 102-107. *From eggs that hatch on beaches, the capelin migrates to the open ocean where it is a most important food for fishes, seabirds, and marine mammals.*

Levine, J.S.: "For Fish in Schools Togetherness is the Only Way to Go," *Smithsonian*, vol. 21, no. 4, July 1990, pp. 88-93. *Communication plays a critical role in schooling, a behavior widespread among many groups of fishes.*

Pietsch, T.W., and D.B. Grobecker: "Frogfishes," *Scientific American*, vol. 262, no. 6, June 1990, pp. 96-103. *Frogfishes are carnivores that use mimicry to blend with their surroundings, a lure to attract prey, and a highly specialized jaw to swallow it.*

Sanderson, S.L., and R. Wassersug: "Suspension-Feeding Vertebrates," *Scientific American*, vol. 262, no. 1, January 1990, pp. 96-101. *How animals, including cartilaginous and bony fishes, filter their food. The different types and mechanisms of filter-feeding, as well as its significance, are discussed.*

Van Dyk, J.: "Long Journey of the Pacific Salmon," *National Geographic Society*, vol. 178, no. 1, July 1990, pp. 2-37. *The Pacific salmon as the star of a spectacular migration and victim of many hazards that threaten its existence.*

Webb, P.W.: "Form and Function in Fish Swimming," *Scientific American*, vol. 251, no. 1, July 1984, pp. 72-82. *The mechanics behind the different techniques fishes use to swim.*

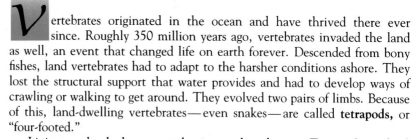

8 Marine Reptiles, Birds, and Mammals

Figure 8-1 A killer whale (*Orcinus orca*) with young.

Vertebrates originated in the ocean and have thrived there ever since. Roughly 350 million years ago, vertebrates invaded the land as well, an event that changed life on earth forever. Descended from bony fishes, land vertebrates had to adapt to the harsher conditions ashore. They lost the structural support that water provides and had to develop ways of crawling or walking to get around. They evolved two pairs of limbs. Because of this, land-dwelling vertebrates—even snakes—are called **tetrapods,** or "four-footed."

Living on land also means having to breathe air. Tetrapods evolved **lungs,** which are internal air sacs that allow absorption of oxygen. Tetrapods also had to evolve ways to keep from drying out. The delicate egg is especially vulnerable, and the first land tetrapods—the **amphibians** (class **Amphibia**),—never really solved this problem. Represented today by frogs, salamanders, and their relatives, amphibians must keep themselves moist and lay their eggs in water. None of them are marine.

Other groups of tetrapods solved the problem of water loss and truly adapted to life on land. **Reptiles** (class **Reptilia**) evolved from amphibians and for a long time were the dominant land vertebrates (see Figure 7-2 and the classification scheme shown in Figure 8-33). The **birds** (class **Aves**) and **mammals** (class **Mammalia**) both evolved from reptiles.

Having adapted to the land, various groups of reptiles, birds, and mammals turned around and re-invaded the ocean. This chapter deals with these marine tetrapods. Some have not fully made the transition and still return to land part of the time. Others, like the killer whale shown in Figure 8-1 spend their entire lives at sea. They have adapted so completely to a marine existence that their streamlined bodies look almost fishlike. This fishlike appearance, however, belies the fact that they evolved from animals that once, about 55 million years ago, walked on land. Their embryos even have the four limbs that characterize all land vertebrates (see Figure 8-17).

The marine animals to be covered in this chapter include some of the most fascinating and awesome creatures on the planet. Unfortunately, many are in danger of disappearing forever because of our own greed. Some already have.

MARINE REPTILES

There are now close to 6,000 living species of reptiles, including lizards, snakes, turtles, and crocodiles. Their dry skin is covered with scales to prevent water loss. The egg has a leathery shell that prevents it from drying out so that reptiles can lay their eggs on land. Like fishes, reptiles are **ecto-therms, (or poikilotherms)**, incorrectly called "cold-blooded." As in other ectotherms, their metabolic rate—and therefore activity level— varies with temperature: They get sluggish in the cold. This tends to keep them out of cold regions, especially on land because the air temperature fluctuates more widely than does the ocean temperature.

ectotherms (or poikilo-therms) Those organisms having a body temperature that varies with that of the environment
Chapter 4, p. 102

Reptiles are air-breathing, "cold-blooded" vertebrates. Their skin is covered with dry scales and they lay their eggs on land.

Since reptiles first appeared more than 300 million years ago, several different groups have invaded the seas. Many are long gone, like the seagoing ichthyosaur (see Figure 8-15) that thrived during the so-called Age of Reptiles. Only a few reptiles still roam the seas. Some are rare and endangered; others are common but have a restricted distribution.

Sea Turtles

Sea turtles belong to an ancient group of reptiles. Their bodies are enclosed by an armorlike shell, or carapace, that is fused to the backbone. Unlike land tortoises and turtles, sea turtles cannot retract their heads into the shell. Their legs, particularly the larger first pair, are modified into flippers for swimming. They usually leave the water only to lay eggs.

There are only seven species of sea turtles, which live primarily in warm water. Green turtles (*Chelonia mydas*; Figures 8-2, A; 8-10, E; and 4-21, A) were once found in coastal waters throughout the tropics. Their shell may grow to 1 m (40 inches) in length. They feed mostly on seagrasses and seaweeds. Like all turtles, green turtles lack teeth, but they have strong biting jaws. The hawksbill turtle (*Eretmochelys imbricata*; Figure 8-2, B) is smaller, and the shell is reddish brown with yellow streaks. It uses its beaklike mouth to feed on encrusting animals (sponges, sea squirts, barnacles) and seaweeds.

The largest sea turtle is the leatherback (*Dermocheiys coriacea*; Figure 8-2, C). Individuals may attain a length of 2.1 m (7 feet) and weigh at least 540 kg (1,200 pounds). Instead of a solid shell, they have a series of small bones buried in the dark skin, forming distinct longitudinal ridges. Leatherbacks are an open-water species and are rarely seen by humans except in nesting beaches. Their diet consists largely of jellyfishes.

All sea turtles must return to land to reproduce. They migrate long distances to lay their eggs on remote sandy beaches, which they have been doing for millions of years before humans appeared on the scene. Green turtles still gather to nest on beaches on the east coast of Central America, northern Australia, Ascension Island (in the middle of the south Atlantic), and a few other locations. Marine biologists have tagged adult sea turtles at As-

A

B

C

Figure 8-2 **(A)** The green turtle (*Chelonia mydas*) gets its name not from the color of its shell, but because of its greenish fat. **(B)** The hawksbill turtle (*Eretmochelys imbricata*) takes its name from the shape of its jaw. It is the source of tortoiseshell. **(C)** The largest of all sea turtles, the leatherback turtle (*Dermochelys coriacea*), may venture into cold waters as far north as Newfoundland and Alaska.

cension and have found that the turtles regularly cross 2,200 km (1,360 miles) of open water to their feeding grounds along the coast of Brazil, a journey that takes a little more than 2 months (see world map in Appendix C). Though we are still not sure how they find their way, evidence suggests that they do it by sensing wave motion and the earth's magnetic field.

Most of what we know about the reproduction of sea turtles is based on the green turtle. They return to their nesting areas every 2 to 4 years, often against prevailing currents. Evidence that females return to the beaches where they were born has been obtained by analyzing the DNA of breeding populations in separate Caribbean and Atlantic sites. The DNA of turtles breeding in one area differs from the DNA of turtles breeding in other sites. It thus appears that turtles keep returning to the same place generation after generation.

Copulating pairs are often seen offshore, but only females venture ashore—usually at night. Therefore biologists have tagged mostly females, since they can most easily be tagged on land. The females congregate on the beach, and each proceeds to excavate a hole in the sand, using both pairs of flippers (Figure 8-3). The large, leathery eggs are laid in this nest. More than 100 eggs, and as many as 160, are laid. The female covers the eggs with sand before she returns to the sea. She may make several trips ashore during the breeding season, laying eggs each time.

The eggs hatch after about 60 days of incubation in the sand. The tiny baby turtles must then dig themselves out of the sand and crawl all the way back to the water, protected by darkness if they're lucky. Green turtles and other sea turtles have many enemies. The eggs are often eaten by ghost crabs, wild pigs, and other animals. The hatchlings are easy prey for land crabs and birds, especially during the day. Even more young turtles are lost in the water, where they are taken by a variety of fishes and sea birds.

To this list of natural enemies we must add the most destructive of all—humans. Many nesting areas have been turned into resorts. Turtles drown in fishing nets, especially drift nets (see the section in Chapter 17, "Other Marine Species Risking Extermination"), and choke to death after swallowing plastic bags they think are jellyfish. Turtles have been used as food for centuries. Their eggs are taken by the bucketful. They are located by pushing a stick into the sand until it comes out yellow. The eggs are eaten or fed to pigs or cattle. Leatherback eggs are said to be an aphrodisiac, a myth that probably arose because adult leatherbacks are seen copulating for long periods at sea.

Sea turtles can live for months without food or water. In the days before refrigeration, sailors kept them alive aboard ship as a source of fresh meat, storing them on their backs for months. Coming ashore by the thousands, females were an easy catch. They were, and still are, immobilized by turning them on their backs to be gathered later without giving them a chance to lay their eggs. The green turtle is especially esteemed for its meat, and its cartilage is used to make turtle soup. Some people consider the oily leatherback meat a delicacy.

The polished shell of the hawksbill is the source of valuable **tortoiseshell** used to make jewelry, combs, and other articles, particularly in Japan.

DNA A complex molecule that contains the cell's genetic information *Chapter 4, p. 88*

Figure 8-3 Nest building in green turtles culminates a long and hazardous trip by females. It is at this time that they are most vulnerable to human poachers. This photograph was taken on Sipadan Island, one of the Turtle Islands off the northeastern coast of Borneo, Malaysia.

The oil and leather of many sea turtles are also of commercial value (see Figure 17-22). Even baby sea turtles are valuable: They are stuffed and sold as souvenirs.

The green turtle, once very common, has disappeared in many areas as the result of relentless overexploitation for eggs and meat. It is the most widely distributed and most common of all sea turtles, but only an estimated half a million individuals are left world-wide. All seven species of sea turtle are endangered, the leatherback, hawksbill, and Kemp's—or Atlantic—ridley (*Lepidochelys kempi*) critically so. Shrimp nets in the Gulf of Mexico have been especially deadly to the Kemp's ridley turtle, once very common but now so rare that only a few hundred breeding females remain. It is the most endangered of all sea turtles. After a lengthy struggle, the U.S. government mandated that shrimp nets be fitted with turtle exclusion devices—or TEDs—that allow sea turtles to escape once caught in the nets. Though sea turtles are protected by many countries, enforcement is difficult. It is impossible to protect all coasts and nesting grounds from fishermen and egg hunters. Stricter world-wide enforcement of conservation practices, the control of pollution, the regulation of trade in sea turtle products, and the restocking of former nesting areas might help save them.

Sea Snakes

Approximately 55 species of **sea snakes** are found in the tropical Indian and Pacific oceans (Figure 8-4). Their body is flattened, and the tail paddle-shaped for swimming. Most are 1 to 1.3 m (3 to 4 feet) long. Practically all sea snakes lead a totally marine existence. They are **ovoviviparous** and do not need to return to land to breed. A few species, however, still come ashore to lay their eggs.

Sea snakes are carnivores. Most feed at the bottom on fish, a few specializing in fish eggs. They are closely related to cobras and their allies, the most venomous of all snakes. Sea snakes are among the most common of all venomous snakes, and their bites can be fatal to humans. Fortunately, they are rarely aggressive, and their mouth is too small to get a good bite. Most casualties have been reported in Southeast Asia—swimmers accidentally stepping on them and fishermen removing them from nets. Sea snakes are also victims of overexploitation. They are hunted for their skins, and some species have become rare.

Other Marine Reptiles

An unusual lizard is among the unique inhabitants of the Galápagos Islands, which lie off the Pacific coast of South America. The **marine iguana** (*Amblyrhynchus cristatus;* Figure 8-5) spends most of its time basking in large groups on rocks along the coast, warming up after swimming in the cold water. It eats seaweeds and can dive as deep as 10 m (33 feet) to graze.

The one remaining marine reptile to be discussed is the **saltwater crocodile** (*Crocodylus porosus;* see Figure 16-21), which inhabits mangrove swamps and estuaries in the eastern Indian Ocean, Australia, and some of

Figure 8-4 Sea snakes are found from the Indian Ocean coast of South Africa to the Pacific coast of tropical America. They sometimes occur under floating debris, feeding on the fish it attracts. The conspicuous coloration of sea snakes may be a warning to potential predators, since many fishes learn to associate the bright colors with danger. There are no sea snakes in the Atlantic, but a sea-level canal across Central America may allow their migration into the Caribbean.

ovoviviparous animals
Livebearing animals in which the eggs develop in the reproductive tract of females
Chapter 7, p. 223

the western Pacific islands. They live mostly on the coast but are known to venture into the open sea. There is a record of an individual 10 m (33 feet) long, but they are rarely over 6 m (20 feet). They are among the most aggressive of all marine animals and are known to attack and eat people. In some areas they are more feared than sharks.

Marine reptiles include the sea turtles, sea snakes, the marine iguana, and the saltwater crocodile.

SEABIRDS

Birds have some significant advantages over reptiles, including the ability to fly. Birds are **endotherms** (or **homeotherms**), also known as "warm-blooded." This has allowed them to live in a wide variety of environments, unlike reptiles. Their bodies are covered with waterproof feathers that help conserve body heat. Waterproofing is provided by oil from a gland above the base of the tail. The birds rub the oil into their feathers with their beaks. Flight is made easier by their light, hollow bones. Furthermore, their eggs are more resistant to water loss than those of reptiles.

Birds are "warm-blooded" vertebrates that have feathers and light bones as adaptations to fly.

Seabirds are those birds which spend a significant part of their lives at sea. Seabirds nest on land but feed at least partially at sea and have webbed feet to aid them in the water. Seabirds descended from several different groups of land birds. As a result, they differ widely in their flying skills, feeding mechanisms, and ability to live away from land.

Seabirds are birds that nest on land but feed entirely or partially at sea.

Though comprising only about 3% of the estimated 8,600 species of birds, seabirds are distributed from pole to pole, and their impact on marine life is significant. Most are predators of fish, squid, and bottom invertebrates, but some feed on plankton. Seabirds have amazing appetites. They need much food to supply the energy required to maintain their body temperatures.

Penguins

Penguins are the seabirds most fully adapted for life at sea. They are flightless, with wings modified as stubby "flippers" (Figure 8-6, A) that come alive underwater. Their bones are heavier than those of other birds to reduce buoyancy and make diving easier.

Penguins are spectacular swimmers, propelling their streamlined bodies with powerful strokes of the wings (see Figure 8-15). They can also jump out of the water and sometimes cover long distances by alternately swimming and jumping. On land it is another story—they are clumsy and awkward, looking like drunk little men in coattails.

Penguins are also adapted for cold temperatures. All but one of the 18

endotherms (or homeotherms) Those animals able to keep their body temperature more or less constant regardless of the temperature of the environment
Chapter 4, p. 102

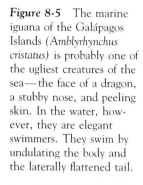

Figure 8-5 The marine iguana of the Galápagos Islands (*Amblyrhynchus cristatus*) is probably one of the ugliest creatures of the sea—the face of a dragon, a stubby nose, and peeling skin. In the water, however, they are elegant swimmers. They swim by undulating the body and the laterally flattened tail.

A

B

C

D

Figure 8-6 Seabirds. **(A)** Emperor penguin *(Aptenodytes forsteri)* and chick. **(B)** A breeding colony of the macaroni penguin *(Eudyptes chrysolophus)* in Antarctica. **(C)** A royal albatross *(Diomedea epomophora)* flying over a breeding colony of cormorants *(Phalacrocorax)* in New Zealand. **(D)** Brown boobies *(Sula leucogaster)*, here with chick, nest in the Caribbean and Gulf of California. They are regular visitors to the Gulf of Mexico. **(E)** The Heermann's gull *(Larus heermanni)* in its nesting site at Isla de la Raza, a small island in the Gulf of California.

E

species of penguins live in Antarctica and other cold and desolate regions of the Southern Hemisphere. The exception is the Galápagos penguin *(Spheniscus mendiculus)*, which lives right on the Equator. Even so, this penguin is confined to regions that are bathed by cold currents. Protection against low temperatures is provided by a layer of fat under the skin, and cold air is also trapped by the dense, waterproof feathers. This air, warmed by body heat, protects from the cold like a down coat.

The larger penguins, like the imposing emperor penguin *(Aptenodytes forsteri)*, hunt for fish and squid. The Adélie *(Pygoscelis adeliae)* and other small penguins feed mostly on krill. Penguins have strong beaks, a characteristic of seabirds that feed on fish and large plankton like krill (Figure 8-7, B). Some species migrate seasonally between feeding grounds at sea and nesting areas on land or ice. They establish breeding colonies (Figure 8-6, B), which in Adélies may number more than a million pairs.

Breeding season and number of eggs laid vary from species to species. Emperor penguins form life-long pairs. The male incubates a single large egg during the dark Antarctic winter. The female leaves to feed as soon as she lays the egg. The male, standing on ice, must keep the egg warm by holding it between his feet and against his body for 64 days of dreadful winter storms and cold.

You may wonder why the penguins lay their eggs at the coldest time of the year. Reproduction is timed so that the egg hatches during the productive Antarctic summer, when food is most plentiful. When the egg hatches, the female finally returns and regurgitates food for the fuzzy chick. After that, both parents take turns feeding the chick. While the parents feed, the fast-growing young are herded into groups guarded by a few adult "babysitters." Returning parents identify their chick among thousands by its voice and appearance. The parents continue to feed the chick for 5 ½ months, until it is strong enough to feed itself at sea.

krill Planktonic, shrimp-like crustaceans
Chapter 6, p. 174; Figure 6-39

Figure 8-7 The beak's shape is related to the kind of food and the feeding style of seabirds. Beaks are relatively short, heavy, and hooked in tubenoses such as petrels (*Pterodroma*), an ideal shape for holding and tearing prey that is too big to be swallowed whole **(A).** Such a beak is best suited for shallow feeding, since its size and shape interfere with fast pursuit underwater. Beaks are heavy but more streamlined in the penguin (*Aptenodytes* and others), the razorbill (*Alca*), and other seabirds that dive deeper to feed on crustaceans and other slow prey **(B).** Boobies (*Sula*), terns (*Sterna*), and other plunge divers have straight and narrow beaks for feeding on fish that are swallowed whole **(C).** Skimmers (*Rynchops*) are the only birds with a lower mandible that is longer than the upper **(D),** which permits feeding while flying. Shorebirds that feed on mud flats have long, thin beaks that allow them to get to prey buried in the mud (see Figure 11-15, A).

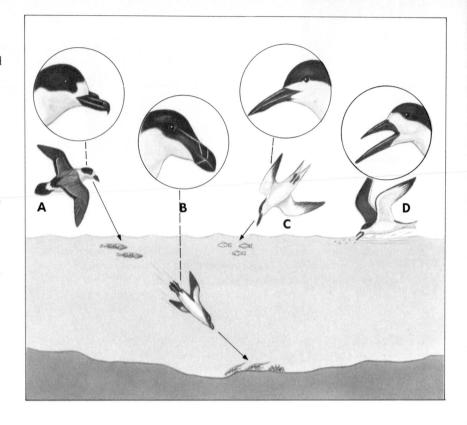

Tubenoses

The **tubenoses** comprise a large group of seabirds with distinctive tubelike nostrils and heavy beaks that are usually curved at the tip (Figure 8-7, A). They spend months and even years in the open sea. Like other seabirds and sea turtles, they have **salt glands** that get rid of excess salts; these empty into the nostrils. Tubenoses include the albatrosses (*Diomedea*), shearwaters (*Puffinus*), and petrels (*Oceanodroma*).

Tubenoses are very skillful fliers. Most catch fish at the sea surface (see Figure 8-7, A), though some scavenge on dead birds or whales. The whalebirds, or prions (*Pachyptila*), feed on krill and other plankton. Albatrosses are magnificent gliders with huge wings they hardly ever seem to flap. Wandering albatrosses (*D. exulans*) and royal albatrosses (*D. epomophora*; see Figure 8-6, C) have wing spans of up to 3.4 m (11 feet), the longest of any bird alive.

Male and female tubenoses remain faithful to each other and perform elaborate courtship and greeting behaviors. Most nest on remote islands. Incubation and care of the single chick takes 8 months, and even longer in some species. Tubenoses make some of the most spectacular migrations of any animal. Many breed on islands around Antarctica, then migrate across the open ocean to summer feeding grounds near the Arctic. The wandering albatross gets its name from the fact that it spends 2 years or more traveling around the Southern Hemisphere before returning to nesting sites near Antarctica. Some nonbreeding individuals wander off and pay visits as far away as California and the Mediterranean!

Pelicans and Allies

Several quite different-looking seabirds are grouped together because they have webbing between all four toes. They are relatively large fish-eaters of wide distribution.

Pelicans (*Pelecanus*) have a unique pouch below their large beaks. Some species, like the brown pelican (*P. occidentalis*), catch their food by plunging into the water and catching fish in the pouch (Figure 8-8). The brown pelican was once common along the coasts of the United States but was decimated by pesticide pollution (see the section in Chapter 17, "Synthetic Chemicals"). It is now making a comeback (see Figure 17-8). **Cormorants** (*Phalacrocorax*) are black, long-necked seabirds that dive and pursue their prey. They can be easily identified by their low flights over water and because they float low in the water, with only the neck above the surface. **Frigate birds** (*Fregata*) have narrow wings and a long, forked tail. They soar majestically along the coast, forcing other seabirds to regurgitate fish in midair or catching prey from the surface (see Figure 8-8). These agile pirates never enter the water, not even to rest, since their feathers are not very waterproof.

Pelicans and their kin nest in large colonies along the coast. They build messy nests of twigs and anything else they can find. The excrement of mil-

Figure 8-8 Feeding strategies vary widely among seabirds. Pelicans (*Pelecanus*) and boobies (*Sula*) plunge into the water, jaegers (*Stercorarius*) pursue other seabirds and force them to regurgitate food, and frigate birds (*Sterna*) take fish from the surface and steal fish from other seabirds. Gulls (*Larus*) rarely dive from the air, and storm petrels (*Oceanodroma*) simply flutter over the waves. Divers such as cormorants (*Phalacrocorax*) pursue prey using their wings or feet. Mud-flat shorebirds also follow various strategies (see Figure 11-15, B).

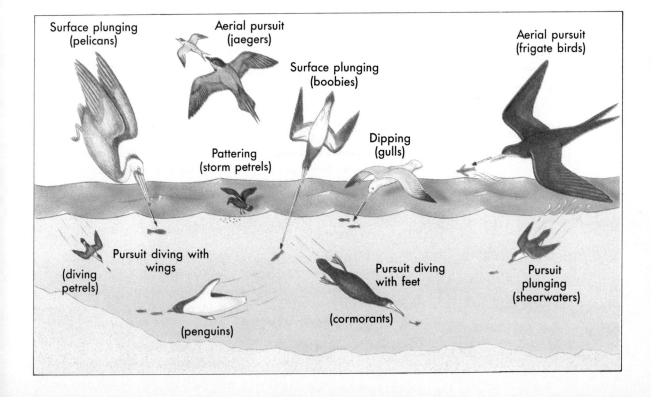

lions of boobies (Figure 8-6, *D*), cormorants, pelicans, and other seabirds accumulates as **guano.** Guano deposits are particularly thick in dry coastal regions and islands near very productive waters, such as the coasts of Perú, Chile, and southwest Africa, where they are mined for fertilizer (see Box 16-1).

Gulls and Allies

Gulls *(Larus;* Figure 8-6, *E)* and their kin make up the largest variety of seabirds. Common and widespread, gulls are predators and scavengers happy to eat just about anything (see Figure 8-8). They are very successful in the company of humans and congregate near piers, garbage dumps, or anywhere else we throw refuse. Jaegers *(Stercorarius)* and skuas *(Catharacta)* are gull-like predators that steal fish from other birds (see Figure 8-8). They nest near the rookeries of penguins and other seabirds and eat their eggs and young.

Terns *(Sterna)* are graceful flyers that hover over their prey before plunging for it. Their slender beaks are specialized to catch small fish, which they swallow whole (Figure 8-7, *C).* The arctic tern *(S. paradisaea)* is another amazing wanderer. It breeds in the Arctic during the northern summer, travels 16,000 km (10,000 miles) to Antarctica for the southern summer, and then returns to the Arctic.

Also related to gulls are several cold-water diving seabirds. Puffins *(Fratercula)* have heavy beaks that make them look like misplaced parrots. The related razorbill *(Alca torda)* is a black and white bird reminiscent of penguins (see Figure 8-7, *B).* In fact, these birds may fill the role of penguins, which are absent in the Northern Hemisphere. Like penguins, they use their wings to swim underwater. Their extinct cousin, the great auk *(Pinguinus impennis),* looked and acted like a penguin. Great auks once lived in great numbers in the north Atlantic but were slaughtered for their eggs and feathers. The last great auk died in 1844.

Shorebirds

Usually included among the seabirds are numerous species of wading **shorebirds.** Since they do not swim much, they are not really seabirds in the strict sense. Many live in inland waters, as well as the sea. Some are common in estuaries and coastal marshes. Plovers, sandpipers, and similar birds are related to gulls. Many other shore birds may live on the coast: rails, coots, herons, egrets, and even ducks. The distribution and significance of shorebirds in estuaries will be discussed in Chapter 11.

MARINE MAMMALS

About 200 million years ago another major group of air-breathing vertebrates, the **mammals** (class **Mammalia**), evolved from the reptiles (see Figure 7-2 and the classification scheme shown in Figure 8-33). For a long time the mammals were overshadowed by the dinosaurs, which were reptiles. About 65 million years ago, however, the dinosaurs disappeared. It was then that mammals thrived by taking the place of the dinosaurs. There are now roughly 4,500 species of mammals, including humans. Fishes, reptiles, and birds all outnumber mammals in number of species.

Like birds, mammals have the advantage of being endotherms, or "warm-blooded." The skin of mammals, however, has hair instead of feathers to retain body heat. With few exceptions, mammals are **viviparous.** The embryo receives nutrients and oxygen through the **placenta,** a membrane that connects it with the womb. It is also known as the "afterbirth." The newborn is fed by milk secreted by the mother's **mammary glands.** Instead of releasing millions of eggs, mammals produce few—but well-cared-for—young.

And then there is the brain—it is larger in relation to body size and far more complex than that of other vertebrates, allowing the storage and processing of more information. This accounts in part for the amazing adaptability of mammals. They live anywhere there is air to breathe and food to eat. This, of course, includes the ocean.

viviparous animals   Live-bearing animals whose embryos develop within their mothers' bodies and are nourished by the maternal bloodstream
Chapter 7, p. 224

Mammals are air-breathing, warm-blooded vertebrates that have hair and feed their young with milk.

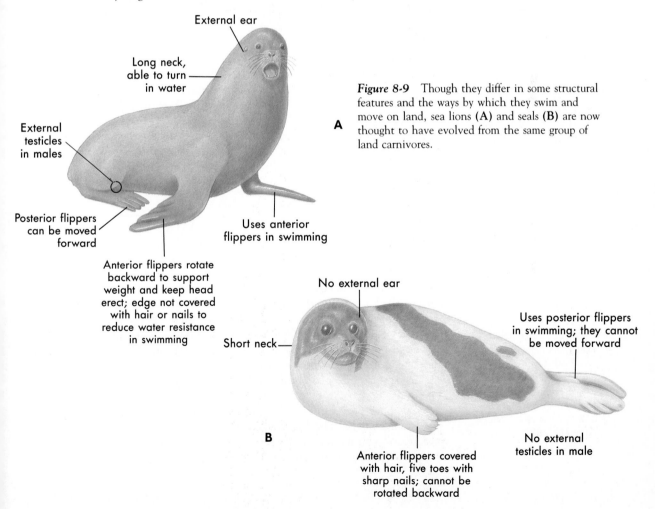

External ear

Long neck, able to turn in water

External testicles in males

Posterior flippers can be moved forward

Anterior flippers rotate backward to support weight and keep head erect; edge not covered with hair or nails to reduce water resistance in swimming

Uses anterior flippers in swimming

A

B

No external ear

Short neck

Uses posterior flippers in swimming; they cannot be moved forward

Anterior flippers covered with hair, five toes with sharp nails; cannot be rotated backward

No external testicles in male

Figure 8-9   Though they differ in some structural features and the ways by which they swim and move on land, sea lions (**A**) and seals (**B**) are now thought to have evolved from the same group of land carnivores.

A   **B**

D

E

Figure 8-10 **(A)** The northern elephant seal (*Mirounga angustirostris*). **(B)** A rare Hawaiian monk seal (*Monachus schauinslandi*) basking together with a green turtle. **(C)** The New Zealand fur seal (*Arctocephalus forsteri*), is characterized by a thick underfur. **(D)** Female harp seals (*Phoca groenlandica*) give birth to white, furry pups on the floating Arctic ice. **(E)** The crabeater seal (*Lobodon carcinophagus*) does not eat crabs, but Antarctic krill.

Types of Marine Mammals

There is something fascinating about mammals that live at sea like fishes. Four or five different groups of land mammals succeeded in invading the oceans. They have followed different paths in adapting to the marine environment. Some are so fishlike that we have to remind ourselves that they have hair and bear live young nourished by their mother's milk.

Seals, Sea Lions, and Walruses. **Seals** and other related forms are marine mammals that have paddle-shaped flippers for swimming but still need to rest and breed on land. They make up one of the 19 or 20 major groups, or orders, of mammals—the **pinnipeds** (order **Pinnipedia;** see the classification scheme shown in Figure 8-33). Pinnipeds evolved from an early form of terrestrial carnivore (order Carnivora)—the cats, dogs, bears, and their kin. The similarities are so close that many scientists classify them with the carnivores. Pinnipeds are predators, feeding mostly on fish and squid. Their streamlined bodies help them swim (Figure 8-9).

Most pinnipeds live in cold water. To keep warm they have a thick layer of fat under their skin called **blubber** (see Figure 4-4). Besides acting as insulation, it serves as a food reserve and it helps provide buoyancy. Pinnipeds also have bristly hair for added protection against the cold. Many of them are quite large, which also helps conserve body heat because large animals have less surface area for their size than small animals, and therefore lose less body heat (see Figure 4-23).

> Pinnipeds, which comprise seals and allies, are marine mammals with flippers and blubber that need to breed on land.

The largest group of pinnipeds, some 19 species, is the seals. Seals are distinguished by having rear flippers that cannot be moved forward (Figure 8-9, B). On land they must move by pulling themselves along with their front flippers. They swim by powerful strokes of the rear flippers.

Harbor seals (*Phoca vitulina*) (see Figure 8-9, B) are common in both the north Atlantic and Pacific oceans. Elephant seals (*Mirounga*, Figure 8-10, A), are the largest pinnipeds. Males, or bulls, reach 6 m (20 feet) in length; they can weigh as much as 3,600 kg (4 tons). Unlike most seals, monk seals (*Monachus*) live in warm regions. The Mediterranean (*M. monachus*) and Hawaiian (*M. schauinslandi;* Figure 8-10, B) monk seals are now very rare. A third species, the Caribbean monk seal (*M. tropicalis*), was last seen in 1952.

Seals have been hunted for their skin, meat, and for the oil extracted from their blubber. The Marine Mammal Protection Act of 1972 extends protection to all marine mammals and restricts the sale of their products in the United States. For some seals, this protection has not been enough.

Sea lions, or eared seals, are similar to seals, except that—unlike seals—they have external ears (Figure 8-9, A). They can also move their rear flippers forward, so they can use all four limbs to walk or run on land. The front flippers can be rotated backward to support the body, permitting the animal to sit on land with its neck and head raised. Sea lions are graceful and agile swimmers, relying mostly on their broad front flippers. Adult males are much bigger than females, or cows, and have a massive head with a hairy mane (see Figure 8-30, A). The head of sea lions looks somewhat doglike, whereas the head of seals has much softer outlines, making it look more like a cat.

There are five species of sea lions, plus nine species of the related **fur seals.** The most familiar of all is the California sea lion (*Zalophus californianus;* Figure 8-25) of the Pacific coast of North America and the Galápagos Islands. They are the trained barking "seals" that do tricks for a fish or two. Fur seals, like the northern fur seal (*Callorhinus ursinus*), were once almost exterminated for their thick fur (Figure 8-10, C). They are now mostly protected around the world, though some species are still hunted. Sea lions were luckier since they lack the underfur of their cousins. Still, both sea lions and fur seals may run afoul of fishermen. They sometimes drown in nets and are accused of stealing fish.

The **walrus** (*Odobenus rosmarus;* Figure 8-11) is a large pinniped with a pair of distinctive tusks protruding down from the mouth. It feeds mostly on bottom invertebrates, particularly clams. It was once thought that the walrus used its tusks to dig up food, but there is no evidence for this. Instead, these pinnipeds apparently move along the bottom sucking up their food. The stiff whiskers of the snout probably act as feelers.

Sea Otters. While there is doubt about the pinnipeds, the **sea otter** (*Enhydra lutris;* Figure 8-12) is definitely a member of the order **Carnivora.** Sea otters are the only marine members of this large order, though the **polar bear** (*Ursus maritimus*) can be considered marine (Figure 8-13). The sea otter is the smallest marine mammal; an average male weighs 25 to 35 kg (60 to 80 pounds). It also differs from other marine mammals by lacking a layer of blubber. Insulation from the cold is provided by air trapped in its dense fur. This splendid, dark brown fur unfortunately attracted hunters. Sea otters were slaughtered to near extinction until they became protected by an international agreement in 1911. The sea otter was then able to slowly expand from the few individuals that had managed to survive in some remote locations.

Sea otters are playful and intelligent animals. They spend all or most of their time in the water, including breeding and giving birth. The furry pup is constantly groomed and nursed by its mother. Sea otters require 7 to 9 kg (15 to 20 pounds) of food every day, so they spend a lot of time looking for it. They satisfy their ravenous appetites with sea urchins, abalone, mussels, crabs, other invertebrates, and even fish. They live in or around kelp beds in the Aleutian Islands and central California, where food is plentiful (see the section in Chapter 12, "Kelp Communities").

Manatees and Dugongs. It is hard to believe that relatives of the elephant live at sea. **Manatees** and the **dugong** are also known as **sea cows,** or

Figure 8-11 Walruses (*Odobenus rosmarus*) typically inhabit the edge of pack ice in the Arctic. They migrate as far south as the Aleutian Islands and Hudson Bay, Canada. They also crowd onto beaches on isolated islands that they use as resting places.

Figure 8-12 Sea otters (*Enhydra lutris*) are remarkable for their use of a tool—a rock for crushing shells. The sea otter floats on its back at the surface, places a flat rock on its chest, and crushes its toughest prey against it.

Figure 8-13 Polar bears (*Ursus maritimus*) are semi-aquatic animals that spend a good part of their life on drifting Arctic ice. They feed primarily on seals, which they stalk and capture as the seals surface to breathe or rest.

Figure 8-14 It has been estimated that approximately 1,000 West Indian manatees (*Trichechus manatus*) remain along the coasts and rivers of Florida. Some concentrate on the warm-water effluents of power plants. They are strictly protected, but collisions with boats take their toll. Manatees have been considered as a possible way to control weeds that sometimes block waterways. Some people have suggested raising manatees for food.

sirenians (order **Sirenia**). They have a pair of front flippers but no rear limbs (Figure 8-14). They swim with up-and-down strokes of the paddle-shaped, horizontal tail. The round, tapered body is well padded with blubber. They have wrinkled skin with a few scattered hairs. Not a pretty sight, even though the group is named after the sea nymphs or mermaids (*sirenas* in Spanish) whose songs drove sailors crazy!

Sirenians are gentle, peaceful creatures that love to be tickled. They usually live in groups. They are the only strict vegetarians among marine mammals. Their large lips are used to feed on seagrasses and other aquatic vegetation. All sirenians are large. Dugongs may reach 3 m (10 feet) in length and 420 kg (930 pounds) in weight. Manatees reach 4.5 m (almost 15 feet) and 600 kg (1,320 pounds) in weight. The largest sirenian of all was the now extinct Steller's sea cow, which supposedly grew to 7.5 m (25 feet) long (see Figure 17-16).

Humans have exploited sirenians for their meat (which supposedly tastes like veal), skin, and oil-rich blubber. Like elephants and other large mammals, they reproduce slowly, typically one calf every 3 years. Only four species remain, and all are in danger of extinction. Three species of manatees (*Trichechus*) live in the Atlantic Ocean. One is restricted to the Amazon; the others inhabit shallow coastal waters and rivers from Florida to West Africa. The dugong (*Dugong dugon*) is strictly marine and survives from East Africa to some of the western Pacific islands. Their numbers are critically low throughout most of their range.

Whales, Dolphins, and Porpoises. The largest group of marine mammals is the **cetaceans** (order **Cetacea**), the **whales, dolphins,** and **porpoises.** No group of marine animals has captured our imaginations like the smiling dolphins and Leviathan the whale. They have inspired countless legends and works of art and literature (see the section in Chapter 18, "Oceans and Cultures"). The rescue of whales stranded on a beach or the birth of a killer whale in an oceanarium bring out strong emotions in all of us.

Of all marine mammals, the cetaceans have made the most complete transition to aquatic life. While most other marine mammals return to land at least part of the time, cetaceans spend their entire lives in the water. Their bodies are streamlined and look remarkably fishlike (Figure 8-15). This is a dramatic example of **convergent evolution,** where different species develop similar structures because they have similar lifestyles. Though they superficially resemble fishes, cetaceans breathe air and will drown if trapped below the surface. They are "warm-blooded," have hair (though scanty), and produce milk for their young.

Cetaceans have a pair of front flippers (Figure 8-16), but the rear pair of limbs has disappeared. Actually, the rear limbs are present in the embryo but fail to develop (Figure 8-17). In adults they remain only as small, useless bones. Like fishes, many cetaceans have a dorsal fin. The muscular tail ends in a pair of finlike, horizontal **flukes.** Blubber provides insulation and buoyancy; body hair is practically absent. Cetacean nostrils differ from those of other mammals. Rather than being on the front of the head, they are on top, forming a single or double opening called the **blowhole** (see Figure 8-16).

LIFE IN THE MARINE ENVIRONMENT

Shark

Dolphin

Ichthyosaur

Penguin

Figure 8-15 Streamlining to reduce water resistance evolved independently in different groups of fast-swimming marine animals: sharks, dolphins, the ichthyosaur (a reptile that became extinct about 65 million years ago), and penguins. Notice that dolphins lack posterior fins and that their flukes are horizontal, not vertical as in fins (see also Figure 8-21).

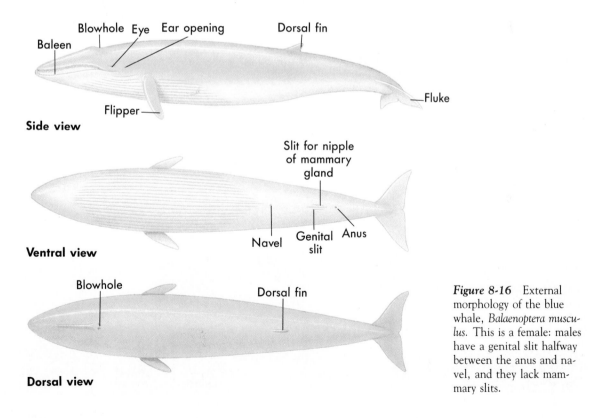

Baleen

Blowhole Eye Ear opening

Dorsal fin

Flipper

Fluke

Side view

Slit for nipple of mammary gland

Navel Genital slit Anus

Ventral view

Blowhole

Dorsal fin

Dorsal view

Figure 8-16 External morphology of the blue whale, *Balaenoptera musculus*. This is a female: males have a genital slit halfway between the anus and navel, and they lack mammary slits.

There are around 90 species of cetaceans. They are all marine except for five species of freshwater dolphins. Cetaceans are divided into two groups: (1) the toothless, filter-feeding whales and (2) the toothed, carnivorous whales, a group that includes the dolphins and porpoises.

The toothless whales are better known as the **baleen whales.** Instead of teeth they have rows of flexible, fibrous plates named **baleen** that hang from the upper jaws (Figure 8-18). Baleen is made of the same material as our hair and nails. The inner edge of each plate consists of hairlike bristles that overlap and form a dense mat in the roof of the mouth. The whale filter feeds by taking a big mouthful of water and squeezing it out through the bristles. The whale then licks the food that is left behind on the bristles and swallows it.

Baleen whales are cetaceans that filter feed with baleen plates.

Baleen whales are not only the largest whales, they are the largest animals that have ever lived on earth. There are 11 species of these majestic creatures. At one time, they were common in all the oceans, but overhunting has brought many species to the brink of extinction (see the section in Chapter 17, "The Case of the Whales"). The blue whale (*Balaenoptera mus-*

Figure 8-17 The fetus of a white-sided dolphin (*Lagenorhynchus*) shows two distinct pairs of limbs; the rear pair will eventually disappear.

Figure 8-18   The filtering apparatus of whales consists of vertical baleen plates. The number and length of plates vary among species, up to an average of 360 on each side in the sei whale (*Balaenoptera borealis*). The plates vary from 30 cm (1 foot) long in the minke whale (*B. acutorostrata*) to 4.5 m (15 feet) in the bowhead whale (*B. mysticetus*). The baleen is also called "whalebone" and was once used to make corset stays. Water is filtered as the mouth closes and the tongue (*large arrow*) pushes up, forcing the water out through the baleen.

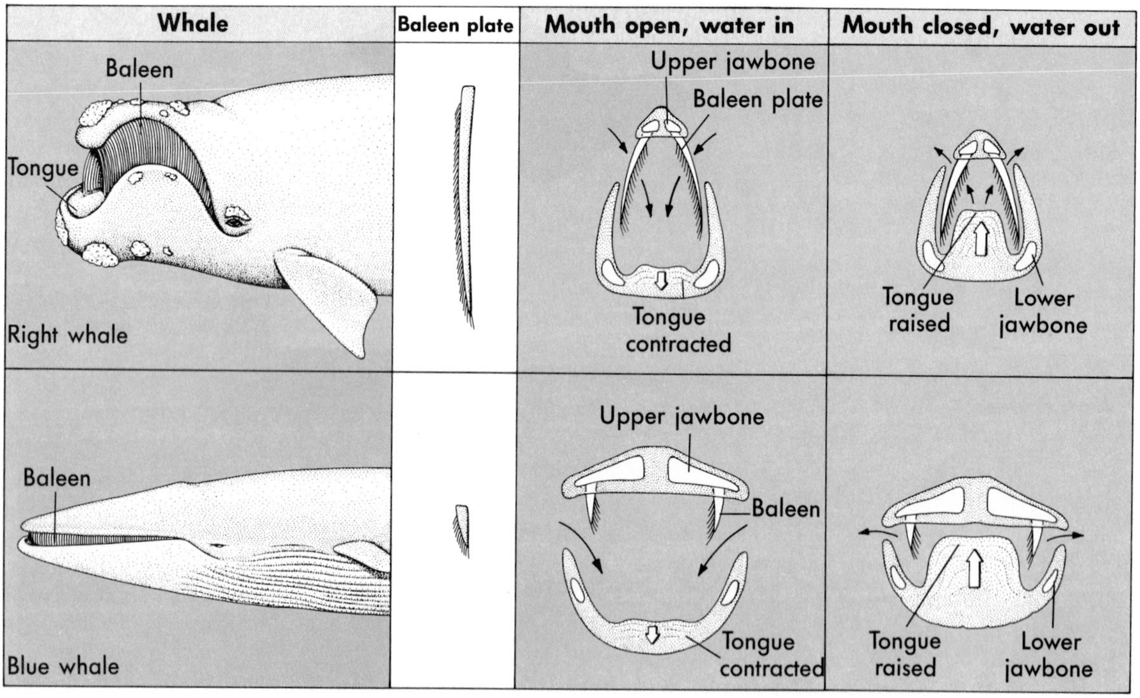

Whale	Baleen plate	Mouth open, water in	Mouth closed, water out
Right whale Baleen / Tongue		Upper jawbone / Baleen plate / Tongue contracted	Tongue raised / Lower jawbone
Blue whale Baleen		Upper jawbone / Baleen / Tongue contracted	Tongue raised / Lower jawbone

LIFE IN THE MARINE ENVIRONMENT

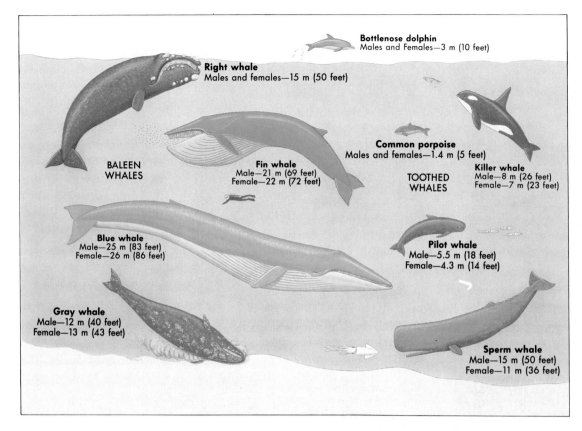

Figure 8-19 Representative baleen and toothed whales.

culus)—which is actually blue-gray—is the largest of all (Figure 8-19). Males average 25 m (80 feet), and there is a record of a female 33.5 m (110 feet) long. How do you weigh a blue whale? Very carefully—they average 80,000 to 130,000 kg (90 to 140 tons), but the record is an estimated 178,000 kg (200 tons)!

The blue whale, the fin whale (*B. physalus*), and the minke whale (*B. acutorostrata*)—together with two other related species—are known as the **rorquals.** They and the humpback whale (*Megaptera novaeangliae*), which is often included among the rorquals, feed by gulping up schools of fish and swarms of krill (see Figure 8-19). The lower part of the throat expands when feeding, hence the distinctive accordion-like grooves on the underside of these whales. Humpback whales often herd fish by blowing curtains of bubbles around them. Krill is the most important part of the rorqual diet, especially in the Southern Hemisphere, but fishes such as herring and mackerel are also eaten.

The right whales (*Eubalaena, Caperaea*) and the bowhead whales (*Balaena mysticetus*) feed by swimming along the surface with their huge mouths open. They have the largest baleen plates of the whales but the finest bristles. This allows them to filter small plankton like copepods and some krill.

amphipods Small crusta-
ceans whose bodies are
compressed from side to
side
Chapter 6, p. 173

Figure 8-20 Dolphins
often ride the bow wave of
boats **(A)** or of even large
whales. They ride without
beating their tails, obtaining thrust from the pres-
sure wave in front of the
ship. This spinner dolphin
(*Stenella longirostris*) from
the eastern Pacific was
photographed swimming
along a ship **(B).**

A

B

The gray whale (*Eschrichtius robustus*) is primarily a bottom feeder.
When examined, their stomachs contain mostly amphipods that inhabit
soft bottoms. Grays stir up the bottom with their pointed snouts and then
filter the sediment (see Figure 8-19), leaving characteristic pits on the bottom. It is thought that they swim on their right side when feeding, since
the baleen on this side is more worn. A 10-week-old female kept in captivity in San Diego, California, ate over 815 kg (1,800 pounds) of squid every
day, gaining weight at the rate of 1 kg (2.2 pounds) an hour!

The roughly 80 remaining species of cetaceans are toothed whales that
lack baleen. Their teeth are adapted for a diet of fish, squid, and other prey.
They use the teeth only to catch and hold prey, not to chew it. Food is
swallowed whole. As in all cetaceans, food is ground up in one of the three
compartments of the stomach. The blowhole has one opening, as opposed
to two in the baleen whales.

**The toothed whales, which include the dolphins and porpoises, lack baleen
and feed on fish, squid, and other prey.**

The largest toothed whale is the sperm whale (*Physeter catodon*), the un-
mistakable blunt-nosed giant of *Moby Dick* fame (see Figure 8-19). Together, the sperm and baleen whales are often called the **great whales.** The
sperm whale is now the most numerous of the great whales, even though it
has been one of the mainstays of the whaling industry for centuries (see Table 17-1). The largest on record weighed 38,000 kg (42 tons).

Sperm whales are fond of squid, including the giant deep-sea ones. Undigested squid beaks and other debris accumulate in the gut as large globs of
sticky material known as **ambergris.** Believe it or not, ambergris is an ingredient in fine perfumes. Sperm whales also eat a wide variety of fishes, including sharks, lobsters, and other marine animals.

The other toothed whales are much smaller than the great whales. One
is the killer whale, or orca (*Orcinus orca;* see Figures 8-1 and 8-19), a magnificent black and white predator with a taste for seals, penguins, large
fishes, and even other whales. It is most common in cold water but is found
around the world. It is often kept in captivity. Killer whales have a nasty
reputation, but there are no confirmed cases of them attacking humans in
the wild.

Though they are all whales, most of the small, toothed whales are called
dolphins or porpoises. Technically speaking, porpoises comprise only a
small group of blunt-nosed whales (see Figure 8-19), but in some places the
name "porpoise" is given to some of the dolphins. Most of the small whales,
however, are called dolphins and some people prefer to call all of them dolphins.

The many species of dolphins typically possess a distinctive snout, or
beak, and a perpetual smile. Playful, highly sociable, and easily trained,
dolphins easily win people's hearts. They often travel in large groups called
pods, herds, or schools. They like to catch rides along the bows of boats
(Figure 8-20, A) or even on great whales. The bottlenose dolphin (*Tursiops
truncatus*) is the dolphin seen in marine parks and oceanaria around the

world. The spinner dolphin (*Stenella longirostris*; Figure 8-20, *B*), is so named because of its spectacular twisting jumps in the air. It is one of the species of dolphins that get caught in the nets of tuna fishermen. This happens because the tuna and dolphins eat the same fish and often occur together (see the section in Chapter 17, "Other Marine Species Risking Extermination").

Biology of Marine Mammals

It is surprising how little we know about marine mammals. Most are difficult or impossible to keep in captivity or even to observe for long periods at sea. Some whales and dolphins are rarely seen, so what little we know about them comes from captive or stranded individuals. What we *do* know about marine mammals, however, is simply fascinating.

Swimming and Diving. Streamlining of the body for swimming is a hallmark of marine mammals. Seals, sea lions, and other pinnipeds swim mostly by paddling with their flippers. Sirenians and cetaceans, in contrast, move their tails and flukes up and down (Figure 8-21). Fishes, you will recall, move their tails from side to side (see Figure 7-12). Cetaceans turn mostly by sideways movements of the tail and flukes. Sea lions have been timed at speeds of 35 kph (22 mph). Blue and killer whales can reach speeds of 50 kph (30 mph). A group of common dolphins (*Delphinus delphis*) was recorded bowriding at a speed of 64 kph (40 mph)!

To avoid inhaling water, marine mammals take very quick breaths. A fin whale can empty and refill its lungs in less than 2 seconds, half the time we take—even though the whale breathes in 3,000 times more air! When swimming fast, many pinnipeds and dolphins jump clear out of the water to take a breath. Cetaceans have the advantage of having the blowhole on top of the head. This allows them to breathe even though most of the body is underwater. It also means, by the way, that cetaceans can eat and swallow without drowning. In the large whales the moisture in their warm breath condenses when it hits the air. Together with a little mucus and seawater, this water vapor forms the characteristic **spout** or **blow**. The spout can be seen at great distances and its height and angle used to identify the whale (Figures 8-22 and 1-33). The blue whale, for instance, has a spout some 6 to 12 m (20 to 40 feet) high.

Marine mammals have mastered the art of diving, and most make prolonged dives to considerable depths for food. There is a wide range in diving ability. Sea otters can dive for only 4 or 5 minutes, to depths of perhaps 55 m (180 feet). Pinnipeds normally dive for up to 30 minutes, and maximum

Figure 8-21 Swimming in cetaceans involves strong up-and-down movements of the tail and flukes. Flip the pages and see for yourself without getting wet.

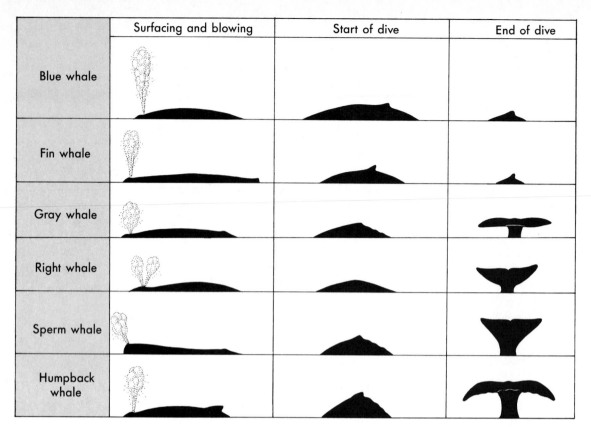

	Surfacing and blowing	Start of dive	End of dive
Blue whale			
Fin whale			
Gray whale			
Right whale			
Sperm whale			
Humpback whale			

Figure 8-22 Great whales may be identified from a distance by their blowing pattern, their outline on the surface, and the way they dive.

depths are roughly 150 to 250 m (490 to 820 feet). The Weddell seal *(Leptonychotes weddelli)* has been recorded diving for as long as 1 hour 13 minutes and as deep as 575 m (1,900 feet). A northern elephant seal *(Mirounga angustirostris)* was recorded at a depth of 880 m (2,900 feet).

The plankton-feeding habits of baleen whales do not require them to dive too deeply for their food, and they seldom venture below 100 m (300 feet). Toothed whales, however, are excellent divers. Dolphins are known to dive as deep as 300 m (990 feet). The champion diver is the sperm whale, which can stay under for least an hour. They are known to dive to 2,250 m (7,380 feet) and can probably go much deeper.

The long, deep dives of marine mammals require several crucial adaptations. For one thing, they must be able to go a long time without breathing. This involves more than just holding their breath, for they must keep their vital organs supplied with oxygen. To get as much oxygen as possible before dives, pinnipeds and cetaceans hold their breaths for 15 to 30 seconds, then rapidly exhale and take a new breath. As much as 90% of the oxygen contained in the lungs is exchanged during each breath, in contrast to 20% in humans.

LIFE IN THE MARINE ENVIRONMENT

Not only do diving marine mammals breathe more air faster than other mammals, they are better at absorbing and storing the oxygen in the air. They have relatively more blood than nondiving mammals. Their blood also contains a higher concentration of red blood cells, and these cells carry more hemoglobin. Furthermore, their muscles are extra rich in myoglobin, which means that the muscles themselves can store a lot of oxygen.

Marine mammals have adaptations that reduce oxygen consumption in addition to increasing supply. When they dive, their heart rate slows dramatically. In the northern elephant seal, for example, the heart rate decreases from about 85 beats per minute to about 12. Blood flow to nonessential parts of the body—like the extremities and the gut—is reduced, but it is maintained to vital organs like the brain and heart. Thus oxygen is made available where it is needed most.

Another potential problem faced by divers results from the presence of large amounts of nitrogen in the air. Nitrogen dissolves much better at high pressures, like those experienced at depth. The blood of scuba divers picks up a lot of nitrogen while they are below the surface. If the pressure is suddenly released, some of the nitrogen will not stay dissolved and will form tiny bubbles in the bloodstream. You can see a similar phenomenon when you open a bottle of soda pop: As long as the top is on, the contents are under pressure. The carbonation, actually carbon dioxide gas, remains dissolved. When you open the bottle the pressure is released and bubbles form. When nitrogen bubbles form in the blood after diving, they can lodge in the joints or block the flow of blood to the brain and other organs. This produces a horribly painful condition known as the **bends.** To avoid the bends, human divers must be very careful about how deep they go, how long they stay underwater, and how fast they come up.

Marine mammals dive deeper and stay down longer than human divers, so why don't they get the bends? The answer is that they have adaptations that prevent nitrogen from dissolving in the blood in the first place. Human lungs work pretty much the same underwater as on land. When marine mammals dive, on the other hand, their lungs actually collapse. They have a flexible rib cage that gets pushed in by the pressure of the water. This squeezes the air in the lungs out of the places where it can dissolve into the blood. Air is moved instead into the central spaces, where little nitrogen is absorbed. Some pinnipeds actually exhale before they dive, further reducing the amount of air—and therefore nitrogen—in the lungs.

> **Adaptations for deep, prolonged dives in marine mammals include efficient exchange of air on the surface, storage of more oxygen in the blood and muscles, reduction of the blood supply to the extremities, and collapsible lungs to prevent the bends.**

Echolocation. Marine mammals depend little on the sense of smell, which is so important to their terrestrial cousins. Their vision remains fairly acute, but they have developed another sensory system—echolocation—based on hearing. Echolocation is nature's version of sonar. The animal emits sound waves, which travel about five times faster in water than in air,

hemoglobin A blood protein that transports oxygen in many animals; in vertebrates it is contained in *red blood cells*
Chapter 7, p. 210

myoglobin A muscle protein in many animals that stores oxygen
Chapter 7, p. 210

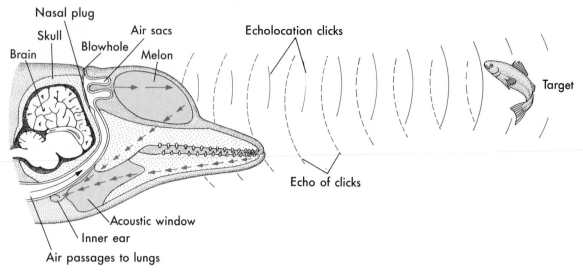

Nasal plug

Skull

Air sacs

Blowhole

Brain

Melon

Echolocation clicks

Target

Echo of clicks

Acoustic window

Inner ear

Air passages to lungs

Figure 8-23 Dolphins echolocate by emitting bursts of sound waves, or "clicks," by pushing air through the air passages. Two muscular nasal plugs act as valves, closing and opening the passages. Flaps of tissue on the plugs probably also produce sound by vibrating in the moving air. The clicks are focused into a beam by the melon. To cover a wider area the dolphins move their heads from side to side. The melon is also known to receive the echoes and transmit them to the ears.

and listens for the echoes that are reflected back from surrounding objects (Figure 8-23). The echoes are then analyzed by the brain. The time it takes the echoes to return tells the animals how far away the object is. Most if not all toothed whales, including dolphins and porpoises, and some pinnipeds are known to echolocate. At least some baleen whales may also use echolocation. Echolocation is not exclusive to marine mammals. Bats, for example, use it to find insects and other prey while flying at night.

Many marine mammals echolocate by analyzing the echo of sound waves they emit. Echolocation is used to find prey and orient to the surroundings.

The sounds used in echolocation consist of short bursts of sharp **clicks** that are repeated at different frequencies. Low-frequency clicks have a high penetrating power and can travel long distances. They reflect from large features and are used to obtain information on the surrounding topography. Low-frequency sound waves may also be used in some toothed whales to stun their prey (see Box 8-1). To discriminate more detail and locate nearby prey, high-frequency clicks that are inaudible to humans are used. Experiments have shown that blindfolded bottlenose dolphins can discriminate between objects of slightly different size or made of different types of materials and even detect wires.

We are not completely sure how echolocation operates in marine mammals. It is a rather complicated arrangement (see Figure 8-23). The clicks, squeaks, and whistles of cetaceans are produced and modulated as air is forced through the air passages and several associated air sacs while the blowhole is closed. Cetaceans do not have vocal cords. The frequency of the clicks is changed by contracting and relaxing muscles along the air passages and sacs. A fatty structure on the forehead of toothed whales, the **melon,** appears to focus and direct the sound waves. The melon gives these

LIFE IN THE MARINE ENVIRONMENT

BOX 8-1

The Other "Big Bang" Theory

It is widely known that toothed whales use sound for echolocation and to communicate with each other. Recently a different use of sound waves by these cetaceans has been suggested.

This new hypothesis developed as a possible explanation for the feeding habits of sperm whales, the largest of the toothed whales. Squid taken from the stomach of captured and stranded whales often show no tooth marks or scars of any kind. In fact, live squid have been known to swim out of the stomachs of freshly caught whales! It seems that sperm whales have a way of catching the squid—including the giant squid—without using their teeth, even if their teeth are actually of little help since they are present only in the lower jaw. Another puzzle is explaining how sperm whales weighing 36,000 kg (40 tons) or more and averaging speeds of just 2 to 4 knots catch squid that can swim at 30 knots.

How about the possibility that whales and dolphins may use powerful blasts of sound to catch their food? This ingenious hypothesis has been dubbed a second "big bang theory," the original being the well-known view proposed to explain the origin of our universe (see the section in Chapter 2, "The Structure of the Earth"). Catching prey whole and still alive could be explained if the whale stuns its prey with a blast of sound and then simply swallows it whole.

Some indirect evidence is provided by the now extinct ancestors of toothed whales. Fossils of the earliest known toothed whales, which lived in the ocean 20 million years ago, indicate the presence of long snouts armed with numerous piercing teeth. Like those of a barracuda, the teeth were probably used to catch small fish and other prey. The long snout, however, has disappeared in practically all modern cetaceans, and the teeth have become wider and shorter. Have modern toothed whales evolved a new technique to catch their food?

Sonic hunting may involve a beam of low-frequency sound waves powerful enough to stun a fish or squid. Though the sophisticated sound-producing mechanism of cetaceans is not fully understood, it is thought to be capable of emitting the required sound waves. It has been suggested that sonic hunting evolved as a by-product of echolocation in the early toothed whales.

It is not easy to obtain the necessary evidence to support this hypothesis. Loud noises can indeed stun fish, but it is difficult to reproduce the exact sounds produced by cetaceans. Dolphins living in the wild have produced gunlike bangs that can be heard by humans. Unfortunately, undertaking detailed studies of sonic hunting in the wild presents many complications. Captive dolphins do not produce loud noises, which is not surprising since the echo of the noise off the tank walls would be very painful to them. The function of big bangs in whales is but one of the many surprising adaptations of cetaceans waiting to be explored.

A

B

Figure 8-24 The beluga (or "white one" in Russian, *Delphinapterus leucas*) is a white Arctic whale with a conspicuous melon **(A).** In their natural environment, belugas live in small groups **(B).**

Figure 8-25 A California sea lion (*Zalophus californianus*) with nursing pup.

whales their characteristic rounded foreheads. To accommodate the melon, the skull is modified to form a pointed, dish-shaped face. The skull is also asymmetric, the right side being slightly different from the left side. Belugas (*Delphinapterus leucas;* Figure 8-24) have a bulging forehead that changes shape as the melon, moved by muscles, focuses the sound. The huge forehead of the sperm whale is filled in part with a massive melon called the **spermaceti organ.** This organ is filled with a waxy oil, **spermaceti,** once much sought for making candles and still used as a lubricant for precision instruments. It has been suggested that this deep diving whale may also use the spermaceti organ to regulate buoyancy.

In toothed whales incoming sound waves are received primarily by the lower jaw (see Figure 8-23). The ear canal that connects the outside with the inner ear is reduced or blocked in most cetaceans. The jawbones, filled with fat or oil, transmit sound to the two very sensitive inner ears. Each ear receives sound independently. The ears are protected by a bony case and embedded in an oily mixture that insulates the ear but allows sound waves to pass from the jaws. Sound information is sent to the brain, which forms a mental "picture" of the target or surroundings.

Behavior. Echolocation is just one indication of the amazing mental capabilities of marine mammals. The mammalian brain has evolved as an association center for complex behaviors in which learning, not instinct, dominates. In contrast to fishes, birds, and other vertebrates, mammals rely mostly on past experience—stored and processed by the brain—to respond to changes in the environment (see Box 8-2).

Most marine mammals are highly social animals that live in groups at least part of the time. Many pinnipeds live in huge colonies during the breeding season. Most cetaceans spend their entire lives in highly organized pods of a few (see Figure 8-24, *B*) to thousands of individuals. Some pods include smaller sub-groups organized by age and sex that often do different things. To keep in contact, many of their highly complex and sophisticated behaviors are directed toward members of their own species.

Sounds, or **vocalizations,** play a prominent role in communication. Sea lions and fur seals communicate by loud barks and whimpers; seals use more sedate grunts, whistles, and chirps. The vocalizations of pinnipeds are especially important in maintaining territories during reproduction (see the section "Reproduction" on p. 255). Females and their pups or calves recognize each other by their "voices" (Figure 8-25).

Cetaceans produce a rich variety of vocalizations that are different from the sounds used for echolocation. Both types of sounds can be produced simultaneously, further evidence of the complexity of sound production in marine mammals. Social vocalizations are low-frequency sounds that humans can hear. The variety of sounds is amazing and include grunts, barks, squeaks, chirps, and even "moos." Different sounds are associated with various moods and are used in social and sexual signalling. Whistles, emitted in a multitude of variations and tones, are characteristic of each species. Some of these sounds serve as a "signature," allowing individuals to recognize one another. Among the more than 70 calls that have been identified among killer whales, some are present in all individuals, whereas others are "dialects" that identify certain pods.

BOX 8-2

How Intelligent Are Cetaceans?

We often hear that whales, dolphins, and porpoises are as intelligent as humans, maybe even more so. Are they really that smart? There is no question that cetaceans are among the most intelligent of animals. Dolphins, killer whales, and pilot whales in captivity quickly learn tricks. The military has trained bottlenose dolphins to find bombs and missile heads and to work as underwater spies.

This type of learning, however, is called **conditioning.** The animal simply learns that when it performs a particular behavior it gets a reward—usually a fish. Many animals—including rats, birds, and even invertebrates—can be conditioned to perform tricks. We certainly don't think of these animals as our mental rivals.

Unlike most other animals, however, dolphins quickly learn by observation and may spontaneously imitate human activities. One tame dolphin watched a diver cleaning an underwater viewing window, seized a feather in its beak, and began imitating the diver—complete with sound effects! Dolphins have also been seen imitating seals, turtles, and even water-skiers.

Given the seeming intelligence of cetaceans, people are always tempted to compare them with humans and other animals. Studies on discrimination and problem-solving skills in the bottlenose dolphin, for instance, have concluded that its intelligence lies "somewhere between that of a dog and a chimpanzee."

Such comparisons are unfair, however. It is important to realize that "intelligence" is a very human concept and that we evaluate it in human terms. After all, not many people would consider themselves stupid because they couldn't locate and identify a fish by its echo. Why should we judge cetaceans by their ability to solve human problems?

Both humans and cetaceans have large brains with an expanded and distinctively folded surface, the **cortex.** The cortex is the dominant association center of the brain, where abilities such as memory and sensory perception are centered. Cetaceans have larger brains than ours, but the ratio of brain to body weight is higher in humans. Again, direct comparisons are misleading. In cetaceans it is mainly the portions of the brain associated with hearing and the processing of sound information that are expanded. The enlarged portions of our brain deal largely with vision and eye-hand coordination. Cetaceans and humans almost certainly perceive the world in very different ways. Their world is largely one of sounds, ours one of sights.

Movies and television to the contrary, the notion of "talking" to dolphins is also misleading. Though they produce a rich repertoire of complex sounds, they lack vocal cords and their brains probably process sound differently from ours. Bottlenose dolphins have been trained to make sounds through the blowhole that sound something like human sounds, but this is a far cry from human speech. By the same token, humans cannot make whale sounds. We will probably never be able to carry on an unaided conversation with cetaceans.

Instead of "intelligence," some people prefer to speak of "awareness." In any case, cetaceans probably have a very different awareness and perception of their environment than do humans. Maybe one day we will come to understand cetaceans on their terms instead of ours, and perhaps we will discover a mental sophistication rivalling our own.

Bottlenose dolphins participating in research to test acoustic communication. The devices on their heads, which are held in place by suction cups, light up every time a dolphin whistles.

Figure 8-26 A humpback whale *(Megaptera novaeangliae)* performing a full spinning breach.

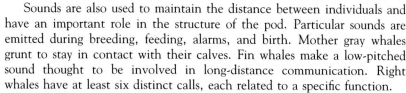

Sounds are also used to maintain the distance between individuals and have an important role in the structure of the pod. Particular sounds are emitted during breeding, feeding, alarms, and birth. Mother gray whales grunt to stay in contact with their calves. Fin whales make a low-pitched sound thought to be involved in long-distance communication. Right whales have at least six distinct calls, each related to a specific function.

The humpback whale is renowned for its soulful **songs.** They are sung by breeding males to attract females by advertising their readiness to mate. The songs consist of phrases and themes repeated in a regular pattern for a half hour or longer. They may be repeated over and over for days! The songs change over time. Males also start each breeding season with the song they were singing at the end of the previous breeding season. Researchers record and catalogue songs to help track whales in their annual migrations.

Communication between cetaceans is not restricted to vocalizations. Researchers have described a variety of postures and movements that may indicate the animal's mood. Dolphins clap their jaws or turn around with their mouths open as a threat. The loud cracking sound made when some marine mammals flap their flukes or flippers on the surface is thought to be a warning signal.

Cetaceans are notorious for their **play behavior,** what appears to be pleasurable activities with no serious goal. Many species—including the great whales—play with food or floating objects like logs, kelp, and feathers, throwing them up in the air or holding and pushing them with their snouts. Individuals may swim head down or on their backs apparently just for the fun of it. Dolphins like to surf; pilot *(Globicephala)* and right whales go sailing with their flukes out of the water to catch the wind. Sex play, the rubbing and touching of the genital opening, is also common.

The sight of a great whale **breaching,** leaping up in the air and loudly crashing on the surface, is awesome (Figure 8-26). Breaching has been interpreted as a warning signal, a way of scanning the surface or the shoreline or of getting rid of external parasites or an ardent lover, or simply as fun. After a deep dive, sperm whales may breach, fall on their backs, and make a splash that can be heard 4 km (2.5 miles) and seen 28 km (17.4 miles) away! Many whales stick their heads out of the water to "spy" on their surroundings (Figure 8-27, A).

The complex behavior of cetaceans is evident in other ways. When one individual is in trouble, others may come to assist (Figure 8-27, B). Members of a pod refuse to leave a wounded or dying comrade. Whalers knew that a harpooned whale was a lure for others, who came to the rescue from miles around. Dolphins will carry injured individuals to the surface to breathe (Figure 8-27, C), and there are records of females carrying the body of a stillborn calf until it rots.

Many of the toothed whales work together when they hunt, some in coordinated pairs. Sometimes whales take turns feeding while their buddies herd a school of fish. An individual may investigate something strange lying ahead while the rest of the group waits for the "report" of the scout.

Marine mammals, particularly cetaceans, use a rich variety of vocalizations and other tactile and visual signals to communicate with each other. Play

252

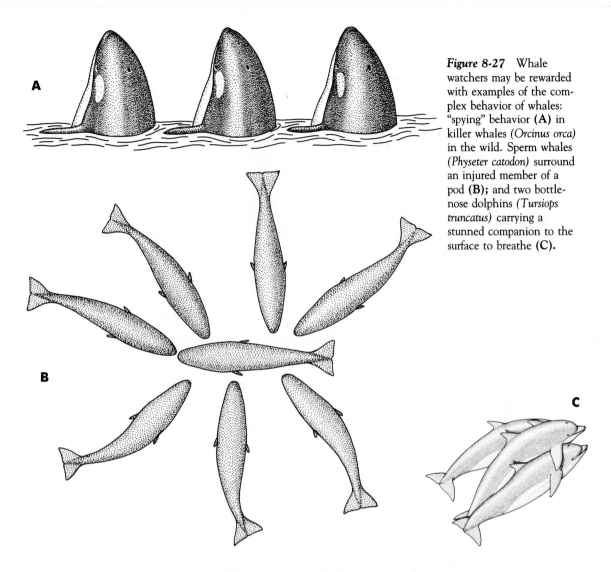

Figure 8-27 Whale watchers may be rewarded with examples of the complex behavior of whales: "spying" behavior **(A)** in killer whales *(Orcinus orca)* in the wild. Sperm whales *(Physeter catodon)* surround an injured member of a pod **(B)**; and two bottlenose dolphins *(Tursiops truncatus)* carrying a stunned companion to the surface to breathe **(C)**.

behavior and mutual assistance are additional evidence of the complexity of their behavior.

The relationship between dolphins and humans is a controversial one. Some people swear of "spiritual inspiration" while swimming among dolphins during the "dolphin encounters" offered by some resort hotels. Others see this as outright exploitation of the captive animals. It has been suggested that stress among captive dolphins reduces their life span. Though exaggerations abound, there are authenticated cases of dolphins approaching human swimmers who appeared to be in trouble. For more than a century, fishermen in southern Brazil have established a unique partnership with dolphins. The dolphins detect fish and deliver them to fishermen waiting with nets. Fishermen have learned to interpret cues given by the dolphins about the location and abundance of fish. Generations of dolphins have learned that a row of fishermen holding a net in shallow water means

Figure 8-28 The body of this gray whale (*Eschrichtius robustus*) was probably washed ashore after it died. Other whales, however, become stranded while still alive.

an easy catch for themselves, even if it has to be shared with funny-looking, air-breathing mammals.

One of the mysteries of the behavior of whales and dolphins is the **stranding** or **beaching** of dozens of individuals on beaches (Figure 8-28). The animals refuse to move, and efforts to move them into deeper water usually fail. Even if they are pulled out to sea, they often beach themselves again. The whales die because their internal organs collapse without the support of the water. Stranding has been described in many species, but some—such as pilot and sperm whales—are stranded more often than others. It appears that whales become stranded when they follow one or more members of their group that have become disoriented by a storm, illness, or injury. This indicates the strong cohesiveness and herd instinct of the group.

Migrations. Many pinnipeds and cetaceans make seasonal migrations, often traveling thousands of miles from feeding to breeding areas. Most toothed whales, on the other hand, do not migrate at all, though they may move about in search of food.

The migrations of the great whales are by far the most remarkable. Many baleen whales congregate to feed during the summer in the productive waters of the polar regions of both hemispheres, where huge concentrations of diatoms and krill thrive in the long days. During the winter they migrate to warmer waters to breed. The seasons are reversed in the Northern and Southern Hemispheres, so when some humpback whales are wintering in the Hawaiian Islands or the West Indies, other humpbacks living in the Southern Hemisphere are feeding around Antarctica during the southern summer (Figure 8-29).

> Most great whales migrate from winter breeding areas in the tropics to summer feeding areas in colder waters.

The migratory route of the gray whale is the best known of any of the great whales (see Figure 8-29). From the end of May to late September the whales feed in shallow water in the northern Bering, Beaufort, and East Siberian seas. They begin moving south in late September when ice begins to form. By November they begin crossing through the eastern Aleutian Islands. They eat less while on the move, burning off close to a quarter of their body weight. The whales cover about 185 km (115 miles) per day. They travel alone or in small groups along the coast of the Gulf of Alaska and down the western coast of North America en route to the Baja California Peninsula in Mexico (see Figure 17-19). Migrating individuals often show spying behavior, pushing their heads out of the water. This raises the possibility that they navigate by using memorized landmarks. They reach Oregon around late November or early December and San Francisco by mid-December. Females generally migrate earlier, and in late February pregnant females are the first to appear in shallow, quiet lagoons in southern Baja California and the southern mainland coast of the Gulf of California. It is here that females give birth and males mate with nonpregnant females.

The northbound migration begins by March, after the birth of the 700- to 1,400-kg (1,500- to 3,000-pound) calves. Females mate every 2 years,

LIFE IN THE MARINE ENVIRONMENT

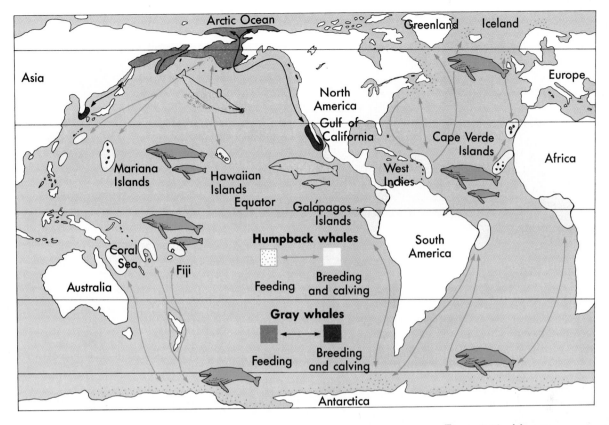

and the first to migrate north are the newly pregnant females that did not give birth. They will return 12 months later to give birth. Mothers with calves leave last. On the way north the whales tend to stay farther from the coast and move slower, an average of 80 km (50 miles) per day, because of the newborn calves and the adverse currents. The last whales leave the coast off Washington state by early May. They start reaching their feeding areas by late May, completing an amazing 8-month trip of up to 18,000 km (11,200 miles), the longest migration of any mammal.

There is still much to be learned about the migrations of the gray whale and other whales. It has been found, for instance, that some isolated groups of gray whales along the migratory route do not migrate at all. This is the case in a group that resides in the Queen Charlotte Islands off the coast of British Columbia. Scientists are using novel ways to investigate the migration of whales. Attaching small radio transmitters to whales and tracking their movements by satellite promises to uncover intriguing details. Analysis of the DNA of humpback whale populations in the Hawaiian Islands suggests that, as in the green turtle, individuals always return to the feeding grounds of their mothers. Another vexing question is how whales navigate. It has been suggested that they use the earth's magnetic field, a possibility that implies that they must carry some type of internal compass to orient themselves.

Reproduction. The reproductive system of marine mammals is similar

Figure 8-29 Migration routes of humpback (*Megaptera novaeangliae*) and gray (*Eschrichtius robustus*) whales. Both species tend to migrate and breed close to shore, which made them easy to hunt. Now under protection, both are on the comeback. The western Pacific population of grays that may still breed south of Korea is close to disappearing forever. Gray whales used to live in the north Atlantic until exterminated in the last century. Also see Figure 17-19.

Figure 8-30 (A) A male Steller sea lion (*Eumetopias jubatus*) guarding his harem on a rocky island off the coast of Alaska. Steller sea lions are the largest of the eared seals: males may weight nearly 900 kg (one ton). (B) A harem of female California sea lions (*Zalophus californianus*) on Santa Barbara Island, southern California. The harem (*center*) is being guarded by a large, darker bull (*top left*). Large female elephant seals (*Mirounga angustirostris*) rest near the harem, oblivious to the occasional fights between the bull and rival males around the harem.

to that of land mammals. They have some unique adaptations to life in the water, however. To keep the body streamlined, male cetaceans and most other marine mammals have an internal penis and testes. The penis, which in blue whales is over 3 m (10 feet) long, is kept rigid by a bone. It is extruded just before copulation through the **genital slit,** an opening anterior to the anus (see Figure 8-16).

Pinnipeds breed on land or ice, some migrating long distances to isolated islands to do so. In most species of seals each adult male breeds with only one female. This is not true of sea lions, fur seals, and elephant seals. During the breeding season the males of these species, who are much bigger and heavier than females, come ashore and establish breeding territories. They stop eating and defend their territories by constant, violent fighting. They herd **harems** of as many as 50 females onto their territories and keep other males away (Figure 8-30, B). Only the strongest males are able to hold territories and breed. The others gather into **bachelor groups** and spend much of their time trying to sneak into harems for a quick copulation. Defending the harem is exhausting work, and dominant males "burn out" after a year or two, making way for newcomers. It nevertheless pays off in the huge number of offspring they leave compared with the males that never reach dominance—even if the submissive males live longer!

Female pinnipeds give birth to their pups on shore. They seem to be indifferent to the birth process but soon establish a close relationship with the pup. Since females continue to go to sea to feed, they must learn to recognize their own pups out of all the others by sound and smell (see Figure 8-25). The pups generally cannot swim at birth. They are nursed for periods of 2 weeks to 2 years, depending on the species. Most pinnipeds have two pairs of mammary glands that produce a fat-rich milk ideal for the rapid development of the pup's blubber.

A female pinniped can become pregnant only during a brief period after **ovulation,** the release of an egg by her ovaries. This occurs just days or weeks after the birth of her pup. Females of most species return to the breeding grounds only once a year. By contrast, **gestation**—the length of time it takes the embryo to develop—is less than a year. This difference would cause the pup to be born too early, before the mother returns to the breeding ground. To prevent this, the newly-formed embryo stops develop-

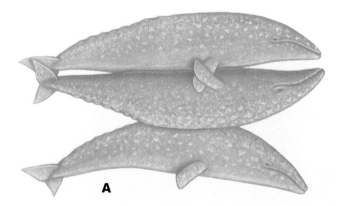

A

B

ing and remains dormant in the female's womb, the **uterus.** After a delay of as long as 4 months, the embryo finally attaches to the inner wall of the uterus and continues its normal development. This phenomenon, known as **delayed implantation,** allows pinnipeds to prolong the embryo's development so that the timing of birth coincides with the female's arrival to the safety of shore.

> **Delayed implantation allows pinnipeds to time the birth of pups to the arrival of pregnant females in breeding areas.**

Our knowledge of the reproductive behavior of cetaceans in their natural environment is limited. We do know that cetaceans are intensely sexual animals. Sex play is an important component of the behavior of captive dolphins. Like humans, they appear to use sex not only for procreation, but for pleasure as well. Sexual behavior appears to have a role in the establishment and maintenance of bonds among all individuals, not just potential mates. The sexes are typically segregated within the pod, and males perform elaborate courtship displays to catch the attention of potentially receptive females. Fights among rival males are common, but cooperation also occurs sometimes. Gray whales are known to copulate with the help of a third party, a male that helps support the female (Figure 8-31, A). Group matings have been observed in humpback and white whales. Considerable touching and rubbing is known to precede copulation (Figure 8-31, B). Actual copulation lasts less than a minute but is repeated frequently.

Gestation lasts for 11 or 12 months in most cetaceans. An exception is the sperm whale, which has a gestation period of 16 months. Development in the large baleen whales is actually accelerated in most species to time it with the annual migration to warm waters. It is remarkable that it takes 9 months for a 3-kg (7-pound) human baby to develop, but a 2,700-kg (3-ton) blue whale calf needs only about 11 months to develop!

The calves of probably all cetaceans are born tail-first (Figure 8-32). This allows them to remain attached to the placenta for as long as possible to prevent drowning. The calf immediately swims to the surface. In captive dolphins, the mother or an attending female may help the calf to the surface. Fat-rich milk is responsible for the rapid growth of calves, particularly in the great whales. They are born without their full complement of blubber and must gain weight before migrating with their mothers to feeding

Figure 8-31 Mating behavior in great whales. **(A)** Gray whales *(Eschrichtius robustus)* often mate with the help of a third party—a male that props the female against the male. Actual copulation is reported to last for just 30 to 60 seconds. **(B)** Courtship in humpback whales *(Megaptera novaeangliae)* includes rolling, slapping of the flukes, and pairs surfacing vertically face to face.

Figure 8-32 A Commerson's dolphin (*Cephalo-rhynchus commersoni*) giving birth in captivity. Not much is known about this dolphin, which is found only in southern South America.

grounds in polar waters. It has been estimated that a typical blue whale calf gains 90 kg (200 pounds), the equivalent of 4 cm (1.5 inches), every day for the first 7 months of its life! The mother's milk is produced by two mammary glands with nipples located on both sides of the genital slit (see Figure 8-16). The milk is squirted right into the calf's mouth, which allows the calf to drink underwater. In at least some of the great whales, females do not seem to feed much while they are nursing. The calves are not weaned until they arrive at the feeding grounds. In some species they continue to nurse for more than a year after birth.

The relationship between mother and calf during the nursing period is very close. Frequent contact and vocalizations are used in communication. Cows are known to defend their calves when there is danger. There is a report of a female gray whale lifting her calf onto her flipper to save it from the attacks of killer whales. The bond between mother and calf probably lasts for several years. Captive young dolphins are known to return to their mothers for comfort in times of danger or stress.

Cetaceans reach sexual maturity relatively early, at age 5 or 10 years in great whales, but most females give birth only to a single calf—rarely twins—every 2 or 3 years. This low birth rate, coupled with extensive hunting, may have already sealed the fate of some of the great whales.

LIFE IN THE MARINE ENVIRONMENT

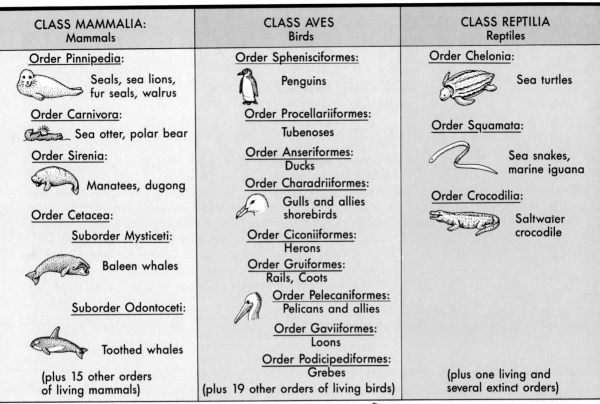

CLASS MAMMALIA: Mammals	CLASS AVES Birds	CLASS REPTILIA Reptiles
Order Pinnipedia: Seals, sea lions, fur seals, walrus	Order Sphenisciformes: Penguins	Order Chelonia: Sea turtles
Order Carnivora: Sea otter, polar bear	Order Procellariiformes: Tubenoses	Order Squamata: Sea snakes, marine iguana
Order Sirenia: Manatees, dugong	Order Anseriformes: Ducks	Order Crocodilia: Saltwater crocodile
Order Cetacea: Suborder Mysticeti: Baleen whales Suborder Odontoceti: Toothed whales	Order Charadriiformes: Gulls and allies shorebirds Order Ciconiiformes: Herons Order Gruiformes: Rails, Coots Order Pelecaniformes: Pelicans and allies Order Gaviiformes: Loons Order Podicipediformes: Grebes	
(plus 15 other orders of living mammals)	(plus 19 other orders of living birds)	(plus one living and several extinct orders)

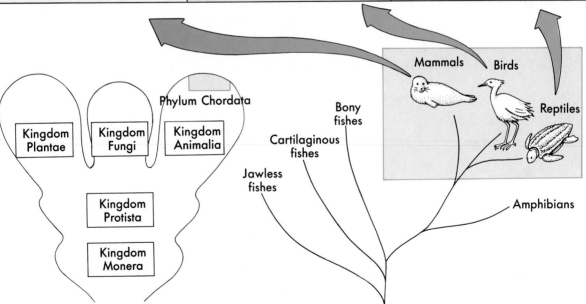

Figure 8-33 Classification scheme of reptiles, birds, and mammals.

Do It Yourself

SUMMARY

1. The two largest groups of marine reptiles are the _____ and _____.
2. In contrast to fishes and reptiles, birds are said to be homeotherms or "warm-blooded," which means they can _____.
3. Mammals are homeotherms like birds, but—save a few exceptions—they are _____ since their embryos develop in the female's womb. Embryos of most mammals receive nutrients and oxygen across the _____, a membrane connecting the embryo to the womb.
4. The _____ include the _____, sea lions, and walruses. Sea lions can be distinguished by having _____.
5. Unlike other marine mammals, sea otters lack _____ and rely on their rich fur for insulation.
6. Sirenians, dugongs, and _____, look like oversized pinnipeds but lack a pair of _____ limbs.
7. Toothless, or _____, whales are filter-feeding cetaceans. Most filter out plankton, especially _____.
8. Most cetaceans are _____ whales, carnivores that include the dolphins and porpoises.
9. Marine mammals are excellent divers as the result of several adaptations. These include the storage of additional oxygen in the blood and muscles, the collapse of the _____, and the direction of the _____ to the brain and other vital organs during diving.
10. _____ in many cetaceans involves producing sound waves and analyzing their _____.
11. Complex behaviors among cetaceans includes communication by means of _____ and _____ signals, play behavior, and _____ of individuals in danger.
12. Many cetaceans and pinnipeds migrate every year from _____ to breeding areas. In most of the great whales, breeding areas are located in _____ regions.
13. Pinnipeds prolong the development of their embryos by _____.

LIFE IN THE MARINE ENVIRONMENT

THOUGHT QUESTIONS

1. Sea turtles have disappeared from many regions, and one way of trying to save them is to reintroduce them to areas where they have been wiped out. This is done by reburying eggs or by releasing newborn baby turtles on beaches. Why are eggs reburied or baby turtles released instead of fully-grown individuals?

2. Most seabirds are specialists that feed on particular types of fish and other prey. In some cases this may reduce the chances of competing with other species of seabirds for limited resources. Sometimes, however, we find two or more species of seabirds feeding on the same type of fish. What type of mechanisms might have evolved to prevent direct competition?

3. Cetaceans give birth to few well-developed calves at well-spaced intervals. They also feed and protect the calves for long periods. This is in sharp contrast to most fishes, in which many eggs are spawned and the parents spend no time feeding and protecting the offspring. What do you think is the best strategy? Has this effort paid off in the case of the great whales?

FOR FURTHER READING

Bruemmer, F.: "Survival of the Fattest," *Natural History*, July 1990, pp. 26-33. *Maternal care and the fast development of blubber determine the survival of hooded seal pups, which are born on floating ice in the Arctic.*

Darling, J.: "Whales: An Era of Discovery," *National Geographic Magazine*, vol. 174, no. 6, December 1988, pp. 872-909. *A profile of the findings of research on the life of whales. Excellent photographs and illustrations.*

Gentry, R.L.: "Seals and Their Kin," *National Geographic Magazine*, vol. 171, no. 4, April 1987, pp. 474-501. *A family album of the pinnipeds. The issue includes "Pinnipeds Around the World," a wall map that illustrates the world distribution of pinnipeds and notes on their biology.*

Jackson, D.D.: "The Bad and the Beautiful: Gulls Remind Us of Us," *Smithsonian*, vol. 20, no. 7, October 1989, pp. 72-85. *Some fascinating aspects of the feeding and reproductive habits of the ubiquitous gulls.*

Nelson, C.H., and K.R. Johnson: "Whales and Walruses as Tillers of the Sea Floor," *Scientific American*, vol. 256, no. 2, February 1987, pp. 112-117. *Gray whales and walruses create pits and burrows on the bottom as a result of their feeding activities. It results in the "reseeding" of the sea floor for food.*

Renouf, D.: "Sensory Function in the Harbor Seal," *Scientific American*, vol. 260, no. 4, April 1989, pp. 90-95. *A report of how the harbor seal navigates, finds prey, and manages to keep track of its pup.*

Rudloe, J., and A. Rudloe: "Shrimpers and Lawmakers Collide Over a Move to Save the Sea Turtles," *Smithsonian*, vol. 20, no. 9, December 1989, pp. 44-55. *Much controversy surrounds the drowning of sea turtles caught in the net of shrimp fishermen.*

Swartz, S.L., and M.L. Jones: "Gray Whales At Play in Baja's San Ignacio Lagoon," *National Geographic Magazine*, vol. 171, no. 6, June 1987, pp. 754-771. *Gray whales are making a comeback and larger numbers of individuals are making it to the breeding grounds in Baja, California.*

"Symphony Beneath the Sea," *Natural History*, March 1991, pp. 36-76. *Ten articles about recent findings on hearing, sound making, and sound communication in cetaceans and pinnipeds.*

Trivelpiece, S.G., and W.Z. Trivelpiece: "Antarctica's Well-Bred Penguins," *Natural History*, December 1989, pp. 28-37. *How three species of penguins share their resources, even if they overlap in their distribution.*

Whitehead, H.: "Why Whales Leap," *Scientific American*, vol. 252, no. 3, March 1985, pp. 84-93. *Breaching in whales: types, distribution among whales, and its role in social behavior.*

"Whither the Whales?" *Oceanus*, vol. 32, no. 1, Spring 1989. *A series of 19 excellent articles covering numerous aspects of the biology of whales—from behavior to conservation efforts.*

Wiley, J.P.: "Manatees, Like Their Siren Namesakes, Lure Us to the Deep," *Smithsonian*, vol. 18, no. 6, September 1987, pp. 92-97. *The unique manatee, mermaids in risk of extinction, makes a last stand in Florida.*

Wursig, B.: "The Behavior of Baleen Whales," *Scientific American*, vol. 258, no. 4, April 1988, pp. 102-107. *Feeding and social behavior among the baleen whales. It includes a useful scheme of the evolution and classification of cetaceans.*

Zapol, W.M.: "Diving Adaptations of the Weddell Seal," *Scientific American*, vol. 256, no. 6, June 1987, pp. 100-105. *The Weddell seals manage to swim deeper and longer than most mammals. How do they do it?*

Part Three

Structure and Function of the Marine Environment

9
An Introduction to Ecology

habitat The place where
an organism lives
Chapter 2, p. 22

Figure 9-1 This oyster-catcher, contrary to its name, does not eat many oysters, but it does eat many of the other inverte-brates that compete for space along the shore. The oystercatcher has its own enemies, which is why its chicks are camouflaged so well that they are almost invisible. Both the hunter and the hunted are af-fected by the waves, tides, and other physical factors of the environment.

E verywhere you look in the ocean there are living things. How many and what kind of organisms there are depends on where you go— that is, on the specific nature of the **habitat.** Every environment has distinct characteristics that determine which organisms live there and which do not. The amount of light, for example, determines whether plants can grow. The type of bottom, the temperature and salinity of the water, waves, tides, currents, and many other aspects of the environment profoundly affect marine life. Some of these physical and chemical features were introduced in Chapters 2 and 3.

Equally important are the ways that organisms affect each other. They eat each other, crowd each other out, and even cooperate. Living things interact in complex and fascinating ways. Part II of this book, Chapters 4 through 8, introduced the organisms that live in the sea. Part III looks a little more carefully at how these organisms live. We will explore the different habitats that make up the ocean world, from the water's edge to the darkest depths. We will consider the special physical and chemical features of each habitat, how the organisms have adapted to each environment, and how they affect each other.

Before examining each individual habitat, we briefly examine a few basic principles that apply to all habitats. No matter where you go, for instance, plants need light and animals need food. These are just two examples, though fundamental ones, of the interactions among organisms and their environment (Figure 9-1). The study of these interactions is called **ecology.** It is important to realize that ecology is an objective science, the branch of biology that studies how and why organisms interact among themselves and with their environment. Contrary to the way the term is used in the media, ecology involves a lot more than defending the environment, as important as that may be.

THE ORGANIZATION OF COMMUNITIES

The nature of life in a particular habitat is determined to a large extent by the nonliving, or **abiotic,** part of the environment—that is, the environment's physical and chemical features. Each environment makes different demands, and the organisms that live there must adapt to those demands. Organisms are also affected by other organisms—the living, or **biotic,** envi-

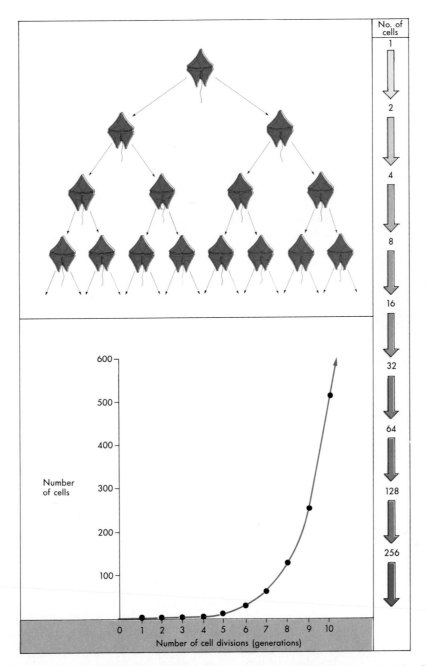

Figure 9-2 All species are capable of undergoing population explosions. When a single-celled dinoflagellate divides, for example, it leaves two daughter cells. Each daughter cell in turn produces two new daughter cells, which also divide, and so on. At each generation the number of cells doubles, and the population grows at an ever-increasing rate. If this population growth were to continue unchecked, the combined weight of the dinoflagellates would soon exceed the weight of the universe! Imagine the growth potential of organisms that can produce thousands of young instead of just two. A graph of the number of cells at each generation produces a J-shaped curve, which is typical of populations undergoing unrestrained growth.

population A group of individuals of the same species that live together *Chapter 4, p. 97*

community All the different populations of organisms that live in the same place *Chapter 4, p. 97*

ecosystem The community or communities in a large area plus the physical environment *Chapter 4, p. 98*

Figure 9-3 Sea urchins such as the purple and red sea urchins (*Stronglyocentrotus purpuratus* and *S. franciscanus*, repectively) of the Pacific coast of North America sometimes undergo population explosions. The effects of such explosions can be devastating (see Figure 12-29).

nutrients Raw materials other than carbon dioxide and water that are needed by plants for primary production *Chapter 4, p. 93*

ronment. Biological **populations** interact in complex ways that make the organisms in a **community** dependent on each other.

> The nature of a community is determined by both biotic, or living, and abiotic, or nonliving, factors.

How Populations Grow

When the conditions are right, organisms can produce many more offspring than it takes to just replace themselves. If every individual has more than one offspring—or each pair has more than two—the total number of offspring increases every generation (Figure 9-2). The population grows faster and faster, and there is a population "explosion." Were its reproduction left unchecked, any species could grow to cover the earth in a relatively short time.

Fortunately, we are not up to our necks in dinoflagellates. Though many organisms occasionally undergo dramatic population explosions (Figure 9-3), reproduction obviously does not continue unchecked forever. Many factors can control, or **regulate,** the growth of populations. For one thing, explosive population growth can occur only under favorable conditions. When the abiotic environment changes, the population may stop growing or even start to decline. Some environmental fluctuations are regular and predictable, like the seasons. A species might thrive during the long, warm summer days, for instance, but die off in the winter. Other changes, like a sudden storm or a drifting log that smashes organisms on a rocky shore, are not so predictable. Predictable or not, large or small, changes in the environment can slow down or even wipe out populations.

There are mechanisms that limit population growth even if the nonliving environment doesn't fluctuate. Some animals slow down or stop reproduction when their habitat becomes too crowded. Others fight among themselves or even cannibalize each other. Natural enemies may be attracted when the population gets large, and diseases often spread faster under crowded conditions. Large populations can pollute the environment with their own wastes.

As more and more individuals join the population, they use up their **resources**—the things they need to live and reproduce, like food, nutrients, and living space. Eventually there are just not enough resources to support any more individuals and the population levels off (Figure 9-4).

Organisms use many different resources. A lack of any of these prevents growth and reproduction. A dinoflagellate living in the middle of the ocean, for example, can be limited by a lack of nitrate (NO_3^{-2}), even if it has plenty of light, water, carbon dioxide (CO_2), and other nutrients. During the winter in polar regions, there may be plenty of nitrate but not enough light. A **limiting resource** is one whose short supply prevents the growth of a population (Figure 9-5).

Because of the drain of resources or other effects of crowding, populations do not grow forever. As the size of the population increases, its *growth rate* goes down. In this sense the population is **self-regulating;** that is, its

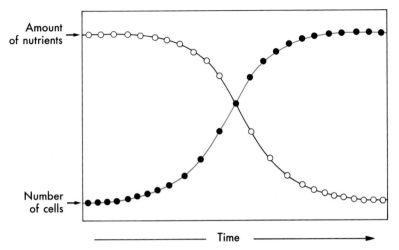

Figure 9-4 The explosive population growth shown in Figure 9-2 would never go on forever. As the dinoflagellates get more and more numerous, they use up the nutrients and other resources that they need to grow.

Figure 9-5 These sea-weeds get plenty of both light and nutrients. They are unable to increase in number because there is no room for more plants to attach themselves. In other words, their limiting resource is space.

growth rate depends on its own numbers. Whereas nonliving factors such as the weather can affect populations of any size, self-regulation only acts when the population is large.

Populations are capable of explosive growth. Growth may be limited, however, by either the nonliving environment or by the activities of the organisms themselves.

As resources run out, there are not enough to go around. Each individual has to vie with other organisms for what resources remain. **Competition** is the interaction that results when a resource is in short supply and one organism uses the resource at the expense of another (Figure 9-6). Those who compete successfully survive to replace themselves by reproducing. The unsuccessful competitors do not reproduce themselves as successfully and eventually disappear. In this way nature favors the members of the population that are best suited to the environment. Because of this natural selection, the population as a whole is a little better adapted each generation. In other words, it evolves.

Ways That Species Interact

Species do not live in a vacuum, isolated from other species. Members of different species may have strong effects on each other. Species can interact in many ways, all of which influence the organization of communities.

Competition. Organisms must compete for resources not only with members of their own species, but with members of other species. When two species use the same resource and when the resource is scarce, the two must compete just as if they were members of the same population. One of the two species almost always turns out to be a little better at the competition. If two species eat exactly the same food, for example, one of the two will be better at catching it. Unless something interferes, the inferior competitor loses out and the competitively superior species takes over. When

natural selection When some individuals are better adapted than others and produce more offspring
Chapter 4, p. 109

evolution The genetic change in a population that results when certain individuals are favored by natural selection
Chapter 4, p. 108

Figure 9-6 These barnacles from a rocky shore in Japan are competing. They have plenty of food, but there is not enough space on the rock where they live to go around.

Figure 9-7 The sargasso fish (*Histrio histrio*) is beautifully adapted to live among fronds of floating sargasso weed (*Sargassum*).

one species eliminates another by out-competing it, **competitive exclusion** is said to occur.

> Competition between species results when both species have the same limiting resource. In competitive exclusion, one species wins the competition and eliminates the other from the community.

Sometimes a competitively superior species is prevented from excluding poorer competitors. The superior species might be affected by bad weather or have natural enemies that keep it from taking over. This gives the lesser competitors a chance. Furthermore, which species is superior might depend on the conditions. In the summer, for example, one species might do better, but the other species might dominate in the winter. As long as the conditions are variable, neither species will be able to exclude the other. Thus there may be a shifting balance between competing species.

Species can also avoid excluding each other if they manage to share the limiting resource, with each species specializing on just part of the resource. For example, two species of fish that both eat seaweeds might divide the resource by specializing on different kinds of seaweed. Animals that eat exactly the same thing might live in different places or feed at different times, staying out of each other's way (see Figure 11-15). Ecologists call this sharing of a resource by specialization **resource partitioning**.

> Species that might otherwise competitively exclude each other sometimes coexist by resource partitioning, or dividing the resource.

Resource partitioning allows species to coexist when they might otherwise exclude each other, but it does have its price. By specializing, each species gives up some of the resource. The fish that specializes on a particular kind of seaweed, for example, will have less food available than if it ate any type of seaweed. With fewer resources available, the size of the population tends to be smaller. On the other hand, the species might be able to use the resource more efficiently by being a specialist than if it were a "jack-of-all-trades" (Figure 9-7).

To be successful in the long run, a species must find the right balance between specialization and generalization. There is no one best answer to the problem—just look at all the different kinds of organisms. Each species has its own special place, or **ecological niche,** in the community. The species' niche is defined by the combination of virtually every aspect of its lifestyle, such as what it eats, where it lives, when and how it reproduces, how it behaves, and so on.

> The role a species plays in the community is called its ecological niche. The niche includes feeding habits, habitat, and all the other aspects of the species' lifestyle.

Eating Each Other. Species don't always compete for resources. Sometimes they use each other as the resource. In other words, they eat one another (Figure 9-8). When an animal eats another organism it is called **predation.** The animal that does the eating is the **predator,** and the organism that gets eaten is the **prey.** The term "predator" is often reserved for carnivores, animals that eat other animals. A special case of predation occurs when animals eat plants. This is usually called **herbivory** rather than predation, and the animal is called a **herbivore.** No matter what you call it, the plant gets eaten just the same.

Predation obviously affects the individual that gets eaten. It also affects the prey population as a whole by reducing the number of prey. If the predators don't eat too much, the prey population can replace the individuals that get eaten by reproducing. If predation is intense, however, it can cut down the prey population greatly.

The interaction between predator and prey is not a one-way street. After all, the predator depends on the prey for its food supply. If bad weather or disease wipes out the prey, the predators suffer as well. Prey populations can also decline if there are too many predators or the predators eat too much. In this case the predator population soon starts to decline, having used up its food supply.

> **Predation, which occurs when an animal eats another organism, affects the numbers of both predator and prey.**

The more successful an individual predator is at catching its prey, the better its chances of surviving to reproduce. Natural selection, therefore, favors the most efficient predators in the population. Each generation will be a little better at capturing its food.

There is more than one way to solve a problem, and a tremendous variety of predatory strategies have evolved. Sharks and tunas are highly efficient killing machines that are swift and powerful. Other predators sneak up on their prey, catching them unaware. Predators such as anglerfishes actually lure their prey. Many snails can drill through the shells of their prey, and some predators, like slime eels, actually climb in and eat their prey from the inside! Many predators eat only part of the prey, leaving the rest alive to continue growing. In the following chapters we will take a closer look at some of the many ways that predators catch their prey.

Just as natural selection favors the best predators, it favors the prey individuals that are most successful at getting away. Organisms have evolved at least as many ways to escape from predators as there are predators. Some are fast and elusive, others use camouflage. Organisms may have protective spines, shells, or other defensive structures. Many plants and animals protect themselves with distasteful or even poisonous chemicals (Figure 9-9).

Thus there is a continual "arms race" between predators and their prey. The predator continually gets better at catching the prey and overcoming its defenses. In response, the prey becomes more adept at escaping or develops better defenses. This interplay, with each species evolving in response to the other, is known as **coevolution.**

Living Together. Coevolution becomes even more important when species interact more intimately. Members of different species may live in very close association, even with one inside another (see Box 9-1). Such close relationships are known as examples of **symbiosis,** which literally means "living together." The smaller partner in the symbiosis is usually called the **symbiont** and the larger one the **host.**

Figure 9-8 A grouper eating a rainbow wrasse in the Red Sea.

Figure 9-9 This sea slug (*Bajaeolis bertschi*) can get away with being so gaudy because its tissues contain stinging structures, so it doesn't have to hide from predators. In fact, the bright colors warn predators away. The sea slug gets the stinging structures—nematocysts—from the hydroid *Eudendrium ramosum* that it lives on and eats.

BOX 9-1

Symbiosis and the Modern Cell

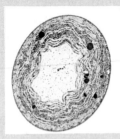

Prochloron.

Symbiosis, the living together of different kinds of organisms, is a fascinating and vital part of many marine ecosystems. Indeed, many organisms, such as the corals that depend on their zooxanthellae (see below), cannot live without their symbionts. Symbiosis may be of even more profound importance than that. Many scientists think that the very cells of higher organisms arose as a result of symbiosis in the distant past.

There are two fundamental types of cells: the simple prokaryotic cells of bacteria and blue-green algae (or cyanobacteria) and the more complex eukaryotic cells of all other organisms. What sets eukaryotic cells apart is the presence of organelles, which are membrane-bound structures within the cell. Mitochondria are the organelles where respiration takes place—the "powerhouses" of the cell (see Figure 4-13). They are 1 to 3 microns long, about the same size as many bacteria. Like many groups of bacteria, mitochondria contain layers of folded membranes. Although most of the cell's DNA is contained in the nucleus, mitochondria contain a small amount of their own DNA in the form of a single, circular molecule similar to the DNA of bacteria. When a cell needs new mitochondria it does not make them from scratch. Instead, the existing mitochondria, divide in two, much as bacteria divide by cell division. Mitochondria also resemble bacteria in several other respects.

The similarities between mitochondria and bacteria have led many biologists to believe that mitochondria originated as disease bacteria that invaded other cells. Most of the host cells died of the infection, but a few survived. The invading bacteria began to live symbiotically with the surviving host cells.

According to this theory, the host cells could not use oxygen in respiration. The symbiotic bacteria did use oxygen and gave this ability to the hosts when they moved inside. This gave the host cells an advantage over cells that did not have the symbiotic bacteria. Over the course of time, the bacteria and the host cells became more and more dependent on each other. Eventually the bacteria became mitochondria.

Chloroplasts, the organelles that perform photosynthesis, are also thought to have arisen from symbiotic bacteria. They contain folded membranes like those of today's photosynthetic bacteria (see Figure 4-12). They also contain a circular DNA molecule, can reproduce themselves, and bear other resemblances to bacteria. In fact, biologists have found a possible "missing link" between photosynthetic bacteria and chloroplasts. Some sea squirts and other invertebrates contain symbiotic, photosynthetic bacteria of the genus *Prochloron*. Although it is prokaryotic, *Prochloron* also has similarities to the chloroplasts of green algae and higher plants. It is thought that the chloroplasts of these plants arose from *Prochloron*-like bacteria when host cells engulfed the bacteria. The few bacteria that managed to resist being digested became chloroplasts.

The hypothesis that mitochondria and chloroplasts arose from some bacteria was once highly controversial and is still not accepted by a few biologists. Evidence supporting the hypothesis continues to accumulate, however. It seems likely that symbiosis has played a profound role in the evolution of all higher organisms, including ourselves.

Biologists often divide symbiosis into different categories according to whether the organisms involved benefit from or are harmed by the relationship. In **commensal** relationships, one species obtains shelter, food, or some other benefit without affecting the other species one way or the other. For example, certain barnacles live only on whales (Figure 9-10). The barnacles get a place to live and a free ride. They feed by filtering the water, and as far as we can tell, the whale is neither harmed nor helped by the barnacles. There are many other examples of symbiotic organisms that live on, or even in, a host without apparently either hurting or helping it (see Figure 6-41).

On the other hand, sometimes the symbiont benefits at the expense of the host (Figure 9-11). This is called **parasitism.** The giant tapeworms that live in the guts of whales (see the section in Chapter 6, "Flatworms") are considered parasites because they derive food and shelter from the host. In the process they may weaken their whale hosts. There are many examples of marine parasites. In fact, probably no marine species escapes having at least one type of parasite.

Not all symbiotic relationships are one-sided. In **mutualism,** both partners benefit from the relationship. In many places, small cleaner fishes and shrimps have mutualistic relationships with larger fishes. The cleaners remove harmful parasites from the fish, getting a meal in return. The fish, which could easily eat the cleaners, allow them to poke and prod over their bodies and even inside their mouths (Figure 9-12). This type of symbiosis is called **cleaning symbiosis** (see Box 12-1).

In the case of cleaning symbiosis, both partners can get by without the other if they have to. Sometimes the partners depend on each other. The colorful crab shown in Figure 6-1, for example, is found nowhere else but on its host coral. The coral supplies not only shelter but food, by producing the mucus that the crab eats. Even though the coral isn't really dependent on the crab, it benefits because the crab helps chase away sea stars and other coral predators.

The coral provides an example of an even closer relationship, one in which neither partner can live without the other. The tiny zooxanthellae that live within its tissues help the coral make its calcium carbonate ($CaCO_3$) skeleton. The zooxanthellae also make food for the coral by photosynthesis. The zooxanthellae, in turn, get both nutrients and a place to live from the deal. Another example of mutualism is shown in Figure 5-4.

> In symbiosis, members of different species live in close association. Symbiosis includes parasitism, in which the symbiont harms the host, commensalism, in which the host is not affected, and mutualism, in which both partners benefit.

Figure 9-10 These barnacles live only on the skin of whales. Because they do not appear to affect their host at all, they are considered to be commensals.

zooxanthellae Photosynthetic dinoflagellates that live in animal tissues *Chapter 5, p. 124; Figure 13-9*

THE FLOW OF ENERGY AND MATERIALS

All living things use energy to make and maintain the complex chemicals necessary for life. **Autotrophs** get this energy from the environment, usually in the form of sunlight. They use the energy to make their own food from simple molecules like carbon dioxide, water, and nutrients. As the autotro-

autotrophs Organisms that can use energy (usually solar energy) to make organic matter *Chapter 4, p. 90*

Figure 9-11 This Cooper's nutmeg snail (*Cancellaria cooperi*) is a parasite of electric rays, which rest on the bottom for long periods. The snail smells the ray from far away and moves up to the ray (*top*). It then extends its long proboscis, makes a tiny cut in the ray's skin with its radula (see Figure 6-27), and sucks the ray's blood.

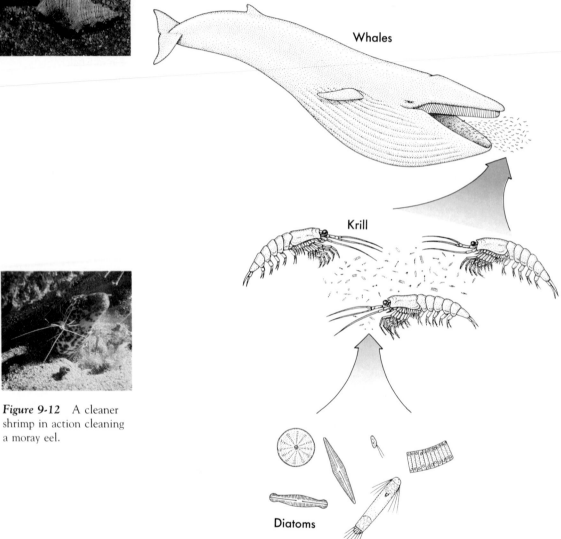

Whales

Krill

Diatoms

Figure 9-12 A cleaner shrimp in action cleaning a moray eel.

Figure 9-13 This three-step food chain is typical of Antarctic waters. The main primary producers are diatoms, which capture sunlight to fix carbon. The energy is transferred from the diatoms to the krill that eat them, and then to blue whales that eat the krill. Though this food chain is greatly oversimplified (see Figure 9-14), it accounted for a great deal of the total productivity of Antarctic waters until whales were hunted nearly to extinction.

phs grow and reproduce they become food for **heterotrophs.** When one organism eats another, both the organic material and the energy stored in it are passed from one to the other. Thus energy and chemical substances flow from the nonliving part of the ecosystem to the organism and from organism to organism. The pathways taken by energy and materials tell us a lot about how the ecosystem works.

Trophic Structure

The flow of energy and matter through an ecosystem can be traced by observing the food, or **trophic,** relationships among its organisms: who makes the food and who eats it. The organisms can be divided into two broad components: **primary producers,** the autotrophs that make the food, and **consumers,** the heterotrophs that eat it. Not all consumers feed directly on producers. Many animals eat other animals rather than plants. Thus the transfer of energy through the system usually takes place in several steps known as a **food chain** (Figure 9-13). Each of the steps in the food chain is known as a **trophic level.**

Most ecosystems have a number of different primary producers. Furthermore, many animals eat more than just one kind of food, and many change their diet as they get older and larger (see Figure 14-29). For these reasons, trophic structure is usually a complex, interwoven **food web** (Figure 9-14) instead of a simple, straight-line food chain. Such food webs may be difficult for biologists to unravel and understand, but their complexity is one reason for the tremendous diversity of life.

> Energy and materials are passed from one trophic level to another along a food chain or food web. The first level is occupied by primary producers, the other levels by consumers.

Trophic Levels. Food chains and food webs are a little easier to understand if we consider how many steps, or trophic levels, the energy has traveled. The first step in the flow of energy, and therefore the first trophic level, is occupied by the primary producers that originally capture the energy and store it in organic compounds. Diatoms are the primary producers in the food web illustrated in Figure 9-14. Consumers that feed directly on the producers are called **first level** or **primary consumers,** and occupy the next trophic level. At the next level are **second level** or **secondary consumers,** predators that eat the primary consumers. Feeding on the secondary consumers are the **third level** or **tertiary consumers,** and so on. Each trophic level relies on the level below for sustenance. At the end of the food web are **top predators** such as the killer whales in Figure 9-14.

The concept of trophic levels helps us understand how energy flows through ecosystems, but the organisms themselves are not especially concerned about which level they feed at. Predators commonly feed on prey from different trophic levels, for example.

The Trophic Pyramid. Instead of being passed on to the next higher level, much of the energy contained in a particular trophic level is used up by the activities of the organisms. Energy and organic matter are also lost as waste. In fact, only about 10% of the energy, on average, is passed from one

heterotrophs Organisms that cannot make their own food and must eat the organic material produced by autotrophs
Chapter 4, p. 90

primary production The conversion of carbon dioxide into organic matter by autotrophs, that is, the production of food
Chapter 4, p. 92

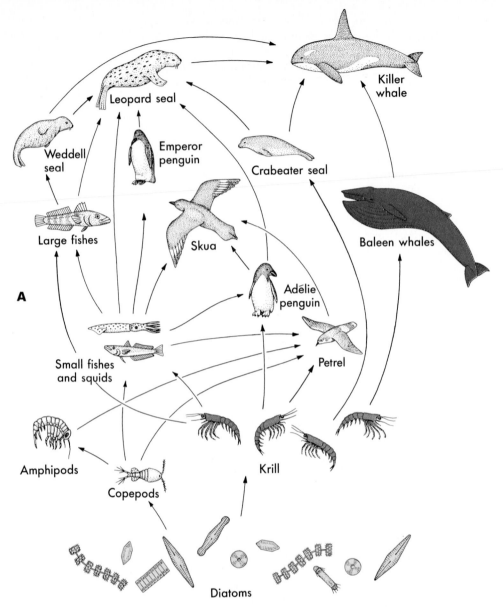

Figure 9-14 **(A)** A simplified Antarctic food web. The simple diatom-to-krill-to-whale chain illustrated in Figure 9-13 is an important part of the food web.

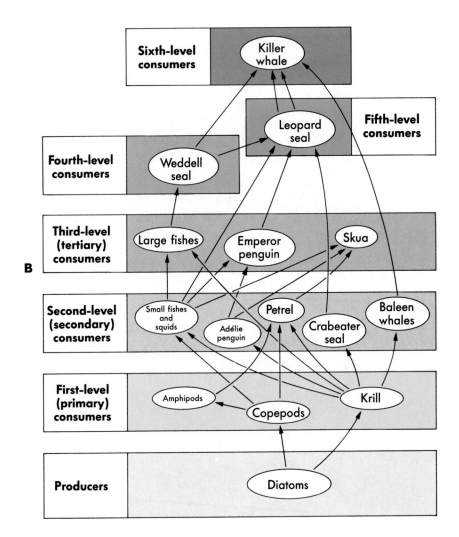

Figure 9-14, cont'd. (B) The food web can be simplified even further by combining organisms into categories that are based on their trophic level.

Sixth-level consumers — Killer whale

Fifth-level consumers — Leopard seal

Fourth-level consumers — Weddell seal

Third-level (tertiary) consumers — Large fishes, Emperor penguin, Skua

Second-level (secondary) consumers — Small fishes and squids, Adélie penguin, Petrel, Crabeater seal, Baleen whales

First-level (primary) consumers — Amphipods, Copepods, Krill

Producers — Diatoms

B

level to the next. Suppose, for example, that the diatoms in the Antarctic food chain illustrated in Figure 9-13 contain a total of 10 million calories of energy. According to the 10% rule, only about 1 million calories will make it to the primary consumers, the krill, and the whales would get only about 100,000 calories. The trophic structure of ecosystems can be represented by a **pyramid of energy** (Figure 9-15), with less energy contained in each succeeding level.

Because there is less energy available at each level, there are also fewer individual organisms. Thus there are fewer primary consumers than producers, and fewer secondary than primary consumers. The trophic pyramid can be pictured in terms of numbers of individuals as well as energy, and it is then called the **pyramid of numbers.** The pyramid can also be expressed in terms of the total weight, or **biomass,** of the organisms at each level. In this

Figure 9-15 (A) Although the figure varies, an average of 90% of the energy that primary producers capture in photosynthesis is lost at each step up the food chain. (B) This relationship is often depicted as a pyramid, the trophic pyramid. The pyramid can represent either the amount of energy, the numbers of individuals, or the biomass at each trophic level. The idea remains the same: it takes a lot of energy, numbers, and biomass at the bottom to support just a little at the top.

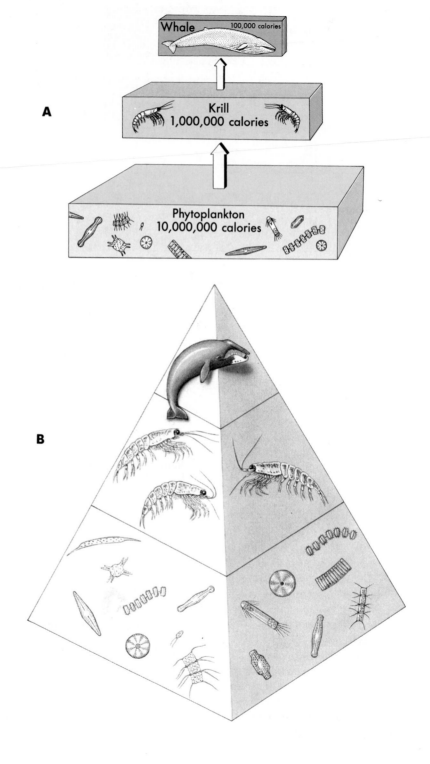

STRUCTURE AND FUNCTION OF THE MARINE ENVIRONMENT

case, the trophic pyramid is called the **pyramid of biomass.** To support a given biomass of primary consumers, primary producers must make about 10 times as much living tissue. To support 1,000 grams of copepods, for example, about 10,000 grams of phytoplankton must be eaten. In turn, only about one tenth of the primary consumer's biomass will make it to the secondary consumers.

On an average, only about 10% of the energy and organic matter in one trophic level is passed to the next higher level.

Because each level has less energy available, there is a limit to how many levels there can be. Eventually the system just runs out of energy. Most marine food webs have three or four trophic levels.

As noted above, some organic material is lost as waste instead of moving through a food web of herbivores and predators. This material is not lost to the ecosystem, however. Bacteria, fungi, and other **decomposers** break down the waste products and dead bodies of organisms. They break the nonliving organic matter down into its original components: carbon dioxide, water, and nutrients. Together, the decomposers and dead organic matter are called **detritus.** Detritus is an important energy pathway in marine ecosystems because many marine organisms feed on it. This funnels the organic matter back into the food web.

Detritus consists of nonliving organic matter and the decomposers that break it down.

Decomposers are a very vital component of all ecosystems. If it were not for decomposers, waste products and dead bodies would accumulate instead of rotting away. Not only would this make quite a mess, nutrients would remain locked up in the organic matter. When decomposers break down organic matter, the nutrients incorporated into the organic matter during primary production are released, making the nutrients available again to photosynthetic organisms. This process is known as **nutrient regeneration.** Without it, nutrients would not be available to autotrophs, and primary production would be greatly limited.

Measuring Primary Productivity. Because primary production supplies the food at the base of the trophic pyramid, it is useful to know how much production occurs in a given area. The rate of primary production, or **productivity,** is expressed as the amount of carbon fixed under a square meter of sea surface in a day or in a year (Figure 9-16). It includes the production of both phytoplankton in the water column and plants that live on the bottom.

To measure primary production, biologists determine either the amount of raw materials used up in photosynthesis or the amount of end products given off. Photosynthesis produces oxygen (O_2) and consumes carbon dioxide, so photosynthesis can be estimated by measuring *how much* oxygen is produced or carbon dioxide used up.

Primary production has traditionally been measured by placing water samples in bottles and determining the amount of oxygen produced. Usually

photosynthesis CO_2 + H_2O + SUN ENERGY $\longrightarrow$ organic matter + O_2
Chapter 4, p. 88

respiration Organic matter + $O_2 \longrightarrow CO_2$ + H_2O + ENERGY
Chapter 4, p. 90

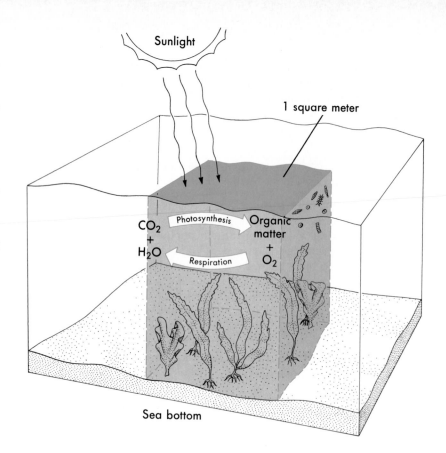

Figure 9-16 To describe primary productivity, marine biologists figure out how much carbon ends up being fixed, or converted from carbon dioxide to organic material, in the water column under 1 m² of sea surface in a given time. Primary productivity is thus expressed in units of grams of carbon per square meter per day or year (gC/m²/day or gC/m²/year).

two bottles are used: a clear one and one that is opaque—painted black, covered with foil, or made of dark glass. To understand the need for this "light–dark bottle" technique, you must remember that all organisms, including phytoplankton, constantly use up energy to stay alive. Even while photosynthesizing in the light bottle, phytoplankton respire some of the organic matter they produce to provide this energy. Thus changes in carbon dioxide reflect the amount taken up in photosynthesis minus the amount given off by respiration (Figure 9-17). To measure photosynthesis, the amount produced by respiration must be known. Respiration is estimated from the amount of carbon dioxide produced in the dark bottle. There is no light in the dark bottle so there can be no photosynthesis. Any change in the carbon dioxide level must be due to respiration. The carbon dioxide used up in the light bottle is then corrected for the rate of respiration, and the rate of photosynthesis is calculated.

A more modern method measures the amount of carbon dioxide used by phytoplankton. This method uses a radioactive form of carbon (^{14}C) that can be measured very accurately.

The rate of primary production is usually measured in light and dark bottles. Changes in the oxygen or carbon dioxide level in the light bottle indicate both photosynthesis and respiration, whereas changes in the dark bottle reflect only respiration.

STRUCTURE AND FUNCTION OF THE MARINE ENVIRONMENT

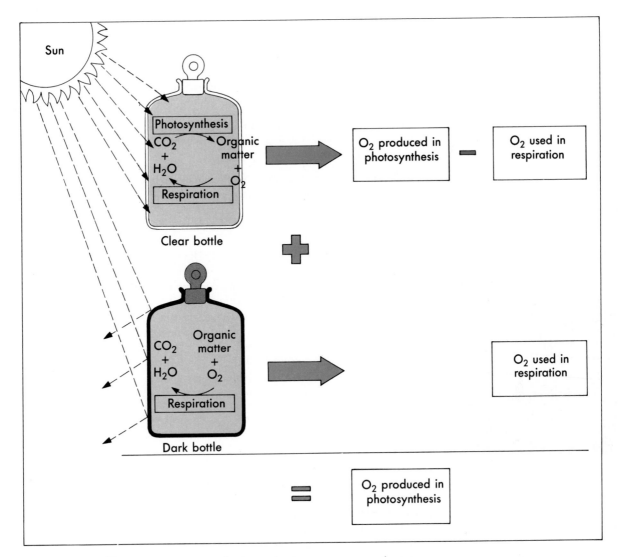

Figure 9-17 To measure primary production, scientists put water samples into both light and dark bottles. In the light bottle both photosynthesis and respiration take place. In the dark bottle there can be only respiration. Combining the results in the two bottles allows the determination of photosynthesis alone.

The total amount of phytoplankton in the water—or the **standing stock** of phytoplankton—is related to the primary productivity, but the two are not the same thing. The standing stock refers to how much phytoplankton is already in the water. Primary productivity refers to the amount of new organic material being created. In a lawn, for example, the standing stock would correspond to the length of the grass. The primary productivity would be the amount of clippings produced if the lawn were mowed every day to keep it the same length.

Like all plants, phytoplankton contain **chlorophyll,** so the amount of chlorophyll in the water is a good indication of the amount of phytoplankton. Indeed, the most common way to measure the phytoplankton standing stock is to measure the concentration of chlorophyll in the water.

chlorophyll A green pigment that absorbs light energy for photosynthesis *Chapter 4, p. 89*

> **The standing stock of phytoplankton is the total amount of phytoplankton in the water column. Standing stock is usually determined by measuring the chlorophyll concentration.**

Chlorophyll can be measured by chemically extracting it from the water, but this is time consuming and expensive. Instead, an instrument called a fluorometer is usually used. Under certain kinds of light, chlorophyll emits a glow, or fluorescence; the amount of chlorophyll can be determined by measuring this fluorescence.

Phytoplankton standing stock can even be measured from space. Satellites equipped with special cameras take color pictures of the sea surface and beam the images down to earth. Using computers, scientists carefully analyze the photos, paying special attention to the characteristic green color of chlorophyll. By applying various correction factors, taking account of the weather at the time the photo was taken, and relating the results to actual chlorophyll measurements made from ships, phytoplankton standing stock can be estimated over vast areas of the sea surface (Figure 9-18). Satellites are the only means to get a large-scale understanding of the distribution of phytoplankton: Ships can measure standing stock in only one place at a time.

Figure 9-18 This satellite image combines data on the concentration of chlorophyll in the Atlantic Ocean from November 1978 through June 1981. Chlorophyll concentrations are low in the central regions of the ocean and high along the coasts and at high latitudes.

Cycles of Essential Nutrients

Once the energy that is stored in organic compounds is used in metabolism or given off as heat, it is lost to the system forever. Unlike energy, the *materials* that make up organic matter can be used over and over again in a repeating cycle. These materials—like carbon, nitrogen, and phosphorus—originally come from the atmosphere or the weathering of rocks. Starting out in the form of simple inorganic molecules, they are converted into other forms and incorporated into the tissues of autotrophs. When this organic material is broken down by digestion, respiration, and decomposition, the raw materials are released back to the environment, and the cycle begins again.

The **carbon cycle** provides a good example of this process (Figure 9-19). The carbon that forms the backbone of all organic molecules starts in the atmosphere as carbon dioxide, then dissolves in the ocean. This inorganic carbon is converted into organic compounds by photosynthesis. Respiration by consumers, decomposers, and the producers themselves breaks down the organic compounds and makes the carbon dioxide available to producers again. Some carbon is also deposited as calcium carbonate in biogenous sediments and coral reefs. Under certain conditions, some of this calcium carbonate dissolves back into the water.

280   STRUCTURE AND FUNCTION OF THE MARINE ENVIRONMENT

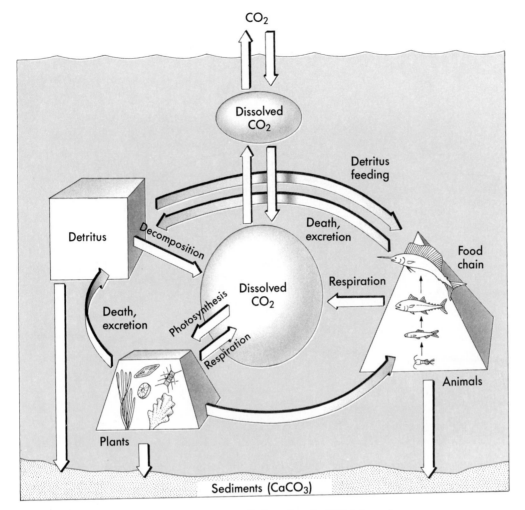

CO₂

Dissolved
CO₂

Detritus
feeding

Detritus

Decomposition

Death,
excretion

Food
chain

Death,
excretion

Photosynthesis

Dissolved
CO₂

Respiration

Respiration

Plants

Animals

Sediments (CaCO₃)

Figure 9-19 The carbon cycle in the ocean. Carbon dioxide (CO_2) from the atmosphere dissolves in the ocean. In photosynthesis, marine plants convert carbon dioxide into organic matter, which is then passed to animals along the food chain. Both plants and animals release some carbon dioxide as a product of their respiration. The rest of the carbon fixed by plants is eventually either excreted as waste or ends up in the form of dead bodies. This non-living organic matter, together with decay bacteria, forms detritus. The bacteria in detritus decompose the organic matter, releasing carbon dioxide and starting the cycle anew.

Nitrogen is also present in the atmosphere. Atmospheric nitrogen, however, is in the form of nitrogen gas (N_2), which most organisms are unable to use. A few types of bacteria and blue-green algae (or cyanobacteria) are able to convert nitrogen gas into forms that can be used by plants and other organisms. This conversion process is known as **nitrogen fixation,** and the organisms that perform it are called **nitrogen fixers.** Without nitrogen fix-

biogenous sediment Made of the skeletons and shells of marine organisms *Chapter 2, p. 40*

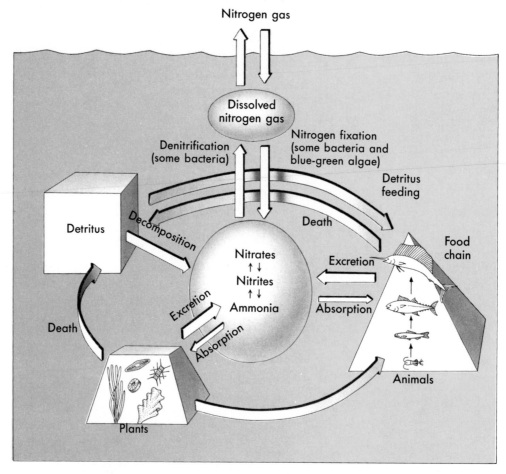

Nitrogen gas

Dissolved nitrogen gas

Denitrification (some bacteria)

Nitrogen fixation (some bacteria and blue-green algae)

Detritus feeding

Detritus

Decomposition

Death

Food chain

Nitrates
↑ ↓
Nitrites
↑ ↓
Ammonia

Excretion

Absorption

Excretion

Absorption

Death

Absorption

Plants

Animals

Figure 9-20 The nitrogen cycle in the ocean. The atmosphere is rich in nitrogen gas (N_2), which cannot be used by most organisms. Some bacteria and blue-green algae (or cyanobacteria) convert N_2 into usable forms of nitrogen in the process of nitrogen fixation. There are several different forms of usable nitrogen; various bacteria and other organisms convert nitrogen from one usable form to another. Plants take up the usable nitrogen to produce organic matter. The nitrogen is passed on to animals when they eat the organic matter. Usable nitrogen may be excreted as waste by plants and animals or regenerated by decomposing bacteria.

ers, plants would be unable to obtain the nitrogen that they need to grow and reproduce.

> **Nitrogen fixation is the conversion of atmospheric nitrogen gas to a form that organisms can use. It is performed by certain bacteria and blue-green algae.**

Once nitrogen has been fixed it cycles within the ecosystem as part of the **nitrogen cycle.** (Figure 9-20). There are a number of nutrient forms of nitrogen, the most important of which is nitrate. Nitrate and other nitrogen compounds are taken up during primary production and regenerated by de-

composition. Most of the nitrogen-containing nutrients used in primary production come from this recycling of nitrogen compounds, not the original fixation of nitrogen.

The other raw materials needed by organisms, like phosphorus and silicate, undergo similar cycles. The tremendous abundance and diversity of living things depends on this recycling. Without it, the living world would soon run out of raw materials, and life on earth would be scarce indeed.

The living world depends on the recycling of materials such as carbon, nitrogen, and phosphorus. Decomposition is a vital part of this cycle.

BIOLOGICAL ZONATION OF THE MARINE ENVIRONMENT

It has already been noted that, because physical and chemical conditions change from place to place, different parts of the ocean harbor very distinct communities. For convenience, marine biologists categorize communities according to where and how the organisms live. Perhaps the simplest classification relates to the lifestyle of the organism: whether it lives on the bottom or up in the water column. **Benthic organisms,** or the **benthos,** are those that live on or buried in the bottom (Figure 9-21). Some benthic organisms are **sessile,** or attached to one place. Others are able to move around. **Pelagic** organisms, on the other hand, live up in the water column, away from the bottom.

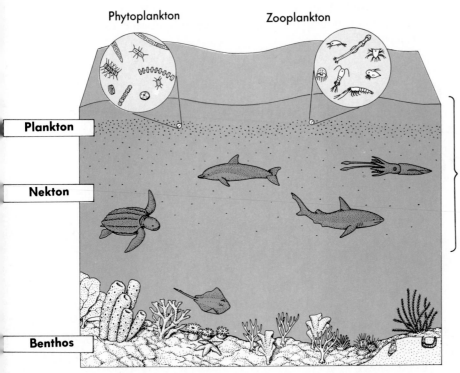

Phytoplankton

Zooplankton

Plankton

Nekton

Benthos

Pelagic

Figure 9-21   Three major groups result when marine organisms are classified by their lifestyles: benthos, plankton, and nekton.

Pelagic organisms are further subdivided according to how well they can swim. Some marine organisms swim only weakly or not at all. These organisms, called **plankton,** are at the mercy of the currents and are carried from place to place. The term "plankton" comes from the Greek word for "drifters." Planktonic plants and other autotrophs are collectively called the **phytoplankton** and are the most important primary producers in many marine ecosystems. The animal plankton are called the **zooplankton.**

Animals that can swim well enough to oppose the currents are called the **nekton.** Most nektonic animals are vertebrates, mainly fishes and marine mammals. There are a few nektonic invertebrates, however, such as squids and large jellyfish. Note that not all nekton are pelagic. The ray in Figure 9-21, for example, is part of the nekton because it can swim. It is not considered pelagic, however, because it spends most of its time on or just above the bottom rather than up in the water column.

Another way to classify marine communities is by where they live. Zonation in benthic organisms, for example, relates to depth and the continental shelf (Figure 9-22). Right at the boundary between land and sea is the shallowest part of the shelf, the **intertidal zone.** This is the area between the tides: exposed to the air when the tide is out but under water at high tide. The intertidal zone makes up only a small fraction of the continental shelf. Most of the shelf is never exposed to the air, even at the lowest of low tides. Benthic organisms that inhabit the continental shelf beyond the intertidal live in the **subtidal zone,** sometimes called the **sublittoral zone.** Away from the shelf the benthic environment is subdivided by depth into the **bathyal, abyssal,** and **hadal** zones (Figure 9-22). For the sake of simplicity, we shall lump these different zones together and just call them the "deep ocean floor."

Figure 9-22 The marine environment is divided into zones according to distance from land, water depth, and whether the organisms are benthic or pelagic.

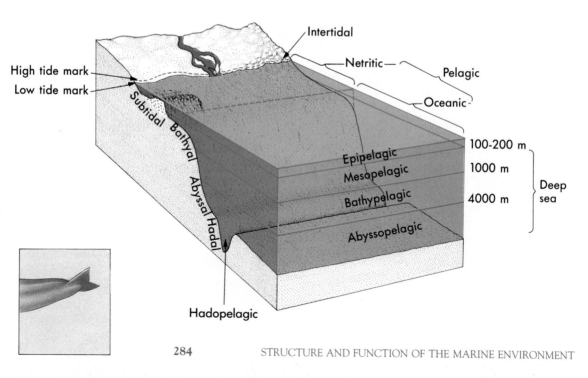

STRUCTURE AND FUNCTION OF THE MARINE ENVIRONMENT

The pelagic environment, too, is divided with reference to the continental shelf. The pelagic environment that lies over the shelf is called the **neritic,** or **coastal, zone.** Pelagic waters beyond the shelf break are the **oceanic zone.**

Besides being divided into the neritic and oceanic zones, the pelagic is divided vertically into depth zones that correspond to the amount of light. In the shallowest, the **epipelagic zone,** there is plenty of light for photosynthesis, at least for part of the year. Given a supply of nutrients, plants thrive, producing food for the rest of the ecosystem. The epipelagic usually extends down to a depth of 100 to 200 meters (350 to 650 feet). Because this is approximately the depth of the continental shelf, nearly all neritic waters lie in the epipelagic.

Other pelagic zones lie in the deep oceanic waters beyond the shelf. Below the epipelagic lies the **mesopelagic zone.** There is not enough light in the mesopelagic to support photosynthesis; thus plants cannot grow. There is, however, enough light to see by. On even the brightest day, the mesopelagic is an area of twilight. This "twilight zone" typically extends to a depth of around 1,000 meters (3,000 feet).

No sunlight at all penetrates to the deepest parts of the ocean, the **bathypelagic, abyssopelagic,** and **hadopelagic** zones. Though each of these zones supports a different community of animals, they share many similarities. We refer to all these zones together as the "deep sea" or "deep pelagic" environment.

shelf break   The increase in steepness that marks the outer edge of the continental shelf
Chapter 2, p. 42

Do It Yourself

SUMMARY

1. _____ is the study of how organisms interact with each other and with their environment.
2. The things that organisms need to survive and reproduce, such as food, water, and space, are called _____. A _____ is one whose short supply prevents the growth of a population.
3. _____ is what happens when there are not enough resources to go around.
4. _____ occurs when a superior competitor eliminates another species from the community.
5. _____ occurs when two species avoid competition by sharing a resource.
6. The _____ of a species is its precise role in the community: its diet, habitat, and other aspects of its lifestyle.

7. _____ occurs when animals eat other organisms. Carnivores are animals that eat _____ and herbivores are animals that eat _____.

8. _____ is the living together of different organisms. In _____, one organism benefits without harming the host. In parasitism, _____. In _____, both partners benefit.

9. From the point of view of food, or _____, relationships, a community can be divided into _____, which make food by photosynthesis, and consumers, which _____. Energy and materials flow from one level to another up the _____.

10. On an average, about _____% of the energy and _____ contained in a given trophic level is passed to the next higher level.

11. Microorganisms that break down nonliving organic matter are called _____. Together with the organic matter, they make up _____.

12. _____ is the transformation of nitrogen gas into a form that organisms can use.

13. Organisms that live on the bottom are called _____. Pelagic organisms live in _____. _____ are organisms that drift in the currents and nekton are organisms that can swim _____ the currents.

THOUGHT QUESTIONS

1. Two species of sea urchins live practically side by side on sandy bottoms. The two species appear to have the same diet: drift seaweeds and other bits of organic matter. They are able to live in the same environment without competing with each other. How might they be able to share their habitat and food resources?

2. The many instances of symbiosis in the marine environment are not always easy to classify under a particular category. Suppose that the relationship of a snail, which is always found living on a coral, is assumed to be an instance of commensalism because there is no evidence that the coral is harmed or weakened. How would you go about to find out whether corals inhabited by this snail are actually "harmed"? Is it possible that this symbiosis may prove to be a case of mutualism? How?

FOR FURTHER READING

Isaacs, J.D.: "The Nature of Oceanic Life." *Scientific American*, vol. 221, no. 3, September 1969, pp. 146-162. *A classic summary of the ecology of the seas.*

McDermott, J.: "A Biologist Whose Heresy Redraws the Tree of Life." *Smithsonian*, vol. 20, no. 5, August 1989, pp. 72-81. *A biologist named Lynn Margulis broke with conventional "knowledge" and first proposed that organelles arose from symbiotic bacteria. Now most biologists think she is right.*

Yanagisawa, Y.: "Strange Seabed Fellows." *Natural History*, August 1990, pp. 46-51. *The symbiotic relationship between a burrowing shrimp and a fish "sentry."*

Between the Tides 10

Of all the vast ocean, the narrow edge that is the **intertidal zone** is by far the most familiar to marine biologists and laypersons alike. The reason for this is simple: the intertidal is the only part of the marine world that we can experience first-hand without leaving our own natural element. We don't need a boat to visit the shore, and at least at low tide, we can see it without goggles and move through it without swim fins. Perhaps more importantly from the scientist's point of view, we can work in the intertidal without cumbersome and expensive equipment and can easily return to the exact same place time after time. For these reasons, intertidal communities are probably the most studied and best understood of all marine communities. The lessons taught in the intertidal have added immensely to our knowledge not just of marine biology, but of biology in general.

The intertidal, sometimes called the **littoral zone,** is defined as the part of the sea floor that lies between the highest high and the lowest low tides (Figure 10-1). It is unique among marine environments in that it is regularly exposed to the air. Every organism that lives in the intertidal must have some way to cope with this daily exposure, even it if means giving up characteristics that would be advantageous below the tides. Being out of the water and exposed to air is called **emersion**—the opposite of **immersion,** or being placed underwater.

The effects of the tide, and thus the nature of the community, depend to a great extent on the type of bottom. The bottom—or, to be more correct, the material on or in which an organism lives—is called the **substrate.** Hard, rocky bottoms pose very different problems from soft bottoms made of mud or sand. In this chapter we shall consider intertidal communities on both kinds of substrate, beginning with the rocky intertidal. The major coastal communities of North America are indicated in Appendix C.

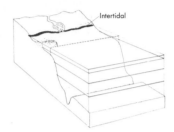

community   All the different populations of organisms that live in an area
Chapter 4, p. 97

> The intertidal zone is the part of the shore between the high and low tide marks. It is the only part of the marine world that faces regular exposure to air, or emersion. Intertidal communities are often divided according to the type of substrate: rocky and soft bottoms.

ROCKY SHORE COMMUNITIES

Rocky shores generally occur on steep coasts that lack large amounts of sediment. Often, such areas have been recently uplifted or are still rising as a result of geologic events. These uplifted coasts have not had much time to erode or accumulate sediments. The west coast of the Americas, for example, is largely rocky because its active margin has been uplifted by geologic

active continental margin
One that is colliding with another plate and therefore has a lot of geologic activity
Chapter 2, p. 42; Figure 2-24

passive continental margin
One that is on the "trailing edge" of a continent and has little geologic activity
Chapter 2, p. 44; Figure 2-24

Figure 10-1 The intertidal zone is alternately submerged (**A**) and exposed to air (**B**) as the tides rise and fall.

processes. Eastern Canada and New England were covered by huge sheets of ice during the last ice age (see the section in Chapter 2, "Climate and Changes in Sea Levels"). The ice sheets scraped the sediments from the continental shelf, exposing the bare rock underneath. Under the tremendous weight of the ice, the coast actually sank partially down into the mantle. When the ice melted, the coast slowly rose, or **rebounded,** leaving a coastline of exposed rock. Sea level, however, began to rise, and eventually the rise in sea level overtook the rebound of the coast. The sea flooded the rocky coast, forming the beautiful, deeply sculptured shoreline we see north of Cape Cod (see Figure 10-27). This area is the only part of the Atlantic coast of North America that has much rocky shore. The southern Atlantic and Gulf coasts of North America are slowly sinking, or **subsiding,** weighed down by the huge amounts of sediment that are accumulating on the passive continental margin. Rocky areas are practically absent from these coasts.

Some rocky shores are not formed by uplift. Waves and currents can carry sediments away, leaving bare rock behind. Much of the coast of the island of Hawaii is rocky because it is geologically very young. Formed by successive flows of lava into the sea, the coast has not had time to accumulate sediment. In fact, the coast is still being formed by the periodic eruptions of the Kilauea volcano (Figure 10-2).

> **Rocky shores occur on recently uplifted or geologically young coasts that have not accumulated sediments or on coasts where erosion removes the sediment. In North America, rocky shores are common on the West Coast and on the East Coast north of Cape Cod.**

It is difficult to burrow through rock, though rock-boring clams (*Zirfaea, Penitella*) can do it in soft rock such as sandstone. Most rocky intertidal organisms live right on the rock's surface. Animals that live on the surface of the substrate—be it rock, sand, mud, or anything else—are called **epifauna.** Some epifauna move about over the rocks, but many are **sessile** and stay put, attaching to the rock.

mantle One of the three main internal layers of the earth. The mantle lies under the *crust,* the outermost layer, and over the *core,* the innermost
Chapter 2, p. 26; Figure 2-4

LIFE IN THE MARINE ENVIRONMENT

Figure 10-2 **A,** Lava flows are creating new shoreline on the island of Hawaii. **B,** Much of Hawaii's coast is black, volcanic rock.

Living on the rock's surface, the plants and animals in the rocky intertidal are fully exposed to the elements. This subjects them to great physical stress.

Exposure at Low Tide

Low tide presents severe problems for the organisms of the rocky intertidal. Left high and dry, they are exposed to the air, which is a much harsher environment than the water. It is important to note that this exposure is more of a problem for those plants and animals that live high in the intertidal. The upper part of the intertidal zone is only submerged at the peak of high tide and then only briefly. The very upper edge may not be submerged every day, but only during exceptionally high spring tides. The highest part of the "intertidal," in fact, is almost never submerged. It is kept wet by wave splash. Organisms that live low in the intertidal, on the other hand, are submerged most of the time and have to cope with emersion only for short periods or, at the lowermost edge, only at the most extreme low tides. The time spent out of the water, then, gets longer moving up the shore.

Emersion time, or time spent out of the water, gets longer the higher in the intertidal you go.

Water Loss. Marine organisms tend to dry out, or **desiccate,** when they are out of the water. To survive in the intertidal, an organism must be able to prevent desiccation, tolerate it, or both. Most intertidal organisms cope with the problem of drying out in one of two basic ways: They either "run and hide" or "clam up."

The "run and hide" strategy is simple enough. When the tide goes out, the organism simply moves somewhere wet and waits for the tide to come back in. It is not unusual to see shore crabs, hermit crabs, snails, and other denizens of rocky shores huddled in moist, shady cavities or crevices in the rocks (Figure 10-3). **Tide pools,** which are depressions in the rocks that hold seawater after the tide goes out, are favorite places to hide (Figure 10-4). Some areas are kept moist by spray from ocean waves or by water that slowly leaks out of tide pools. The hiding place may even be provided by other organisms (Figure 10-5).

spring tides Tides with a large range that occur around times of the full or new moon

neap tides Tides with a small range that occur when the moon is in quarter

tidal range The difference in height between a high tide and the next low tide
Chapter 3, p. 75

Figure 10-3 These periwinkles (*Littorina cincta*) get through low tide by clustering in a moist, shady crevice. They also seal against the rock to retain moisture.

Figure 10-4 When the tide goes out, a little bit of the ocean stays in this tide pool. Without the tide pool, many of its inhabitants would dry out and die. Even in tide pools, however, life can be tough. The water undergoes drastic changes in temperature, salinity, and oxygen content.

Figure 10-5 Mussels (*Mytilus*) form dense clumps that retain moisture. They provide a place to live for a variety of smaller organisms, like crabs, sea anemones, and barnacles, among a host of others.

Seaweeds and sessile animals cannot move, but many adopt the "hide" part of the "run and hide" approach. Instead of moving to moist areas when the tide goes out, they live in moist areas all the time (Figure 10-6). This may be because the larvae settle only in moist, shady places or because larvae that settle elsewhere dry out and die.

Organisms that use the "clam up" strategy have some sort of protective covering, such as a shell, that they can close to hold in water. Some animals, like barnacles and mussels, are completely enclosed and can seal in moisture simply by closing their shell. Others, like limpets, have an opening that cannot be completely shut. These organisms typically clamp themselves tightly to the rock to seal the opening (Figure 10-7). Some use mucus to make a better seal. They may also carve out shallow depressions, or "home scars," in the rock that make the seal more effective. They do this by scraping the rock with their shell or radula, slowly wearing the rock away.

Some organisms use a combination of strategies. The periwinkles shown in Figure 10-3, for example, clamp themselves against the rocks to seal in moisture. They can also seal off the opening of their shell by closing the

radula A rasping ribbon of small teeth that is possessed by most molluscs
Chapter 6, p. 165; Figure 6-27

LIFE IN THE MARINE ENVIRONMENT

Figure 10-6 These seaweeds flourish in a slight depression in the rock that stays wet when the tide goes out. The surrounding rock dries out, and the seaweeds cannot grow there.

Figure 10-7 The common European limpet (*Patella vulgata*) clams up at low tide but grazes around a home territory at high tide. Bare patches among seaweeds like this and a "home scar" reveal their home bases.

operculum, a stiff plate that fits the opening like a door. Periwinkles are still not immune to desiccation, however, so at low tide they congregate in moist, shady places, especially on hot, sunny days.

Finally, some plants and animals use neither the "run and hide" nor the "clam up" approach. Instead, they simply allow themselves to dry out. Some intertidal chitons can survive the loss of 75% of the water in their tissues. Some intertidal seaweeds, such as rockweeds (*Fucus*; see Figure 5-15), can withstand a water loss of as much as 90%—becoming almost completely dry and practically crunchy. They quickly recover when the tide comes in and wets their tissues.

> **Some intertidal organisms avoid drying out by moving to or living in wet spots. Others can close a shell to retain water. Still others are able to dry out and recover when the tide returns.**

Temperature and Salinity. Emersion creates other problems besides desiccation for marine organisms. Ocean temperatures are relatively constant and mild because of the high heat capacity of water, but air temperatures can be much more extreme. At low tide, intertidal organisms are at the mercy of the sun's heat and the freezing cold of winter. Because tide pools are shallow, they too experience extreme temperatures, though not as extreme as air temperatures.

Most intertidal organisms can tolerate a wide temperature range. Tide-pool fishes, for example, are much more tolerant of extreme temperatures than their relatives that live below the tides. Certain species of periwinkle (*Littorina*) that live high in the intertidal show an amazing heat tolerance. One is known to survive temperatures as high as 49° C (120° F) under laboratory conditions.

chitons Molluscs whose shells consist of eight overlapping plates on their upper, or dorsal, surface
Chapter 6, p. 169

heat capacity The ability of a substance to hold heat energy; water has one of the highest heat capacities of any natural substance
Chapter 3, p. 55

Figure 10-8 The ridges on the shell of this tropical snail, the pleated nerite (*Nerita plicata*), help it to stay cool by radiating heat. The white color of the snail probably also helps by reflecting sunlight.

Organisms may deal with temperature extremes in ways besides simply being "tough." Those that move to moist "hiding places" to avoid drying out, for example, also avoid high temperatures, because such places tend to be cool. Some snails, especially in the tropics, have pronounced ridges on their shells (Figure 10-8). These ridges, like the fins on a car radiator, help the snail give off heat.

The color of its shell can also help a snail to tolerate high temperatures. Snails that are regularly subjected to extreme high temperatures tend to be light in color. The light color helps reflect sunlight so that the snail does not absorb the energy and heat up. On Atlantic shores, for example, the dog whelk, or winkle (*Nucella lapillus*), has two color forms: a white-shelled form and a brown-shelled form. On shores that are exposed to heavy wave action the brown form of the whelk predominates. On sheltered shores the white form is most common. The difference appears to be related to temperature. Brown whelks absorb more heat than white ones and die in conditions that do not appear to bother the white snails at all. On exposed coasts where the brown snails live, there are dense mussel beds, which stay moist (see Figure 10-5) and therefore tend to remain cool. Wave spray also helps cool the snails. On sheltered coasts there are few mussels and little wave spray. Temperatures are therefore higher, so white whelks have an advantage.

Salinity also fluctuates widely in the intertidal. When it rains, exposed plants and animals have to endure fresh water, which is fatal to most marine organisms. Many simply keep the fresh water out by closing their shells—another benefit of the "clam up" strategy. Even so, rainstorms during low tide sometimes cause the mass mortality of intertidal organisms.

Tide-pool residents, too, face extreme variation in salinity. At low tide the pool may be diluted by rain, lowering the salinity. On hot, dry days, salinity goes up as a result of evaporation. To cope with this variation, tide-pool organisms can usually withstand a wide range of salinity, as well as temperature. They may also burrow or reduce their activity to "ride out" the extreme salinity and wait for high tide.

The intertidal faces more extremes of temperature and salinity than other marine environments because it is exposed to the air. Intertidal organisms have evolved various mechanisms to avoid or endure these extremes.

Restriction of Feeding. Because little sediment accumulates in the rocky intertidal, **deposit feeders** are rare. Most of the sessile animals are **filter feeders** and cannot feed when the tide is out. Obviously, they can't filter water while they're not in the water. Furthermore, many of them "clam up" during low tide to avoid water loss and can't extend their filtering or pumping apparatus with their shell closed.

Even animals that are not filter feeders have trouble feeding at low tide. Many mobile animals in the rocky intertidal are grazers that scrape algae, bacteria, and other food from the rocks. Others are predators and move over the rocks in search of prey. At low tide these animals seek shelter or clamp to the rocks to avoid water loss. This prevents them from moving around to find food.

Being unable to feed when the tide is out is not much of a problem for animals that live low in the intertidal and are submerged most of the time.

deposit feeders Animals that eat organic matter that settles to the bottom
Chapter 6, p. 158; Figure 6-20

suspension feeders Animals, including *filter feeders*, that eat particles suspended in the water column
Chapter 6, p. 147; Figure 6-20

LIFE IN THE MARINE ENVIRONMENT

Higher in the intertidal, however, animals may not have enough time to feed. This may cause them to grow more slowly than they would with more ample feeding times or may prevent them from living high in the intertidal at all.

Many intertidal animals are unable to feed when exposed at low tide.

The Power of the Sea

Desiccation, temperature, salinity, and restriction of feeding are all problems that intertidal organisms face while the tide is out. Even when the tide is in, however, life in the intertidal is not exactly easy. The shoreline is where ocean waves expend their tremendous energy as they crash on the shore. Anyone who has been knocked around by the surf appreciates how much energy waves can carry. Rocky intertidal organisms are exposed to the full power of the sea (see Figure 3-1).

The Distribution of Wave Energy Along the Shore. The impact of the waves varies along the shoreline. Some areas are sheltered from the surf, whereas others are fully exposed. Enclosed bays, for example, are usually protected from wave action—which is why they are used for harbors. It is not always so easy to predict which areas will be sheltered and which exposed, however. To understand the distribution of wave energy along the shore, we must learn a little more about the behavior of waves.

Recall from Chapter 3 (see the section, "Waves") that, when a wave enters shallow water, it "feels" the bottom and slows down. Thus the same wave will travel faster in deep water than in shallow water.

Waves almost never approach the shore straight on, but instead come in at an angle (Figure 10-9). As a result, one "end" of the wave reaches shallow water before the other. The end in shallow water slows down, but the end in deep water continues to travel at its original speed. As a result, the wave bends, just as a two-wheeled cart will turn to one side if the wheel on that side sticks. This bending of the wave is called **refraction,** and causes the waves to become more nearly parallel to the shore (Figure 10-10). They never quite get perfectly parallel, however.

Figure 10-9 Like nearly all waves, these waves are approaching the coast at an angle. As a result, they refract, or bend toward the coast. By the time the waves break, they have bent until the surf is almost parallel to shore.

Figure 10-10 Wave refraction. Wave crests, which appear as white bands in Figure 10-9, can be represented by lines on a diagram. In this diagram the lines indicate the crest of a single wave at successive times. At time 1, the entire wave is in deep water. At time 2, the left end of the wave, which is closer to shore, has entered shallow water and slows down, but the right end is still in deep water. As a result, the wave bends to the right, becoming more parallel to the shore. This process continues so that the wave is nearly, but not quite, parallel to shore when it breaks (time 5).

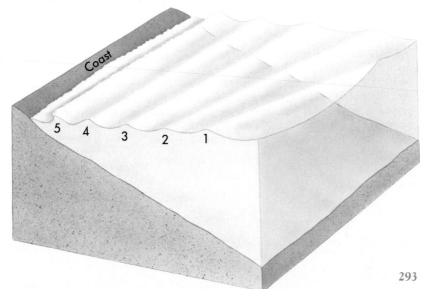

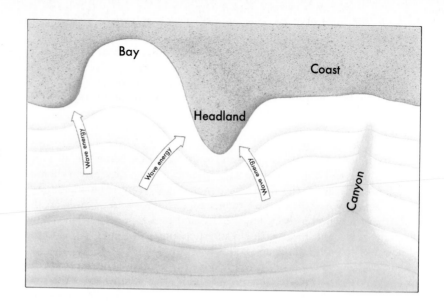

Figure 10-11 The coastline and bottom both affect wave refraction. For example, an incoming wave first hits shallow water straight out from a headland. This part of the wave slows, lagging behind the sections on either side. As a result the wave "wraps around" the headland, which gets most of the wave shock. Just the opposite occurs in bays or over submarine canyons: the central part of the wave is in the deepest water, and the wave turns off to both sides. This deflects wave energy away from the bay or the shore behind the canyon.

Refraction can produce especially complicated wave patterns when the coast is not a straight line. In particular, wave action tends to focus at headlands (Figure 10-11). Bays, even if they are not physically sheltered from incoming waves, tend to get less wave energy because of refraction.

> **Incoming waves refract or bend to become more parallel to the shore. This increases wave impact at headlands and decreases it in bays.**

Offshore bottom features can also influence the effect of waves on the coast. Submarine canyons, for example, may cause wave refraction. Also, waves often break on reefs or sand bars and expend their energy before they reach the shore.

The result of all this is that there is tremendous variation in the intensity of wave impact, or **wave shock,** from place to place along the shore. Exposure to waves strongly affects intertidal organisms (Figure 10-12).

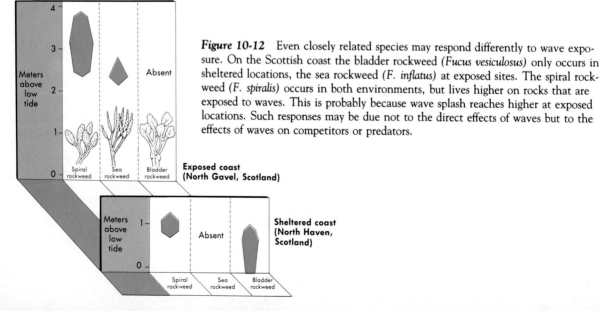

Figure 10-12 Even closely related species may respond differently to wave exposure. On the Scottish coast the bladder rockweed (*Fucus vesiculosus*) only occurs in sheltered locations, the sea rockweed (*F. inflatus*) at exposed sites. The spiral rockweed (*F. spiralis*) occurs in both environments, but lives higher on rocks that are exposed to waves. This is probably because wave splash reaches higher at exposed locations. Such responses may be due not to the direct effects of waves but to the effects of waves on competitors or predators.

Coping With Wave Shock. Some intertidal organisms simply cannot bear the full brunt of the waves and are found only in sheltered locations. Such organisms are often relatively delicate or cannot keep a very firm grip on the rocks. On the other hand, they often tolerate sediment better than organisms that live on exposed coasts. Sediment accumulates mainly in sheltered areas.

One way or another, exposed organisms must cope with wave shock. Sessile organisms anchor themselves firmly to the rocks to keep from being washed away. Seaweeds use their **holdfasts** or simply encrust on rocks. Barnacles secure themselves with a glue so strong that several companies have tried to duplicate it, without success. Mussels hold on with their **byssal threads** (see Figure 10-18), strong fibers made of protein that the mussel produces with a special gland in its foot. It is these threads that form the mussel's "beard." Though we usually think of them as sessile, mussels can slowly move from place to place by putting out new threads and detaching old ones.

More mobile animals may also cling strongly to the rocks. Limpets and chitons, for example, use their muscular foot like a powerful suction cup. Gobies (*Gobius*) and clingfishes (*Gobiesox*), which are common tide-pool fishes, also have suction cups (Figure 10-13). The fishes don't stick as strongly as limpets, but unlike limpets, they can always swim home if they are dislodged. Intertidal fishes also tend to lack swim bladders, so they sink and stay on the bottom.

holdfast   **A structure used by many seaweeds to anchor themselves to the bottom**
Chapter 5, p. 128

swim bladder   **A gas-filled sac that provides bony fishes with buoyancy**
Chapter 7, p. 198; Figure 7-13

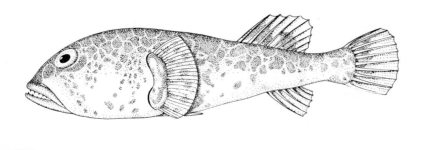

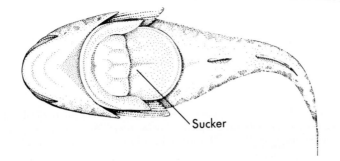

Sucker

Figure 10-13 This clingfish (*Chorisochismus*) uses its modified pectoral fins as a suction cup to hold on to the bottom of tide pools.

There are always those who prefer to hide from trouble. Many intertidal animals move to sheltered spots when wave action gets too strong. This is especially true of highly mobile animals—shore crabs *(Pachygrapsus, Hemigrapsus, Grapsus)*, for example. In fact, there may be a trade-off between being able to move fast and having a strong grip on the rocks. Holding on tightly tends to slow an animal down. Many intertidal snails are easier to knock from the rocks when they are moving around than when they stop and clamp down. Not surprisingly, they usually stop and hold on when wave action is heavy. This, of course, prevents them from feeding.

Intertidal organisms have other adaptations that help them withstand wave shock. Exposed animals tend to have thicker shells than sheltered ones. A compact shape can help reduce the impact of the waves (Figures 10-14 and 10-15). Many intertidal animals— including barnacles, mussels, limpets, and chitons—have low profiles close to the rocks. Colonies of the tubeworm *Phragmatopoma californica* (Figure 10-16) provide another example of the importance of shape. These colonies are not especially strong and are easily crushed if stepped on, but they can withstand considerable wave shock because water flows easily over their domed surface. Some organisms, especially seaweeds, are flexible and can "go with the flow" (Figure 10-17). There may also be some safety in numbers (Figure 10-18).

Organisms on exposed coasts deal with wave shock in several ways, including strong attachments, thickened shells, low profiles, and flexibility.

Figure 10-14 The giant green sea anemone *(Anthopleura xanthogrammica; see Figure 10-32)* lives low in the intertidal zone on the Pacific coast of North America. In sheltered locations (left), the anemone stands relatively tall, catching food particles with its tentacles. Anemones from areas with heavy wave action (right) are much shorter. This reduces the body area that is exposed and allows more water to slide over the top of the anemone.

 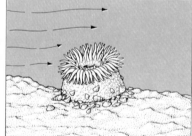

Figure 10-15 This sea urchin *(Colobocentrotus atratus)* from the tropical Indian and Pacific oceans is beautifully adapted to rocky shores. Its compact shape helps it withstand wave stress. Its flattened spines form a protective shield and help retain water.

LIFE IN THE MARINE ENVIRONMENT

Figure 10-16 The sand-castle worm (*Phragmatopoma californica*) is common in California. Each of the many worms in the colony constructs a tube of sand grains that are glued together.

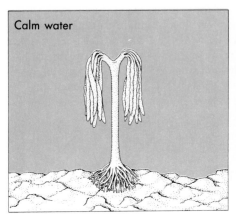
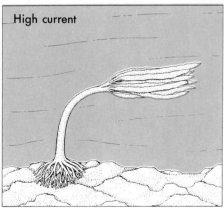

Calm water

High current

Figure 10-17 Organisms like this brown alga, the kelp *Eisenia arborea*, beat the waves by being flexible. In calm water (left) the plant stands upright. When a wave hits (right), the plant bends over and the fronds fold up. This streamlines the plant and reduces water resistance.

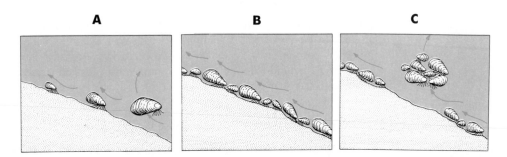

A   B   C

Figure 10-18 Some biologists have suggested that intertidal animals can gain protection from wave shock by growing in dense groups. When mussels (*Mytilus*), for example, grow as isolated individuals (**A**), they get the full impact of the waves. Small individuals, who are closer to the rock, may be able to hang on but larger mussels, with more exposed surface area, can be torn away. When the mussels grow in dense clusters (**B**) the wave does not get a direct "shot" at any of them, but "slides" over the top. This strategy can backfire, however. If the clumps get *too* dense (**C**) many of the mussels attach to other mussels instead of to the rock. A wave can cause more strain than the mussels on the bottom can bear, and entire clumps may be ripped away. This opens up space for other organisms, including other mussels.

Even the firmest grip on the rocks is of little help if the wave smashes a drifting log or something into the rocks, crushing whatever happens to live there. Waves can also turn rocks over, even large boulders. When this happens the organisms on top of the rock may be smashed or buried, and sunlight for the plants is blocked. The animals living underneath fare no better. Once nestled comfortably under the rock, they are suddenly exposed to the sun and waves, not to mention hungry predators. When waves overturn a rock, therefore, the organisms on both top and bottom usually die. People can have the same effect. If you happen to look under rocks at the shore, don't forget to turn them back over the way you found them.

Not only does wave action vary from place to place, but organisms differ in how they are affected by waves. The result of this is that sheltered and exposed areas have very distinct communities. Quiet bays, for example, harbor plants and animals much different from the ones at unprotected headlands. The effect of wave action can also be seen on a smaller scale: The organisms in a sheltered crevice usually differ from those on exposed rocks nearby.

The Battle for Space

For organisms that have adapted to its physical extremes, the rocky intertidal can be a good place to live. The shallow coastal waters provide lots of light and nutrients—the basic requirements for photosynthesis. The vigorous plant growth that results produces abundant food for animals. Furthermore, high tide bathes the intertidal in plankton-rich water, and the waves and tides bring in even more food in the form of drifting seaweeds and detritus (Figure 10-19). In general, food is not a limiting resource for intertidal animals, though high on the shore, they might not be underwater long enough to eat it.

Despite its harsh nature, the intertidal can be hospitable to both plants and animals—provided they can stay put. In even the calmest areas, the organisms will drift away or be smashed on the rocks if not attached to the substrate. Sessile organisms—animals and plants alike—need a permanent place to hold on to. Unfortunately, there is often not enough room to go around. In fact, it is the availability of space that most often limits rocky intertidal populations. Nearly all the space in the intertidal is occupied (Figure 10-20), and colonizers quickly take over what open space there is. Space is so scarce that plants and animals may attach to each other instead of directly to the rock.

Intertidal populations are usually limited by space, not food or nutrients.

It should come as no surprise that competition for space is a dominant biological factor in the rocky intertidal. There are many ways to compete for space. One way is to be the first to get to open spots. For intertidal plants and animals, this means having an effective means of **dispersal**— that is, organisms that depend on being the first to occupy new patches of open space must be good at getting themselves or their offspring from place to place. Having taken over the space, the organism either must be good at holding on to it or must be able to rapidly reproduce and disperse its young to the next opening. Both strategies are used in the intertidal.

photosynthesis $CO_2 +$ H_2O + SUN ENERGY $\longrightarrow$ organic matter + O_2
Chapter 4, p. 88

detritus   Dead organic matter and the microscopic decomposers that live on it
Chapter 9, p. 277

limiting resource   A resource that is so scarce that it prevents a population from growing
Chapter 9, p. 266

LIFE IN THE MARINE ENVIRONMENT

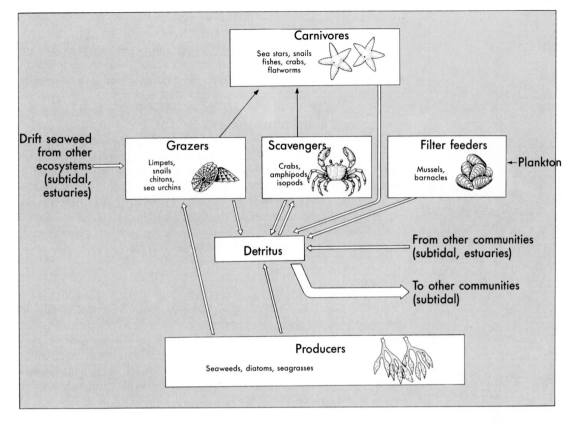

Figure 10-19 A generalized food web on a rocky shore.

Rather than colonizing open patches, some intertidal organisms take over space that is already occupied. Barnacles, for instance, undercut their neighbors, loosening them from the rock (see Figure 9-6). Owl limpets (*Lottia*) keep a territory for themselves by "bulldozing" out intruders. Less forceful methods may also work. Many intertidal organisms simply grow over their competitors, making them vulnerable to waves (Figure 10-21), smothering them, or—in the case of seaweeds— blocking precious sunlight.

Figure 10-20 Most organisms in the intertidal zone live attached to the rocks, and most of the available space is occupied. This little patch of rock supports two species of mussels and two of barnacles, plus limpets and sea anemones. What little open space there is doesn't last for long—the small bare patch (left) has been colonized by newly settled barnacles.

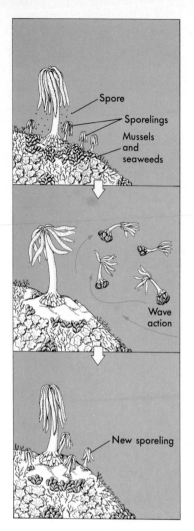

Figure 10-21 The sea palm (*Postelsia palmae-formis*; see Figure 5-17) is a kelp that takes advantage of wave action to compete for space. Adult plants literally drip spores, which develop into young sporelings near the parent. When the space below is occupied, the sporelings grow on these competitors, who are eventually pulled from the rocks. This clears the way for a new generation of young plants.

Two species of mussel provide an example of how competition and physical factors can interact. On the west coast of North America the blue mussel (*Mytilus edulis*) is found mainly in sheltered locations, whereas the California mussel (*M. californianus*) is found on the open coast. As you might expect from this, the California mussel has a thicker shell and attaches more strongly to the rocks than does its cousin. On the other hand, the blue mussel does just fine on open coasts, as long as the California mussel is absent. The blue mussel is apparently rare on open coasts not because it cannot take the waves but because it cannot compete with the California mussel. The California mussel, with its thicker shell, crushes the blue mussel when the two grow together. On the other hand, the blue mussel can live in calm bays because, unlike the California mussel, it can tolerate a lot of silt. These two mussels provide just one example of how physical and biologic factors can interact to determine the distribution of organisms.

Vertical Zonation

The exact plants and animals that make up rocky intertidal communities vary from place to place. There is, however, one feature that is remarkably universal. Nearly everywhere, the community is divided into distinct bands, or **zones,** at characteristic heights in the intertidal (Figures 10-22 and 10-23). Thus a given species is usually not found throughout the intertidal, but within only a particular vertical range. Biologists call this pattern of banding **vertical zonation.**

When the shore consists of an evenly sloping rock face, the zones are often sharply defined belts that can be easily distinguished by the colors of the organisms within (Figure 10-22). Many shores are much more uneven, with scattered boulders, channels, and gullies. Zonation may not be as obvious in areas such as this, but it usually exists nonetheless.

Figure 10-22 Intertidal organisms often form distinct bands, or zones, at different heights on the shore.

LIFE IN THE MARINE ENVIRONMENT

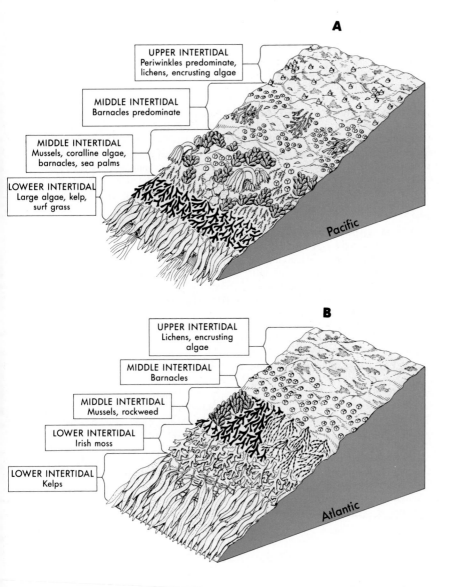

Figure 10-23 "Typical" patterns of zonation on rocky shores on the Pacific **(A)** and Atlantic **(B)** coasts of North America. In any particular place the exact pattern may differ.

A

UPPER INTERTIDAL
Periwinkles predominate, lichens, encrusting algae

MIDDLE INTERTIDAL
Barnacles predominate

MIDDLE INTERTIDAL
Mussels, coralline algae, barnacles, sea palms

LOWEER INTERTIDAL
Large algae, kelp, surf grass

Pacific

B

UPPER INTERTIDAL
Lichens, encrusting algae

MIDDLE INTERTIDAL
Barnacles

MIDDLE INTERTIDAL
Mussels, rockweed

LOWER INTERTIDAL
Irish moss

LOWER INTERTIDAL
Kelps

Atlantic

Marine biologists have spent a tremendous amount of time and energy trying to find out what causes vertical zonation. These efforts have not been wasted. Though there is still much to be done, we now know that zonation is produced by the complex interaction of physical and biologic factors. A general rule is that the upper limit at which a species occurs is usually determined mainly by physical factors, whereas the lower limit is usually determined by biologic factors, especially predation and competition.

Most rocky shores have a distinct pattern of vertical zonation. The upper limit of a zone is often set by physical factors, the lower limit by biologic factors.

BOX 10-1

Transplantation, Removal, and Caging Experiments

Experiments using transparent plexiglass cages provide valuable information on the behavior of fishes. A caged oval butterflyfish (*Chaetodon trifasciatus*) elicits an aggressive response from an individual chevron butterflyfish (*C. trifasialis*), an indication that the caged fish has intruded in the territory of the latter. This technique can be used to map the boundaries of butterflyfish territories. Territorial butterflyfishes that live in pairs, like *C. trifasciatus* (see Figure 7-30), fight with other pairs, but fights occur between individuals of the same sex of each pair. Therefore, by using a caged fish of known sex, experimenters can determine the sex of the wild fish, which is normally impossible in the field.

A general principle that is often applied to the intertidal is that physical factors, especially drying in air, usually determine how high in the intertidal a particular organism can live, whereas biologic factors such as competition and predation set the lower limit. Where does this principle come from? How can you tell why a certain species of barnacle or a seaweed is found only in a particular zone, not higher or lower?

Like most scientific questions, this one is best answered by performing experiments—altering natural conditions to see what happens. In the intertidal, three general types of experiment have been particularly effective. In one of these, organisms are **transplanted** from one place to another to see whether they can survive under the new conditions. In one such transplantation experiment barnacles were grown on panels that could be placed at various heights on pier pilings. The barnacles soon died if they were moved higher than they normally lived. Measurements of their body fluids showed that they had lost a lot of water, so the cause of death was probably drying out. The experiment showed that the barnacles cannot survive the conditions above their normal range. Thus their upper limit is set by a physical factor—desiccation.

Transplantation can also tell us when physical factors are *not* important. When barnacles are transplanted lower in the intertidal than they normally live, they do just fine. The same is true of other intertidal species, including certain mussels, snails, and seaweeds. In fact, some species may actually grow faster when submerged longer than they would be at their normal position. Thus most intertidal organisms can live below their normal range—their lower limit is not set by the physical environment.

If organisms *can* live lower in the intertidal, then why don't they? Experiments suggest that other organisms, either competitors or predators, often set a species' lower boundary. In **removal experiments,** one species is removed from an area, which is then compared to an untouched **control** area.

A classic experiment of this type examined competition between two species of barnacles. Little gray barnacles (*Chthamalus*) normally live higher on the shore than rock barnacles (*Semibalanus*). Rocks with newly settled larvae of both species were collected and placed at different heights. The rock barnacles were removed from some of the rocks with a needle, whereas other rocks were not disturbed.

This rule is a useful place to begin when trying to understand the causes of vertical zonation; however, as with all generalizations, there are exceptions. For example, zonation can be produced if the larvae of a species settle at a particular height on the shore. One cannot just *assume* that the upper limit of a certain species is set by physical factors. Instead, this must be tested by *experiment* (see Box 10-1). Furthermore, different factors often *interact* to determine the limits, and sometimes the line between physical and biologic factors is fuzzy. As noted above, for example, some filter feeders do not live high in the intertidal because they are not submerged long enough to feed. Is this a physical factor—emersion by the tide—or a biologic one—feeding?

At high levels, little gray barnacles survived better than rock barnacles, so they apparently tolerate drying better. But why are the little gray barnacles rare or absent at lower levels? The answer seems to be that they are out-competed by the rock barnacles. At lower tidal levels, little gray barnacles survived much better when the rock barnacles were removed. On the control rocks, where the rock barnacles were not removed, the little gray barnacles were crushed, undercut, or smothered by the bigger, faster-growing rock barnacles. In other words, little gray barnacles do not live lower on the shore, not because they *can't* live there, but because they lose the competition for space. At even lower levels, little gray barnacles don't settle at all.

Removal experiments have also shown the importance of grazing and predation. When limpets are removed from test areas, for example, the result is usually a profusion of seaweeds. Because the limpets scrape the young algae off the rocks, the algae can never get started. Once the seaweeds get a foothold, however, they are often able to hold on even if the limpets return. As adults they can stand up to the grazers—only the young stages are susceptible. Another removal experiment, this one involving a starfish *(Pisaster)*, is summarized in Figure 10-28.

Sometimes removal experiments are just not practical. There may not be enough time to remove all the predators or competitors, or they may move back into an area too fast after removal. In these situations, **caging experiments** are often used. The basic idea is that some sort of barrier is used to stop animals from getting into the study area. Sometimes the barriers are actual cages made of wire, but they can be anything that keeps out the animal in question. Plastic rings screwed to the rocks will keep out sea stars and limpets, for example, who cannot climb over the rings. Similarly, limpets, snails, and other molluscs can't crawl on the artificial turf that is used in stadiums—the surface is too rough. Strips of the turf can be used to keep these animals out of an experimental site.

Cages can also be used to hold animals inside an area. Similar to removing them, *adding* certain animals to an area can have consistent effects on a community. In these cases the cages are used to prevent the animals from moving away. Whether the animals are kept in or kept out, changes in the community provide good evidence that the animals in question play an important role in the structure of the community.

Rocky shores around the world frequently show a pattern of vertical zonation that is quite similar in its general nature, but highly variable in its particulars. Biologists studying a certain area sometimes name the zones after the "dominant" organism: for instance, the "limpet zone" or the "mussel zone." These zones actually contain a tremendous variety of organisms. The mussel zone, for example, is also home to worms, crabs, snails, seaweeds, and many others. Also, the specific organisms found in the intertidal vary from place to place. On the east coast of North America, for example, the mussel zone is often replaced by a rockweed zone.

Instead of naming the zones after the animals and plants, we shall simply divide the intertidal into the *upper, middle,* and *lower zones.* These names may not be especially informative or creative, but at least they apply everywhere! Bear in mind that the following sections are general descriptions of "typical" intertidal communities and probably don't apply exactly to any particular place. Furthermore, the boundaries between zones are not absolute: Some species occur in more than one zone. The next time you go to the shore, think about how these descriptions apply. Perhaps more importantly, think about how the place you visit is different. What factors might cause the differences?

The Upper Intertidal. Seldom submerged, the inhabitants of the upper intertidal must be well adapted to withstand exposure to air. This zone actually lies mostly above the high-tide mark, and the organisms are wetted mostly by wave splash and spray. It is often called the "splash zone." On exposed, stormy coasts with more wave splash, the upper intertidal extends higher above the high-tide line than on sheltered coasts.

In most places, lichens (*Verrucaria*) form black, tarlike blotches on rocks in the upper intertidal. The fungus part of the lichen soaks up water like a sponge, storing it for long dry periods. Dark green mats of blue-green algae (*Calothrix*) are abundant. These are protected from drying out by a jellylike coating and have the advantage of being able to fix nitrogen from the air. Small tufts of a filamentous green alga (*Ulothrix*), which are also resistant to drying out, may spring up here and there. Various other green (*Prasiola, Enteromorpha;* see Figure 5-11), brown (*Pelvetia*), and red (*Porphyra, Bangia;* see Figure 5-21) algae are occasionally found in the upper intertidal, usually in moist spots.

Large numbers of periwinkles graze on the algae, scraping it from the rocks with their radula. Periwinkles are so abundant that the upper intertidal is often called the "*Littorina* zone" after the scientific name of the snails. Not all of the many species of periwinkles live high in the intertidal, but those that do are well adapted for it. They can "breathe" air like land snails and live out of water for months. They can also tolerate extreme temperatures. They do have limits, however, and may have to seek shelter on the hottest days (see Figure 10-3).

Limpets (*Colisella, Acmaea, Lottia*) may also be found in the high intertidal. Like periwinkles, limpets are hardy grazers. Shore crabs occasionally venture into the upper intertidal, scurrying over the rocks. They eat mostly algae that they scrape off the rocks with their claws, but they may also eat animal matter, dead or alive. Sea lice or sea roaches (*Ligia;* see Figure 10-24), which are also crustaceans, breathe air and live above the water's edge, moving into the upper intertidal at low tide.

Most of the upper intertidal lies above the high-tide line and is kept wet by wave splash. The dominant plants are lichens and blue-green algae. Periwinkles are the most common animal.

Few marine predators can reach the upper intertidal. Shore crabs occasionally eat periwinkles and limpets. From time to time predatory snails, such as unicorn shells (*Acanthina*), may venture into the splash zone to do the same. On the other hand, the upper intertidal is visited by predators from land. Birds, such as oystercatchers (*Haematopus*), may eat large num-

lichens Close, *mutualistic* associations between fungi and microscopic algae; mutualism is a type of *symbiosis*
Chapter 5, p. 125, and Chapter 9, p. 269

nitrogen fixation Conversion of nitrogen gas (N_2) into nitrogen compounds that can be used by plants as nutrients
Chapter 9, p. 281

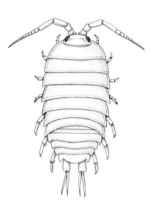

Figure 10-24 The sea louse, or sea roach (*Ligia oceanica*), is neither a louse nor a roach but an isopod, a flattened crustacean with several similar pairs of legs. It feeds mainly on decaying seaweeds carried in by the waves.

LIFE IN THE MARINE ENVIRONMENT

bers of limpets and snails. Raccoons, rats, and other land animals are also known to enjoy these tasty molluscs. People in many parts of the world agree!

The Middle Intertidal. Unlike the upper intertidal, which is affected only by wave splash and the highest spring tides, the middle intertidal is submerged and uncovered by the tides on a regular basis. A diurnal tide exposes the organisms once a day, a semidiurnal tide twice. If the tide is mixed, the lower of the two successive high tides may not submerge the high part of the zone, and the higher of the two lows may not expose the bottom part. Thus, even in places with a semidiurnal tide, some parts of the intertidal may only be submerged or exposed once a day. With so much variation in the time spent out of the water, different heights within the middle intertidal may support very different plants and animals. In other words, the middle intertidal is often made up of several separate vertical bands.

The upper boundary of the middle intertidal is almost always marked by a band of acorn barnacles. In many places at least two separate species of barnacles, such as little gray (*Chthamalus*) and rock barnacles (*Balanus, Semibalanus*), form distinct bands, with the little gray barnacle occurring higher up.

Experiments (see Box 10-1) have shown that this zonation results from a combination of the barnacles' larval settlement pattern, how well they tolerate desiccation, and competition and predation (Figure 10-25). The larvae of both species settle over a wider range than that occupied by adults.

diurnal tide   A tidal pattern with one high and one low tide each day

semidiurnal tide   A tidal pattern with two high and two low tides each day

mixed tide   A tidal pattern in which two successive high tides are of different heights
Chapter 3, p. 76, Figure 3-33

Figure 10-25 Larvae of both little gray barnacles (*Chthamalus*) and rock barnacles (*Semibalanus*) settle out over a wider range than occupied by adults. At the upper edge, young barnacles dry out and die. Little gray barnacles are more resistant so they can live higher on the shore. Little gray barnacles that settle where the rock barnacles can survive, however, are out-competed. The lower edge of the rock barnacle zone occurs where predatory snails, called dog whelks (*Nucella*), can easily reach them. In some places, competition from mussels also helps determine the lower edge.

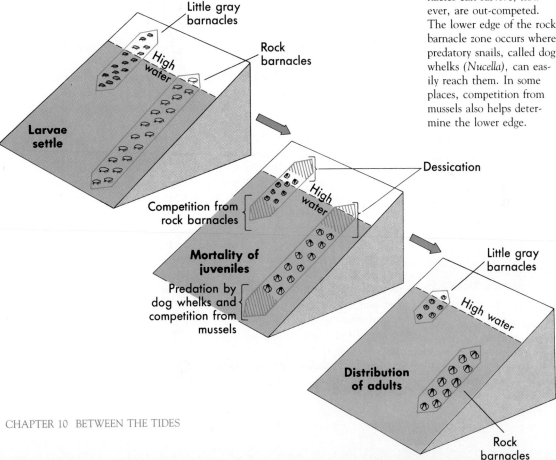

Figure 10-26 The Pacific gooseneck barnacle (*Pollicipes polymerus*) forms dense clusters in the middle intertidal zone. A similar species is much esteemed as food by many Europeans.

Figure 10-27 Rockweeds (*Fucus*) often dominate the middle intertidal zone on north Atlantic rocky shore communities, like this one on the coast of Maine.

The upper limit of both species appears to result from emersion: Larvae that settle too high in the intertidal die out. Little gray barnacles appear to tolerate drying better than rock barnacles and can therefore live higher on the shore. At lower levels, where the rock barnacles can live, they out-compete the little gray barnacles. Adult little gray barnacles, then, live perched above their competitors in the narrow band where they can survive emersion.

The lower limit of rock barnacles, such as that of little gray barnacles, is determined by biologic factors—in this case both predation and competition. Dog whelks are major predators of barnacles. These snails use their radula to drill holes in barnacles' shells. They also secrete chemicals that soften the shells. If protected from the snails by wire cages (see Box 10-1), rock barnacles flourish below their normal limit. Unprotected barnacles, however, suffer heavy losses to snail predation. The only thing that saves the rock barnacles is that the dog whelks can find and eat the barnacles only when the tide is in. In the upper intertidal, the snails do not have enough time at high tide to eat the barnacles. Sea stars, which also eat barnacles, have the same problem as the snails. Rock barnacles may also face competition from mussels. Similar to the little gray barnacles, the rock barnacles persist in a narrow zone between the level where they die of "thirst" and the level where their enemies can reach them.

> Acorn barnacles usually occupy the top of the middle intertidal. Their upper limit is determined by how high they can live without drying out. Their lower limit is set by competition with other barnacles or mussels or by snail or sea star predators.

Many other plants and animals live in the middle intertidal below or with the barnacles. Which organisms are present, how many there are, and where they occur all depend on the unique combination of physical and biologic factors at that site. The pattern of the tides, steepness of the shore, exposure to waves, and the local weather all exert a profound influence. Predation, competition, and larval settlement patterns are almost always involved. If you consider all the possible ways in which these factors might interact, you may get some idea of how complicated the story can be.

Depending on all these factors, mussels (*Mytilus*), gooseneck barnacles (*Pollicipes*; Figure 10-26), and brown seaweeds, particularly rockweeds (*Fucus, Ascophyllum, Pelvetia, Hesperophycus*) often dominate the middle intertidal below the barnacles (Figure 10-27). Many rockweeds have air bladders called **pneumatocysts** that float their fronds closer to the light. They may form a luxuriant thicket, or **algal turf,** that shelters many small animals. Mussels are especially common on stormy, exposed shores. They cannot live as high on the shore as barnacles, both because they dry out and because they do not get enough time to filter feed at high levels. At levels where they *can* live, however, mussels are the dominant competitors, smothering or crowding out other organisms. In some places, in fact, the lower limit of barnacles is set not by dog whelk predation, but by competition with mussels.

The main predators of mussels are sea stars (*Pisaster, Asterias*). Contrary to popular belief, sea stars do not force open the mussel's shell to eat it.

Instead, they insert their stomachs into the shell, and begin digesting the mussel from the inside. The shell opens after the mussel is weakened. The sea star can insert its stomach through a very narrow crack—only 0.2 mm (0.008 inch) for the ochre sea star (*Pisaster ochraceus*).

These sea stars do not tolerate drying very well and need to be underwater to look for food, though they may finish a meal after the tide goes out. In the lower part of the middle intertidal this is no problem, and sea stars take a heavy toll on the mussels. Thus mussels are found only above the level of heavy sea star predation. As with other intertidal species, the lower limit of mussels is set by a biologic factor: predation by sea stars.

Mussels are the dominant competitors for space on many rocky shores. Their upper limit is set by desiccation and filtering time, their lower limit by predatory sea stars.

Sea stars and other mussel predators can have a strong effect, even above the level where they completely wipe out the mussels. They venture into the mussel bed during high tide and eat some of the mussels before retreating with the tide. This opens up space for other organisms—such as acorn barnacles, gooseneck barnacles, and seaweeds—that would otherwise be crowded out. In addition, the sea stars eat dog whelks, doing the barnacles another favor. If the barnacles can live long enough, they get too big for the dog whelks to eat. By reducing the number of dog whelks, the sea stars give the barnacles more of a chance.

The sea stars are absent from areas with extremely strong wave action. In such areas, or if the predators are experimentally removed (Figure 10-28), the mussels take over. They out-compete other organisms and form dense beds. Thus many of the species in the middle intertidal would be crowded out if the sea stars did not keep the number of mussels down. Be-

Figure 10-28 Removing predatory sea stars—or keeping them out with cages—shows how they affect the middle intertidal community. Below a certain level, the sea stars can easily reach the mussels, their favorite food, and eat them all. Above this level the sea stars cannot eat all the mussels, but they eat enough to make a space available for other species. When there are no sea stars around, the mussels are able to live lower and they monopolize the available space by overgrowing and crowding out other species.

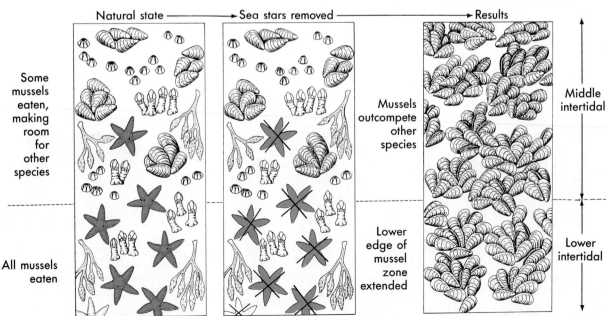

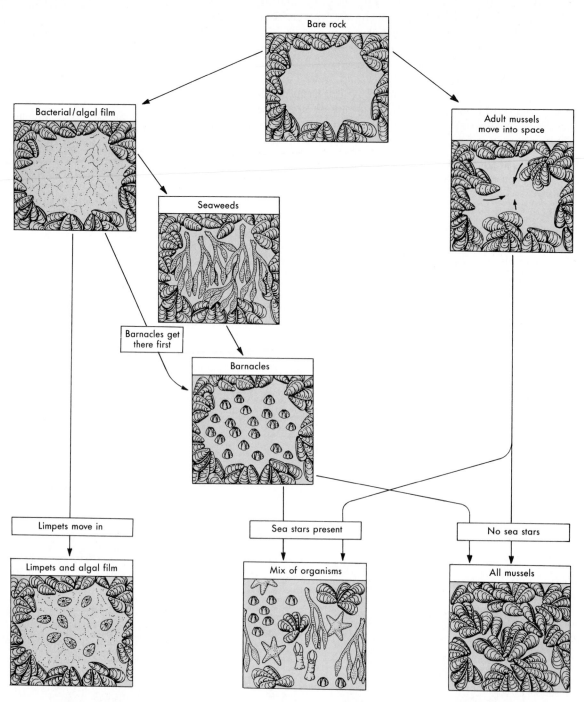

Figure 10-29 Ecological succession following the clearing of a patch in a mussel bed. The pathway taken and the end result depend on the size of the patch when it opens up and just plain luck—which organisms get there first. Some biologists think that the term succession should only be used when the steps are completely predictable: they do not think that, in this particular case, the rocky intertidal zone is a true example of ecological succession.

cause much of the structure of the community depends on the sea stars, they are known as **keystone predators.**

Note that many intertidal species are more abundant than the sea stars, but the sea stars are of central importance in the community. Removing them would profoundly affect the other species. This demonstrates a common feature of biological communities: just because a species is relatively uncommon does not mean that it is unimportant. Though not the most abundant species, keystone predators such as sea stars are important because they structure the community.

Natural disturbances can have effects similar to those of predation. When mussel beds get too dense, clumps of mussels can be torn away by waves (see Figure 10-18), exposing the bare rock. Drifting logs that batter the rocks and, in cold places, scouring ice have the same effect: opening up new space. This prevents mussels from monopolizing the available space, allowing other organisms to persist.

When a patch of space is cleared, new organisms often move into the patch and get replaced by others in a regular sequence (Figure 10-29). The term **ecological succession** is used to refer to such regular patterns of regrowth. In the rocky intertidal the first stage is usually a thin film of bacteria and microscopic algae, such as diatoms, that cover the rock. This film may "condition" the rock surface, since some species prefer surfaces with this film better than bare rock for settling larvae. Seaweeds often move in next, followed by barnacles, and finally mussels, the dominant competitor. The final stage in ecological succession is called the **climax community.**

The typical steps of ecological succession on many rocky shores are, first, a bacterial and algal film, then seaweeds, barnacles, and finally the climax mussel community.

The succession may deviate from the "typical" pattern for many reasons. For example, grazers such as limpets and chitons may remove newly settled animal larvae and seaweed spores. If grazers move into a patch early on, the succession may never get past the bacterial and algal film stage. Predators and other disturbances help determine whether the final stage is a solid mussel bed or a mixture of species.

Furthermore, steps in the succession may be skipped. For example, seaweeds may never be able to colonize a patch if barnacle larvae get there first. The time of year that the patch is created is important. If the patch opens up when barnacle larvae are abundant and seaweed spores are rare, the barnacles have the upper hand. Thus there is an element of random chance—"luck"— in the development of the community. Stages may also be skipped if the bare patch is small. A small patch cleared in the middle of a mussel bed may be taken over by other mussels that move in from the sides before the succession gets started.

Without predation and other disturbances, the best competitors take over and many species in an area disappear from the area. Disturbances can prevent this. Mussels, for instance, are often unable to completely cover middle intertidal rocks because sea stars eat them and waves tear them away. This gives other species a chance. Disturbance can thus increase the

competitive exclusion Occurs when one species out competes and eliminates another
Chapter 9, p. 268

number of species that live in an area by interfering with **competitive exclusion.** The rocky intertidal in some places can be thought of as a mosaic of patches that were cleared at different times and are going through different stages of succession. Because each patch supports a separate set of organisms, the number of species in the whole area increases (Figure 10-30). On the other hand, if predation and other disturbances happen too often, the community will keep getting knocked back to the starting point. It never gets a chance to develop, and not many species will be able to live there.

> The number of species that live in an intertidal area is strongly affected by predation and other disturbances. Without such disturbances, a few dominant competitors, especially mussels, take over. Occasional disturbance removes the mussels and gives other species a chance. Too much disturbance removes most of the species.

The Lower Intertidal. The lower intertidal is submerged most of the time. This makes it easy for predators such as sea stars and dog whelks to feed, so mussels and barnacles are rare. The lower intertidal is dominated by seaweeds, which form a thick turf on the rocks. These seaweeds—including species of red, green, and brown algae—cannot tolerate drying out, but in the intertidal, they grow in profusion. Not surprisingly, grazing and competition are also important in the lower intertidal. Light, as well as space, is an important resource. Seaweeds often compete by overgrowing each other, blocking the light.

The lower intertidal is dominated by red, green, and brown seaweeds.

A good example of the importance of the interaction of competition and grazing comes from the coast of New England. Two common seaweeds there are a green alga, *Enteromorpha,* sometimes called the green thread alga, and a red alga (not a moss), called "Irish moss" (*Chondrus crispus;* see Figure 5-22), which, by the way, is edible (see Box 5-2). When the two seaweeds grow together in tide pools, *Enteromorpha* dominates by overgrowing the Irish moss (Figure 10-31). Surprisingly, however, Irish moss is the most common seaweed in many tide pools.

The reason for this is that grazing sometimes prevents *Enteromorpha* from taking over. A common grazer in the area is the common periwinkle (*Littorina littorea*), which is closely related to the periwinkles that live in the upper intertidal. Though *Enteromorpha* is one of its favorite foods, the snail hardly touches Irish moss. If a tide pool has a lot of snails, they eat the *Enteromorpha* and make way for Irish moss. The importance of the periwinkles has been shown in transplantation and removal experiments (see Box 10-1). If snails are transplanted into an *Enteromorpha* pool, *Enteromorpha* quickly disappears and Irish moss takes over. On the other hand, *Enteromorpha* soon moves in after periwinkles are removed from an Irish moss pool.

The story doesn't end here. If the periwinkles like *Enteromorpha* so much, you might think that they would move into *Enteromorpha* pools, but they don't. For one thing, large periwinkles do not move around a lot and prefer to stay in one tide pool. Young periwinkles are eaten by green crabs (*Carcinus maenas*), which also live in tide pools. The crabs, in turn, are

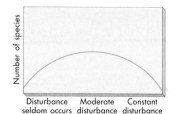

Figure 10-30 Other things being equal, the number of species depends on how often disturbances occur. In this context, predation on the dominant competitors counts as a "disturbance." When disturbance is rare, the dominant competitors take over and exclude other species. Intermediate levels of disturbance prevent this and give other species a chance. Thus the number of species is highest when disturbance is moderate. If disturbance happens too often, most species cannot get a foothold and the number of species drops.

LIFE IN THE MARINE ENVIRONMENT

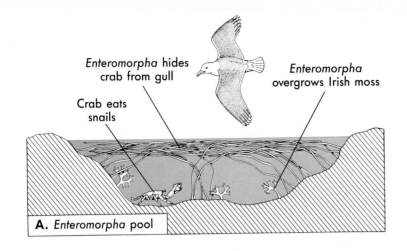

Enteromorpha hides
crab from gull

Crab eats
snails

Enteromorpha
overgrows Irish moss

A. Enteromorpha pool

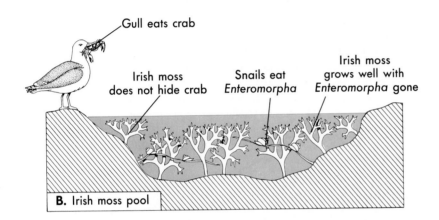

Gull eats crab

Irish moss
does not hide crab

Snails eat
Enteromorpha

Irish moss
grows well with
Enteromorpha gone

B. Irish moss pool

Figure 10-31 Biological interactions determine what type of seaweed predominates in New England tide pools (see text). Unless something interferes both types of pool are self-sustaining. As with the chicken and the egg, you might ask yourself which part of each system comes first.

eaten by gulls *(Larus). Enteromorpha* provides cover for the crabs, protecting them from gulls. The crabs then eat young periwinkles, preventing *them* from colonizing the pool. In Irish moss pools, on the other hand, the crabs lack cover and are eaten by gulls. Without the crabs, young periwinkles can survive in the tide pools. Thus both types of tide pool perpetuate themselves.

The lower intertidal supports many other seaweeds besides *Enteromorpha* and Irish moss. These range from delicate reds and greens to large, tough kelps *(Egregia, Laminaria;* see Figure 12-27). The kelps mark the lower limit of the intertidal and continue down into the subtidal zone. Thus there is no precise boundary between the lower reaches of the intertidal and the upper edge of the subtidal. Coralline red algae *(Corallina, Lithothamnion)* can also be abundant. Surf grass *(Phyllospadix),* a flowering plant, is common in the lower intertidal on the Pacific coast of North America (see Figure 5-26, C and E).

A host of small animals live among the seaweeds, hiding from predators and staying moist during very low tides. Sea urchins *(Arbacia, Strongylocentrotus, Echinometra;* see Figure 12-21) are common grazers on the seaweeds.

subtidal zone The part of the continental shelf that is never exposed by low tide
Chapter 9, p. 284

coralline red algae Red algae that deposit calcium carbonate in their tissues
Chapter 5, p. 132, Figure 5-23

Figure 10-32 The aggregating sea anemone (*Anthopleura elegantissima*) not only catches small prey with its tentacles, but also gets nutrients from symbiotic algae, or zooxanthellae, in its tissues. See also Figure 4-24, B.

Also present in the lower intertidal are sea anemones (*Anthopleura, Metridium, Tealia;* Figure 10-32), polychaete worms (*Spirorbis, Phragmatopoma;* see Figure 10-16), snails (*Tegula, Nucella*), sea slugs (*Aplysia, Dendronotus*), and many others.

Most intertidal fishes live in the lower intertidal or in tide pools. Gobies, clingfishes, sculpins (*Oligocottus*), pricklebacks (*Cebidichthys*), and gunnels (*Xererpes*) are among the most common. These are all small fishes that are adapted to the environmental extremes of the intertidal. Most are carnivorous.

SOFT-BOTTOM INTERTIDAL COMMUNITIES

Marine biologists refer to any bottom that is composed of sediment, as opposed to rock, as a **soft bottom.** The dividing line isn't always clear. Boulder fields are usually thought of as rocky, but how small the rocks have to be before the bottom is considered "soft" is not precisely defined. In this text, we will refer to a bottom as soft if organisms are able to burrow in it easily.

Soft bottoms occur where large amounts of sediment have accumulated. In North America, soft bottoms dominate on the east coast south of Cape Cod and on virtually all of the Gulf coast. The rocky west coast is interrupted here and there by sandy beaches and mud flats, especially in and near river mouths. The kind of sediment that accumulates in an area and whether sediment is deposited at all depends on how much water motion there is, as well as on where the sediment comes from. In turn, the type of sediment strongly influences the community.

The Shifting Sediments

Soft bottoms are unstable and constantly shift in response to waves, tides, and currents. Thus soft-bottom organisms do not have a solid place for attachment. Very few seaweeds have adapted to this. Seagrasses are about the only large plants that live on soft bottoms, and they live only in certain places. Under the right conditions, however, sea grasses can form thick beds (see the section in Chapter 12, "Seagrass Beds"). In this chapter we will deal only with intertidal soft-bottom communities that lack seagrasses.

Like plants, the animals that live on soft bottoms lack solid attachment sites. Though a few of them are epifauna and live on the sediment surface, most burrow in the sediment to keep from being washed away. Animals that burrow in the substrate are called **infauna.** This is easy to remember: Infauna live *in* the sediment.

The kind of sediment on the bottom, especially the size of the grains, is one of the most important physical factors affecting soft-bottom communities. Most people use the terms "sand," "silt," "clay," and "mud" without much thought. To a geologist these terms refer to sediments with specific particle sizes (Figure 10-33).

Careful analysis is needed to determine grain size precisely, but a "quick and dirty" test will literally give you a feel for the different sediment types. Rub a pinch of sediment or soil between your fingers. Any grittiness that

LIFE IN THE MARINE ENVIRONMENT

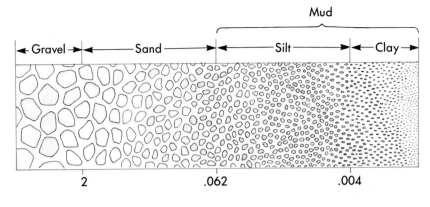

Figure 10-33 Sediments are classified by the size of the particles. Sand is relatively coarse, clay is very fine. Together silt and clay are called *mud.*

you feel is caused by sand. Silt and clay, which together are called mud, feel smooth. The finest sand also corresponds to the smallest particles that you can see. To tell silt and clay apart, rub a little between your teeth. The silt feels gritty, but the clay still is smooth.

The terms "sand," "silt," and "clay" refer to sediments of particular grain sizes. Sand is the coarsest, followed by silt, and then clay. "Mud" refers to silt and clay combined.

Actually, most sediments are made up of a mixture of different particle sizes. The terms for the sediments refer to the most common grain size. "Sand," for instance, has mostly sand-sized grains, but it may also have small amounts of mud or a few large rocks.

The sediment composition is directly related to the degree of water motion. Imagine or, better yet, actually *perform* the following experiment. Take a handful of sediment or soil that contains a wide range of particle sizes, from large pebbles down to clay. Put it in a container with some water and shake it up. As soon as you stop shaking, the pebbles will sink to the bottom. As the water gradually stops moving, first the coarse sand and then finer particles will begin to sink. Very fine material will remain suspended for a long time, making the water look muddy. You might have to let the container sit still for weeks before the finest clay settles out completely. If the container is stirred even gently, the fine material will never settle out.

This demonstrates a general rule: Fine sediments remain suspended with even a small amount of water motion, whereas coarse sediments settle out unless there is considerable flow. Bottom sediments thus reflect the prevailing conditions at the site. Calm, sheltered areas have muddy bottoms because fine sediments can settle out. Particles of organic matter sink at about the same rate as clay particles, so the two tend to settle out together; clay sediments are rich in organics. Places that experience waves and currents have coarser bottoms. If the water motion is strong enough, it may carry away *all* the loose material, leaving bare rock.

Fine sediments are found in calm areas such as bays and lagoons. Coarser sediments are found in areas that are affected by waves and currents.

In the imaginary experiment described, all the particles would eventually settle out if the container were kept still. If there was a constant flow of water through the container, however, the fine material would be carried away and only certain particles, namely the coarse ones, would remain. The coarse particles would have been sorted out from the fine ones.

Living in the Sediment

Living in sediment has advantages in the intertidal. Soft bottoms stay wet after the tide is out, so the problem of desiccation is not as important as in the rocky intertidal. This depends on the grain size, however. Coarse sands drain, and therefore dry out, quite rapidly (Figure 10-34). Partially because of this, coarse sand beaches have relatively very little animal life.

Oxygen Availability. The amount of organic matter in bottom sediments is particularly important to deposit feeders. Because there are no plants to speak of, detritus is the main source of food for soft-bottom animal communities. Deposit feeders extract this organic matter from the sediments. The amount of detritus depends on grain size (see Figure 12-12). Coarse sands contain very little organic matter. This is why we think of sand as being "clean." Silt and clay, on the other hand, are usually rich in detritus, which is why they are often smelly.

The grain size also affects the amount of oxygen that is available in the sediments. Oxygen in sediments is used up by the respiration of animals and, more importantly, decay bacteria. Below the sediment surface there is no light and therefore no photosynthesis, so the infauna depend on the circulation of water through the sediments to replenish the oxygen supply. Grain size and the degree of sorting strongly affect how porous a sediment is (see Figure 10-34). Any backyard gardener knows that water flows more easily through sand than clay. Water circulation through fine sediments is greatly restricted.

Muddy bottoms, then, have a double problem. First, they have more organic material to decay and use up oxygen. Second, the flow of water that brings in new oxygen is reduced. Except for the upper few centimeters of mud the **interstitial water,** or water between the grains, is deficient in oxygen. If you go very far down at all, in fact, the oxygen is completely used up. Sediments that have absolutely no oxygen are said to be **anoxic.**

respiration   Organic matter + $O_2 \longrightarrow CO_2 + H_2O$ + ENERGY
Chapter 4, p. 90

Figure 10-34 "Sorting" refers to the uniformity of the grain size in the sediment. In well-sorted sediments (**A** and **B**) the grains are all about the same size. There is a lot of space between the grains, so water can flow through. The grain size also reflects how well water circulates: larger grains have larger spaces and are thus more porous. Poorly-sorted sediments (**C**) have many different-sized grains. The smaller grains fill in the spaces between the larger ones, making water flow especially difficult.

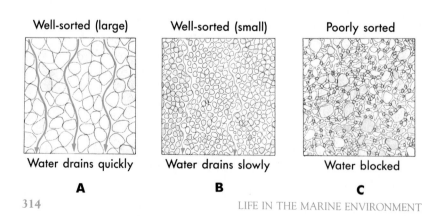

Well-sorted (large)   Well-sorted (small)   Poorly sorted

Water drains quickly   Water drains slowly   Water blocked

A   B   C

LIFE IN THE MARINE ENVIRONMENT

Surprisingly, anoxic conditions are no problem at all for bacteria. Many bacteria are capable of **anaerobic respiration,** in which they break down organic matter without oxygen. A noxious gas called **hydrogen sulfide** (H$_2$S) is produced as a by-product. If you have ever had to unclog the drain of your kitchen sink, you have probably run into hydrogen sulfide: It turns things black and smells like rotten eggs. In the sediments, just as in your drain, hydrogen sulfide is produced when bacteria continue to decompose organic material after all the oxygen has been used up. A distinct black band shows where anoxic conditions begin in the sediments (Figure 10-35).

Figure 10-35 Mud snails (*Cerithidea californica*) have left their tracks by exposing the black band in the sediment that is caused by hydrogen sulfide. It shows the level where the interstitial water becomes anoxic.

> The decay of organic matter in the sediments uses up oxygen. Because of this, the interstitial water beneath the sediment surface is often depleted in oxygen, especially in fine sediments.

The infauna, unlike bacteria, must adapt to the short supply of oxygen, especially in muddy bottoms. Many animals avoid the problem by pumping oxygen-rich water from the sediment surface with siphons or through their burrows (see Figure 11-13). Animals such as these are never actually exposed to low oxygen levels, though the sediments around them may be oxygen deficient.

Other animals are completely buried and have adapted to low-oxygen environments. They often have special hemoglobins and other adaptations that allow them to extract every last bit of oxygen from the interstitial water. Some are sluggish, which reduces their consumption of oxygen. They may even have a limited capability to carry out anaerobic respiration. A few animals, in fact, have symbiotic bacteria that help them live in low-oxygen sediments. Even so, hydrogen sulfide is highly toxic, and anoxic sediments have very little animal life.

hemoglobin   Red protein in the blood of vertebrates and some invertebrates that transports oxygen
Chapter 7, p. 210

Getting Around.   Soft-bottom animals use a variety of methods to burrow through the sediments. Clams (*Macoma, Mya, Donax*) and cockles (*Cardium, Clinocardium*) take advantage of being able to change the shape of their muscular foot. First, they make the foot thin and reach forward with it (Figure 10-36). Then, the end of the foot thickens and acts as an anchor while they pull the rest of the body along. Many soft-bodied animals do something similar, using their bodies the way clams use the foot (Figure 10-37). Heart urchins (*Lovenia, Echinocardium;* see Figure 12-8) burrow or plough through the sediment with their spines and tube feet. The many crustaceans that live on soft bottoms—such as amphipods (*Talitrus, Coro-*

Figure 10-36 Mechanism of burrowing in clams and cockles.

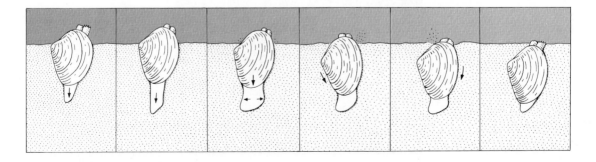

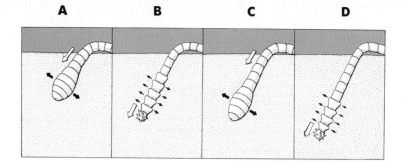

Figure 10-37 Burrowing in the lugworm (*Arenicola*). **A,** The worm expands the end of its body (*black arrows*). The expanded end acts as an anchor and the rest of the body is pulled along behind (*white arrow*). **B,** The worm then flares its segments (*black arrows*), which prevents it from sliding backward when it pushes its proboscis forward (*white arrow*). As the process is repeated (**C** and **D**), the worm moves forward through the sediment.

diatoms Single-celled, photosynthetic organisms that have a shell, or test, made of silica
Chapter 5, p. 121; Figure 5-5

plankton Small plants (*phytoplankton*) and animals (*zooplankton*) that drift with the currents
Chapter 9, p. 284; Figure 9-21

phium); sand, or mole, crabs (*Emerita*); and ghost (*Callianassa*) and mud (*Upogebia*) shrimps—use their jointed appendages to dig.

Quite a few of the deposit feeders solve two problems at once by eating their way through the sediment. Sea cucumbers, certain worms, and other animals use the same "inchworm" style of locomotion as do the clams and worms described above, with one important exception. Rather than push through the sediments, they push the sediments through *them*. They digest the organic matter and leave the rest of the sediment behind (Figure 10-38).

Some soft-bottom animals are so small that instead of burrowing through the sediments they move between the grains. These animals are collectively called the **meiofauna.** There are many different types of animals in the meiofauna, but most have independently evolved a wormlike shape (see Box 11-1). This is clearly an adaptation that makes it easier to move in the spaces among the grains.

Feeding. Because it lacks large plants, the main source of food in the soft-bottom intertidal is detritus (Figure 10-39). Diatoms at the sediment surface sometimes form mats that can be very productive. Diatoms generally do not account for much production, however. In any case most animals don't distinguish between the diatoms and detritus. Plankton brought in by the tides also make a small contribution to the food supply.

Many different methods of deposit feeding have evolved. One of the more common methods has already been described: Animals such as sea cucumbers and various worms take in sediment as they burrow, digest out the detritus and small organisms, and leave the rest of the sediment in their

Figure 10-38 After digesting out the organic matter, the dark sea cucumber (*Holothuria nobilis*) on the right has deposited the pile of undigested sand pellets seen on the left.

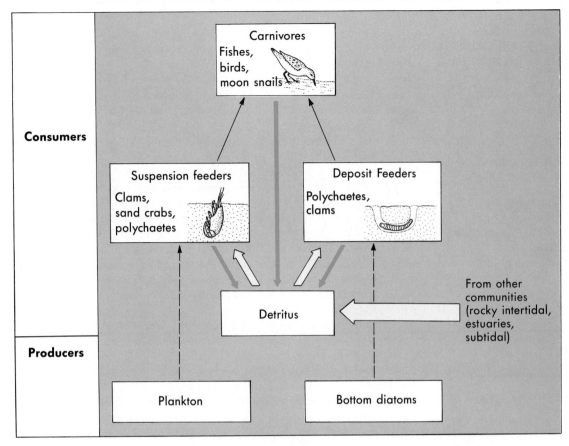

Figure 10-39 A generalized food web on a sandy beach.

wake. This technique is more common in mud bottoms than sand, probably because mud contains more organic matter. Sand is also abrasive and hard on the digestive system.

Sand dollars (*Dendraster, Mellita*; see Figures 6-51, C and 12-9) are more selective. They use their tube feet to pick up organic particles. Sand dollars burrow near the sediment surface, where detritus is still accumulating and is therefore more abundant. The bent-nosed clam *(Macoma)* also concentrates on the upper layer of sediment. It uses a long siphon to suck up food particles from the surface (Figure 10-40). A number of animals catch particles as they settle out and are thus on the line between suspension feeding and deposit feeding. Some polychaete worms *(Terebella)* have long, sticky tentacles that they spread out on the bottom to gather food (see Figure 6-21, A). Other polychaetes produce a mucous net that they use instead of tentacles.

There are also suspension feeders who do not wait for the detritus to settle out. An olive shell *(Olivella)* from Central America makes a mucous net, but instead of spreading the net on the bottom, the snail holds the net up in the water to filter food. Sand, or mole, crabs use a somewhat similar method. They have a pair of large, bushy antennae that they hold up into the water to trap food. By quickly burying and uncovering themselves, they

Figure 10-40 The bent-nosed clam (*Macoma*) uses its long incurrent siphon to feed.

combine feeding and locomotion. When the tide is coming in and they want to move up the beach, they uncover just before a wave hits, ride the wave up, and raise their antennae. They then burrow to keep from being washed back down. As the tide recedes, they do the opposite, burrowing before the wave arrives, then popping out to ride the backwash down the beach (Figure 10-41).

Soft bottoms also have their share of predators. Moon snails (*Polinices*) burrow through the upper sediments looking for clams. When they find one, they drill a hole in its shell and eat it. Several polychaetes and other worms are also important predators. Birds can be major predators during low tide (see Figure 11-15). At high tide, fishes can come in. Fishes often don't eat an entire animal. Instead they just nip off clam siphons or other bits that stick out.

Zonation. Because the organisms live in the sediments and can't be seen, zonation on soft bottoms is not as obvious as in the rocky intertidal. Zonation does exist, however, especially on sandy beaches (Figure 10-42). Water drains rapidly from the sand and, because the beach slopes, the upper part is drier than the lower part.

The upper beach is inhabited by sand fleas, which are really amphipods, and by isopods. In warmer areas, these small crustaceans are replaced by ghost (*Ocypode*) and fiddler crabs (*Uca;* see Box 11-2), who scurry about catching smaller animals, scavenging bits of dead meat, and gathering detritus. Polychaetes, clams, and other animals appear lower on the beach.

Zonation is practically absent from muddy areas. The bottom in such places is flat, and fine sediments retain water. The habitat therefore does not change very markedly between the high- and low-water marks.

Figure 10-41 The sand crab (*Emerita analoga*) uses its long, featherlike antennae to capture its food (left). Another pair of appendages then proceeds to remove the food particles from the antennae (right).

LIFE IN THE MARINE ENVIRONMENT

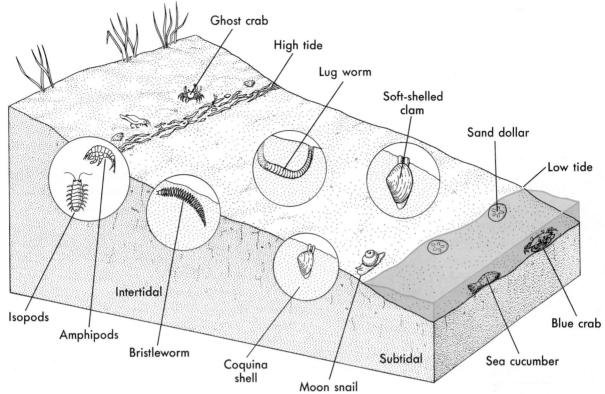

Ghost crab

High tide

Lug worm

Soft-shelled clam

Sand dollar

Low tide

Isopods

Amphipods

Bristleworm

Intertidal

Coquina shell

Moon snail

Subtidal

Sea cucumber

Blue crab

Figure 10-42 Typical zonation pattern on sandy beaches of the Atlantic coast of North America.

Do It Yourself

SUMMARY

1. The _____ is the material on which an organism lives.
2. On the east coast of North America, rocky shores are common only north of _____. They dominate the _____ and are practically absent on the _____.
3. Animals that live on the substrate surface are called _____. Those that burrow are called _____.
4. Zonation in the rocky intertidal is related to the difference in _____ time at different heights. Usually the upper limits of a zone is set by _____, the lower limit by _____.
5. Wave _____ causes waves to bend and become more parallel to the shore. Wave energy is focused at _____.
6. _____ attach mussels to the rock. Molluscs such as _____ use a muscular foot.
7. _____ is usually the limiting resource for rocky intertidal populations.

8. Typically, the most common inhabitant of the high intertidal are _____.

9. Barnacles usually occupy the uppermost part of the _____. Their upper limit is set by _____; predation by _____ largely determines their lower limit.

10. Sea stars are called _____ because they play a central role in the structure of rocky intertidal communities.

11. The occurrence of a regular pattern of recovery after a community is disturbed is called a _____. Typical stages of this in the rocky intertidal are a _____ followed by seaweeds, then _____, and finally _____. The final stage in the process is called the _____.

12. The lower intertidal is dominated by _____. _____, as well as space, is an important limiting resource here.

13. _____ is relatively coarse and contains little organic matter. It is found in _____ areas. Silt and _____ together are called _____, which is fine and organic rich. Muddy bottoms are found in _____ areas.

14. The most important food source for soft-bottom intertidal animals is _____. As a result, most are _____ feeders.

FOR FURTHER READING

Horn, M.H. and R.N. Gibson: "Intertidal Fishes." *Scientific American*, vol. 258, no. 1, January 1988, pp. 64-70. *The ways in which fishes have adapted to the rigorous demands of the intertidal are described.*

Koehl, M.A.R.: "The Interaction of Moving Water and Sessile Organisms." *Scientific American*, vol. 247, no. 6, December 1982, pp. 124-134. *A look at the different ways that attached marine organisms withstand currents and waves.*

Mangin, K.: "A Pox on the Rocks." *Natural History*, June 1990, pp. 50-53. *In Mexico a recently discovered hydroid, a relative of jellyfishes, kills young barnacles. This makes way for seaweeds and limpets.*

Winston, J.E.: "Intertidal Space Wars." *Sea Frontiers*, vol. 36, no. 1, January-February 1990, pp. 46-51. *An examination of the battle for space among intertidal invertebrates.*

Wolcott, T.G. and D.L. Wolcott: "Wet Behind the Gills." *Natural History*, October 1990, pp. 46-55. *Land and shore crabs carry salt water with them to breathe.*

THOUGHT QUESTIONS

1. There are marked differences in the type of organisms found at four different locations at the same tidal height along a rocky shore. What might account for this? Mention at least three possible explanations.

2. Most marine biologists hypothesize that space, not food, limits populations in the rocky intertidal. What kind of experiments would be performed to test this hypothesis?

11

Estuaries: Where Rivers Meet the Sea

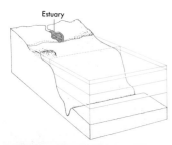

Estuary

Figure 11-1 Mangroves, such as the red mangrove (*Rhizophora mangle*), thrive along the shores of many estuaries in the tropics and subtropics.

A unique environment develops where fresh water from rivers enters the sea. **Estuaries** are more or less enclosed areas where fresh water and seawater meet and mix. Estuaries are environments inhabited by far fewer species than are rocky shores. Nevertheless, they are among the most productive environments on earth. Salt-marsh grasses or mangroves thrive along the shore (Figure 11-1) and seaweeds live farther out. A multitude of worms, clams, and shrimps burrow in the muddy bottom. Snails and crabs crawl along the shore. Fishes swim in the murky, plankton-rich water.

Estuaries also rank among the environments most affected by humans. Estuaries are natural harbors, and as such, many of the world's great cities— New York, London, and Tokyo among others—developed along them. The consequences of human intrusion in estuaries have been disastrous. Estuaries are dredged or filled and transformed into marinas, seaports, industrial parks, cities, and garbage dumps. Many have been obliterated, and most surviving ones are endangered (see the section in Chapter 17, "Modification and Destruction of Habitats").

ORIGINS AND TYPES OF ESTUARIES

Estuaries are scattered along the shores of all the oceans and vary widely in origin, type, and size. They may be called lagoons, sloughs, or even bays, but all share the mixing of fresh water with the sea in a partially enclosed section of the coast. Some oceanographers go as far as classifying enclosed, low-salinity seas, such as the Baltic and Black seas, as estuaries.

Estuaries are partially enclosed coastal regions where fresh water from rivers meets and mixes with seawater.

Many estuaries were formed when sea level rose due to the melting of ice at the end of the last ice age, about 18,000 years ago. The sea invaded

322

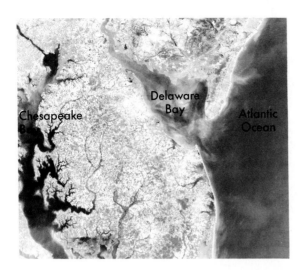

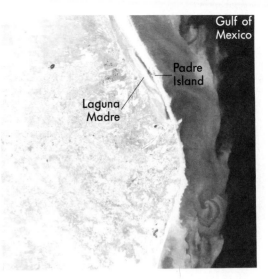

Figure 11-2 Satellite view of the mouth of the Delaware River and the northern margin of Chesapeake Bay on the eastern coast of the United States. Both are drowned river estuaries that formed as the sea flooded river valleys at the end of the last ice age.

Figure 11-3 Satellite view of the southern end of Laguna Madre, Texas, a bar-built estuary. Padre Island is a barrier island that limits access to the open ocean. The small amount of fresh water that flows into the lagoon, its shallow depth, and the high rate of evaporation increase the salinity of the lagoon.

lowlands and river mouths in the process. These estuaries are called **drowned river valleys** or **coastal plain estuaries** (Figure 11-2). They are probably the most common type of estuary. Examples are Chesapeake Bay and the mouths of the Delaware and St. Lawrence rivers on the eastern coast of North America and the mouth of the Thames in England.

A second type of estuary is the **bar-built estuary** (Figure 11-3). Here the accumulation of sediments along the coast builds up **sand bars** and **barrier islands** (see Box 17-2) that act as a wall between the ocean and the fresh water coming from rivers. Bar-built estuaries are found, for instance, along the Texan coast of the Gulf of Mexico, the section of the North Carolina coast protected by the Cape Hatteras barrier islands, and along the North Sea coast of the Netherlands and Germany.

Other estuaries, such as San Francisco Bay in California, were created not because sea level rose but because the land sank, or **subsided,** as the result of movements of the crust. These are known as **tectonic estuaries.**

Another type of estuary was created when retreating glaciers cut deep, often spectacular valleys along the coast. The valleys were partially submerged when sea level rose, and rivers now flow into them. These estuaries, or **fjords,** are common in southeastern Alaska, British Columbia, Norway, southwestern Chile, and the South Island of New Zealand (Figure 11-4).

plate tectonics The process in which large sections of the earth's crust move about
Chapter 2, p. 28

> **Estuaries can be classified into four basic groups based on their origin: drowned river valleys, bar-built, tectonic, and glacier-carved estuaries, or fjords.**

Figure 11-4 Milford Sound, on the southwestern coast of New Zealand's South Island, is an awesome example of a fjord. It is a fingerlike inlet surrounded by sheer walls that rise 1,200 m (3,900 feet) above sea level and plunge to depths of 500 m (1,640 feet). Its entrance is only 55 m (180 feet) deep. As in other fjords, the shallow entrance restricts the exchange of water between the fjord and the open sea.

passive continental margin One that is on the "trailing edge" of a continent and has little geologic activity
Chapter 2, p. 43; Figure 2-24

active margin One that is colliding with another plate and therefore has a lot of geologic activity
Chapter 2, p. 43; Figure 2-24

Broad, well-developed estuaries are particularly common along regions with flat coastal plains and wide continental shelves, a condition typical of **passive margins.** Such is the case of the Atlantic coast of North America. The opposite is true for the steep coasts and narrow continental shelves of the Pacific coast of North America and other **active margins.** Here narrow river mouths carved along the steep coast have restricted the development of estuaries.

PHYSICAL CHARACTERISTICS OF ESTUARIES

Influenced by the tides and the mixing of fresh and salt water, estuaries possess a unique combination of physical and chemical characteristics. These factors govern the lives of the organisms that live there. Fjords are in some ways more similar to coastal waters than to other estuaries. Much of the rest of this chapter focuses on the three remaining types of estuaries.

Salinity

The salinity of estuaries fluctuates dramatically both from place to place and from time to time. When seawater, averaging about 35‰ salinity, mixes with fresh water, nearly 0‰, the mixture has a salinity somewhere in between. The more fresh water that is mixed in, the lower the salinity. Salinity therefore decreases moving upstream (Figure 11-5).

Salinity also varies with depth in the estuary. The salty seawater, being denser, tends to sink. It flows in along the bottom in what is frequently known as a **salt wedge.** Meanwhile, the fresher, less dense water tends to flow out on the surface.

The salt wedge moves back and forth with the daily rhythm of the tides (Figure 11-6). It moves up the estuary on the rising tide, then recedes as the tide falls. This means that organisms that stay in one place are faced with dramatic fluctuations in salinity. They are submerged under the salt wedge at high tide and under low-salinity water at low tide. If the area has a diurnal tide, the organisms are subject to two shifts in salinity every day: one as

diurnal tide A tidal pattern with one high and one low tide each day

semidiurnal tide A tidal pattern with two high and two low tides each day

mixed tide A tidal pattern in which two successive high tides are of different heights
Chapter 3, p. 76; Figure 3-33

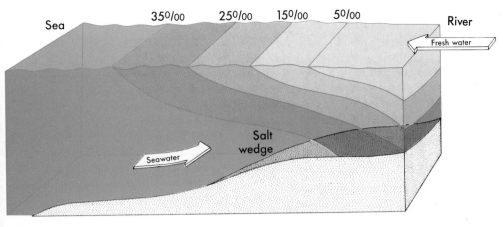

Sea   35⁰/oo   25⁰/oo   15⁰/oo   5⁰/oo   River

Fresh water

Salt wedge

Seawater

Figure 11-5 Profile of a typical, idealized estuary. The lines across the estuary connect points of similar salinity and are known as *isohalines*.

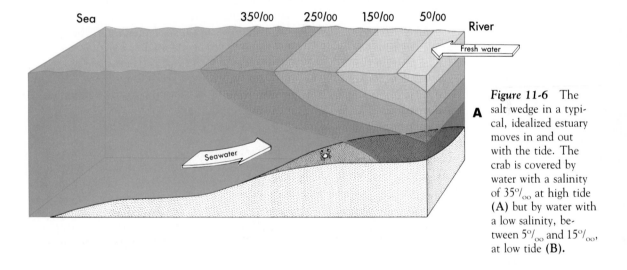

Sea   35⁰/oo   25⁰/oo   15⁰/oo   5⁰/oo   River

Fresh water

Seawater

A

Figure 11-6 The salt wedge in a typical, idealized estuary moves in and out with the tide. The crab is covered by water with a salinity of 35⁰/oo at high tide (A) but by water with a low salinity, between 5⁰/oo and 15⁰/oo, at low tide (B).

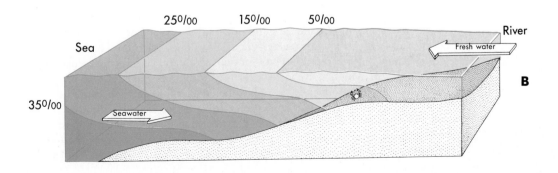

25⁰/oo   15⁰/oo   5⁰/oo   River

Sea   Fresh water

35⁰/oo   Seawater

B

the tide moves upstream and the second as it retreats. In an estuary with semidiurnal tides, salinity changes four times a day.

Estuaries are subject to wide fluctuations in salinity.

The behavior of water masses in estuaries is not always as simple as outlined previously. The shape of the estuary and its bottom, the wind, evaporation of water from the surface, and changes in the tide and freshwater runoff all influence the distribution of salinity. Currents are often especially important. Because most estuaries are long and narrow, the tide doesn't just rise, it rushes in, creating strong **tidal currents.** In a few places the tide actually comes in as a nearly vertical wall of water known as a **tidal bore.** Huge tidal bores in eastern Canada, the People's Republic of China, and a few other places attract surfers from all over the world! Such strong water movements greatly affect the pattern of salinity in an estuary.

Another factor that affects circulation in estuaries is the **Coriolis effect.** North of the Equator, the fresh water that flows from rivers toward the sea is turned toward the right, and south of it, the flow is to the left. This means that, in estuaries located in the Northern Hemisphere, marine organisms can penetrate farther upstream on the left side, whereas this is true of the right side in the Southern Hemisphere.

In regions of little freshwater runoff and high evaporation, the salinity of the water increases. This is the case of Laguna Madre, a shallow bar-built estuary with limited access to the open ocean that parallels the Texas coast for 185 km (115 miles; see Figure 11-3). Average salinity is over 50‰ in some areas, but it may reach 100‰ and over during periods of little rain.

Substrate

Rivers carry large amounts of sediment and other material, including pollutants, into most estuaries. The sand and other coarse material settle out near the river mouth, as soon as the current slows. The fine, muddy particles, however, are carried out into the main body of the estuary. There they eventually settle out when the current slows even more. Therefore, the **substrate,** or type of bottom, of most estuaries is soft mud. Mud, which is actually a combination of silt and clay (see Figure 10-33), is rich in organic material.

As in other organic-rich sediments, respiration by bacteria involved in decomposition of organic matter uses up oxygen that is present in the **interstitial water,** the water between sediment particles. Water cannot easily flow through the fine sediments to replenish the oxygen supply. As a result, the sediments in estuaries are often devoid of oxygen, or are **anoxic,** below the first few centimeters (see Figure 10-35). They have the black color and "rotten-egg" smell typical of oxygen-poor sediments in which **hydrogen sulfide** (H_2S), which is toxic to most organisms, accumulates. Anoxic sediments are not completely devoid of life. **Anaerobic bacteria,** which do not need oxygen to carry out respiration, thrive under these conditions.

In estuaries that have unimpeded tidal flow, which includes most shallow ones, there is plenty of oxygen dissolved in the water. Some deep-water estuaries such as fjords (Figure 11-4), however, have a shallow "sill" at the

Coriolis effect As a result of the earth's rotation, anything that moves large distances on the earth's surface tends to bend to the right in the Northern Hemisphere and to the left in the Southern Hemisphere
Chapter 3, p. 65

entrance that restricts water circulation. Low-salinity water flows out unimpeded on the surface. The sill, however, prevents seawater from flowing in along the bottom. The stagnant deep water may become depleted in oxygen because of bacterial respiration associated with the decomposition of organic matter that sinks and accumulates on the bottom.

Fine, muddy sediments brought into estuaries by rivers settle out in the relatively quiet waters. Bacterial respiration in these organic-rich sediments depletes the oxygen in them.

Other Physical Factors

Besides extreme salinity fluctuations and muddy substrates, other physical factors help make estuaries one of the least inviting of all marine environments. The water temperature in estuaries, except fjords, varies markedly because of their shallow water and large surface area. Organisms that are exposed at low tide may have to face even more drastic daily and seasonal temperature fluctuations.

There is a lot of suspended sediment in estuaries, greatly reducing the water clarity. This permits very little light to penetrate through the water column. The particulate material in the water can also clog the feeding apparatus of some filter feeders and even kill some organisms, such as corals, that are sensitive to sediment.

ESTUARIES AS ECOSYSTEMS

To the uninitiated, estuaries may at first look desolate and almost lifeless. This is far from the truth. Estuaries are tremendously productive and are home to large numbers of organisms, some of which are of commercial importance. Estuaries also provide vital breeding and feeding grounds for many birds, fishes, and other animals. Estuarine ecosystems actually consist of several distinct communities, each with its own characteristic assemblage of plants and animals.

Living in an Estuary

Life in an estuary revolves largely around the need to adapt to extremes in salinity, temperature, and other physical factors. Though other marine environments may be more extreme—they may be colder or more saline, for instance—none change so rapidly or in so many ways as estuaries. Living in an estuary is not easy, so relatively few species have successfully adapted to estuarine conditions.

Coping with Salinity. Maintaining the proper salt and water balance of cells and body fluids is one of the greatest challenges that estuarine organisms must deal with (see the section in Chapter 4, "Regulation of Salt and Water Balance"). Most estuarine organisms are marine species that have developed the ability to tolerate low salinities (Figure 11-7). How far they get up the estuary depends on just how tolerant they are. Most estuarine organisms are **euryhaline** species, those that tolerate a wide range of salinities. They evolved from marine species. The relatively few **stenohaline**

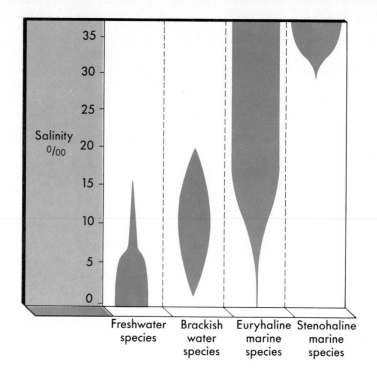

Figure 11-7 Type of species living in an idealized estuary in relation to salinity. The thickness of the bars represents the relative number of species.

species, those that only tolerate a narrow range of salinities, may be either freshwater or marine; most are limited to the outer ends of the estuary and rarely penetrate into the estuary proper. There are also some brackish water species that are adapted to live in intermediate salinities. Some of these species are stenohaline; others are euryhaline.

Because of their marine background, estuarine organisms face the problem of the water in estuaries being diluted with fresh water. Those which have an internal salt concentration higher than that of the surrounding water tend to take on water through **osmosis**. Some animals adapt by simple changes in behavior. They may hide in their mud burrows, close their shells, or swim away if the salinity drops. This "run or hide" strategy is not widespread, however, and most organisms rely on more complex but reliable mechanisms.

Soft-bodied estuarine animals, such as many molluscs and polychaete worms, often maintain osmotic balance simply by allowing their body fluids to change with the salinity of the water. They are called **osmoconformers** (Figure 11-8). Many crabs and fishes, as well as some molluscs and polychaetes, are instead **osmoregulators**. They keep the salt concentration of their body fluids more or less constant regardless of the water salinity. Basically, they pump excess water out and absorb solutes from the surrounding water to compensate for those lost in the elimination of water. The kidneys, gills, salt glands, and other structures accomplish this.

Organisms in estuaries have adapted to salinity fluctuations in various ways. Many are osmoconformers and let the salinity of their body fluids vary with that of the water. Others are osmoregulators and keep the salt concentration of their body fluids constant.

osmosis   The movement of water from high to low concentrations across a membrane
Chapter 4, p. 99

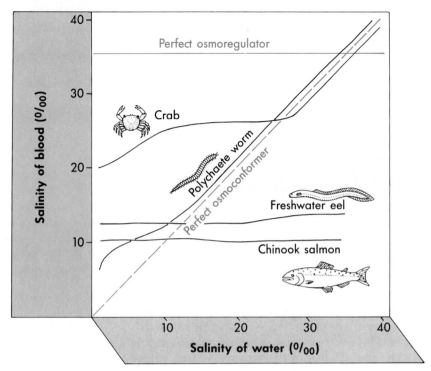

Figure 11-8

- 40 — Perfect osmoregulator
- 30 — Crab
- 20 — Polychaete worm / Perfect osmoconformer
- Freshwater eel
- 10 — Chinook salmon

Salinity of blood (°/oo)

Salinity of water (°/oo): 10   20   30   40

Figure 11-8 The concentration of salts and other solutes, known as the *salinity,* in the body fluids of estuarine animals responds in various ways to the salinity of the surrounding water. In a perfect osmoconformer, the salinity of the blood exactly matches that of the water. In a perfect osmoregulator, blood salinity stays the same no matter what the water salinity is. We have drawn the line for an imaginary perfect osmoregulator at 35°/oo. The salmon and freshwater eel are nearly perfect osmoregulators even though their bloods are more dilute. The important point is not the actual salinity of the blood, but the fact that it remains relatively constant.

Fishes that inhabit estuaries need to osmoregulate even if their blood is less salty than seawater (see Figure 4-22 and the section in Chapter 7, "Regulation of the Internal Environment"). Salmon and freshwater eels are able to migrate back and forth between rivers and the sea and still maintain a stable internal environment thanks to their kidneys.

Not all animals can be neatly classified as perfect osmoconformers or perfect osmoregulators. Many invertebrates, for instance, osmoregulate at low salinities and osmoconform at higher salinities (see Figure 11-8). Even efficient osmoregulators such as salmon and freshwater eels do not keep *exactly* the same concentration of salts and other solutes in their blood as salinity changes.

Estuarine plants must also handle salinity variations. Grasses and other salt-marsh plants are land plants that have developed a high salt tolerance. These plants actively absorb salts to match the outside concentrations and prevent water from leaving their tissues. Notice that this is opposite to the situation in marine organisms, which have to adapt to *reduced,* not increased, salinities.

Cord grasses (Figure 11-9) other salt-marsh plants, and some mangroves excrete excess salts with salt glands in their leaves (see Figure 4-23, B). Some estuarine plants, such as pickle weed, instead accumulate large amounts of water to dilute the salts they take up (Figure 11-10). Fleshy plants such as these are known as **succulents.**

Adapting to the Mud. As we learned in Chapter 10 (in the section, "Living in the Sediment"), living in mud has its problems. There is nothing to hold on to, so most animals either burrow or live in permanent tubes beneath the sediment surface. Because it is difficult to move through mud,

Figure 11-9 Cord grass (*Spartina*) is an important component of salt marshes on the Atlantic and Pacific coasts of North America.

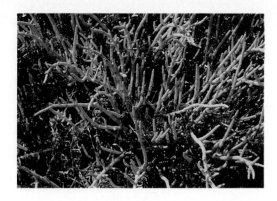

Figure 11-10 Numerous species of pickle weed (*Salicornia*), also known as saltwort or glasswort, are common succulent plants in salt marshes around the world.

the inhabitants tend to be stationary or slow movers. Living in mud, however, has a side benefit: Salinity fluctuations are less drastic than in the water column.

The depletion of oxygen caused by the decay of organic matter in the mud presents another challenge. This is no problem to burrowers that pump oxygen-rich water into their burrows. Burrowers without this luxury have special adaptations to low-oxygen environments. Some have blood that is packed with hemoglobin. Furthermore, the hemoglobin itself is especially "hungry" for oxygen: It can hold and carry oxygen even when only minute quantities are available. Some clams and a few other mud-dwellers can even survive without oxygen for days.

hemoglobin Red protein in the blood of vertebrates and some invertebrates that transports oxygen *Chapter 7, p. 210*

Types of Estuarine Communities

Several distinct communities are associated with estuaries. One consists of the plankton, fishes, and other open-water organisms that come in and leave with the tide. Several other communities are permanent parts of the ecosystem.

plankton Small plants (*phytoplankton*) and animals (*zooplankton*) that drift with the currents *Chapter 9, p. 284; Figure 9-21*

Estuarine communities consist of relatively few species. These species, however, are typically represented by numerous individuals. A surprising number of estuarine species, particularly those inhabiting temperate estuaries, are widely distributed around the world.

Open Water. The type and density of plankton inhabiting estuaries varies tremendously with the currents, salinity, and temperature. The murky water restricts the penetration of light and may limit **primary production** by phytoplankton. Most of the phytoplankton and zooplankton in small estuaries are marine species flushed in and out by the tides. Larger, more stable estuaries may also have their own, strictly estuarine, species.

primary production The conversion of carbon from an inorganic form, carbon dioxide, into organic matter by autotrophs—that is, the production of food *Chapter 4, p. 92*

A rich variety of fishes live in most estuaries. Many are the juveniles of marine species that breed at sea but use estuaries as nurseries, taking advantage of the abundant food and relative safety from predators. Many are of enormous commercial importance world wide. Examples are the menhaden (*Brevoortia*), anchovies (*Anchoa*), mullets (*Mugil*), many flatfishes (*Solea, Platichthys, Pseudopleuronectes*), and croakers (*Micropogonias*). Some fishes move through the estuary during their migrations. Migrating fishes are ei-

Figure 11-11 Mud flats are characteristic of many estuaries. The great majority of their inhabitants are burrowers that are exposed to wide fluctuations in temperature, salinity, and oxygen. Cord grass *(Spartina)* typically limits the upper edge of mud flats in temperate estuaries like this one.

ther **anadromous**—such as salmon *(Oncorhynchus, Salmo)*, smelts *(Osmerus)*, and shad *(Alosa)*—or **catadromous**—such as freshwater eels *(Anguilla)*. Relatively few fishes spend their entire lives in estuaries. One of these, at least through part of its range, is the striped bass *(Morone saxatilis)*.

Shrimps are also found in estuaries. Some commercially valuable species *(Penaeus)* use estuaries as nurseries.

Mud Flats. The bottoms of estuaries that become exposed at low tide often form **mud flats** (Figure 11-11). Mud flats are especially extensive in estuaries where there is a large difference in height between high and low tide. Though it all looks pretty much the same, the mud flat can vary widely in the size of particles. Sand accumulates near river mouths and in the tidal creeks that form as the tide changes. In the calmer central part, the mud flat contains more fine, silty material.

Mud-flat communities in estuaries are similar to those on muddy shores discussed in Chapter 10 (in the section, "Soft-Bottom Intertidal Communities"). Low tides expose organisms to desiccation, wide variations in temperature, and predation, just as in any other intertidal community. In estuaries, however, mud-flat organisms must also withstand regular variations in salinity.

Plant life is usually not very evident on mud flats. A few hardy seaweeds—such as the green algae *Enteromorpha* (Figure 11-12) and *Ulva*, the sea lettuce, and the red alga *Gracilaria*—manage to grow on bits of shell. These and other **primary producers** may be particularly common during the warmer months. Large numbers of benthic diatoms grow on the mud and often undergo extensive blooms, forming golden-brown patches. In tide pools left by the receding tides these patches become coated with oxygen bubbles as intense photosynthesis takes place in the sunlight.

Bacteria are extremely abundant on mud flats. They decompose the huge amounts of organic matter brought in by rivers and tides. When the oxygen is used up by decay, some bacteria produce hydrogen sulfide. This, in turn, is used by **sulfur bacteria,** chemosynthetic bacteria that derive energy by breaking down sulfur compounds such as hydrogen sulfide. Diatoms and bacteria actually account for most of the primary production in mud flats.

The dominant animals in mud flats burrow in the sediment and are

anadromous fishes Those which migrate from the sea to spawn in fresh water

catadromous fishes Those which migrate from fresh water to spawn at sea
Chapter 7, pp. 217-218

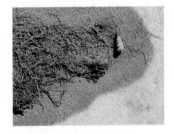

Figure 11-12 *Enteromorpha* is a green alga that tolerates wide fluctuations in salinity, temperature, and pollution. It can be found on rocky shores (see Figure 5-11), as well as in estuaries.

chemosynthetic bacteria Bacteria that use energy stored in chemical compounds to make their own food
Chapter 5, p. 119

BOX 11-1

Life in the Mud

The space between sediment particles is largely invisible to us, but it is an active and important part of the marine environment. This microscopic world, which extends from sandy beaches and estuaries to deep water, is inhabited by highly specialized interstitial organisms, often called the meiofauna.

Some of the meiofauna glide freely among the sediment particles, whereas others attach to them. Though detritus is probably the most important source of food, many of the meiofauna are predators and scavengers. There also are grazers who eat the diatoms and other tiny plants that live in the upper few millimeters of the sediment. The meiofauna, in turn, are part of the diet of many deposit feeders.

The underground world of the meiofauna is packed with bizarre creatures. There are monstrous protozoans, wiggling nematodes, tiny attached hydroids that capture their food with snakelike tentacles, and weird vacuum-cleaning worms. Some members are just miniature versions of flatworms, polychaetes, copepods, and other more familiar animals that live elsewhere. Most, however, are uniquely adapted to the interstitial habitat, and some are found nowhere else.

Only a little over 100 species of **kinorhynchs (phylum Kinorhyncha)** are known, and all are restricted to marine muds. Their body is divided into a series of segments armed with spines. The head, which is also surrounded by spines, can be retracted into the body.

Gastrotrichs (phylum Gastrotricha) use suction to eat detritus and smaller members of the meiofauna. They have short, hairlike cilia on the head and the ventral surface of the body. About half of the over 500 known species are marine. Though they look much like protozoans, gastrotrichs have many cells and true tissues and organ systems.

Rotifers (phylum Rotifera) are a group of nearly 2,000 species, of which only about 100 are marine. Of these, several species are part of the meiofauna. They are known as "wheel animals" because of a crown of cilia on their heads. This structure is used in locomotion and feeding.

Oligochaetes are segmented worms **(phylum Annelida)** that are widely distributed on land and in fresh water. They include the earthworms and related forms. Oligochaetes are less common at sea. They are present in the meiofauna, however, especially in polluted bays and harbors.

Collecting and extracting live meiofauna is not particularly complicated. The easiest procedure is to take moist mud or sand and let it stagnate in a container for an hour or so. The meiofauna will move to the top layer of sand as they run out of oxygen. Samples are taken close to the sediment surface with an eye dropper and observed under a dissecting or, even better, a compound microscope. The show is about to begin.

Examples of interstitial fauna, or meiofauna.

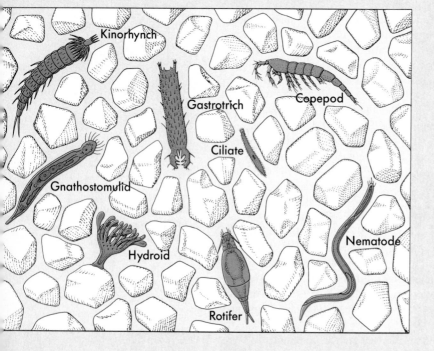

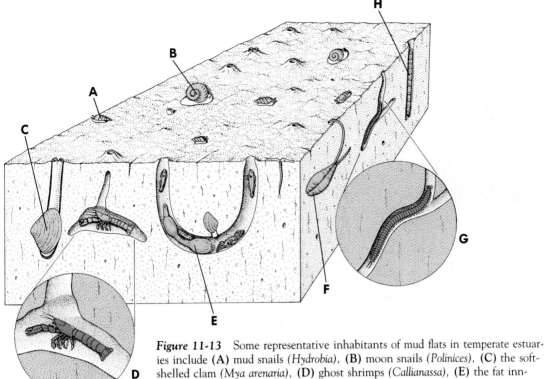

A
B
C
D
E
F
G
H

Figure 11-13 Some representative inhabitants of mud flats in temperate estuaries include **(A)** mud snails (*Hydrobia*), **(B)** moon snails (*Polinices*), **(C)** the soft-shelled clam (*Mya arenaria*), **(D)** ghost shrimps (*Callianassa*), **(E)** the fat innkeeper (*Urechis caupo*) and guests, **(F)** bent-nosed, or small, clams (*Macoma*), **(G)** sandworms (*Neanthes*), and **(H)** bamboo worms (*Clymenella*). Many mud-flat organisms can be found in muddy bottoms outside estuaries (see Figure 12-6).

known as **infauna** (Figure 11-13). Very few can be classified as **epifauna,** those that either live *on* the sediment or are attached to a surface as **sessile** forms. Though there are not very many species of these burrowing animals, they often occur in immense numbers. At low tide their presence may be revealed only by small hills topped by a hole or piles of feces and other refuse. They feed on the abundant **detritus** suspended in the water and deposited in the sediment. Most of the food for these animals is brought in by the rivers and tides and is not actually produced on the mud flat.

Mud flat inhabitants feeding on detritus are **deposit** and **suspension feeders.** Deposit feeders are more common than suspension feeders in mud flats and other muddy bottoms (see Figure 12-12). Suspension feeders, which include **filter feeders,** are actually excluded, or eliminated, by deposit feeders in these soft bottoms. They are at a disadvantage because their filtering mechanisms tend to be clogged by the higher amounts of detritus that rains on soft bottoms. Suspension feeders, on the other hand, are more common in bottoms where sediment particles are more sandy. The wider interstitial spaces between these larger sand particles hold less of the detritus that feeds deposit feeders. Suspension feeders, many of which stay in one place while filtering the water, find it easier to live in sandier sediments than do deposit feeders, who find softer sediments easier to dig into. The disturbance of sediments caused by deposit feeders further helps in the ex-

detritus Dead organic matter and the decomposers that live on it
Chapter 9, p. 277

deposit feeders Animals that feed on organic matter that settles in the sediment
Chapter 6, p. 158; Figure 6-20

suspension feeders Animals, including *filter feeders,* that feed on particles suspended in the water column
Chapter 6, p. 147; Figure 6-20

Figure 11-14 The California horn snail *(Cerithidea californica)*, a deposit feeder, is extremely abundant in mud flats.

clusion of suspension feeders by clogging filtering surfaces and burying newly settled larvae.

The dominant producers on mud flats are diatoms and bacteria. Most of the animals are burrowing deposit and suspension feeders that feed on detritus.

Protozoans, nematodes, and many other minute—or **interstitial**—animals that compose the **meiofauna** (see Box 11-1) also thrive on detritus. The larger burrowing animals include numerous polychaetes (see Figures 11-13 and 10-37). Most are deposit feeders. Other polychaetes are suspension feeders that filter water or extrude tentacles to collect the detritus that falls from the water column. Yet another detritus-feeding strategy is to switch back and forth between suspension and deposit feeding, depending on the amount of suspended material in the water.

A number of bivalves also thrive in mud flats. Many are filter feeders that are also found on muddy and sandy shores outside estuaries (see Figure 10-36). Examples from temperate waters are the quahog, or hard clam *(Mercenaria mercenaria)*, the soft-shelled clam *(Mya arenaria;* Figure 11-13, C), and razor clams *(Ensis, Tegulus)*. Some of these are of considerable commercial importance. Bent-nosed, or small, clams *(Macoma;* Figure 11-13, F) are deposit feeders that use their long incurrent siphon to vacuum clean the surface (see Figure 10-40).

The ghost *(Callianassa;* Figure 11-13, D) and mud *(Upogebia)* shrimps make elaborate burrows that, as a side effect, help oxygenate the sediment. These shrimps feed on detritus that they filter from the water and sift from the mud. Fiddler crabs *(Uca)* also live in burrows but are active on the mud flats at low tide (see Box 11-2). They eat the mud and extract the detritus.

Figure 11-15 Differences in the bill length of wading shorebirds allow them to feed on particular mud-flat animals **(A).**

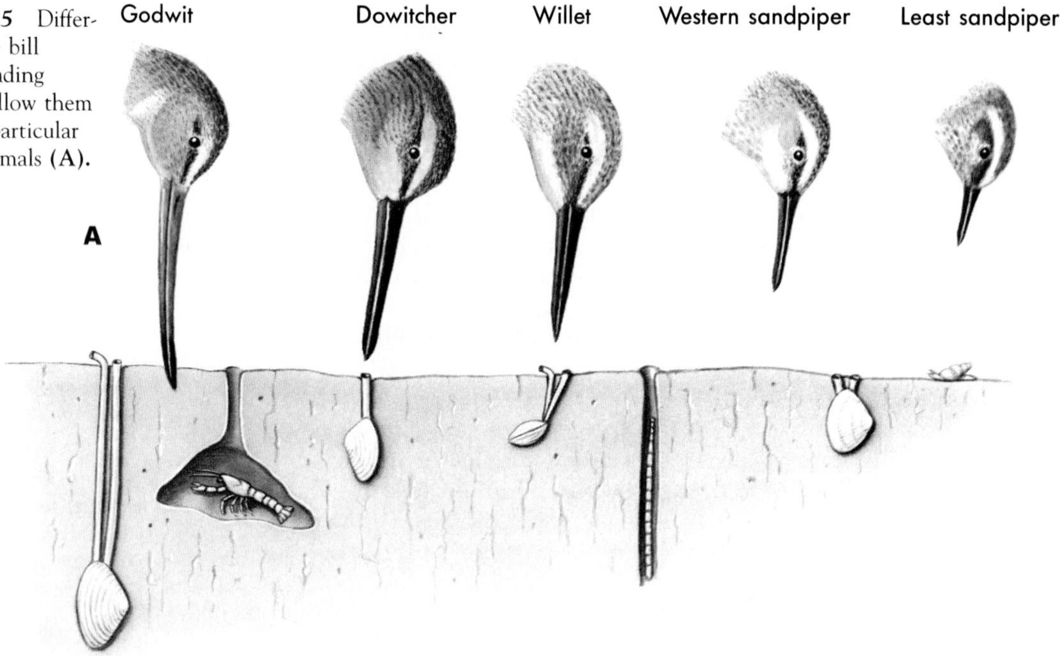

Godwit Dowitcher Willet Western sandpiper Least sandpiper

A

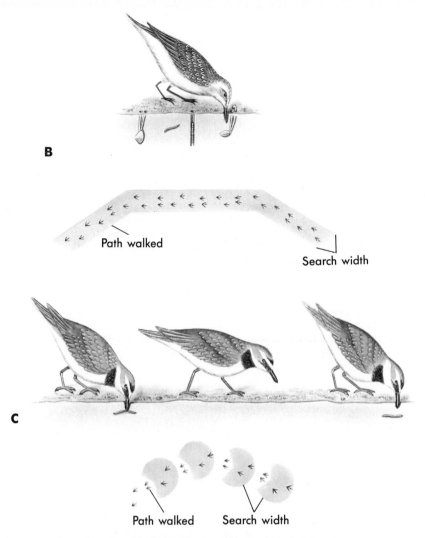

B

Path walked

Search width

C

Path walked   Search width

Figure 11-15, cont'd
Feeding behavior also varies among these shorebirds: sandpipers use their bills in searching for food and follow a roughly straight path **(B)**, whereas plovers rely mostly on their vision, turning their heads sideways as they move **(C)**.

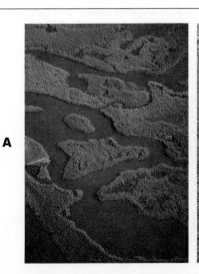

A

B

Figure 11-16 Aerial view of one of the extensive salt marshes typical of the eastern coast of the United States, this one in North Carolina **(A)**. A salt marsh crossed by a tidal creek **(B)**. Creeks are formed as water rushes in and out with the tides. Turbulence sorts out sediment particles so that the bottom of tidal creeks often consists of larger particles.

BOX 11-2

Fiddler on the Mud

What lives in mud, feeds on mud, and find mates by calling and waving across the mud? But, of course—fiddler crabs, creatures remarkable in many ways but best remembered as the ultimate experts in mud.

The many species of fiddler crabs (*Uca*) are inhabitants of mud and sand flats in estuaries and other sheltered coasts. They are found mostly in the tropics and subtropics, but some species are found as far as southern California and Boston Harbor. Fiddlers are deposit feeders that feed on detritus. They feed at low tide, using their pincers to scoop mud up into the mouth. The detritus in the mud is extracted with the help of brushlike mouthparts. Water is pumped from the gill chambers into the mouth to make the lighter detritus float and thus help separate it from the mud. The detritus is swallowed, and the clean mud is spat out on the substrate in neat little balls.

Fiddlers are active at low tide and retreat into their burrows at high tide. Each burrow has an entrance (with neat little balls of mud around it!) that the occupant can plug when the tide comes in. At the next low tide, crabs emerge from home to feed and do whatever healthy, active fiddlers like to do. The tidal cycle means everything to them. Tides are so crucial that if crabs are taken away and isolated in an environment of constant light and temperature, they will continue to be active at the times that correspond to low tide and sit quietly when they were expected to be at home! It is as if they can tell time with little watches they carry around. Not only activity patterns, but color changes that normally take place in their natural environment (darker color during the day, lighter at night), continue to be observed in isolation. These are examples of **biological clocks,** repeated rhythms that are syn-chronized with time. In the case of fiddlers, activity patterns and color changes are synchronized with tides—which depend on lunar cycles—and also with day-night changes—which depend on the sun.

Fiddlers also have an interesting sex life. Males feature a tremendously-enlarged claw, either right or left. It is brightly colored or highlighted with markings in many species. Females have a much smaller pair, which is used in feeding, as is the case of the males' small pincer. Males use their big claw to advertise their sex—to tell females they mean business and to threaten any other males that may be on their way. They wave their claw at low tide on territories established around their burrows. In one species, males build a mud "shelter" around the burrow to heighten the effect. Claw-waving gets so heated sometimes that a whole mud flat seems to move up and down with hundreds of displaying males. Males entice any interested females into their bachelors' burrows, and a female may visit a few pads before deciding on a particular one. Males often fight for prospective mates. They fight very carefully—a lost claw means disaster. You may as well retire. It takes many molts to regenerate one that is big enough to get you back in action, and here size is the name of the game.

In those areas coinhabited by several species of fiddlers, waving is used to prevent a male from attracting females of the wrong species. Some species wave up and down; others sideways. The angle and frequency of waving also varies, and bowing, fancy steps, and other body movements may form part of the ritual. Some beat the claw on the ground. Males of many species even produce sound by vibrating a joint of the large claw. Whatever turns you on!

Signaling in a male fiddler crab.

On the Pacific coast of North America the fat innkeeper (*Urechis caupo*), an echiuran worm, secretes a funnel-shaped net of mucus through which it pumps water to filter out food (Figure 11-13, *E*). It gets its common name because it shares its U-shaped burrow with a polychaete (*Hesperonoe adventor*), a crab (*Scleroplax granulata*), one or more fish (*Clevelandia ios*), and other guests.

Some animals live on the surface of the mud or move in and out with the tide. These include deposit-feeders such as mud snails (*Ilyanassa, Cerithidea, Batillaria,* and *Hydrobia;* Figures 11-3, *A* and 11-14), amphipods, and shrimps. Carnivores include polychaete worms (*Glycera, Neanthes;* Figure 11-13, *G*), moon snails (*Lunatia, Polinices;* Figure 11-13, *B*) and other predatory snails (*Busycon, Urosalpinx*), and the blue crab (*Callinectes sapidus*).

By far the most important predators in the mud-flat community are fishes and birds. Fishes invade mud flats at high tide, whereas birds congregate at low tide to feed. Estuaries are important stop-over and wintering areas for many species of migratory birds. The open spaces offer them safety from natural enemies and food is plentiful. The most significant predators in mud flats are wading shorebirds (Figure 11-15). These include the willet (*Catoptrophorus semipalmatus*), godwits (*Limosa*), dowitchers (*Limnodromus*), the knot (*Calidris canutus*), and many species of plovers (*Charadrius*) and sandpipers (*Erolia, Ereunetes*). They feed on polychaetes, ghost shrimps and other small crustaceans, clams, and mud snails. Oystercatchers (*Haematopus*) specialize on clams and other bivalves.

These many birds do not all exploit the same type of prey. The varying lengths of their bills may represent a specialization in prey, because different types of prey live at different depths in the mud (Figure 11-15, *A*). In addition, shorebirds use different strategies to locate their food. Birds such as sandpipers rely mostly on their bills, probing in the mud as they walk around (Figure 11-15, *B*). Others, such as plovers, use their eyesight to detect slight movements on the surface of the mud (Figure 11-15, *C*). These differences in feeding habits are an example of **resource partitioning.**

Herons (*Ardea, Hydranassa*) and egrets (*Casmerodius, Leucophoyx*) compose yet another group of wading birds. They specialize in catching fishes, shrimps, and other small, swimming prey. Birds that feed by swimming or diving in the estuary, such as ducks (*Aythya*), terns (*Sterna*), and gulls (*Larus*), are often seen resting on mud flats.

Salt Marshes. Estuaries in temperate and subarctic regions are invariably bordered by extensive grassy areas that extend inland from the mud flats (see Figure 11-11). These areas are partially flooded at high tide and are known as **salt,** or **tidal, marshes** (Figures 11-16, *A* and 11-17). Sometimes they are grouped with freshwater marshes and collectively referred to as **wetlands.** Though mostly associated with estuaries, salt marshes can also develop along sheltered open coasts. Tidal creeks, freshwater streams, and shallow pools frequently cut through the marsh (Figure 11-16, *B*).

Salt marshes are particularly extensive along the Atlantic and Gulf coasts of North America (Figure 11-16, *A*). The broad estuaries and shallow bays of these gently sloping coastlines provide optimal conditions for the development of salt marshes. The Pacific coast of North America, on the other hand, is generally steeper, and most of the estuaries have formed

resource partitioning The sharing of a resource by two or more species to avoid competition
Chapter 9, p. 268

Figure 11-17 The daily tides have a crucial role in salt marshes. They help circulate detritus and nutrients and expose mud flats to predation by shorebirds and other animals.

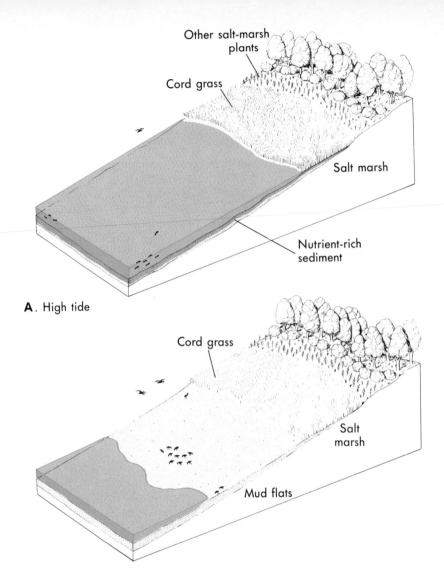

Other salt-marsh plants

Cord grass

Salt marsh

Nutrient-rich sediment

A. High tide

Cord grass

Salt marsh

Mud flats

B. Low tide

Figure 11-18 A salt marsh near Atlantic Beach, North Carolina. As in other Atlantic marshes the salt marsh cord grass (*Spartina alterniflora*) occupies the edge of the marsh that is flooded the most by sea water. It is replaced higher up in the marsh by the salt marsh hay (*Spartina patens*), a shorter and finer grass that may form extensive meadows. It grows where the marsh is flooded only at high tide.

along narrow river valleys. This has resulted in the development of less extensive salt marshes.

Salt marshes are subject to the same extremes in salinity, temperature, and tides that affect mud flats. They also have a muddy bottom, but it is held together by the roots of marsh plants and thus is more stable.

The salt-marsh community is dominated by a few hardy grasses and other salt-tolerant land plants. These plants thrive in the marsh, though the environment is too harsh for most other land plants. There is a pronounced zonation of plants in salt marshes. This zonation is related to the height relative to the tide, but it varies according to geographic location, type of substrate, and other factors.

Cord grasses (*Spartina alterniflora* on the Atlantic coast; *S. foliosa* on the Pacific coast) are typically the most common plants along the seaward limit of the salt marsh, where it meets the mud flats (Figure 11-18; see also Figures 11-9 and 11-11). These grasses invariably occupy the fringe above the mean low-tide level. The tops of their tall leaves remain exposed to the air even when the bottom is covered at high tide. The plants have extensive horizontal stems that stretch out underground. The leaves and roots develop from the stems.

Cord grass may gradually invade mud flats, because the plants slow down the tidal flow and thus increase the amount of sediment that is trapped between the roots. The extension of the salt marsh is eventually slowed down by the height of the highest tide.

Cord grass gives way to other plants in the higher parts of the marsh. On the Atlantic coast a second species of cord grass (*S. patens*) dominates, but rushes (*Juncus*), pickle weed (*Salicornia*), and several other plants often form distinct zones (see Figure 11-18). The higher levels of salt marshes on the Pacific coast are usually dominated by pickle weed (Figure 11-10). The landward limit of salt marshes, a transition zone with adjacent terrestrial— or land—communities, is characterized by a large variety of plants resistant to salt spray, such as salt grasses (*Distichlis*) and several species of pickle weed.

The muddy salt-marsh substrate is home to bacteria, diatoms, and thick mats of filamentous blue-green and green algae (Figure 11-19). Bacteria play a crucial role by breaking down, or decomposing, the large amounts of dead plant material produced in the salt marsh. These bacteria and the partially broken-down organic matter are a major source of the detritus that feeds many of the inhabitants of the estuary. Some bacteria and blue-green algae are **nitrogen fixers** that enrich the water.

Salt marshes are dominated by grasses and other marsh plants. Bacteria in the mud decompose dead plant material and contribute a large portion of the detritus in the estuary.

Some of the burrowing animals of mud flats also inhabit the salt marshes. In addition, nematodes, small crustaceans, larvae of land insects, and other small invertebrates live among the algal mats and decaying marsh plants. Crabs are conspicuous inhabitants of salt marshes. Fiddler crabs build burrows along the mud-flat edges. Other marsh crabs (*Sesarma*,

Figure 11-19 The Atlantic horse, or ribbed, mussel (*Geukensia demissa*) is a mussel adapted to live half buried in mud.

nitrogen fixation Conversion of nitrogen gas (N_2) into nitrogen compounds that are used as nutrients by plants
Chapter 9, p. 281

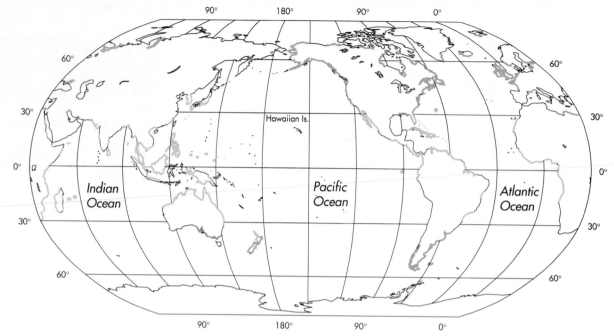

90°   180°   90°   0°

60°   60°

30°   Hawaiian Is.   30°

0°   *Indian Ocean*   *Pacific Ocean*   *Atlantic Ocean*   0°

30°   30°

60°   60°

90°   180°   90°   0°

Salt marshes   ———————

Mangrove forests   ———————

Mixed salt marshes and   ———————
mangrove forests

Figure 11-20 The world distribution of salt marshes and mangrove forests. Mangroves replace salt marshes in tropical regions. Both overlap in such areas as the Gulf of Mexico and southern Australia. Mangroves were introduced by humans to the Hawaiian Islands.

Hemigrapsus, Pachygrapsus) are scavengers that eat dead organic matter. They also live in burrows.

Marsh plants provide shelter and food to many marine and land animals. Coffee bean snails (*Melampus*) and marsh periwinkles (*Littorina*) are air-breathing snails that feed on detritus and minute algae that grow on marsh plants. The horse, or ribbed, mussel (*Geukensia demissa*) is a suspension feeder that lives half buried in the mud among the cord grass (Figure 11-19). Killifishes (*Fundulus*) and juvenile silversides (*Menidia*) are examples of fishes that inhabit tidal creeks and pools in the marsh. Rails (*Rallus*) and the American coot (*Fulica americana*) are among the birds that feed and nest here. Many other land birds and mammals, from ospreys to raccoons, are occasional visitors.

Mangrove Forests. Although not limited to estuaries, mangrove forests are in many ways the tropical equivalents of salt marshes. **Mangroves** are flowering land plants adapted to live in the intertidal (see the section in Chapter 5, "Flowering Plants"). These shrubs and trees may form thick forests. The forests are sometimes called **mangals** to distinguish them from "mangroves," the actual plants themselves (see Figures 5-27 and 5-28). Mangroves are typical of tropical and subtropical regions, where they replace the temperate salt marshes (Figure 11-20). It has been estimated that from 60% to 75% of all tropical shores are fringed with mangroves, a suggestion of their tremendous importance.

Mangrove forests, or mangals, are the tropical equivalents of salt marshes. They are dominated by mangroves, which are bushes or trees adapted to be partially immersed at high tide.

Mangroves grow along protected coasts where muddy sediments accumulate. Though mangroves grow in estuaries, they are not restricted to river mouths. As in salt-marsh plants, the various species of mangroves have different tolerances to immersion by high tide and thus show a distinctive zonation in the intertidal, from a marine to a progressively terrestrial environment.

The red mangrove (*Rhizophora mangle*) is the most common species of mangrove along the shores of southern Florida, the Caribbean, and the gulfs of California and Mexico. It lives right on the shore (Figure 11-21) and is easily identified by its peculiar prop-roots, which branch downward and support the treelike stilts (see Figure 5-27). Flexible air roots drop down from the higher branches, helping to extend the tree laterally. The trees can be as high as 9 m (30 feet). Under optimal conditions they form dense forests noted for their high primary production.

Other species of *Rhizophora* are found on tropical coasts around the world. Along the Caribbean and Atlantic coasts of the Western Hemisphere, the black mangrove (*Avicennia germinans*) and the white mangrove (*Laguncularia racemosa*) live inland from the red mangrove (Figure 11-21). Throughout their range the species of the black mangrove develop conspicuous, unbranched aerial roots that grow upward from the oxygen-poor mud to help aerate the plant tissues (Figure 11-22). The white mangrove typically grows along the landward edge of the mangrove forest. The base of its leaves has two clearly visible salt glands for the excretion of salts. Salts are also excreted by the leaves of the black mangrove (see Figure 4-21, B).

Most species of mangroves are found along the shores of the western Pacific and the Indian Ocean. In southern New Guinea and some of the islands of Indonesia, where the influence of the tides extends far up estuaries, mangrove forests may reach as far inland as 320 km (200 miles).

Many marine and land animals live in mangrove forests. Several species of fiddler crabs (see Box 11-2) excavate burrows in the mud. Mudskippers

Figure 11-21 Aerial view of a mangrove forest on the southern coast of Puerto Rico. The outer seaward edge of the forest is dominated by the red mangrove (*Rhizophora mangle*). As in Florida and most of the Western Hemisphere, the coastal red mangrove is followed by a broad belt of the black mangrove (*Avicennia germinans*). Further inland is the white mangrove (*Laguncularia racemosa*).

Figure 11-22 The aerial roots of a black mangrove (*Avicennia marina*) in Palau, western Pacific. Here, as in most other mangrove forests in the western Pacific, black mangroves dominate the seaward fringe of mangrove forests.

Figure 11-23 Mudskippers (*Periophthalmus*), from mud flats of mangrove forests in New Guinea. Their protruding eyes are adapted to see in air. Each eye can be retracted into a moist pocket to keep it from drying out.

(*Periophthalmus*) are unique fishes found in Indian Ocean and western Pacific forests. They have burrows in the mud but spend most of their time out of the water, skipping over the mud and crawling up mangrove roots to catch insects and crabs (Figure 11-23). Their gills get oxygen not from the water, but from air trapped in the mouth. Many seaweeds, oysters, snails, and other animals attach to or take shelter among the submerged mangrove roots (see Figure 11-1). Several species of crabs feed on detritus or vegetation. The muddy bottom around mangroves is inhabited by a variety of deposit and suspension feeders as in temperate mud flats. These include polychaetes, mud shrimps, and clams. Birds make their homes in the branches and feed on fishes, crabs, and other prey. The channels that cross mangrove forests are rich nurseries for many species of shrimps, spiny lobsters, and fishes.

Large amounts of detritus accumulate as leaves and other dead plant material are broken down by bacteria. This makes the mud among the roots black and oxygen-poor, much like those in salt marshes. As sediment and detritus accumulate among the roots, mangroves gradually extend the coastline seaward—actually creating new land. When enough material has accumulated, the mangroves may be replaced by land plants. Thus mangrove forests can be considered a stage of **ecological succession** between marine and terrestrial communities.

Other Estuarine Communities. The muddy bottoms below low-tide levels are sometimes covered by beds, or meadows, of grasslike flowering plants known as **seagrasses.** They include eel grass (*Zostera*; see Figure 5-26, *B*), which is restricted to temperate waters, and turtle grass (*Thalassia testudinum*; see Figure 5-26, *D*), a warm-water species often found around mangrove forests. The roots of seagrasses help stabilize the sediment, and their leaves provide shelter and food to many organisms, as well as an additional source of detritus. Seagrass beds are not restricted to estuaries and are discussed in more detail in Chapter 12 (in the section, "Seagrass Beds").

Oysters (*Ostrea, Crassostrea*) may form extensive beds on the muddy bottoms of estuaries in temperate waters. These **oyster reefs** gradually develop as successive generations of oysters grow on the shells of their predecessors (Figure 11-24). The oyster-reef community includes seaweeds, sponges, tube worms, barnacles, and other organisms that attach to the hard shells. Other animals take shelter among or even inside the shells. A similar estuarine community develops in association with mussel (*Mytilus*) beds, though the mussels require a hard substrate for attachment.

Feeding Interactions Among Estuarine Organisms

Although relatively few species live in estuaries, they reap the benefits of living in a very productive ecosystem. They are adapted to exploit the abundant food resources of the estuary. The generalized food web shown in Figure 11-25 summarizes the feeding relationships among different organisms in estuarine ecosystems.

Nutrients brought in by the tide and rivers, together with those generated by nitrogen-fixing organisms and the decomposition of detritus, are

ecological succession
Successive changes of communities brought about by modifications of the biologic and physical environment
Chapter 10, p. 309

Figure 11-24 An oyster reef formed by the eastern oyster (*Crassostrea virginica*) near Beaufort, North Carolina, exposed at low tide.

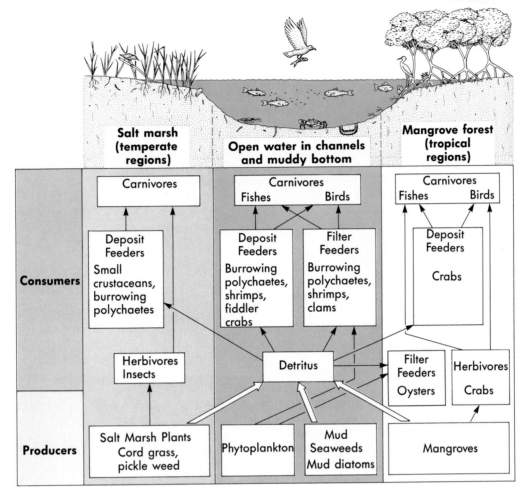

Figure 11-25 Generalized food webs in estuarine ecosystems.

used by plants, the primary producers. Primary production is especially high in the communities that surround estuaries. The dense stands of cord grass and other salt-marsh plants (or mangroves in the tropics), as well as seagrasses, thrive in the high concentration of nutrients. The diatoms and seaweeds in the mud and the phytoplankton in the water also contribute significantly to primary production.

Primary production by these plants varies geographically and seasonally, as does their relative contribution to the ecosystem as a whole. Estimates of primary production range from 130 to as high as nearly 6,000 g dry wt/m^2/year for cord grasses in salt marshes on the Atlantic coast of the United States. See Table 14-2 for a summary of the typical rates of primary production for salt marshes, mangroves, and seagrass beds.

The organic material manufactured by plants is made available to consumers mainly in the form of detritus. A distinctive feature of estuarine ecosystems is that most of the animals feed on dead organic matter. Except insects and other land animals on the fringes, there are few **herbivores** that graze on living plants. There is evidence that many detritus feeders obtain more energy from the bacteria and other decomposers in the detritus than from the dead organic matter itself. They excrete any detritus that remains undigested, however, and return it to the detritus pool. The surplus detritus is exported to the open ocean and neighboring ecosystems, a process known as **outwelling.** The exported detritus serves as a valuable source of food and nutrients to other ecosystems. Outwelling is therefore a most important role of estuaries, an additional reason why they must be preserved and protected.

Estuaries include plant-dominated communities with very high primary production. A significant amount of the food manufactured by these plants is made available to consumers by way of detritus.

Do It Yourself

SUMMARY

1. Estuaries are coastal regions defined by two main features: They are _____, and seawater mixes with _____.
2. Although many estuaries were formed by the flooding of river valleys, others developed as _____ formed along the coast or by the _____, or subsidence, of land.
3. Typical estuaries show wide fluctuations in the _____ of their water and the deposition of _____ sediments along their bottoms.
4. The depletion of _____ in muddy substrates results from a high concentration of _____ and limited _____ of water.
5. Organisms that withstand changes in salinity by keeping the salt concentration of their body fluids constant are known as _____. Another strategy is that of _____, who allow the salt concentration of their body fluids to change with the salinity of the water.
6. Most animals living in mud flats are _____ and suspension feeders. Both feed primarily on _____.
7. The zonation of salt-marsh plants is basically controlled by their tolerance to _____.
8. _____ are the temperate equivalents of mangrove forests.
9. Primary production in the communities that surround estuaries is very high because of the high density of _____ in them.
10. Most of the food manufactured by salt-marsh plants and mangroves finds its way to consumers in the form of _____. This means that relatively few animals are _____.

THOUGHT QUESTIONS

1. A proposal is made to deepen the entrance and main channel of an estuary. What do you think will happen to the salt marshes that surround the channel? What do you predict will happen to the primary production of the estuary as a whole?
2. Some of the organic material manufactured in estuarine communities is exported to other ecosystems. What type of ecosystems receive this material? How is this material transported?

FOR FURTHER READING

Coutant, C.C.: "Thermal Niches of Striped Bass." *Scientific American*, vol. 255, no. 2, August 1986, pp. 98-104. *The distribution of the striped bass in estuaries is affected by responses to temperature. These responses are influenced by the age of the fish.*

Hoy, M.: "Researching the Slough." *Audubon*, vol. 91, no. 2, pp. 99-105. *The National Estuarine Reserve Research System is involved in investigating and protecting the relatively few unspoiled estuaries that remain in the United States.*

Palmer, J.D.: "Biological Clocks of the Tidal Zone." *Scientific American*, vol. 232, no. 2, February 1975, pp. 70-79. *Some inhabitants of the intertidal zone, including estuarine mud flats, are able to keep time. The consequences of this phenomenon are most fascinating.*

Rutzler, K., and C. Fuller: "Mangrove Swamp Communities." *Oceanus*, vol. 30, no. 4, Winter 1987, pp. 16-24. *The structure and function of a typical mangrove forest in the Caribbean.*

Wiley, J.P.: "On Tropical Coasts Mangroves Blend the Forest into the Sea." *Smithsonian*, vol. 15, no. 12, March 1985, pp. 122-135. *A look at the surprisingly rich diversity of land and marine life that converges in a mangrove forest.*

Life on the Continental Shelf

12

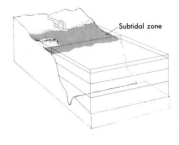

For ages the submerged edge of the continents, the **continental shelf,** was considered the beginning of the unknown blue sea. But when the shelf came under scrutiny during the last century, the unknown began to give up her secrets. We have learned even more in just the last few years, since the use of scuba and underwater habitats has permitted first-hand study and observation (Figure 12-1). Biologically, the shelf is the richest part of the ocean, with the world's most important fishing grounds, which yield about 90% of the total global fish catch. Oil and minerals have been found on the shelf, and nations have extended their international borders to protect their new-found, or yet-to-be found, resources. Unfortunately, not everybody understands the value of the continental shelf, and it remains a cheap dump for garbage and sewage around the world.

This chapter takes a look at life on the continental shelf, paying particular attention to the **benthos.** Life in the water column above the shelf is discussed in Chapter 14; the shelf's fisheries in Chapter 16. Coral reefs, important shelf communities in the tropics, are covered separately in Chapter 13.

benthos Organisms that live on the bottom

plankton Those that drift in the water

nekton Those able to swim against the currents
Chapter 9, p. 283; Figure 9-21

PHYSICAL CHARACTERISTICS OF THE SUBTIDAL ENVIRONMENT

The part of the continental shelf that is never exposed at low tide constitutes the **subtidal zone** of the marine environment. The subtidal zone, which is sometimes called the **sublittoral zone,** extends from the low tide level on shore to the **shelf break,** the outer edge of the continental shelf where depth suddenly increases (see Figure 9-22). The depth of the shelf break varies, but averages around 150 m (490 feet). The width of the shelf is also highly variable, from less than 1 km (0.6 mile) to over 100 km (60 miles), with an average of approximately 80 km (48 miles). The benthos of the continental shelf live in the subtidal zone, whereas the **plankton** and **nekton** of the water column over it inhabit the **neritic** zone (see Figure 9-22).

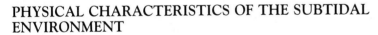

> The subtidal zone consists of the bottom of the ocean from the low tide level to the shelf break, the outer edge of the continental shelf.

347

Figure 12-1 Scuba diving has dramatically expanded the possibilities for the study of the subtidal by marine biologists.

The physical factors that affect subtidal organisms are linked to two of the shelf's fundamental characteristics: its relatively shallow water and its proximity to land. Because the bottom is shallow, temperature varies more from place to place in the subtidal zone than on the deeper bottom beyond the shelf. This is significant because temperature is one of the most important factors influencing the distribution of marine organisms. The kinds of organisms found in the subtidal change dramatically from the Equator to the poles. There are also *more*, not just different, species in the tropics than in temperate or polar waters. The bottom below the Arctic and the Antarctic ice, however, is far from being a desert (see Box 12-1). The benthos in deep water change much less, because the temperature there is uniformly cold and does not vary as much as in shallow water.

The bottom in shallow water is also much more affected by waves and currents than is deep water. The rise and fall of tides can produce particularly strong **tidal currents** on the shelf, especially in bays and narrow straits. Wind waves can affect the bottom even at depths of 200 m (650 feet). Water motion, or **turbulence,** tends to stir up the water column and prevents **stratification.** Thus, except for seasonal variations in temperate waters, the temperature and salinity of the water at the bottom are not much different from those at the surface. More importantly, nutrients do not concentrate in the bottom layer and become unavailable to primary producers at the surface, as they do in deep water. Nutrients are also brought in by rivers, often as a by-product of the highly productive salt marshes or mangrove forests that fringe the shore. As a result, the water over the continental shelf is far more productive and plankton-rich than in the open ocean, so there is much more food available. The high concentration of phytoplankton, plus decaying organic matter brought in by rivers, gives coastal water a greenish tint, as opposed to the deep blue of open-ocean water.

Rivers, obviously, also bring in large amounts of fresh water. Freshwater runoff dilutes the seawater, reducing its salinity. The effects of large rivers such as the Mississippi and the Amazon extend many miles offshore. Away from the influence of rivers, however, the salinity of coastal water is about the same as that beyond the shelf.

The combination of proximity to land and shallow water greatly affects **sedimentation,** the settling of sediment particles from the water. Most of the sediment on the shelf is **lithogenous,** and rivers bring in huge quantities of sediment from the continents. Water motion sorts out the particles by size and density, especially on the shallower parts of the shelf. Large-grained material such as gravel and sand settles out even in areas with strong waves and currents. On the other hand, turbulence keeps fine particles such as silt and clay in suspension, so these are deposited only in quiet areas or in deeper water, where turbulence does not reach the bottom.

The abundant phytoplankton and the sediment that is brought in by rivers and stirred up by waves and currents make shelf water murkier than that in the open ocean. Light does not penetrate as deeply, which reduces the depth at which plants and other primary producers—both phytoplankton and attached seaweeds—can carry out photosynthesis and grow.

> **The type of substrate, depth, turbulence, temperature, salinity, and light are among the most important physical factors that influence life on the continental shelf.**

stratification Separation of the water column into layers, with the densest, coldest water at the bottom, thus preventing the mixing of the nutrient-rich deep water with the less dense, warmer upper layer
Chapter 3, p. 78; Figure 3-35

lithogenous sediment Sediment that comes from the physical or chemical breakdown of rocks on land

biogenous sediment Sediment composed of the skeletons and shells of marine organisms
Chapter 2, p. 40

BOX 12-1

Antarctica: The Last Frontier

Antarctica, one of the last unspoiled wildernesses on earth, is for many of us a vision of awesome whales, smiling seals, and frolicking penguins. Surprisingly, life on the Antarctic shelf is as fascinating and pristine as on the cold waters above it.

Ice is a major physical factor for shelf organisms. A layer of ice several meters thick forms during winter, freezing and eventually crushing any organisms living on the bottom. Winds and currents push the ice, scouring the bottom and crushing even more organisms. Ice may also form around the organisms to depths of 30 m (100 feet) or more. The ice, less dense than the water, often floats to the surface, carrying with it these trapped, often frozen, organisms. Save for winter ice at shallower depths, the physical environment here is remarkably constant. Year-round low temperatures mean that life processes, including growth, proceed at a slow pace in Antarctica.

Though kelps are present on several of the islands offshore, they are surprisingly absent in Antarctica proper. Here they are replaced by *Desmarestia* and a few other brown seaweeds. These are not considered to be kelps, though they are large and look similar.

Sessile organisms are rare at the shelf's shallower depths because of the scouring effect of winter ice. Below this depth, however, benthic invertebrates become plentiful. An unbelievably rich, diverse, and colorful fauna is unveiled to the diver.

Sponges are very common and cover most of the bottom in many places. Sponges are not only represented by many different species, but by a multitude of growth forms. Huge volcano-like sponges may grow 2 m (close to 7 feet) or more in height. These large individuals are probably hundreds of years old. Others are fan shaped or resemble bushes or branching corals. These and other sponges provide refuge and perches for many fishes and other mobile animals, very much as do kelps or corals elsewhere. Sponges are so common that the bottom below 30 m (100 feet) or so is nothing but a thick mat of siliceous, or glass, sponge spicules—truly a glass bottom.

Other invertebrates abound: sea anemones, soft corals, sea stars, nudibranchs, sea urchins, and others. Some of these are suspension feeders such as sponges, whereas many feed on the sponges. Others are predators of sponge-feeders and thus help keep the slow-growing sponges from being wiped out. Not all of these subtidal invertebrates are unique to Antarctica. Some are found elsewhere but in progressively deeper water as one gets closer to the Equator.

Fishes are remarkable in their own right; they must remain active at temperatures that are close to the freezing point of seawater. Many have a chemical "antifreeze" in their blood. The cold water holds so much dissolved oxygen that some, like the icefishes, have colorless blood that lacks hemoglobin.

The wonders of life in Antarctica, both below and above the ice, are unfortunately being threatened. Cruise-ship tourists, oil and sewage pollution, left-over garbage, the potential exploitation of oil and mineral resources, and an ozone hole over the continent (see Box 17-3) may eventually overwhelm our most desolate continent and harm its unique life.

A treelike soft coral, sponges, and a crinoid from Antarctica.

THE CONTINENTAL SHELF AS AN ECOSYSTEM

The type of substrate is very important in determining which particular organisms inhabit the floor of the continental shelf. It also dictates how they are to be sampled by marine biologists (Figure 12-2). In fact, subtidal communities are often classified according to the type of substrate. The most common of these are communities associated with soft bottoms. Hard, or rocky, bottom communities constitute the second type.

Soft-Bottom Subtidal Communities

Sandy and muddy substrates dominate the world's continental shelves. Large areas covered by soft sediments stretch from the shore to the edge of the shelf (Figure 12-3). Even along rocky shores, rocks eventually give way to sand or mud. The organisms that inhabit these often flat and seemingly homogeneous bottoms are not distributed by chance. It has long been recognized that these organisms form distinct communities whose distribution is greatly influenced by such factors as the particle size and stability of the sediment, light, and temperature.

The shelf's soft-bottom communities share many traits with the communities of sandy beaches and mud flats covered in Chapters 10 and 11. In all of these communities, there is a predominance of **infauna,** benthic animals that burrow or dig in the sediment. There are also some **epifauna,** animals that live on top of the sediment. Because there is nothing to hold on to, **sessile**—or attached—forms are virtually absent.

Most of the continental shelf is covered by mud or sand. The subtidal communities in these areas are dominated by infauna.

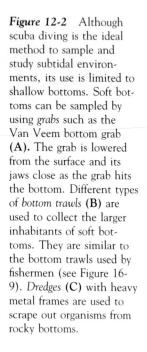

Figure 12-2 Although scuba diving is the ideal method to sample and study subtidal environments, its use is limited to shallow bottoms. Soft bottoms can be sampled by using *grabs* such as the Van Veem bottom grab **(A).** The grab is lowered from the surface and its jaws close as the grab hits the bottom. Different types of *bottom trawls* **(B)** are used to collect the larger inhabitants of soft bottoms. They are similar to the bottom trawls used by fishermen (see Figure 16-9). *Dredges* **(C)** with heavy metal frames are used to scrape out organisms from rocky bottoms.

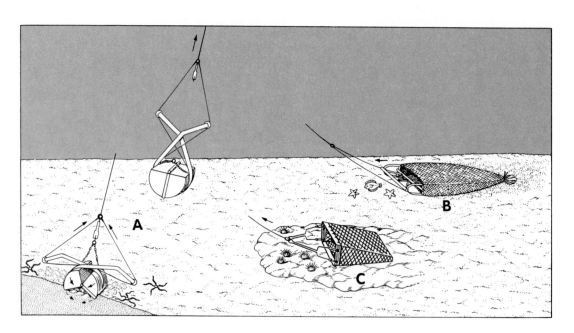

The number of species living on subtidal soft bottoms is usually higher than the number of species on soft bottoms in the intertidal zone. This is largely because the conditions below the low-tide mark are more stable. Desiccation is not a problem. Furthermore, there are not the drastic temperature changes brought about by exposure at low tide or the salinity variations of estuarine mud flats. This allows a wider variety of organisms to live in the subtidal.

The close relationship between particle size and the distribution of organisms is particularly evident in the infauna, which for the most part are very selective as to where they live. Different species may also partition their environment and thus reduce competition by living at different depths in the substrate. Because only the upper layer of sediment contains enough oxygen, however, the depth at which the infauna can live is limited. In organic-rich mud, oxygen is quickly used up by bacteria associated with decomposition. Sand usually contains less organic matter. It is also more porous, so water can circulate through the sediment and replenish the oxygen (see Figure 10-34). Thus the infauna can burrow deeper in sand than in mud. Even in mud, however, burrows aid circulation, helping organisms to obtain oxygen.

Soft substrates often contain patches of different types of sediment. As a result, the distribution of organisms along the bottom is typically **patchy;** that is, the organisms occur in distinct clumps or patches (Figure 12-4). Some species have a patchy distribution because their planktonic larvae are able to select a particular environment on which to settle and undergo metamorphosis. Some larvae are known to postpone metamorphosis and "taste" the bottom until they find a certain type of substrate. Metamorphosis

Figure 12-3 Conspicuous life in shallow sandy bottoms includes occasional clumps of seaweeds that manage to grow on small rocks or shells and large predators like flatfishes.

metamorphosis The transformation of one stage into another, as when a larva changes into a juvenile
Chapter 6, p. 148

Figure 12-4 The spacing, or distribution, of organisms can be classified into three basic types: (A) *random*, in which animals are scattered with no particular pattern; (B) *regular*, where they are spaced at even intervals; and (C) *patchy*, or *clumped*, where they cluster together. Many marine species show a patchy distribution. Can you think of a reason that organisms such as clams may show a regular distribution?

A. Random

B. Regular

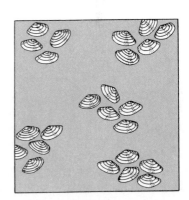

C. Patchy

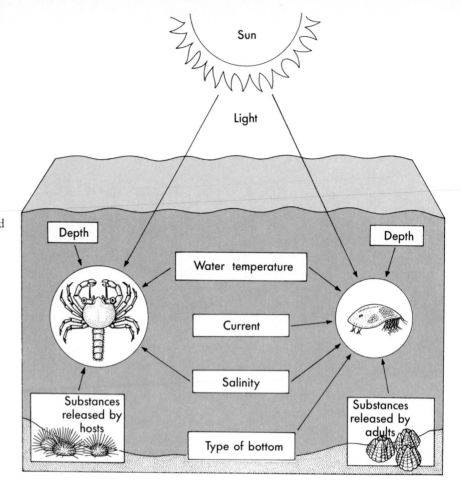

Figure 12-5 Many factors are known to influence the settlement and metamorphosis of planktonic larvae. The larva illustrated on the left, a megalopa (see Figure 6-45), belongs to a crab (*Echinoecus pentagonus*) that as an adult lives on tropical sea urchins. Females live in the rectum, while males wander around the urchin's test.

may be triggered by specific chemical, physical, and biologic factors (Figure 12-5). The larvae of some species can sense adults and prefer to settle near them. This may cause the species to be distributed in clumps even if there is a uniform bottom.

Many marine biologists now think that the establishment of subtidal and other marine communities is influenced to a certain extent by a random, or chance, element. This view, often called the **"lottery" hypothesis,** maintains that, whenever an empty space becomes available on the substrate—because of the death of adults or disturbances caused by, say, a feeding ray—larvae will settle on a "first come first served basis." In other words, there is a an element of unpredictability, or luck, associated with the development of communities: It depends on what kinds of larvae happen to be in the neighborhood. This new approach has opened up some exciting possibilities for future research.

Unvegetated Soft-Bottom Communities. The great majority of the shelf's soft-bottom communities lack significant amounts of seaweeds or seagrasses. They are therefore known as **unvegetated** communities. The absence of large plants is the defining feature of these communities. Sea let-

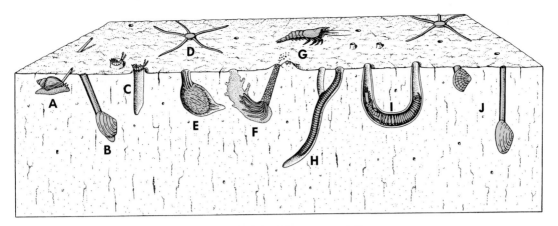

Figure 12-6 The infauna and epifauna of subtidal soft bottoms in different parts of the world include **(A)** snails such as whelks (*Nassarius*), **(B)** clams (the soft-shelled clam *Mya arenaria*), **(C)** amphipods (the tube-building *Haploops*), **(D)** brittle stars (*Ophiura*), **(E)** heart urchins (*Spatangus, Lovenia*), **(F)** trumpet worms (*Pectinaria*), **(G)** shrimps (*Crangon*), **(H)** sandworms (*Neanthes*), **(I)** parchment worms (*Chaetopterus*), and **(J)** cockles (*Cardium*). See also Figure 11-13 for a sample of mud-flat inhabitants.

tuce (*Ulva*) and other green algae such as *Enteromorpha* may grow in the shallows, but only on hard surfaces such as rocks and shell fragments (see Figure 12-3). The main primary producers are diatoms and a few other microscopic algae that grow on sand or mud particles in shallow water. This almost complete absence of plants means that, as in sandy beaches, the primary production is practically nil.

Because there is little primary production, **detritus** is a very important food source for many of the inhabitants. Detritus is brought in by currents from estuaries, rocky shores, and other more productive coastal communities. It is also present in the form of fecal material, dead individuals, and other debris from the plankton and nekton in the water column. Detritus is also generated by the bottom inhabitants themselves as they die and are decomposed by bacteria.

The detritus is used by bacteria and by the many types of microscopic animals that live among the sediment particles, the **interstitial** animals, or **meiofauna** (see Box 11-1).

Larger benthic invertebrates also feed on detritus. They are mostly burrowing **deposit feeders** (Figure 12-6). Examples are numerous species of polychaetes. Trumpet (*Pectinaria*; Figure 12-6, *F*) and bamboo worms (*Clymenella, Axiothella*; see Figure 11-13, *H*) are deposit-feeding polychaetes that inhabit tubes they build from sediment particles. Lugworms (*Arenicola*) and others live in burrows. These polychaetes and other types of worms eat detritus and other organic matter by collecting it with their tentacles or by ingesting sediment and extracting food from it (Figure 12-7).

Some sea urchins are uniquely adapted to live as deposit feeders in soft bottoms. Heart urchins (*Spatangus, Lovenia*) have abandoned the round

primary producers Organisms that can create organic matter from CO_2, usually by photosynthesis
Chapter 4, p. 92

detritus Dead organic matter and the decomposers that live on it
Chapter 9, p. 277

deposit feeders Animals that feed on organic matter that settles in the sediment
Chapter 6, p. 158; Figure 6-20

suspension feeders Animals, including *filter feeders*, that feed on particles suspended in the water column
Chapter 6, p. 147; Figure 6-20

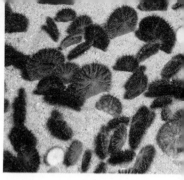

Figure 12-7 The apparent uniformity of soft bottoms is often interrupted by piles of castings such as this. They are made by deposit-feeding worms that swallow large amounts of sediment, remove organic matter in the gut, and then deposit the residue at the posterior opening of their burrow or tube.

Figure 12-8 As in all heart urchins, *Spatangus purpureus* is specialized for living in soft bottoms. Heart urchins burrow below the surface and use modified tube feet for feeding (see Figure 12-6, *E*). This photo was taken in the Mediterranean Sea at night, when some heart urchins emerge from the sediment.

Figure 12-9 *Dendraster excentricus* is a sand dollar that forms dense beds in sandy bottoms along the California coast. Individuals lie partially buried in the sand, "standing up" at an angle perpendicular to the current.

Figure 12-10 Some peanut worms, such as *Sipunculus nudus*, are deposit feeders that burrow in mud or sand and ingest sediment as they burrow.

shape of most urchins, becoming more streamlined, and their shorter spines lie flat (Figures 12-6, *E* and 12-8). Sand dollars (*Dendraster, Mellita*) are almost completely flat and have *very* short spines (see Figure 6-51, *C*). They typically feed on detritus and use mucus to carry the particles to the mouth (Figure 12-9). Other deposit feeders members of the infauna include peanut worms (*Sipunculus;* Figure 12-10), sea cucumbers (*Molpadia*), and ghost shrimps (*Callianassa;* see Figure 11-13, *D*).

The infauna also include invertebrates that are **suspension feeders** that eat drifting detritus and plankton from the water column. Many of these suspension feeders are **filter feeders** that actively filter the water to obtain suspended particles. Soft-bottom suspension feeders include many types of clams (Figure 12-6, *B* and *J*): razor clams (*Ensis, Siliqua*), the quahog (*Mercenaria mercenaria*), cockles (*Cardium, Clinocardium*), the soft-shelled clam (*Mya arenaria*), and others. Though most clams are filter feeders, some—such as the bent-nosed, or small, clams (*Macoma*) and a few others (*Nucula, Tellina, Yoldia*)—are deposit feeders. They collect detritus and even microscopic organisms with their siphons or specialized appendages (Figure 12-11; see also Figure 10-40). Also among the filter feeders are amphipods (Figure 12-6, *C*) and polychaetes such as parchment worms (Figure 12-6, *I*).

The distribution of burrowing deposit and suspension feeders is influenced by several factors. Already mentioned in the previous section is the type of substrate. Deposit feeders tend to predominate in muddy bottoms because more detritus is retained in the smaller interstitial spaces between mud particles (Figure 12-12). Suspension feeders, on the other hand, are more common in sandy bottoms. The types of organisms present also affect the establishment of others. Deposit feeders, for instance, are known to exclude suspension feeders. The constant reworking of the sediment by deposit feeders makes the bottom loose and unstable, conditions that facilitate the clogging of the filtering surfaces of suspension feeders and the burying of their newly settled larvae. Nevertheless, tube builders, many of which are deposit feeders, help stabilize the substrate and make it more suitable for suspension feeders. Their tubes may also interfere with the activities of burrowing deposit feeders.

Many of the surface-dwelling invertebrates are deposit feeders. These include most amphipods and other small crustaceans, as well as many species of brittle stars (see Figure 12-6). Many of these brittle stars are deposit feeders that use their tube feet to collect detritus from the bottom, whereas others are suspension feeders that raise their arms up into the water to catch suspended particles with their tube feet. Still others are scavengers and feed on dead animals. Shrimps, including commercially valuable species (*Penaeus*), can be very abundant (Figure 12-6, G). Shrimps and other large crustaceans are mostly **scavengers** that feed on dead plants and animals. Suspension feeders such as sea pens (*Ptilosarcus, stylatula*; Figure 12-13) and sea pansies (*Renilla*) can often be found in dense stands.

Most soft-bottom subtidal communities lack large plants. They are mostly inhabited by deposit and suspension feeders.

Some members of the soft-bottom epifauna are predators. They may burrow through the sediment to get their prey or catch it on the surface. As in other communities, predators are important in regulating the number and type of bottom inhabitants. They remove individuals and cause disturbances in the sediment that allow recolonization by different types of organisms. Snails such as whelks (*Buccinum, Busycon*; Figure 12-6, A) and moon snails (*Polinices*; see Figure 11-13, B) feed on clams by drilling a hole in the shell

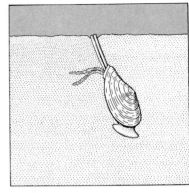

Figure 12-11 The file yoldia, *Yoldia limatula,* is an inhabitant of mud and muddy sand from shallow water to depths of 30 m (100 feet) in the north Atlantic. The siphons bring water in, but are not used for feeding, unlike most other clams. Instead, this clam uses two grooved, tentacle-like appendages to feed. Each groove contains cilia that bring small organisms to the mouth.

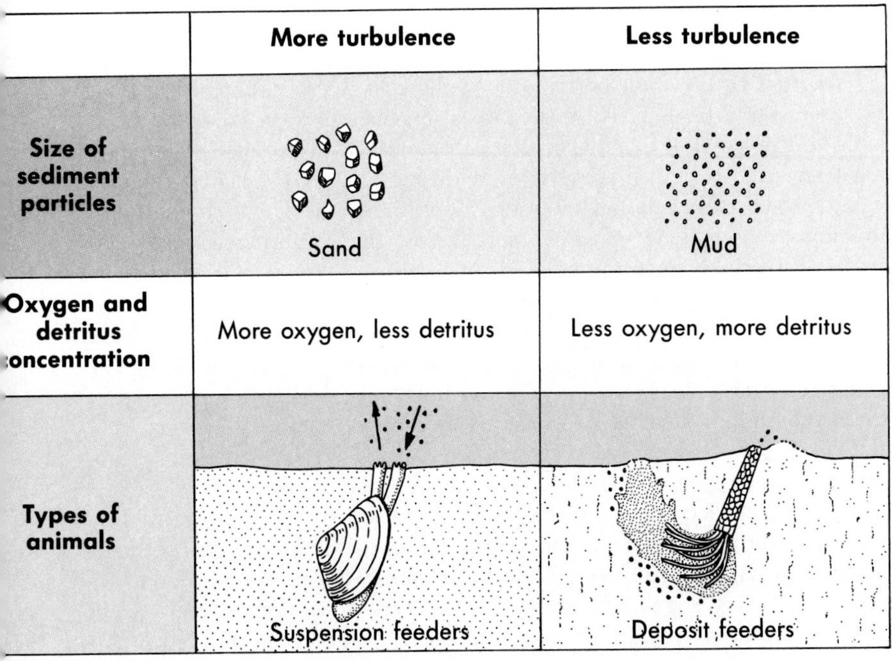

	More turbulence	Less turbulence
Size of sediment particles	Sand	Mud
Oxygen and detritus concentration	More oxygen, less detritus	Less oxygen, more detritus
Types of animals	Suspension feeders	Deposit feeders

Figure 12-12 The distribution of suspension and deposit feeders in soft bottoms is largely influenced by the size of the sediment particles. This relationship, however, is not that clear-cut. The animals already established in the sediment, for instance, influence colonization by others.

Figure 12-14 Sea stars such as *Astropecten* prey on clams and other burrowing invertebrates.

Figure 12-13 Sea pens are suspension-feeding cnidarians. Each individual is a colony that includes numerous minute polyps and a fleshy stalk that is anchored in soft substrates. This is *Stylatula elongata* from the Pacific coast of North America.

pelagic Organisms that live in the water column *Chapter 9, p. 283*

Figure 12-15 Flounders and other flatfishes lie on the bottom, usually covered by sediment with only their eyes protruding. They will literally leap from the bottom and catch small crustaceans, other invertebrates, and small fishes.

and rasping away the flesh. Sea stars (*Astropecten, Luidia;* Figure 12-14) prey on clams, brittle stars, polychaetes, and other animals. Crabs are common predators and scavengers in soft bottoms. Examples are the blue crab (*Callinectes sapidus*) and other swimming crabs (*Portunus*), the lady crab (*Ovalipes ocellatus*), and the many species that partially bury themselves in the sand or mud. Other predatory crabs, hermit crabs, lobsters, and octopuses take shelter in rocky bottoms but move to soft bottoms to feed.

An enormous number of fishes are also noteworthy predators. The great majority of bottom-dwelling, or **demersal,** fishes in these soft-bottom communities are carnivores. Most rays and skates scoop up clams, crabs, and other infauna and epifauna. Flatfishes such as flounders, halibuts, soles, and turbots lie camouflaged or covered (Figure 12-15) on the bottom and forage for a wide variety of prey. Pelagic fishes and squids also feed on the inhabitants of the shelf's soft bottoms. A very large predator, indeed, is the gray whale (*Eschrichtius robustus*), which filters the sediment for amphipods and other small animals (see Figure 8-19).

Seagrass Beds. Soft bottoms along the coast are occasionally carpeted by **seagrasses.** These are flowering plants, grasslike in appearance but unrelated to true grasses, that have become established in the marine environment (see the section in Chapter 5, "Flowering Plants"). Seagrass beds develop best in sheltered, shallow water along the coast. They are also found in estuaries and along mangrove forests (see the section in Chapter 11, "Other Estuarine Communities").

Only about 50 species of seagrasses are known. Most are tropical and subtropical, but several species are common in colder waters. Most seagrasses are restricted to muddy and sandy areas below the low-tide level. Different species of seagrass vary in maximum depth, but all are limited by the penetration of light through the water column.

Turtle grass (*Thalassia testudinum;* see Figure 5-26, D) is the most common seagrass in the tropical and subtropical Atlantic, the Caribbean, and the Gulf of Mexico. Extensive meadows are found to depths of about 10 m (30 feet), but turtle grass can grow deeper in clear water. Another seagrass, **manatee grass** (*Syringodium filiforme;* see Figure 5-26, A) is often found together with turtle grass. **Eelgrass** (*Zostera marina;* see Figure 5-26, B) is widely distributed in temperate and cold waters of the Pacific and North At-

Figure 12-16 Eelgrass (*Zostera marina*) exposed at low tide along the coast of North Carolina, eastern United States.

lantic. It is most common in shallow water, sometimes exposed at low tide (Figure 12-16), but has been found at depths of up to 30 m (100 feet).

Seagrasses can grow into thick, luxuriant beds. Their roots and a network of underground stems keep them anchored in the face of turbulence. The roots and stems also help stabilize the soft bottom, and the leaves cut down wave action and currents. This decrease in turbulence causes more and finer sediment to be deposited, which in turn affects colonization by other organisms. It also improves water clarity because less sediment remains suspended in the water column.

Seagrass beds contain a very high plant biomass, as high as 1 kg/dry wt/ m^2 for turtle grass. With such a high density of photosynthesizing plants, seagrass beds have a higher primary production than anywhere else in soft bottoms. In fact, they rank among the most productive communities in the entire ocean, with estimates as high as 16 $gC/m^2/day$, or over 5,000 $gC/m^2/$ year, for turtle grass. Part of the reason for such high primary production is the fact that seagrasses have true roots, unlike seaweeds, and are thus able to absorb nutrients from the sediment. Phytoplankton and seaweeds, by contrast, must depend on nutrients dissolved in the water. Some typical rates of primary production in seagrass beds and other marine environments are summarized in Table 14-2.

Many small algae grow on the surface of seagrass leaves. These algae, known as **epiphytes,** further increase primary production in seagrass communities. Microscopic diatoms are particularly abundant. Some epiphytic blue-green algae also carry out nitrogen fixation and thus release nutrients in the form of nitrogen compounds.

Though the number of grazers, or **herbivores,** varies with geographic location, surprisingly few animals eat seagrasses. Those that do include sea turtles, manatees, a few diving birds such as ducks, sea urchins (*Diadema, Lytechinus*), and some parrotfishes (*Sparisoma*).

Animals take advantage of the high primary production of seagrasses in several ways, even when they do not actually eat the seagrasses themselves (Figure 12-17). Many feed on the large amounts of detritus produced by the decomposition of leaves and other plant material. Deposit-feeding polychaetes, clams, and sea cucumbers live in or on the detritus-rich sediment. Detritus is also exported to other communities, such as the unvegetated soft bottoms of deeper water.

biomass The total weight of living organisms
Chapter 9, p. 275

nitrogen fixation Conversion of nitrogen gas (N_2) into nitrogen compounds that can be used by plants as nutrients
Chapter 9, p. 281

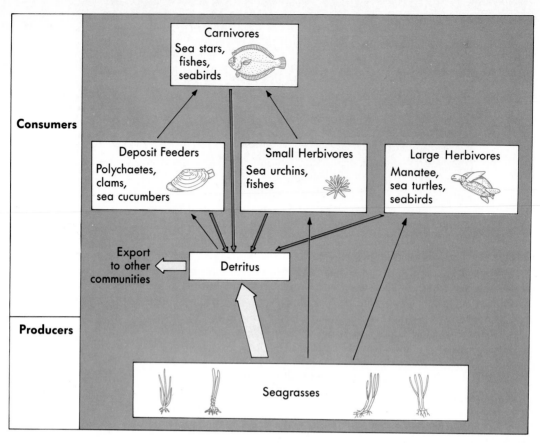

Figure 12-17 Generalized food chain in a seagrass community. As in estuaries (see Figure 11-25), detritus generated from dead plants is an important source of food to many animals.

The dense seagrasses also offer shelter to many animals that do not feed on detritus. A variety of small sessile or crawling animals live on the leaves—hydroids, snails, tiny tube-dwelling polychaetes, amphipods, shrimps, and the like (Figure 12-18). Larger animals live among the plants. Turtle-grass beds, for instance, support a rich fauna (Figure 12-19). These include a pen shell (*Pinna carnea;* see Figure 6-30, B), which is a filter-feeding bivalve, and carnivores such as fishes and the West Indian sea star (*Oreaster reticulatus*). Seagrass beds also serve as nurseries for commercially valuable species such as the Atlantic bay scallop (*Argyropecten irradians*) and shrimps (*Penaeus*).

> **Highly productive seagrass communities are restricted to soft bottoms in shallow, sheltered water. The plants are not heavily grazed, but produce a lot of detritus that is used by deposit feeders and exported to other communities.**

Hard-Bottom Subtidal Communities

Hard bottoms make up a relatively small portion of the continental shelf. Usually they are just the submerged extensions of rocky shores. There are also some subtidal rocky outcrops of varying size. In some cases a significant

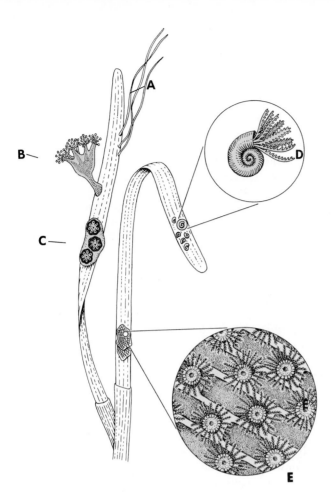

Figure 12-18 Eelgrass (*Zostera marina*) leaves are often inhabited by sessile, or attached, organisms. These include (**A**) green and other algae such as *Enteromorpha*; (**B**) the trumpet-stalked jellyfish, a non-swimming jellyfish (*Haliclystus*); (**C**) colonial sea squirts; (**D**) tube-dwelling polychaetes (*Spirorbis*); and (**E**) bryozoans.

Figure 12-19 The Queen conch (*Strombus gigas*) is a large snail commonly associated with turtle grass beds in Florida and the Caribbean. It is highly prized as food by the locals.

component of the hard substrate is provided by calcareous **coralline algae,** by the tubes of polychaete worms, or the shells of oysters (see Figure 11-24). All of these hard bottoms are often called reefs, though they should not be confused with the live coral reefs of warm waters.

Our discussion of the hard-bottom communities of the continental shelf highlights those found on rocky bottoms. Lush kelp beds, which develop on rocky bottoms in temperate water, are discussed separately because of their particular interest and importance.

Rocky Bottoms. Unlike those in the intertidal, the communities that develop on subtidal rocky bottoms are never subject to desiccation. This means that a wider variety of organisms can live there than in the intertidal.

Rocky-bottom communities in shallow water are rich and productive. Seaweeds are the most conspicuous inhabitants of hard substrates in shallower water. They occur in an awesome array of colors, growth forms, textures, and sizes. Most are brown and red seaweeds. They can be filamentous (*Chordaria, Ceramium*), branched (*Agardhiella, Desmarestia*), thin and leafy

encrusting seaweeds Those
that grow as a thin layer
over rocks, such as some of
the *coralline red algae*
Chapter 5, p. 132

*Sunlight penetrates
through the water column
to depths that depend on
water transparency. Not
all colors of light penetrate
to equal depths. In clear
ocean water, blue light
penetrates the deepest,
red, the least.*
Chapter 3, p. 63

Figure 12-20 Practically
all available space of this
subtidal rocky bottom is
occupied by sponges, sea
squirts, bryozoans, hy-
droids, algae, and barna-
cles. Sea stars and sea urchins are among the most
common examples of the
non-sessile benthos.

(*Porphyra, Gigartina*), or encrusting (*Lithothamnion*). They may even have
different growth forms at different times of their lives (see Figure 5-24).
Many have a **holdfast** (see Figure 12-22) to anchor them to the substrate to
help withstand turbulence.

As in the intertidal, one of the main problems for seaweeds and sessile
animals is to find a place to attach. Every bit of space is occupied, even if it
seems clean to our naked eye. Thus there is intense competition for living
space on the rocks. Different species of seaweeds have different competitive
abilities. They also vary in their tolerances to wave action, grazing by sea
urchins and other herbivores, temperature, light, and the stability of the
substrate. The effect of light is particularly intriguing. As depth increases,
the amount of light available for photosynthesis decreases, making it harder
for seaweeds to live there. Some species are better adapted to deep water
than others, however. Several can live at depths of over 200 m (650 feet)
in clear water. Deep-water seaweeds have increased levels of chlorophyll
and other pigments (see Table 5-1) that capture light energy. These adaptations can be found in all groups of seaweeds and are not restricted to the
reds, once thought to be the best adapted to deep water. In any case, depth
zonation is not determined by light alone. It is affected by grazing, competition for space, temperature, and other factors.

Seaweeds also vary in their life history strategies. Some are **perennial**
and are found year-round, whereas others are found only at certain times.
Although some grow fast and are short lived, others grow slowly but are
sturdy and long lived. Fast growers are typically the first to colonize surfaces
recently disturbed by grazers, turbulence, or other phenomena. Some seaweeds use both strategies by alternating a fast-growing, short-lived stage
with a slow-growing, perennial one. Both stages belong to the same species
but look and function differently.

Seaweeds must compete for space not only with each other, but with a
huge assortment of sessile animals. Hard substrates provide a good place for
organisms to attach. It is much harder, however, to burrow in rock than in
sand or mud. As a result, hard-bottom communities tend to have a rich epifauna and a poor infauna, the reverse of soft-bottom communities. Sponges,
hydroids, sea anemones, soft corals, bryozoans, tube-dwelling polychaetes,
barnacles, and sea squirts (Figure 12-20) are among the most frequently
found groups. Many form thin colonies that outcompete other species simply by growing over them. A few, such as the rock-boring clam (*Pholas*),
live embedded in rocks.

Grazers on subtidal rocky bottoms are usually small, slow-moving invertebrates. Perhaps most important are sea urchins (*Arbacia, Diadema,
Strongylocentrotus*). Chitons, sea hares, limpets, abalones (*Haliotis*), and
other gastropod molluscs are also important grazers. Grazing by parrotfishes
(*Scarus, Sparisoma*), surgeonfishes (*Acanthurus*), and damselfishes (*Pomacentrus*) is particularly heavy in tropical waters. Some seaweeds have evolved
defenses against grazing. These include chemicals that make them unpalat-
able and the ability to rapidly grow back tissue lost to grazers. Some seaweeds even have life history stages that bore into mollusc shells, a good de-
terrent to being eaten by a hungry sea urchin! Upright seaweeds tend to be

tough and leathery. Coralline algae, which include red (*Lithothamnion, Clathromorphum*) and green (*Halimeda;* see Figure 5-13) algae, deposit calcium carbonate in their cell walls, making them particularly resistant to grazing.

Grazers and predators strongly influence the composition of hard-bottom communities (Figure 12-21). They remove residents from the rocks, opening up space for other organisms. Sea urchins feed not only on seaweeds, but on some of the flimsier attached invertebrates as well. Carnivores—including seastars, nudibranchs, and other snails—also feed on these invertebrates. The patches cleared by grazers and carnivores are colonized by settling larvae and seaweed spores. Because these planktonic stages are often seasonal, patches formed at different times are colonized by different species. This increases the overall number of species in a given area.

Carnivores such as crabs, lobsters, and many types of fishes prey on grazers. Other carnivores feed on smaller carnivores or even on the external parasites of other fishes (see Box 12-2). These predators act as a check on the grazers, helping maintain a relatively stable situation. This balance can be easily upset, however. For example, explosive increases in sea urchin populations have been reported in a number of subtidal communities (see p. 367), as have mass mortalities. Like all biological communities, subtidal communities are constantly changing.

The species composition of rocky-bottom subtidal communities is influenced by factors such as light, competition for space, grazing, and predation.

Kelp Communities. One of the most fascinating and important, not to mention beautiful, types of marine community are **kelp communities.** Kelps

Figure 12-21 Sea urchins, like *Strongylocentrotus purpuratus* from the Pacific coast of North America, are most important in influencing the composition of hard-bottom subtidal communities. They graze on seaweeds and on some of the encrusting animals and thus uncover new surfaces where other sessile organisms can settle.

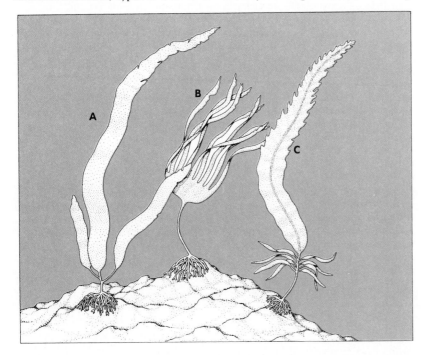

Figure 12-22 Examples of North Atlantic kelps include **(A)** hollow-stemmed kelp (*Laminaria longicruris*), **(B)** horsetail kelp (*Laminaria digitata*), and **(C)** edible kelp (*Alaria escuelenta*). The hollow-stemmed kelp is known to reach lengths of 12 m (40 feet) in deeper water.

BOX 12-2

Cleaning Symbiosis

A cleaning wrasse doing its job on a moray eel.

Living in congested marine environments has resulted in some novel strategies in the art of surviving while getting along with neighbors—especially the big, hungry ones. Such is the case of fishes and shrimps specializing in the art of **cleaning symbiosis.** These small, colorful marine animals make a living by picking up external parasites and diseased tissue from larger fishes. Though especially common in very crowded environments such as coral reefs, cleaning symbiosis has evolved in many other subtidal communities.

Cleaning behavior has been reported in several species of wrasses, butterfly-fishes, gobies, and other fishes. The banded coral shrimp (*Stenopus hispidus*) is among the most well-known of the cleaning shrimps (see Figure 9-12). Cleaning behavior can be a full-time or part-time job. Some are cleaners all of their lives, whereas others are cleaners only as juveniles.

Having the outer body cleaned of parasites and dead skin is a much sought-after service. Prominent topographic features such as a rock, coral head, or large sponge announce that the spot is a "cleaning station" serviced by cleaners. Hosts, as the larger fishes are usually known, regularly congregate in cleaning stations. Fishes learn the location of stations and will continue to congregate there even if cleaners are experimentally removed from the area. Cleaners advertise their services by showing characteristic bright colors and distinctive stripes.

Hosts, many of which are carnivores that can easily swallow a cleaner in one gulp, are willing to cooperate once they reach a cleaning station. They signal the seriousness of their intentions in many ways. Swimming slowly and hovering in mid-water is a common behavior. Some change color, whereas others "pose" by assuming a bizarre posture: "standing" head-up, tail-up, or with the body turned at an angle. They remain motionless and allow cleaners to pick their skin and fins. Gill covers are opened to allow the inspection and cleaning of gills. Cleaners are even allowed to safely move *inside* the mouth! Cleaner fishes approach the host and inspect it by swimming close to the host's surface. Inspection by cleaner shrimps involves tapping with the antennae. Some cleaner fishes "dance" as part of their ritual, perhaps to remind the host that they are cleaners, not food.

Communication gaps between hosts and symbionts are surprisingly rare. Confusing cleaning stations with fast-food stands could prove harmful to hosts, even if the experimental removal of cleaners does not appear to harm resident fishes. The same can be said about cleaners—a continuous supply of food that is used by relatively few other noncleaners outweighs the risks, even for part-time cleaners that can feed on something else. Furthermore, being recognized as a cleaner reduces the chances of being eaten by another fish. A small Caribbean fish, a blenny, has evolved a color pattern that imitates, or mimics, that of a local cleaner wrass. The blenny is so close in appearance and behavior to the cleaner that host fishes are tricked and approach the blenny, only to lose a bit of skin to the fake cleaner!

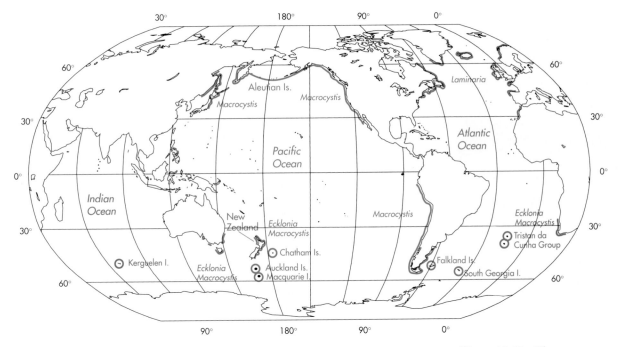

Figure 12-23 The geographic distribution of kelps is greatly affected by surface temperatures, which are influenced by surface circulation of the ocean. Currents along the west sides of the continents transport cold water from polar regions; warm water from the Equator is transported by currents on the east sides. As a result, kelps extend more toward the Equator along the west side of the continents more than on the east side. The opposite is true for reef-building corals, which need warm water (see Figure 13-13).

are a group of large brown seaweeds that live in relatively cold water. Kelp plants are true giants compared with other subtidal seaweeds or seagrasses. They form luxuriant forests that are home to a vast assortment of organisms.

There are several species of kelp. In the North Atlantic and the Asiatic coast of the North Pacific, various species of *Laminaria* (Figure 12-22, A and B) predominate. Their simple or cleft blades may be 3 m (10 feet) long. The giant kelp *(Macrocystis)* dominates kelp communities on the Pacific coasts of North and South America and some other parts of the Southern Hemisphere (Figure 12-23). Each plant is attached to the rocky bottom by a large holdfast (see Figure 5-9). Several long **stipes,** intertwined to form a trunklike foundation, grow from a single holdfast. **Fronds,** a stipe with its many leaflike blades, can reach 20 to 30 m (65 to 100 feet) or more. In some Southern Hemisphere communities, *Ecklonia*—rather than *Macrocystis*—is the main species. Kelp communities on the Pacific coast of North America are particularly diverse. A number of species, including the bullwhip kelp *(Nereocystis;* see Figure 5-18), the elk kelp *(Pelagophycus;* Figure 12-24), *Alaria,* and *Pterygophora,* may be important or may even replace *Macrocystis. Alaria* (Figure 12-22, C), which is edible, is also important in the North Atlantic.

Large, dense patches of kelp plants are known as **kelp beds.** They are called **kelp forests** when the fronds float on the surface (Figure 12-25) in a thick mat. This floating **canopy** is characteristic of the giant kelp and the bullwhip kelp. *Ecklonia* and some species of *Laminaria* and *Alaria* also form a canopy.

Physical factors have a major influence on kelp communities. Temperature is of particular importance because kelps are restricted to cold water. This is caused partially because kelps don't do well in warm water and par-

Figure 12-24 The elk kelp *(Pelagophycus porra)* is a large kelp that grows on the outer, deeper edges of some giant kelp beds (see Figure 12-27). It has two impressive, antlerlike branches from which large blades hang at the mercy of currents.

Figure 12-25 Giant kelp beds in the San Benito Islands, Baja California, Mexico.

Figure 12-26 The life history of the giant kelp *(Macrocystis)* and other kelps includes a large spore-producing sporophyte. Spores settle on the bottom and develop into minute male or female gametophytes. Each gametophyte releases male or female gametes, which—after fertilization—develop into the sporophyte, thus completing the cycle. See also Figure 5-24, A.

tially because, as is explained in Chapter 14 (in the section, "Patterns of Production"), warm waters tend to lack the rich supply of nutrients that kelps need. This dependence on cold water is reflected in the geographic distribution of kelps (Figure 12-23). The surface waters of the oceans flow in great counterclockwise **gyres** (see Figure 3-23). The currents that flow toward the poles on the western sides of the oceans carry warm water from equatorial regions. Because of this, kelps are restricted to high latitudes on the western sides of the oceans. On the other hand, kelps extend well down eastern shores, where cold, nutrient-rich currents flow down from high latitudes.

All but a few species of kelp are limited to hard surfaces, which may include substrates such as worm tubes and the holdfasts of other seaweeds. Given a suitable place to attach, kelps will grow in water as deep as light allows. This can be quite deep—as deep as 40 m (130 feet) in some species. Their fronds float at the surface, basking in sunlight, while their stipes connect them to the bottom far below.

Kelps are actually quite fragile for their giant size—in fact, they are fragile largely *because* of their size. The fronds cause a lot of drag, but the long, thin stipe breaks easily, and the plants are often torn from the bottom. Drifting plants cause more damage by tangling up with other plants. Kelps don't do well where there is heavy wave action, and storms can be disastrous. Many species prefer relatively deep water, where wave action is reduced. Thus the very adaptations that allow the plants to live in deep water, large size and floating fronds, also tend to restrict them there.

Kelps can grow very fast, with the giant kelp growing as fast as 50 cm/day (20 inches/day). Not surprisingly, kelp communities are also very productive. Primary production values of 1,000 gC/m^2/year have been mea-

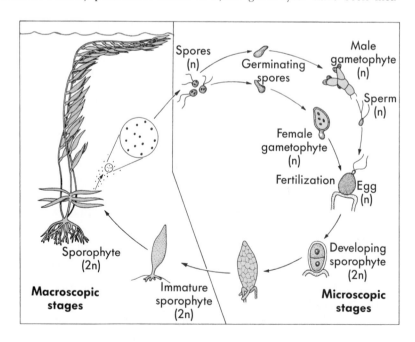

STRUCTURE AND FUNCTION OF MARINE ECOSYSTEMS

sured in *Ecklonia* in Australia and South Africa, up to around 1,500 gC/m²/year in California giant kelp, and close to 2,000 gC/m²/year in *Laminaria* in the North Atlantic (also see Table 14-2).

Kelp communities are restricted to hard substrates in cold, nutrient-rich water.

The large plants we see are only one stage in the life history of kelps. All kelps go through two separate stages: a large spore-producing **sporophyte** and a microscopic **gametophyte,** which produces male and female gametes (Figure 12-26). The sporophyte, the plant we see, is by far the dominant stage.

In some kelps the sporophyte is annual. The giant kelp and others are perennial and can live for several years. These long-lived plants often lose fronds to grazers, storms, and waves. They are able to grow them back because, unlike many seaweeds, growth takes place at the holdfast as well as the end of the stipes.

Kelp communities are arranged in distinct layers, each made up of species that grow at a characteristic height above the bottom (Figure 12-27). This structure results from the interaction of several physical and biologic factors. The floating canopy of the giant kelp, for instance, develops only where the water is deep enough to manage wave action but shallow enough for light to reach the bottom, permitting growth from the holdfasts. Other kelps may contribute to the surface canopy. These include the bullwhip kelp, found closer inshore, and the feather-boa kelp *(Egregia),* which lives in shallower water subject to wave action. *Pelagophycus* forms a midwater canopy under the giant kelp canopy.

Figure 12-27 Distribution of the major types of kelps and other seaweeds in a generalized giant kelp *(Macrocystis)* forest on the Pacific coast of North America. The complex distribution of algae results from the effects of factors such as light, type of substrate, wave action, depth, number and type of grazers, and even time of the year, since some of the seaweeds are annuals.

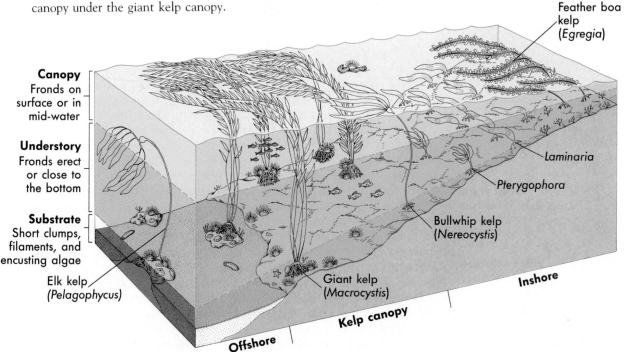

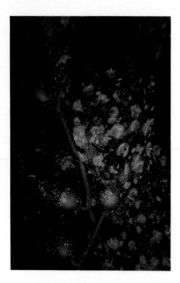

Figure 12-28 *Membra-nipora* is a bryozoan that forms thin, lacelike colonies on the blades of seaweeds, particularly the giant kelp (*Macrocystis*). Although the calcareous encrustations weigh the blades down and cover photosynthetic tissues, their effect appears to be minimal.

The dense canopy of kelp forests cuts down the amount of light underneath. Diving under the canopy is like walking into a dense forest on land: it takes a few moments for your eyes to adjust to the dim light. The plants that live there also have to adjust to the reduced light level, but life under the canopy is surprisingly rich and varied. Smaller kelps exploit the **understory,** the area below the canopy. They include *Laminaria, Pterygophora,* and other kelps that have either erect fronds that stand above the bottom or fronds that lie right on it.

A variety of shorter algae, mostly reds, live on the bottom under the two overlying layers, even when light is greatly reduced. Branching and encrusting coralline red algae are particularly common. Some seaweeds are more abundant in shallow water, where increased wave action reduces the canopy.

North Atlantic kelp beds are dominated by species of *Laminaria* that typically do not form a canopy. They are similar to Pacific kelp forests, however, in that they include many species of seaweeds and are arranged in layers.

The complex three-dimensional structure of kelp communities is exploited by many different animals. An amazing assortment of polychaetes, small crustaceans, brittle stars, and other small invertebrates live on the holdfasts, particularly those of the giant kelp. Tube-dwelling polychaetes, lacelike bryozoans (Figure 12-28), and other sessile organisms are common on the blades and stipes. Like the animals that live associated with the holdfast, they are mostly suspension and deposit feeders. The rocky bottoms around kelps are inhabited, among others, by sponges, sea squirts, lobsters, crabs, hermit crabs, sea stars, abalones (see Figure 6-28, B), and octopuses.

Fishes are very common in kelp communities. They use the food resources and shelter provided by the kelp plants in many different ways. The fishes often partition the available resources by feeding and taking shelter in different areas within the forest. Some species feed close to the bottom. In the Pacific kelp beds, bottom feeders include many species of rockfishes (*Sebastes*) and the kelp bass (*Paralabrax clathratus*). The California sheephead (*Semicossyphus pulcher*) uses its doglike teeth to crush sea urchins, crabs and other bottom invertebrates. Surfperches (*Rhacochilus, Brachyistius*) and others may feed in different parts of the canopy, around the holdfasts, or in the open water between the plants. Topsmelt (*Atherinops*), for example, are plankton feeders that take advantage of large swarms of opossum shrimps, or mysids, and other planktonic animals found around kelps. Fishes may be further differentiated by being active at different times of the day or night.

Small algae are grazed by snails, crabs, sea urchins, and fishes, but surprisingly, few grazers eat the large kelp plants. One large kelp grazer, the Steller's sea cow, is now extinct (see Figure 17-16). Isopods (*Phycolimnoria*) are small crustaceans that burrow into the holdfast, weakening it. A few fishes graze on kelps, but they do not appear to cause much mortality. Instead of feeding on the attached, actively growing plants, animals use most of the huge production of kelps in the form of drift kelp, pieces of the plants that break loose and sink to the bottom or are washed ashore (see Figure

5-18, B). As in seagrasses, salt-marsh plants, and mangroves, much of this material is converted into detritus and exported to other communities.

Kelp beds form a multistoried, complex environment. Drift kelp and understory seaweeds are a major food source, not the kelp itself.

Sea urchins are by far the most important grazers in kelp communities. Of special importance are the red (*Strongylocentrotus franciscanus*) and purple (*S. purpuratus*) sea urchins on the Pacific coast of North America and the green sea urchin (*S. droebachiensis*; see Figure 6-51, *A*) in both the North Atlantic and Pacific oceans. Sea urchin populations sometimes "explode." These explosions have had devastating impacts on kelp communities in several different parts of the world. Sea urchins normally feed on drift kelp. During population explosions or "plagues," however, the urchins eat live kelps and other seaweeds (Figure 12-29). When the sea urchins eat their holdfasts or stipes, the kelp plants break loose, float away, and die. The sea urchins may even climb up plants, weighing down the fronds and allowing other urchins to reach them. The urchins may completely clear large areas, which are then known as "urchin barrens" or "urchin deserts." Encrusting coralline algae are virtually the only plants left on these barrens.

The reasons for such outbreaks of urchins remain unclear. One suggestion is the decline in sea otters (*Enhydra lutris*; see Figure 8-12), an urchin predator that has disappeared from most of its former range in the north Pacific. In a study of several islands in the Aleutian chain, kelp forests were found to be healthy where sea otters were common. In contrast, there were many sea urchins and few kelps on islands where sea otters were absent. These results, however, do not explain the situation in southern California, where sea otters were wiped out more than 150 years ago, but the destruction of kelps by sea urchins was not observed until the 1950's. Heavy fishing on other urchin predators, including lobsters, crabs, and fishes, may play a role.

Another possibility is that the amount of drift algae has decreased, causing the urchins to switch to feeding on live plants. Sewage pollution, temperature increases, and a decrease in plant nutrients are factors that may have caused the availability of drift kelp to drop. Some of these factors may also stimulate the growth and survival of sea urchins. Some organic compounds released in sewage, for example, are used as nutrients by juvenile sea urchins.

The harvesting of abalones for food has also been proposed as a possible cause of sea urchin plagues. These large molluscs compete with sea urchins for shelter in rock crevices. Removal of the abalones may provide more space for urchins.

It is also possible that fluctuations in the number of sea urchins is caused by the higher survival of their planktonic larvae, perhaps the result of more favorable temperatures or more abundant food. It is likely that a combination of factors causes urchin plagues. To some extent, "plagues" may really be natural fluctuations in population size.

Kelp is harvested for food additives and other products (see the section

Figure 12-29 A "sea urchin barren" that has resulted from grazing after a population explosion of the red sea urchin (*Strongylocentrotus franciscanus*) in southern California. See also Figure 9-3.

in Chapter 5, "Economic Importance"). At the peak of sea urchin outbreaks in southern California during the 1970's, swarms of sea urchins were crushed by divers or killed with lime. The situation was complicated when the strong El Niño of 1983 (see the section in Chapter 14, "The El Niño–Southern Oscillation") produced severe storms and unusually warm currents, which caused high kelp mortality. Pollution, especially from sewage, is another complicating factor (see the section in Chapter 17, "Sewage").

Kelp communities may be severely disrupted by strong wave action, grazing by sea urchins, and warm currents.

The recovery of the southern California kelp forests has progressed well in some areas. It has been aided in part by transplanting healthy holdfasts tied to blocks into depleted areas (see the section in Chapter 17, "Restoration of Habitats"). As an experiment, a few sea otters were transplanted to one of the Channel Islands off southern California. Most transplanted animals disappeared, however, so the experiment was called off. The red sea urchin is now being harvested for human food in growing numbers. Will these measures have an effect? Nobody really knows. One thing is for sure—some kelp forests have not recovered and perhaps never will.

Evidence from other parts of the world seems to indicate that catastrophic disturbances of kelp communities are eventually followed by a recovery, all in recurring cycles. Sea urchins and kelps are apparently kept in a delicate balance that can be tripped one way or the other by several factors. The dynamics of kelp communities remain one of the fascinating questions in marine biology. They will ultimately be explained by the complex interplay of the biologic and physical factors that govern kelp communities.

Do It Yourself

SUMMARY

1. The subtidal zone extends from the lower edge of the intertidal zone to the _____.
2. Turbulence in the water above the continental shelf influences the concentration of nutrients in the water column by preventing _____. Turbulence also influences the deposition of sediment. For example, particle size along the coast is _____ where there is more turbulence.
3. Most of the inhabitants of unvegetated soft bottoms are _____ that feed primarily on _____, though some are suspension feeders.
4. Seagrass communities are typical of _____ substrates in _____, _____ waters. Most of the plant material in these communities is slowly decomposed into _____ because relatively few animals graze on seagrasses.

5. _____ bottoms in shallow water are typically dominated by sea-weeds. In contrast to seagrasses, most of the primary production is used by _____.
6. Kelps, which belong among the _____ algae, are fast-growing, highly productive plants. Kelp communities develop on hard bottoms along coasts where _____ currents rich in _____ flow.
7. Technically speaking, kelp _____ consist of kelps whose fronds do not float on the surface. In kelp _____, by contrast, the fronds float on the surface, forming a _____.
8. Kelps are known to be disrupted by grazing, primarily by _____. They are also harmed by _____ currents, heavy wave action, and pollution.

THOUGHT QUESTIONS

1. Eelgrass communities in Europe and the eastern coast of North America were severely affected by a still-unknown disease. The so-called eelgrass blight or wasting disease of the 1930's caused many eelgrass beds to disappear. What changes would you expect to take place if an eelgrass community vanishes from a given area? Consider changes to the substrate, the benthos, and other type of marine animals. Are there any possible changes among animals living on land? What type of community do you think re-placed the eelgrass communities after they disappeared?
2. The life history of kelps consists of a very large sporophyte and a tiny gametophyte. Sea lettuce and some other seaweeds, however, have a gametophyte that is as big as their sporophyte (see Figure 5-24, A). Do you see any advantages for kelps having such an inconspicuous, puny gametophyte?

FOR FURTHER READING

Eastman, J.T., and A.L. Devries: "Antarctic Fishes." *Scientific American*, vol. 255, no. 5, November 1986, p. 106-114. *A group of unique fishes have evolved adaptations to survive in the frigid waters of Antarctica.*

Gore, R.: "Between Monterey Tides." *National Geographic Magazine*, vol. 177, no. 2, February 1990, p. 2-43. *The luxuriant marine life, from kelp beds to deep canyons, of Monterey Bay in central California.*

Iversen, E.S., and D.E. Jory: "Arms Race on the Grass Flats." *Sea Frontiers*, vol. 35, no. 5, September-October 1989, p. 304-311. *Many predators inhabit turtle grass beds. They use different strategies to get their food.*

Martin, G.: "Otter Madness." *Discover*, vol. 11, no. 7, July 1990, p. 36-39. *Sea otters and their role on the structure of kelp beds and other subtidal communities in the Pacific coast of North America.*

Reed, C.G.: "Turbulent World of Moss Animals." *Natural History*, April 1991, pp. 40-47. *Bryozoans, or moss animals, are found practically everywhere in the subtidal.*

Winston, J.E.: "Life in Antarctic Depths." *Natural History*, September 1990, p. 70-75. *A portrait of the spectacular marine life in Antarctica.*

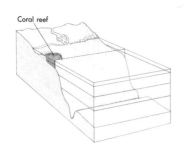

Coral reef

T here is something special about coral reefs. The warm, clear water, the spectacular colors, and the multitude of living things captivate almost everybody who sees a reef (Figure 13-1). Coral reefs rival that other great tropical community, the rain forest, in their beauty, richness, and complexity. Tropical rain forests and coral reefs are also similar in that the basic physical structure of both communities is produced by organisms. Both reef-building corals and the giant trees of a rain forest create a three-dimensional framework that is home to an incredible assortment of organisms. Coral reefs are such massive structures, in fact, that they must be considered not only biological communities but geological features, the largest geological features built by organisms.

A few organisms other than corals, like oysters and polychaete worms, can create reefs (see Figure 11-24). Such reefs are important in some places, but they are minor features compared to coral reefs. This chapter looks exclusively at coral reefs. Coral reefs occur in the subtidal zone, so they could have been covered in Chapter 12. They are so important and interesting, however, that they deserve a chapter all their own.

THE ORGANISMS THAT BUILD REEFS

Coral reefs are made of vast amounts of **calcium carbonate (CaCO$_3$),** limestone, that is deposited by living things. Of the thousands of species in coral reef communities, only a fraction produce the limestone that builds the reef. The most important of these reef-building organisms, as you might guess, are corals.

Reef Corals

Corals are cnidarians. Most of them are in the class Anthozoa, making them closely related to sea anemones. Unlike many cnidarians, they lack a medusa stage and live only as **polyps.** The massive reef is built of the calcium carbonate skeletons of billions of these polyps. Corals that build reefs are called **hermatypic** corals. There are also many **ahermatypic** corals, those that do not build reefs. **Soft corals** (order Alcyonacea; see Figure 13-34) are

cnidarians Animals in the phylum Cnidaria; they have radial symmetry, tissue-level organization, and tentacles with *nemato-cysts,* specialized stinging structures

Life stages of cnidarians:
polyp   the saclike stage, with the mouth and tentacles on top
medusa   the bell-like stage, with the mouth and tentacles underneath
Chapter 6, pp. 150-151

370

abundant on coral reefs but do not help build the reef because they have no hard skeleton. **Black corals** (order Antipatharia; see Figure 6-14) and **gorgonians** (order Gorgonacea), including **sea fans** (see Figure 6-13), **sea whips,** and many **precious corals,** are hard, but their skeletons are made mostly of protein and contribute little to reef formation. This chapter deals mainly with the reef-building, or hermatypic, corals.

Reef-building, or hermatypic, corals are the primary architects of coral reefs.

The Coral Polyp. You have to look closely to see the little polyps that build coral reefs. Coral polyps are not only small but deceptively simple in appearance. They look much like little sea anemones: an upright cylinder of tissue with a ring of tentacles on top (Figures 13-2 and 13-3). Like anemo-

Figure 13-1 A magnificent reef and associated sand island in Papua New Guinea.

Figure 13-2 Cutaway view of one of the polyps in a coral colony and of the calcium carbonate skeleton underneath. The polyps are interconnected by a very thin layer of tissue.

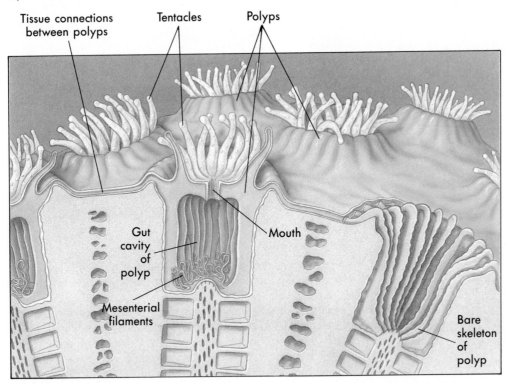

A   B   C

Figure 13-3   Coral polyps. **(A)** A few corals, like the mushroom coral (*Fungia*), consist of a single polyp. Most are colonies of many individual polyps. **(B)** In some of these, like the large-cupped boulder coral (*Montastrea cavernosa*), each poylp has its own individual cup, called a corallite. **(C)** In coral colonies like the brain coral (*Diploria strigosa*), meandering lines of polyps lie in a common corallite.

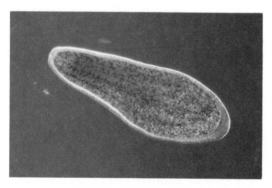

Figure 13-4   A coral colony starts when a planula larva like this settles, develops into a single polyp, and starts building its skeleton.

Figure 13-5   The polyps in a coral colony are interconnected. If you touch an extended polyp it will contract, and so will its neighbors, followed by their neighbors, and so on, so that the polyps contract in a wave that passes over the colony. In this photo the polyps on the left have retracted, but those on the right remain extended.

plankton   Small plants (*phytoplankton*) and animals (*zooplankton*) that drift with the currents
Chapter 9, p. 284; Figure 9-21

nes and other cnidarians, they use their nematocyst-armed tentacles to capture food, especially zooplankton. The tentacles surround the mouth, the only opening to the saclike gut.

Most reef-building corals are colonies of many polyps, all connected by a thin sheet of tissue. The colony starts when a planktonic coral larva, called a **planula** (Figure 13-4), settles on a hard surface. Coral larvae generally do not settle on soft bottoms. Immediately after settling, the larva transforms itself—or **metamorphoses**—into a polyp. This single "founder" polyp, if it

Figure 13-6 This coral (*Lobophyllia hemprichii*) shows especially well how the skeleton is built up. You can see the column of calcium carbonate that was laid down by each large polyp. Only the grey part is living tissue—the rest is calcium carbonate skeleton.

Figure 13-7 The calcium carbonate skeleton makes up most of a coral colony. A live colony is shown on the right. On the left is a colony with the living tissue removed. You can see that the main difference is the color; the live part is only a thin layer on the surface. The "corals" sold in shell and aquarium shops are actually only the skeletons of live corals that were taken from the reef and bleached. Some reefs have been devastated by the collectors who supply these shops in order to feed their families.

survives, divides over and over to form the colony. Thus all the polyps in a coral colony are genetically identical copies, or clones, of the founder polyp. The digestive systems of the polyps usually remain connected, and they share a common nervous system (Figure 13-5).

Most corals are colonies of many identical polyps.

Coral polyps lie in a cuplike skeleton of calcium carbonate that they make themselves. Over the years a series of polyps, each laying down a new layer of calcium carbonate, builds up the skeleton (Figure 13-6). The skeleton forms the bulk of the colony (Figure 13-7), which can take many different shapes (Figure 13-8). The actual living tissue is only a thin layer on the surface. It is the calcium carbonate coral skeletons, growing upward and outward with the coral, that create the framework of the reef.

Nearly all reef-building corals contain symbiotic **zooxanthellae** (Figure 13-9). Without zooxanthellae, corals produce their skeletons very slowly, not nearly fast enough to build a reef. The zooxanthellae enable the coral to deposit calcium carbonate much faster. It is the zooxanthellae as much as the corals that construct the reef framework; without them, corals could not build their skeletons and there would be no reefs.

Coral Nutrition. The zooxanthellae also provide vital nourishment to the coral. They perform photosynthesis and pass on some of the organic matter they make to the coral. Thus the zooxanthellae feed the coral from

zooxanthellae Dinoflagellates (single-celled, photosynthetic organisms) that live within animal tissues *Chapter 5, p. 124*

symbiosis The "living together" of two different species, often divided into *parasitism*, where one species benefits at the expense of the other; *commensalism*, where one species benefits without affecting the other; and *mutualism*, where both species benefit *Chapter 9, pp. 269-271*

photosynthesis CO_2 + H_2O + SUN ENERGY $\longrightarrow$ organic matter + O_2 *Chapter 4, p. 88*

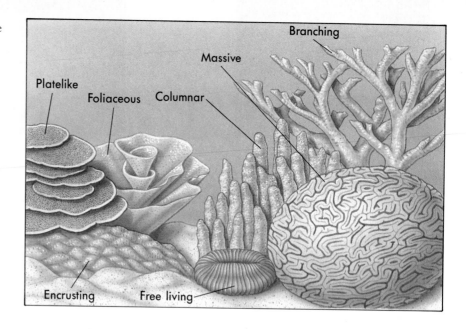

Figure 13-8 Corals come in a multitude of shapes, or growth forms.

Platelike

Foliaceous   Columnar

Massive

Branching

Encrusting   Free living

Figure 13-9 The microscopic zooxanthellae that pack these polyp tentacles help corals build reefs. They also occur in sea anemones (see Figure 10-32), giant clams (see Figure 13-42), and other animals.

the inside. In fact, many corals can survive and even grow without eating, as long as the zooxanthellae have enough light.

Reef-building corals contain symbiotic zooxanthellae. The zooxanthellae nourish the corals and help them produce their skeletons.

Although their zooxanthellae will support them, corals eat when they get the chance. They prey voraciously on zooplankton. The billions of coral polyps on a reef, along with all the other hungry reef organisms, are very efficient at removing zooplankton brought in by currents. Indeed, the reef has been called a "wall of mouths."

In addition to using their tentacles, coral polyps catch zooplankton in sheets of mucus that they secrete along the colony surface. Tiny, hairlike cilia gather the mucus into threads and pass them along to the mouth. Some corals hardly use their tentacles and rely on the mucus method. A few corals have even lost their tentacles altogether.

Corals have still other ways of feeding themselves. There are a number of long, coiled tubes called **mesenterial filaments** attached to the wall of the gut (see Figure 13-2). The mesenterial filaments secrete digestive enzymes. The polyp can extrude the filaments through the mouth or body wall to digest and absorb food particles outside the body. If you drop a bit of fish near a coral, the food will soon be covered by a tangled mass of these white filaments. Corals also use the mesenterial filaments to digest organic matter from the sediments.

As if all these ways of catching food were not enough, corals can actually absorb organic matter directly from the water. Seawater contains a substantial amount of organic material that is dissolved in the water instead of being in particles. This **dissolved organic matter (DOM)** comes from many

BOX 13-1

Coral Reproduction

Corals are amazingly adaptable animals. They come in all shapes and sizes and have many ways to feed themselves. It should come as no surprise, then, that they also have more than one way to reproduce.

At one level, growth and reproduction are the same thing in corals. The coral colony grows as its individual polyps divide to form new polyps. Thus the colony grows as the polyps reproduce. The process crosses the fine line between growth and reproduction when a piece of coral breaks off and continues to grow. It is now a separate colony, though it is a genetically identical clone of its "parent." Certain species of coral may depend a lot on this form of reproduction and actually may be adapted to break easily. After a reef is damaged by a severe storm, an important part of its recovery is the growth of the pieces of shattered coral colonies.

Corals can also reproduce sexually. Like other animals, they produce eggs and sperm, which fuse and eventually develop into a planula larva. Some corals are hermaphrodites, which make both eggs and sperm, whereas other species have separate sexes. The method of fertilization also varies among corals. In some, whether or not they are hermaphrodites, the egg is fertilized and develops inside the polyp. Other corals, probably the majority, are broadcast spawners and release the eggs and sperm into the water.

One of the most spectacular discoveries about coral reefs in recent years is the annual coral spawning event that takes place on the Great Barrier Reef. Once a year, for a few nights between October and early December, a whole variety of coral species spawns at once. The event takes place just after a full moon. At a given place, its occurrence can be predicted down to the night. It may seem surprising that this mass spawning was only recently discovered, but no one was really looking!

On the nights of mass spawning, many of the Great Barrier Reef's corals release their eggs and sperm all at once. The eggs and sperm may be released directly into the water or enclosed in little bundles that are released through the mouth. The bundles float to the surface and break up, allowing the eggs and sperm to mix.

Nobody knows why the corals all spawn at once. Maybe the egg predators get so full that most of the eggs go uneaten. Maybe it has something to do with the tides. Maybe there is an explanation that no one has thought of. Another interesting thing is that although the mass spawning happens in a few other locations, it does not occur in most of the rest of the world. Is there something different about the Great Barrier Reef, or has mass spawning just gone unnoticed elsewhere? The answers to these questions are likely to occupy coral reef biologists for some time to come.

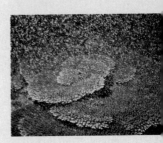

Mass spawning in the Great Barrier Reef: the release of egg and sperm bundles by corals.

sources. It is excreted as waste by animals and plants and released when bacteria break down organic particles. DOM also "leaks" out of many organisms by diffusion. Many marine bacteria and plants can absorb DOM from the water, but corals are among the few animals that can do so.

> **Corals can nourish themselves in a remarkable number of ways. Zooxanthellae provide the most important source of nutrition. Corals can also capture zooplankton with tentacles or mucus nets, digest organic material outside the body with mesenterial filaments, or absorb dissolved organic matter (DOM) from the water.**

Other Reef Builders

Though they are the main architects, corals cannot build a reef alone. Many other organisms help make a coral reef. The most important of these are not animals, as you might expect, but plants. These plants are essential to reef growth. In fact, some marine biologists think that "coral reefs" should be called "algal reefs" or, to be fair to both, "biotic reefs." One reason for this is that zooxanthellae, which can be considered algae because they perform photosynthesis, are essential to the growth of corals. There are other algae, however, that also have key roles in building the reef. Like corals, **coralline algae** produce a "skeleton" of calcium carbonate. Encrusting coralline red algae (*Porolithon, Lithothamnion, Lithophyllum*) grow in rock-hard sheets over the surface of the reef. They deposit considerable amounts of calcium carbonate, sometimes more than corals, and thus contribute to reef growth. Coralline algae are more important on Pacific reefs than Atlantic ones.

Encrusting coralline algae not only help build the reef, they also help keep it from washing away. The stony "pavement" formed by these algae is tough enough to withstand waves that would smash even the most rugged corals. The algae form a distinct ridge on the outer edge of many reefs, especially in the Pacific. This **algal ridge** absorbs the force of the waves and prevents erosion from destroying the reef (Figure 13-10).

Encrusting algae do yet another job that is vital to reef growth: Coral skeletons and fragments create an open network, full of spaces, that traps coarse carbonate sediments (Figure 13-11). Sediment, especially fine sedi-

Figure 13-10 The algal ridge at Enewetak Atoll in the Marshall Islands exposed at low tide.

STRUCTURE AND FUNCTION OF MARINE ECOSYSTEMS

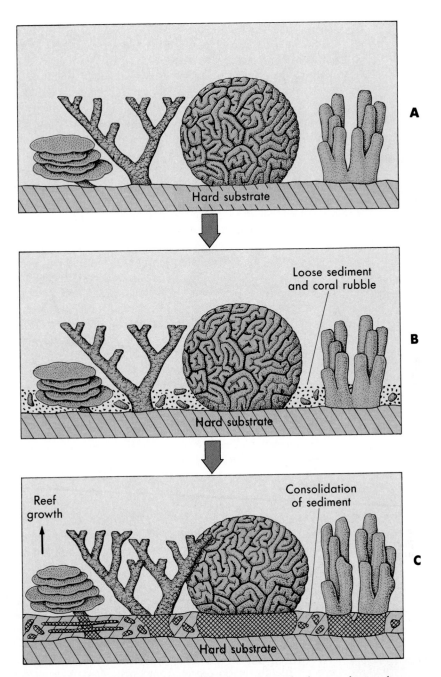

Figure 13-11 Reef growth involves several processes. The framework is made (**A**) when reef-building corals settle and grow on some hard surface—usually a preexisting reef. The spaces in this framework are partially filled in by coarse carbonate sediments (**B**). When the sediments are glued together by encrusting organisms (**C**), new reef "rock" is formed and the reef has grown. On a real reef all three steps go on at the same time.

ment, kills corals when it settles directly on them, but the buildup of coarse sediment in the reef framework is an essential part of reef growth. The structure of a reef is formed as much by the accumulation of coarse calcium carbonate sediment as by the growth of corals. Encrusting algae grow over the sediment as it builds up, cementing the sediment in place. Thus encrusting coralline algae are the glue that holds the reef together. Some invertebrates, notably sponges and bryozoans, also form encrusting growths that help bind the sediments.

Encrusting coralline algae help build the reef by depositing calcium carbonate, by resisting wave erosion, and by cementing sediments.

Nearly all the sediment that accumulates to help form the reef comes from coral fragments, or **coral rubble,** and the shells or skeletons of other organisms. In other words, nearly all the sediment is **biogenous.** The most important of all the sediment-forming organisms is a coralline green alga called *Halimeda* (Figure 13-12). *Halimeda* deposits calcium carbonate within its tissues to provide support and to discourage grazers—a mouthful of limestone is pretty unappetizing. The remnants of *Halimeda* accumulate on reefs in huge amounts, to be bound together by encrusting organisms.

Many other organisms make calcium carbonate sediments and thus contribute to the growth of the reef. The shells of forams (see Figure 6-2), snails, clams, and other molluscs are very important. Sea urchins, bryozoans, crustaceans, sponges, and a host of other animals add their shells or skeletons as well. Reef growth is truly a team effort.

The accumulation of calcium carbonate sediments plays an important role in reef growth. A coralline green alga, *Halimeda,* and coral rubble account for most of the sediment, but many other organisms also contribute.

Conditions for Reef Growth

Other organisms may be important, but coral reefs do not develop without corals. Corals have very particular requirements that determine where reefs develop. Reefs are rare on soft bottoms, for example, because coral larvae need a hard place to settle.

Light and Temperature. Corals can grow only in shallow water, where light can penetrate, because the zooxanthellae on which they depend need light. Coralline algae, being plants, also require sunlight. Particular types of coral and algae have different depth limits—some can live deeper than others—but reefs rarely develop in water deeper than about 50 m (165 feet). Because of this, coral reefs are found only on the continental shelves, around islands, or on top of seamounts. Many types of coral live in deep water and do not need light, but these corals do not contain zooxanthellae or build reefs. Corals also prefer clear waters, since water clouded with sediment or phytoplankton does not allow light to penetrate very well.

Reef-building corals are limited to warm water and are unable to grow and reproduce if the water temperature averages below about 20° C (68° F).

bryozoans (moss animals) Small, colonial, encrusting animals that make delicate calcium carbonate skeletons
Chapter 6, p. 163; Figure 6-25

forams (foraminiferans) protozoans, often microscopic, with a calcium carbonate shell
Chapter 6, p. 144; Figure 6-2

Figure 13-12 The coralline green alga, *Halimeda,* (the green, segmented organism in the center of the photo) is one of the main sediment-forming organisms of the reef. About 95% of the plant's weight is calcium carbonate, with a relatively thin layer of live tissue on the outside. When the tissue dies the segments separate, each leaving a piece of limestone.

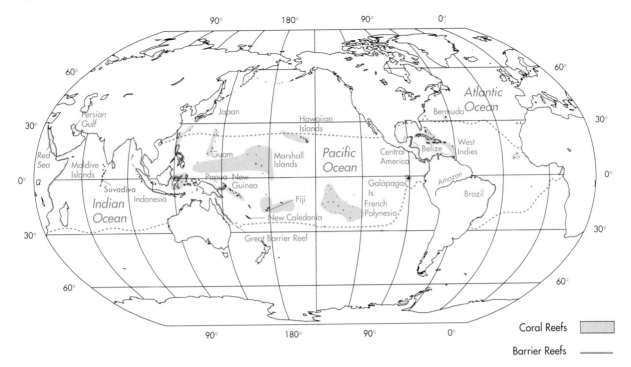

Coral Reefs ▭
Barrier Reefs ▬
Average 20° C Isotherm ┈┈┈

Most reefs grow in considerably warmer areas. Figure 13-13 illustrates the relationship between coral reefs and water temperature.

Corals need light and warm temperatures, so reefs only grow in shallow, clear, warm waters.

Water that is *too* warm is also bad for corals. The upper temperature limit varies, but is usually around 30° to 35° C (86° to 95° F). The first sign of heat stress, or stress of many other kinds, is **bleaching,** in which the coral expels its zooxanthellae (see Figure 17-11). It is called bleaching because the greenish or golden-brown zooxanthellae give the coral most of its color: Without them, the coral is almost white. Corals also slough off large amounts of slimy mucus when stressed. Above a certain temperature the coral never recovers and dies.

The exact temperature range preferred by corals differs from place to place because corals from a particular location adapt to the normal temperatures there (Figure 13-14). Corals from places with very warm water, for example, are able to tolerate higher temperatures. In some places, corals must adapt to drastic fluctuations in temperature. For instance, reefs grow in parts of the Persian Gulf where the water temperature ranges from 16° to 40° C (60° to 104° F).

Corals suffer when exposed to temperatures outside their normal range. This sometimes happens during extreme low tides, when shallow pools on the reef may be cut off from circulation. Heated by the sun, the water can warm up to fatal temperatures. By discharging heated water, electric power plants can also kill corals (see Figure 17-11).

El Niño, a change in surface currents described in Chapter 14 (see the section "The El Niño–Southern Oscillation"), brought unusually warm wa-

Figure 13-13 The geographic distribution of reef-building corals, like that of kelps, is related to temperature. Reef corals require warm water, however, whereas kelps need cold water. Compare the distribution of corals with that of kelps shown in Figure 12-23. Note the effects of warm surface currents: reefs extend farther north and south on the east side of continents than on the west side.

Figure 13-14 The upper temperature limit that corals can stand is related to the temperatures at their home site. For example, the water in the Marshall Islands is warmer than that in Hawaii; during the hottest month of the year the average high temperature is several degrees higher in the Marshalls. Corals from the Marshalls can tolerate correspondingly higher temperatures. Why do you think the highest temperature that the corals can tolerate would be *higher* than the average high temperature?

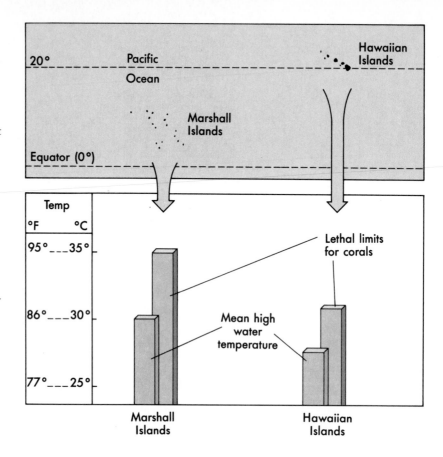

Figure 13-15 Corals often flourish where there is a lot of wave action. The water motion keeps sediment from settling on the corals and brings in food, oxygen, and nutrients.

ter to many parts of the ocean in 1982 and 1983. Afterward, coral bleaching and mortality were observed on a number of reefs, including some from both coasts of Central America, the Florida Keys, the Bahamas, and widespread parts of the Pacific Ocean. Thus changes in atmospheric and ocean circulation patterns can affect coral reefs on a global scale. Some scientists fear that coral bleaching is becoming more frequent, possibly because the earth is getting warmer (see Box 17-1). This global warming could severely threaten the world's coral reefs.

Sediments, Salinity, and Pollution. Fine sediment like silt is very harmful to corals. For one thing, it clouds the water and cuts down the light for the zooxanthellae. Worse, even a thin layer of sediment on the colony surface smothers the coral. To remove the sediment, corals use mucus opposite to the way that they use it to feed. Instead of being brought to the mouth, the mucus is sloughed off, carrying sediment with it. Even with this defense, corals don't do well in places where there is a lot of sediment, unless there is enough wave action to wash the sediment away (Figure 13-15). Reefs tend to be poorly developed, for example, near river mouths. This is not only because of the sediment brought in by the river, but because corals are also quite sensitive to fresh water. Human activities like mining, logging, construction, and dredging can greatly increase the amount of sediment and freshwater runoff, which may have very harmful effects on the local reefs (see the section in Chapter 17, "Coral Reefs").

Corals are also sensitive to pollution of many kinds. Even low concentrations of chemicals like pesticides and industrial wastes can kill them. In high concentrations, nutrients also can be harmful to reef growth. Humans release tremendous amounts of nutrients in sewage and in fertilizers that are washed from farmland and carried to the sea. The nutrients do not harm the corals directly. Instead, they alter the ecological balance of the community. Coral reefs usually grow in water that is very low in nutrients. In such nutrient-poor water, seaweeds do not grow very rapidly and are kept under control by grazers. This allows corals to compete successfully for space and light. When nutrients are added, seaweeds grow much faster, and they can shade and choke out the slow-growing corals.

Corals are very sensitive to fine sediment, fresh water, and pollution, including pollution by high nutrient levels.

The Kaneohe Bay Story. One of the best-known examples of the harmful effects of nutrient enrichment, or **eutrophication,** occurred in a partially enclosed bay in the Hawaiian Islands. Kaneohe Bay, located on the northeast shore of the island of Oahu (Figure 13-16), once had some of the most luxuriant reefs in Hawaii. Until the 1930's the area around the bay was sparsely populated. In the years leading up to World War II, with the military buildup of Oahu, the population began to increase. This increase continued after the war as the shores of the bay were developed for residential use.

The sewage from this expanding population was dumped right into the bay, mainly the south end. At first only small amounts of sewage were released, but major discharges began in the 1950's and increased during the 1960's and 1970's. By 1978 about 20,000 m³ (over 5 million gallons) of sewage were being dumped into the bay every day. Long before then, by the mid-1960's in fact, marine biologists began to notice disturbing changes in the south end of the bay. Loaded with nutrients, the sewage acted as a fertilizer for seaweeds. A green alga, the bubble alga (*Dictyosphaeria cavernosa*; Figure 13-17), found the conditions particularly to its liking and grew at a tremendous rate, literally covering the bottom in many parts of the bay. Bubble algae rapidly took over, overgrowing and smothering the corals. Phytoplankton also multiplied with the increase in nutrients, clouding the water. Kaneohe Bay's reefs began to die.

There is a happy ending to the story, however, at least for the time being. As the once beautiful reefs smothered, scientists and the general public began to cry out. It took a while, but in 1978 public pressure finally managed to stop the discharge of sewage into Kaneohe Bay almost completely, diverting the sewage offshore. The result was dramatic. Bubble algae started to die out, and the bay's corals began to recover much faster than anyone had expected. By the early 1980's, bubble algae were fairly scarce and corals had started to grow again. The reefs were still not what they once were, but they were on the track to recovery.

Figure 13-16 Kaneohe Bay, on the northeastern shore of the Hawaiian island of Oahu.

Figure 13-17 This green seaweed, the bubble alga (*Dictyosphaeria cavernosa*), nearly wiped out coral reefs in Kaneohe Bay, Hawaii. Notice the long detritus-feeding sea cucumber (*Opheodesoma spectabilis*).

Then the ghost of pollution reared its ugly head. In November 1982, Hurricane Iwo struck Kaneohe Bay. During the years of pollution a layer of the coral skeleton had weakened, becoming fragile and crumbly. When the hurricane hit, this weak layer collapsed and many reefs were severely damaged. Fortunately, the corals were already beginning to recover, and the broken pieces were able to grow back. If the hurricane had hit during the years of pollution, the reefs of Kaneohe Bay—and the benefits of fishing, tourism, and recreation—might have disappeared forever.

KINDS OF CORAL REEFS

Coral reefs occur in many different shapes and sizes. Reefs are usually divided into three main categories: **fringing reefs, barrier reefs,** and **atolls.** Some reefs do not fit neatly into any particular category, and some fall between two categories. Still, the division of reefs into the three major types works well for the most part.

The three main types of reefs are fringing reefs, barrier reefs, and atolls.

Fringing Reefs

Fringing reefs are the simplest kind of reef, and the most common. They develop near shore throughout the tropics, wherever there is some kind of firm substrate for the settlement of coral larvae. Rocky shorelines provide the best conditions for fringing reefs. Fringing reefs also grow on soft bottoms if there is even a small patch of hard surface that lets the corals get a foothold. Once they get started, the corals create their own hard bottom and the reef slowly expands.

As their name implies, fringing reefs grow in a narrow band or fringe along the shore (Figure 13-18). Occurring close to land, they are especially

Figure 13-18 Typical structure of a fringing reef. *Inset:* A fringing reef along an island in the Bahamas.

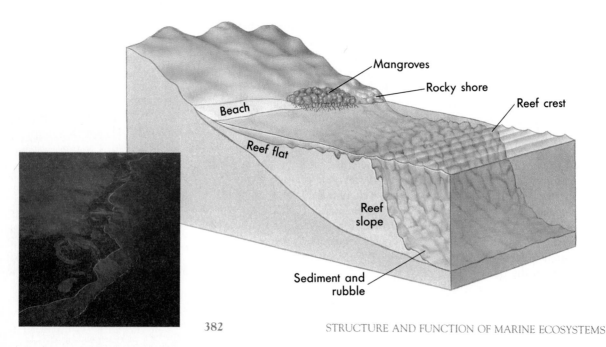

Mangroves

Rocky shore

Reef crest

Beach

Reef flat

Reef slope

Sediment and rubble

STRUCTURE AND FUNCTION OF MARINE ECOSYSTEMS

vulnerable to sediment, freshwater runoff, and to human disturbance. Under the right conditions, however, fringing reefs can be impressive. In fact, the longest reef in the world is not the famous Great Barrier Reef in Australia but a fringing reef that runs some 4,000 km (2,500 miles) along the coast of the Red Sea. Part of the reason that this reef is so well developed is that the climate is dry and there are no streams to bring in sediment and fresh water.

The typical structure of a fringing reef is shown in Figure 13-18. Depending on the area, the shore may be steep and rocky or have mangroves or a beach. The reef itself consists of an inner **reef flat** (Figure 13-19) and an outer **reef slope** (Figure 13-20). The reef flat is the widest part of the reef. It is shallow, often exposed at low tide, and slopes very gently toward the sea. Being closest to land, it is the part of the reef most strongly affected by sediments and freshwater runoff. The bottom is primarily sand or mud, or coral rubble in some places. There are some living corals, but there are neither as many nor as many different kinds as on the reef slope. Seaweeds, seagrasses, and soft corals also may occupy the reef flat, sometimes in dense beds.

The reef slope can be quite steep, nearly vertical in fact. It is the part of the reef with the densest cover and the most different species of coral, because the slope is away from shore and therefore away from the effects of sediment and fresh water. Also, the waves that bathe the slope provide good circulation, bring in nutrients and zooplankton, and wash away fine sediments. The **reef crest** is the shallow edge of the reef slope. The crest usually has the most luxuriant coral growth of all. If there is intense wave action, however, there may be an algal ridge, with the richest coral growth just below the crest. Because there is less light in deep water, the deep part of the reef slope has less live coral and fewer coral species.

> Fringing reefs grow close to shore and consist of an inner reef flat and an outer reef slope.

Large amounts of sediment and coral rubble tumble down the reef slope and settle at the base. As this material builds up, reef organisms may begin to grow on it, depending on the water depth and other factors. Thus the reef can grow outward, as well as upward. Beyond the base of the slope, the bottom is usually fairly flat and covered with sand or mud. On many Caribbean reefs, turtle grass (*Thalassia testudinum*) dominates the bottom beyond the slope.

Barrier Reefs

The distinction between barrier reefs and fringing reefs is sometimes unclear, because the two types grade into one another. Like fringing reefs, barrier reefs lie along the coast, but barrier reefs occur considerably farther from shore, occasionally as far as 100 km (60 miles) or more. Barrier reefs are separated from the shore—which may have a fringing reef—by a relatively deep **lagoon**. Largely protected from waves and currents, the lagoon usually has a soft sediment bottom. Inside the lagoon there are columns of coral called **coral knolls**, or **pinnacles**, that grow up nearly to the surface.

Figure 13-19 The upward growth of reefs is limited by the tides. When extreme low tides occur, shallow parts of the reefs are exposed, as in the case of this reef flat at Phuket Island in the Andaman Sea. If the corals are exposed for only a short time they might survive, but they will die if this is an extreme low tide and they are exposed for too long. It is this occasional exposure at extreme low tides that keeps reef flats flat, because all the corals above a certain depth are killed.

Figure 13-20 The elkhorn coral (*Acropora palmata*) is a dominant coral in the fore-reef slopes of fringing reefs in the Caribbean and Florida. Its broad branches rise parallel to the surface to collect light.

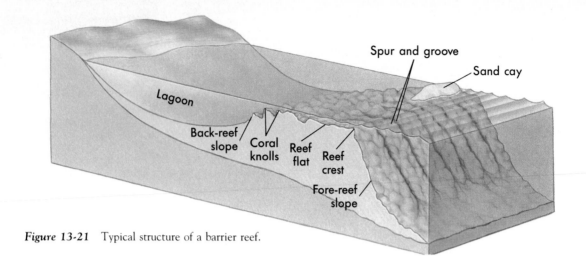

Figure 13-21 Typical structure of a barrier reef.

Figure 13-22 A rich back-reef slope on a Pacific barrier reef.

Figure 13-23 A barrier reef flat in Papua New Guinea.

The barrier reef consists of a **back-reef slope,** a reef flat, and a **fore-reef slope,** which corresponds to the reef slope of a fringing reef (Figure 13-21). The back-reef slope may be gentle or as steep as the fore-reef slope. It is protected from waves by the rest of the reef, but waves wash large amounts of sediment from the reef down the slope. As a result, coral growth is often not as vigorous on the back-reef slope as on the fore-reef slope. This is not always true; some back-reef slopes, especially gentle ones, have luxuriant coral growth (Figure 13-22). Coral pinnacles may be quite common.

The reef flat, like that on fringing reefs, is a shallow, nearly flat platform (Figure 13-23). Sand and coral rubble patches are interspersed with seagrass or seaweed beds, soft corals, and patches of dense coral cover. Waves and currents may pile up sand to form small sand islands called **sand cays** or, in the Caribbean, **keys** (see Figure 13-1).

The richest coral growth is usually at the outer reef crest. There may be a well-developed algal ridge if the reef is exposed to wave action, with coral growth most luxuriant just below the crest. Exposed fore-reef areas often have a series of fingerlike projections alternating with sand channels (Figures 13-21 and 13-24). There is still considerable debate about what causes these formations, known as **spur-and-groove** formations or **buttresses.** The wind, waves, or both are definitely involved, because spur-and-groove formations develop only on reefs that are exposed to consistent strong winds. These formations are also found on atolls and some fringing reefs.

Fore-reef slopes vary from relatively gentle to nearly vertical. The steepness depends on the action of wind and waves, the amount of sediment flowing down the slope, the depth and nature of the bottom at the reef base, and other factors. As with other types of reef, the abundance and variety of corals generally decreases with depth on the fore-reef slope. The growth form of the corals also changes down the slope. At the crest, under the pounding of the waves, the corals are mostly stout and compact; many are massive (see Figure 13-8). Below the crest there is great variety in form. Whether they form branches, columns, or whorls, corals in this zone often grow vertically upward. This may be an adaptation for competition. Corals

Figure 13-24 A portion of the spur-and-groove formations at Enewetak Atoll, Marshall Islands, at low tide.

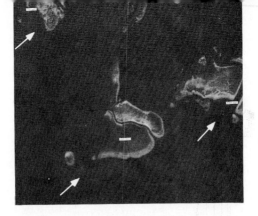

Figure 13-25 Australia's Great Barrier Reef is so large that it can be seen from space. This is a satellite image of the central section of the Reef taken by a Landsat satellite that is used to photograph large-scale coastal features. White arrows indicate reefs; the "bars" the name of islands.

that grow upward like skyscrapers rather than outward need less space to attach. They are also less likely to be shaded, and if they spread out at the top, can shade out other corals. Deeper on the reef slope, corals tend to grow in flat sheets, a form which probably helps them collect more light.

The largest and most famous barrier reef is the Great Barrier Reef (Figure 13-25). It runs more than 2,000 km (1,200 miles) along the northeastern coast of Australia, varying in width between about 15 and 350 km (10 to 200 miles) and covering an area of over 225,000 km² (80,000 square miles). Though not the longest reef in the world, it covers such a large area and is so complex and well developed that it is generally regarded as the largest reef structure in the world. Actually, the Great Barrier Reef is not a single reef but a system of more than 2,500 smaller reefs, lagoons, channels, islands, and sand cays (Figure 13-26).

The world's second largest barrier reef is in the Caribbean, off the coast of Belize, Central America. Major barrier reefs are also associated with New Caledonia and Fiji. There are many other smaller barrier reefs, especially in

Figure 13-26 The Great Barrier Reef is a complicated system of thousands of small reefs, sand cays, and lagoons. Shown here are four reefs that are part of the Great Barrier Reef Marine Park.

Figure 13-27 An atoll in the Palau (Belau) Islands in the western Pacific Ocean.

the Pacific. Like the Great Barrier Reef, these usually are not single reefs but complex systems of smaller reefs.

Atolls

An atoll is a ring of reef, and often islands or sand cays, that surrounds a central lagoon (Figures 13-27 and 13-28). The vast majority of atolls occur in the **Indo—west Pacific region,** that is, the tropical Indian and western Pacific oceans. Atolls are rare in the Caribbean and the rest of the tropical Atlantic Ocean. Unlike fringing and barrier reefs, atolls can be found far from land, rising up from depths of thousands of meters or more. With practically no land around, there is no river-borne silt and very little freshwater runoff. Bathed in pure blue ocean water, atolls display spectacular coral growth and breathtaking water clarity. They are a diver's dream.

Atoll Structure. Atolls range in size from small rings less than a mile across to systems well over 30 km (20 miles) in diameter. The two largest atolls are Suvadiva, in the Maldive Islands in the Indian Ocean, and Kwajalein, one of the Marshall Islands in the central Pacific. These atolls cover

Figure 13-28 Typical structure of an atoll.

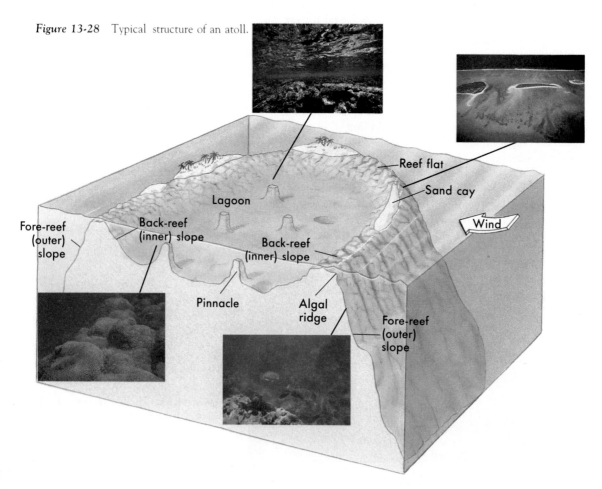

areas of more than 1200 km² (700 square miles). Atolls may include a dozen or more islands and be home to thousands of people.

An atoll's reef flat is much like the reef flat on a fringing or barrier reef: a flat, shallow area. Unlike those of barrier reefs, however, the atoll's reef flat is rarely more than a kilometer or so wide. The fore-reef and back-reef slopes can now be thought of as outer and inner slopes, respectively, since they extend all the way around the ring-shaped atoll.

The reef crest of an atoll is strongly influenced by wind and waves. Since most atolls lie in the zone of the trade winds, the wind usually comes from a consistent direction. Consequently, the wind affects various parts of the atoll in different ways. Encrusting coralline algae, which can stand up under the constant pounding of the waves, build a distinct ridge on the reef crest of the **windward** side of the atoll, the side that faces the prevailing wind. On the few Caribbean atolls, the coralline algae may be replaced by a few species of especially wave-resistant corals. The **leeward,** or sheltered, side of the atoll has no algal ridge. Spur-and-groove formations, too, develop only on the windward side.

The trade winds, the steadiest winds on earth, blow from latitudes of about 30 degrees toward the Equator
Chapter 3, p. 66; Figure 3-21

The fore-reef, or outer, slope is nearly vertical, though there is usually a series of ledges and overhangs. The reef wall continues down to great depths. The water may be hundreds, even thousands, of meters deep just a stone's throw from the reef.

The lagoon, on the other hand, is relatively shallow, usually about 60 m (195 feet) deep. The bottom of a lagoon is very uneven, with many depressions and coral pinnacles. Some pinnacles rise almost to the surface, where they may form "mini-atolls": small rings of coral within the lagoon.

Atolls are rings of reef, with steep outer slopes, that enclose a shallow lagoon.

How Atolls Are Formed. When atolls were discovered, scientists were at a loss to explain them. It was known that corals can grow only in shallow water, yet atolls grow right in the middle of the ocean, out of very deep water. Therefore the atoll could not have grown up from the ocean floor. If the reefs grew on some kind of shallow structure that was already there, like a seamount, why is there no sign of it? The islands on atolls are simple sand cays that have been built by the accumulation of reef sediments and would not exist without the reef. They are *products* of the reef and could not provide the original substrate for reef growth. Finally, why do atolls always form rings?

The puzzle of atoll formation was solved by Charles Darwin in the mid-nineteenth century. Darwin is most famous, of course, for proposing the theory of evolution by natural selection, but his theory of atoll formation is also an important contribution to science.

Darwin reasoned that atolls could be explained by reef growth on a subsiding island. The atoll gets its start when a deep-sea volcano erupts to build a volcanic island. Corals soon colonize the shores of the new island, and a fringing reef develops (Figure 13-29, A). As with most fringing reefs, coral growth is most vigorous at the outer edge of the reef. The inner reef is strongly affected by sediment and runoff from the island.

subsidence   The slow sinking into the *mantle* of a part of the earth's crust that contains a land mass
Chapter 10, p. 288

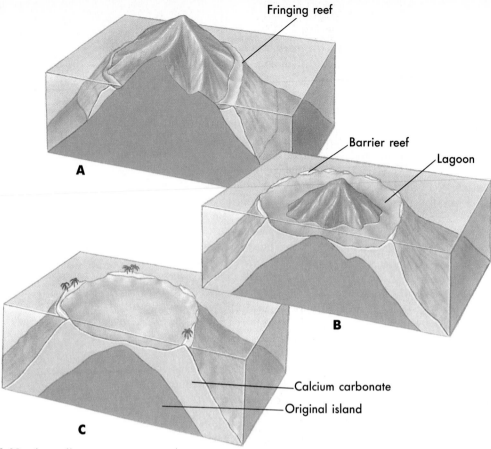

Fringing reef

A

Barrier reef

Lagoon

B

Calcium carbonate

Original island

C

Figure 13-29 An atoll begins as a fringing reef around a volcanic island **(A)**. As the island slowly sinks, the reef flat gets wider and deeper and eventually becomes a lagoon. At this stage the fringing reef has become a barrier reef **(B)**. Eventually the island sinks altogether, leaving only a ring of living, growing reef—an atoll **(C)**.

As time passes, the tremendous weight of the island makes the crust beneath it sink slowly into the mantle. If the island sinks too fast, reef growth cannot keep up and the reef eventually "drowns": It sinks to a depth where the corals can no longer grow. Often, however, the reef grows upward as fast as the island sinks. Thus the living part of the reef always stays in the shallow water that it needs to stay alive.

As the island sinks, the part of it left above water gets smaller and smaller. The reef, however, follows the original outline of the island as it grows upward. The back part of the reef flat, inhibited by sediment and run-off, cannot grow as fast as the outer reef. Therefore it sinks and gets deeper, eventually becoming a lagoon. Thus the reef gradually changes from a fringing reef into a barrier reef (Figure 13-29, B).

As the process continues, the island eventually sinks altogether. The structure is now a full-fledged atoll (Figure 13-29, C). As the island below continues to subside, the reef grows upward, keeping itself in the light. The atoll forms a thick calcium carbonate cap on top of the original island.

Atolls result from the slow sinking of volcanic islands. As an island sinks, the fringing reef around it becomes a barrier reef and eventually an atoll.

STRUCTURE AND FUNCTION OF MARINE ECOSYSTEMS

Surprisingly, Darwin's explanation of how atolls are created was pretty much ignored for a century while scientists proposed various other hypotheses, none of which held up. Finally scientists found conclusive evidence that Darwin was right. Unlike other hypotheses for atoll formation, Darwin's hypothesis predicted that, below the thick calcium carbonate cap, there should be volcanic rock, the original island. In the 1950's the United States Geological Service drilled several deep holes on Enewetak atoll in the Marshall Islands. These cores revealed exactly what Darwin predicted: volcanic rock far beneath the calcium carbonate of the reef. The thickness of the carbonate cap is impressive: The volcanic island that underlies Eniwetak is covered by more than 1,400 m (4,600 feet) of calcium carbonate!

Scientists now believe almost unanimously that Darwin's hypothesis of atoll formation is correct. There are, of course, a few details to be added to the picture, in particular the effects of changes in sea level (see the section in Chapter 2, "Climate and Changes in Sea Level"). When the sea level is low, atolls may be left above the surface. The corals die out and the reef is eroded by the wind and rain. If the sea level rises rapidly, the atoll may be drowned, unable to grow in deep water. In either case, corals recolonize the atoll when the sea level returns to normal.

THE ECOLOGY OF CORAL REEFS

Coral reefs may be impressive to geologists, but to the biologist they are simply awesome. They are easily the richest and most complex of all marine ecosystems. Literally thousands of species may live on a reef. How do all these different species live? How do they affect each other? What is their role in the ecosystem? These and a million other questions fascinate coral reef biologists.

Our ability to answer the questions, however, is surprisingly poor. This is partly because reefs are so complicated. Just keeping track of all the different organisms is hard enough; the task of figuring out what they all *do* is mind-boggling. Furthermore, until the last quarter of a century or so most marine biologists lived and worked in the temperate regions of the Northern Hemisphere, far from the nearest reef. As a result not many biologists studied reefs. Tremendous progress has been made in recent years, but there is still much to learn. The rest of this chapter summarizes what *is* known about the ecology of coral reefs and points out some of the important questions that remain.

The Trophic Structure of Coral Reefs

The tropical waters where coral reefs are found are nearly always very poor in nutrients (see the section in Chapter 14, "Patterns of Production") and therefore have practically no phytoplankton and very little **primary production.** In these barren waters, coral reefs are oases of abundant life. How can such rich communities grow when the surrounding sea is so unproductive? As with most aspects of coral reef biology, we are just beginning to learn the answer.

primary production The conversion of carbon from an inorganic form, carbon dioxide, into organic matter by autotrophs; that is, the production of food
Chapter 4, p. 91

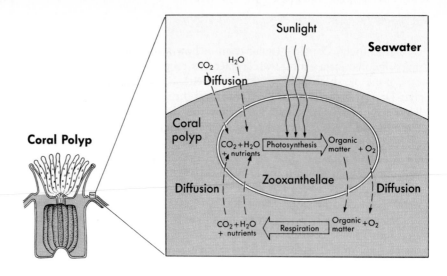

Figure 13-30 Carbon dioxide and nutrients are recycled between a coral polyp and its zooxanthellae.

Figure 13-31 Fishes often take shelter among the branches of reef corals, in this case the elkhorn coral (*Acropora palmata*) in the Caribbean. This is one of many examples of mutualism on the reef. The fish get a hiding place, and their waste products supply vital nutrients to the coral and its zooxanthellae.

A good part of the answer lies in the mutualistic relationship between corals and their zooxanthellae. We have already learned what the zooxanthellae do for the coral: They provide food and help make the calcium carbonate skeleton. In return the zooxanthellae get not only a place to live, but a steady supply of nutrients such as **nitrogen** and **phosphorus.** Most of the coral's waste products are not released into the water. Instead, they are taken up and used as nutrients by the zooxanthellae. Using sunlight, the zooxanthellae incorporate the nutrients into organic matter, which is passed on to the coral. When the coral breaks down the organic matter, the nutrients are released and the whole process begins again (Figure 13-30). The nutrients are **recycled,** used over and over, so that far fewer nutrients are needed than would otherwise be the case. Nutrient recycling is thought to be one of the main reasons that reefs are able to grow in nutrient-poor tropical waters.

Nutrient recycling occurs not just between corals and their zooxanthellae, but among all the members of the coral reef community. When fishes graze on plants, for example, they excrete nitrogen, phosphorus, and other nutrients as waste. These nutrients are then taken up by the plants. Many corals provide shelter to schools of small fish (Figure 13-31). The fish leave the coral at night to feed and return during the day. The waste products of the fish can be an important source of nutrients and help the coral grow faster. Thus nutrients are cycled from whatever the fish feed on to the coral. Nutrients pass through the community again and again in this cycle of feeding and excretion.

Coral reef communities use nutrients very efficiently as a result of recycling. The recycling is not perfect, however, and some nutrients are lost, carried away by the currents. Thus the reef still needs a continual supply of new nutrients. Recycling alone is not enough to account for the high productivity of reefs.

One of the most important discoveries about reefs in recent years is that the reef is able to provide some of its own nutrients. Coral reefs are now known to have among the highest rates of **nitrogen fixation** of any natural community. The main nitrogen fixers are blue-green algae (or cyanobacteria), especially a free-living one called *Calothrix* and a group that lives symbiotically in sponges. There is evidence that corals, too, have symbionts that can fix nitrogen, providing nutrients for the zooxanthellae. Just what the symbionts are is not known. All these nitrogen fixers provide a substantial source of nitrogen nutrients. Nitrogen, therefore, probably does not limit coral reef communities, though not all reef biologists agree about this.

Ocean currents bring in additional nitrogen and, more importantly, phosphorus and other nutrients that are not produced on the reef. Corals, bacteria, algae, and other organisms can absorb nutrients directly from the water. Even though the water contains few nutrients, if enough water washes over the reef it can bring in a lot of nutrients. More importantly, the water carries zooplankton, a rich source of nutrients. When the zooplankton are captured by the "wall of mouths," the nutrients in the zooplankton are passed on to the reef community. In fact, many biologists think that corals eat zooplankton not so much to feed themselves as to get nutrients for their zooxanthellae.

Coral reefs are very productive even though the surrounding ocean water lacks nutrients because nutrients are recycled extensively, nitrogen is fixed on the reef, and the zooplankton and nutrients that occur in the water are used efficiently.

The production and efficient use of nutrients by coral reef communities result in high primary productivity. This is reflected in the overall richness of the community. It may surprise you to find out, however, that scientists don't really know just how *much* primary production there is on coral reefs, or even which particular plants are the most important producers. There is no doubt that zooxanthellae are very important, but since they live inside corals, it is very hard to measure exactly how much organic matter they produce. For a time it was thought that very few animals eat coral, since there is so little live tissue on a coral colony. Therefore it was believed that, even though zooxanthellae produce a lot of organic matter, most of it was consumed by the coral and not much was passed on to the rest of the community. As biologists looked closer, however, they found more and more animals that eat corals or the mucus they shed (Figure 13-32). Thus primary production by coral zooxanthellae may be important not only to corals but to the community at large. Exactly how much production corals and their zooxanthellae contribute is still unknown.

Seaweeds are important primary producers on the reef, especially the small, fleshy, or filamentous types that are called **turf algae** because they often grow in a short, thick "turf" on the reef flat. A great many fishes, sea urchins, snails, and other animals graze on these seaweeds. The turf algae may perform more photosynthesis on the reef than the zooxanthellae, but biologists are not sure.

Zooxanthellae and turf algae are probably the most important primary producers on coral reefs.

nitrogen fixation   Conversion of nitrogen gas (N_2) into nitrogen compounds that can be used by plants as nutrients
Chapter 9, p. 281

Figure 13-32 A number of animals eat coral directly, and many others feed on the mucus that corals produce. The primary production of coral zooxanthellae is thus passed on to coral feeders, and then to the animals that eat them.

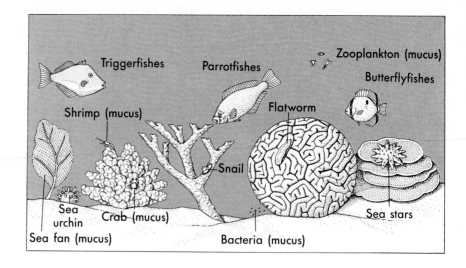

Figure 13-33 When different species of coral come in contact, they attack each other. The white band separating the two coral species is a dead zone where coral in the top of the picture, a species of *Goniastrea*, has killed the other coral, a species of *Acropora*.

Blue-green algae, coralline algae, and some bacteria are also primary producers. They probably account for much less primary production than do zooxanthellae and turf algae; however, biologists have been wrong about coral reefs before.

Coral Reef Communities

With so many species on coral reefs, the interactions among them are exceedingly complex. What is known about these interactions is fascinating; what remains to be learned will be even more so.

Competition. Space is at a premium on coral reefs, as it is in the rocky intertidal community (see the section in Chapter 10, "The Battle for Space"). Corals, seaweeds, and many others need a hard place on which to anchor themselves. Corals and seaweeds need not just space, but space in the sunlight. The reef is crowded, and most of the available space is taken. As a result the sessile animals and plants, those that stay in one place, must compete for space.

Sessile coral reef organisms must compete for space. Corals and seaweeds compete for light as well.

Corals compete for the space they need in different ways. The fast-growing ones tend to grow upward and then branch out, cutting their neighbors off from the light. Other corals take a more direct approach and actually attack their neighbors (Figure 13-33). Some use their mesenterial filaments for this. When they contact another coral, they extrude the filaments and digest away the tissue of the other coral. Still other corals develop special long tentacles, called **sweeper tentacles,** that are loaded with nematocysts and sting neighboring colonies. Corals differ in their aggressive abilities. The most aggressive corals tend to be slow-growing, massive types, whereas the less aggressive forms tend to be fast-growing, upright, and branching.

STRUCTURE AND FUNCTION OF MARINE ECOSYSTEMS

Both strategies have their advantages, and both kinds of coral are able to thrive on the reef.

The two main ways in which corals compete for space are by overgrowing or directly attacking their neighbors.

Corals compete for space and light not only with each other, but with seaweeds and sessile invertebrates. Like corals, encrusting algae have to produce a calcium carbonate skeleton and therefore grow relatively slowly. They tend to be found in places where corals don't do well because of sedimentation, wave action, or predation.

Under the right conditions, nonencrusting algae can grow much faster than either corals or encrusting algae. Even with the nitrogen fixation and nutrient cycling that occur on reefs, seaweeds are probably somewhat nutrient-limited most of the time. Hence they grow fairly slowly. The reef also has an abundance of hungry grazers that eat the seaweeds. The combination of nutrient limitation and grazing keeps the seaweeds in check. If the nutrient levels increase or the grazers disappear, seaweeds rapidly take over, overgrowing and choking out corals and other organisms.

Soft corals are also important competitors for space on reefs, and in some places they may make up almost half of the living tissue (Figure 13-34). Like seaweeds, soft corals lack a calcium carbonate skeleton and are able to grow faster than hard corals. Some soft corals contain sharp little limestone needles, or **spicules,** that discourage predation. Many of them also contain various chemicals—some of which have potential medical value (see the section in Chapter 16, "Drugs from the Sea")—that are toxic or taste bad to predators. Because of these defensive mechanisms, only a few specialized predators are able to eat soft corals. The defensive chemicals can also be released into the water, where they kill hard corals that come too close. Another competitive advantage enjoyed by some soft corals is that they are not completely sessile. Though they stay in one place most of the time, they can move about slowly. This helps them invade and occupy available space on the reef.

Soft corals are important competitors for space on reefs. They grow rapidly, are resistant to predators, and can move about on occasion.

With all these competitive weapons at their disposal, why don't soft corals take over? Not much is known about how soft corals compete with reef-building corals and other reef organisms or what determines the winner. Soft corals do appear to have much shorter lives than reef-building corals and to be much more easily torn away by storm waves. Perhaps these factors keep soft corals in check.

Like soft corals, sponges often have spicules and nasty chemicals that protect them from predators. They can be important users of space on reefs, much more so in the Caribbean than in the Pacific and Indian oceans. Part of the reason for this seems to be that there are far fewer species of coral in the Caribbean than in the Indo–west Pacific. This is a result of geologic history. During the most recent series of ice ages, the sea surface was cooler. Corals survived in the heart of the Indo–west Pacific region, around Indo-

Figure 13-34 Soft corals can form dense patches. This photo is from Madang in Papua New Guinea.

Figure 13-35 Diving on a coral reef such as the Great Barrier Reef is like taking a swim in a tropical fish aquarium—brightly colored fishes seem to be everywhere. The competitive relationships among the various species are poorly understood.

nesia and New Guinea, but many coral species became extinct in other parts of the ocean. When the ice age ended, corals spread out again through the Pacific, recolonizing areas where they had died out. The Caribbean, however, was not recolonized because the Isthmus of Panamá blocked their dispersal. It is thought that the Caribbean contains only those coral species that managed to survive the ice ages.

Fishes are another group in which competition may be important. Along with the corals, fishes are probably the most conspicuous and abundant animals on the reef (Figure 13-35) Many of these fishes have similar diets; for example, many species eat coral, many graze on algae, and many are carnivores. Different species of fishes of the same feeding type at least potentially compete with each other. There is a lot of debate about how competition acts on fishes. Some biologists think that many species overlap very broadly in their feeding habits and lifestyles and that such species therefore compete strongly. According to this view, the outcome of the competition—who wins and who loses—is largely a matter of luck. When space on the reef becomes available, it is occupied by whichever species happens to get there first. Whether a certain species is present at a particular place and time on a reef often depends on whether its larvae or juveniles happen to be at the right place at the right time. This **"lottery" hypothesis** contradicts the more traditional view of competition, in which each species of fish has its own particular niche. Those who hold this view think that each type of fish does something a little different from other fishes, thereby avoiding competition and the risk of competitive exclusion. It is uncertain which of these two schools of thought is correct; the answer probably lies somewhere in between.

> There are two schools of thought about how competition affects coral reef fishes. One holds that every species has a unique niche. The other holds that species may use the same resources and that the outcome of competition is determined largely by chance.

Predation on Corals.   As in other communities, predation and grazing are important in structuring coral reef communities. Though a variety of animals eat corals, instead of killing the coral and eating it entirely, most coral predators eat individual polyps or bite off pieces here and there (Figure 13-36). The coral colony as a whole survives and can grow back the portion that was eaten. In this respect, coral predation is similar to plant grazing by herbivores.

Coral predation can strongly affect both the number and the type of corals that live on a reef, as well as how fast the reef as a whole grows. The 1982-1983 El Niño killed many of the corals in the Galápagos Islands, but before that predation by coral-eating sea urchins limited how fast the reef grew. Fishes, too, may have an effect on reef growth. In Kaneohe Bay, for example, a butterfly fish (*Chaetodon*) slows the growth of a particular coral that it likes to eat. When the coral is protected from the butterfly fish by a cage, it grows much faster. If it were not for the fish, this fast-growing coral would probably dominate other corals in the bay. A similar situation is known in the Philippines, where a snail (*Drupella*) that eats corals prefers the fast-growing types. This gives the slow growers a better chance to compete.

ecological niche   The combination of what a species eats, where it lives, how it behaves, and all the other aspects of its lifestyle
Chapter 9, p. 268

Competitive exclusion occurs when one species out-competes and eliminates another
Chapter 9, p. 268

Figure 13-36   A single chevron butterflyfish *Chaetodon trifascialis* (foreground) and a pair of oval butterflyfish (*C. trifasciatus*; background) are among the many reef fishes that feed on corals without killing colonies. Their mouths are adapted for cleanly nipping off coral polyps. Also see Figure 7-30.

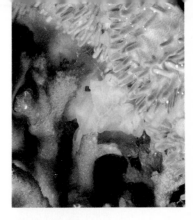

Figure 13-37 The crown-of-thorns sea star (*Acanthaster planci*) is an important and controversial coral predator.

Figure 13-38 The crown-of-thorns sea star (*Acanthaster planci*) feeding on coral (*Pocillopora*) from the Pacific coast of Panamá. Notice the stomach covering the coral.

The Crown-of-Thorns Problem. Another example of the effect of coral predators is the case of the **crown-of-thorns sea star** (*Acanthaster planci*; Figure 13-37). The crown-of-thorns feeds by pushing its stomach out through the mouth (Figure 13-38), covering all or part of the coral colony with the stomach, and digesting away the live coral tissue. The crown-of-thorns has distinct preferences for certain types of coral and avoids other types. Other corals harbor symbiotic crabs (see Figure 6-1), shrimps, and fishes that drive away the sea stars by pinching and biting their tube feet.

The crown-of-thorns can have a major influence on coral reefs. It is relatively abundant, for instance, on some reefs on the Pacific coast of Panamá. The most common coral on these reefs is a branching type called *Pocillopora*. The coral has symbionts that protect it from the sea stars, so the sea stars concentrate on other species of coral. In fact, *Pocillopora* is probably so common because most other types of corals get eaten.

In other parts of the Pacific the crown-of-thorns has had devastating effects on reefs. Beginning in the late 1950's, people began to notice large groups of the sea star, sometimes numbering in the thousands (Figure 13-39), on reefs scattered across the Pacific. The sea stars in these large aggregations move in a mass across the reef, consuming almost every coral in their path. These crown-of-thorns "plagues" have severely damaged a number of reefs; many reefs, fortunately, appear to recover rapidly.

The first response to the problem was panic. Coral reefs are important economic resources: they support fisheries, tourism, and recreation. Reefs also protect many coastlines from wave erosion. With the crown-of-thorns apparently threatening reefs, people decided to take action and control the sea star. The first attempt backfired. Having limited knowledge of the animals' biology, some people cut the sea stars into pieces and dumped them back in the sea. Unfortunately, sea stars can regenerate: the pieces grow into new sea stars! More sophisticated methods, such as injecting the sea stars with formaldehyde, were tried, but these methods were time-consuming and expensive, and did not work well. Fortunately, the outbreaks mysteriously went away by themselves. Did the sea stars starve? Did they move

tube feet Water-filled tubes, possessed only by echinoderms, many of which end in a sucker and can be extended and contracted to grip things and to move around
Chapter 6, p. 179

Figure 13-39 An outbreak of crown-of-thorns sea stars (*Acanthaster planci*) in Australia. The white branches in this colony of staghorn coral (*Acropora*) show the bare calcium carbonate skeleton left behind after feeding by several sea stars.

one knows. Crown-of-thorns outbreaks are still appearing, and disappearing, without explanation.

No one knows what causes the plagues. The subject has been the cause of considerable, sometimes quite emotional, dispute. Some are convinced that the plagues must be the result of some human activity that has altered the ecological balance of reefs. Perhaps dredging, increased runoff from land, or pollution are to blame. Some biologists speculate that the outbreaks happen because shell collectors remove triton shells (*Charonia tritonis*), large snails that eat the crown-of-thorns (Figure 13-40).

At first it seemed obvious that crown-of-thorns outbreaks were unnatural and that humans were to blame. After all, the plagues had never occurred before. Or had they? Some biologists have pointed out that people only began using scuba a short time before the sea star plagues were noticed. Even if the plagues had been occurring for a long time, there were no scientists around to see them. Even though the triton does eat crown-of-thorns sea stars, the snail has probably always been too rare to act as a control on the sea star. Furthermore, geologists have found fossil evidence of crown-of-thorns outbreaks dating back at least 8,000 years. The outbreaks, then, may

Figure 13-40 A triton shell (*Charonia tritonis*) eating a crown-of-thorns sea star (*Acanthaster planci*) on a coral reef in Australia.

be a natural part of the reef ecosystem and have nothing to do with humans. It has been suggested that they are associated with unusually wet years or other natural cycles.

The argument about the causes of crown-of-thorns plagues has practical implications for the management and protection of reefs. If the plagues are caused by humans and threaten reefs, then perhaps humans should do something to stop the outbreaks. On the other hand, the plagues may be a natural, potentially important, part of the coral reef ecosystem. We might do more harm than good by interfering in a system that we don't understand. The best answer, of course, is further study.

The crown-of-thorns sea star has undergone population explosions on many Pacific reefs. There is still debate about what causes the outbreaks and what should be done about them.

Grazing. Grazing on algae by herbivores is at least as important in coral reef ecosystems as is predation on corals. An abundance of fishes, especially surgeonfishes (*Acanthurus;* see Figure 7-11, G), parrotfishes (*Scarus, Sparisoma;* see Figure 7-11, B), and damselfishes (*Pomacentrus, Dascyllus;* see Figures 7-21 and 7-22) graze intensively on reefs. Among the invertebrates, sea urchins (*Diadema, Echinometra*) are especially important. There are also many **microherbivores,** small invertebrates like snails, crustaceans, and polychaete worms that eat algae.

Though non-coralline seaweeds grow rapidly and have the potential to overgrow corals, normally they are checked by a combination of nutrient limitation and grazing. If nutrients are added, as happened in Kaneohe Bay, the algae may grow explosively. The same thing happens if the grazers disappear. In 1983 an unknown disease wiped out populations of the long-spined black sea urchin (*Diadema antillarum*) all over the Caribbean. Within a few months, or even weeks, seaweeds became much more common on the reefs, taking over space from corals, coralline algae, and sponges.

Herbivorous fishes probably have similar effects. Seaweeds are abundant on sand flats next to many Caribbean reefs but relatively scarce on the reef itself. This could be because the seaweeds cannot grow or compete on the reef flat. To test this hypothesis, biologists transplanted seaweeds from the sand flat to the reef. If left unprotected, they were soon eaten by fishes; when protected by cages, they grew even faster than on the sand flat! The seaweeds are rare on the reef not because they are unable to live there, but because they get eaten. If the fishes were removed, the seaweeds would not only survive, but probably take over space from corals and other organisms. Though the evidence is not conclusive, some biologists fear that major changes in reef communities may be occurring as a result of over-fishing.

Grazers help prevent fast-growing seaweeds from overgrowing other sessile organisms on the reef.

In addition to controlling how many seaweeds there are, grazers affect which particular types of algae live on the reef and where. Coralline algae, for example, are abundant because the calcium carbonate in their tissues discourages grazers. Other seaweeds produce noxious chemicals that are poi-

Figure 13-41 Like damselfishes, some surgeon-fishes *(Acanthurus lineatus)* have territories. There are a number of surgeonfish territories on this reef in French Polynesia. The fish keep the large brown alga *(Turbinaria)* out of their territories, so it only grows in the space between territories.

Aristotle's lantern   A complicated set of calcium carbonate teeth and associated muscles that is found in sea urchins
Chapter 6, p. 181

Figure 13-42 The giant clam *Tridacna gigas* does not deserve its reputation as a killer: the legends of people getting trapped in them and drowning are untrue. The colors of the clam come from the zoo-xanthellae that live inside it. The zooxanthellae provide the food that allows the clam to get so big.

sonous or taste bad; these seaweeds also tend to be abundant. The tasty sea-weeds, not surprisingly, are most heavily grazed and thus tend to be rare. They also tend to grow rapidly, however, and are therefore an important food source.

Damselfishes provide some interesting examples of the effects of grazing on reefs. Many damselfishes graze on seaweeds inside territories that they vigorously defend, chasing away other fishes that happen to venture inside. Many such damselfishes actually "farm" their territories. They weed out un-palatable algae, pulling up the undesirable plants and carrying them outside the territory. What is left in the territory is a dense mat of tasty seaweeds, usually fine, filamentous types. Protected by the damselfish, these algae can grow very rapidly and out-compete corals and coralline algae. Outside the territory, parrotfishes and surgeonfishes gobble up the algae, clearing space for other organisms. Thus the community inside the territory is very differ-ent from that outside (Figure 13-41). One interesting point is that blue-green algae, which are nitrogen fixers, are much more common inside dam-selfish territories than outside. Thus damselfishes indirectly may have an im-portant role in the nutrient balance of the reef.

Herbivores obviously affect coral reef plants, but they may have indirect effects on animals as well. Many reef grazers scrape their food off the reef with some kind of specialized hard structure. Parrotfishes, for instance, get their name because their fused teeth form a parrotlike beak (see Figure 7-14, D). Sea urchins use their Aristotle's lantern to scrape off surfaces. If grazing is intense, as it often is, the surface of the reef flat may be scoured smooth; the marks made by herbivores are often visible. In the process of scraping off the seaweeds that are their food, herbivores incidentally remove settling larvae and other small animals. It may be very difficult for new coral colo-nies to establish themselves, for instance, unless the larva happens to settle in a crevice or other protected spot.

Living Together.   Among the vast number of species that live on coral reefs, many have evolved special symbiotic relationships. There are far too many cases of symbiosis on the reef to describe here. In fact, coral reefs most probably have more different symbiotic relationships than any habitat on earth. The few examples discussed here will give you some idea of how fascinating these relationships are.

Symbiotic relationships are very important in coral reef communities. Coral reefs probably have more examples of symbiosis than any other biological community.

We have already seen how mutualism between corals and their zooxan-thellae is the essential feature of reef formation. Many other organisms also have plants that live inside their tissues. Sea anemones, snails, and giant clams *(Tridacna)* all harbor zooxanthellae. The "deal" between the two part-ners is the same as in corals: The zooxanthellae get nutrients and a place to live, and the host gets food. Giant clams are able to grow so large (Figure 13-42) because their zooxanthellae provide a constant food supply.

There are other primary producers that live inside reef animals. As men-tioned previously, some sponges have blue-green algae that fix nitrogen in

BOX 13-2

"Must Have Been Something I Ate"

Ahhhhh! You are relaxing on a warm tropical night after a delicious fish dinner. The palm trees sway, a warm tropical breeze blows, and suddenly your bowels begin to churn. Before long your lips begin to burn and tingle, and your hands and feet get a sensation of pins and needles. You go into the kitchen for a drink, but the cool water feels warm. The cool floor and the cold compress you put on your forehead are also warm to the touch: The sensations of hot and cold are reversed. As you stagger to the bathroom your arms and legs are heavy and weak. Now feeling really sick, you mumble "Must have been something I ate."

You're right. What you have is a case of **ciguatera**: tropical fish poisoning. Maybe it will comfort you to know that you're not alone. Tens of thousands of people get ciguatera every year. And relax, you probably won't die, though you might be a little sick for months or even years. Unfortunately, there is no known cure, unless you care to try a local folk remedy designed to purge your system.

So what *is* ciguatera, anyway? The disease has been known for hundreds of years. The name comes from the Spanish word for a Caribbean snail—you can also get ciguatera from eating molluscs, and perhaps sea urchins, though fish dinners are the most common cause. It is probably most common in the top predatory fishes on the reef, like jacks, barracudas, and groupers, but it can also be caused by eating herbivorous fishes like parrotfishes and surgeonfishes. A particular kind of fish may be perfectly safe in most places but poisonous in particular spots. To make matters even more confusing, ciguatera may disappear from one area and pop up in another. All of which makes things difficult for lovers of fresh reef fish.

It is still uncertain, but ciguatera is thought to be caused by poisonous dinoflagellates that live on the reef. The idea is that herbivorous fish eat the dinoflagellates, and the poison is passed on to the predatory fishes that eat them. As blooms of the dinoflagellates arise and die out, ciguatera comes and goes in the area. Predatory fishes range over larger areas and eat many fish, who may all contain small amounts of the toxin. Thus they are most likely to cause ciguatera. Herbivorous fishes probably carry the disease only when they feed in an area with a bloom.

The big problem is trying to tell when a fish carries the poison. You could just avoid reef fish altogether, but if you are in the tropics that means missing out on a lot of tasty meals. There are all sorts of supposed tests for ciguatera, but most are unreliable. The best-known test is to feed a bit of the fish to your cat—but don't get too attached to your pet. Maybe the neighbor's cat is a better idea. If any mongooses are handy, they're good tasters too. Otherwise, you have to go hungry or take your chances. Ciguatera is a risk, but the authors have been eating reef fishes for years without any problems. Of course, you never know about that next barracuda fillet . . . .

addition to performing photosynthesis. Certain sea squirts house a primitive alga called *Prochloron.* Actually, *Prochloron* is more like a bacterium than a plant. It is especially interesting because some scientists think it is similar to the symbiotic organisms that eventually became the chloroplasts of plants (see Box 9-1).

Another important example of mutualism on the reef is the relationship between corals and the crabs, shrimps, and fishes that protects them from the crown-of-thorns seastar and other predators. Most corals host a number of symbionts, especially crustaceans and snails. Some of these are only "casual" symbionts and can live as well off the coral as on it. Others are much more specialized and are found only on their host coral. These are called **obligate symbionts.** Among the obligate symbionts, some are parasites and harm the coral, some are commensals and do not affect the coral one way or the other, and some are mutualists that benefit the coral. It is hard to tell which is which, because for most symbionts the nature of the relationship between coral and symbiont is not well understood. Those who study such organisms must frequently revise their ideas as more information is obtained. The crabs that protect their coral from predators, for instance, were once thought to be parasites.

Clownfishes (*Amphiprion*), so named because of their spectacular bright colors (Figure 13-43), have an interesting mutualistic relationship with several kinds of sea anemone that inhabit coral reefs. The anemones inhabited by clownfishes have a powerful sting and are capable of killing the clownfish. To understand why the clownfish isn't stung, you have to remember that anemones have no eyes and no brain, but lots of tentacles. If the anemone simply stung everything it touched, it would end up stinging itself every time the tentacles bumped into each other. The anemone recognizes itself by the "taste" of its own mucus. When one tentacle touches another, the anemone detects the mucus coating on the tentacles and refrains from stinging. Clownfishes take advantage of this system. They spend a good part of their time rubbing against and nipping at the anemone's tentacles to coat themselves with the anemone's mucus. When the anemone touches the clownfish, it "tastes" its own mucus and refrains from stinging. If you clean the mucus off a clownfish, the anemone will sting and devour it quite happily.

The relationship between clownfishes and anemones is mutually beneficial. The fish gets a nice safe place to live; predatory fishes who stray too close get stung by the anemone. In return, the anemone gets food. The clownfish may act as lures, attracting prey for the anemone. If the fish have had their fill, in fact, they will actually carry food to the anemone and feed it!

Figure 13-43 A clownfish (*Amphiprion*) and its sea anemone host.

Do It Yourself

SUMMARY

1. The most important reef architects are the _____ corals, which deposit skeletons made of _____. Most corals are colonies of many individual _____.
2. Reef-building corals contain symbiotic _____ that provide _____ and enable the coral to rapidly build its _____.
3. _____ algae help reef growth by depositing _____, by resisting erosion by _____, and by cementing together coarse _____.
4. Coral reefs need _____ and _____, clear water, so they are only found in the _____. Corals are very sensitive to _____, fresh water, and _____ of many kinds.
5. The main types of reefs are _____, _____, and _____.
6. When a _____ island slowly sinks, leaving only living, growing reef near the surface, an _____ is formed.
7. Though the surrounding water is poor in _____, coral reefs have high _____ production. This is because in reef communities nutrients are _____, nitrogen is _____, and the few nutrients that are brought in are used efficiently.
8. Seaweeds are important primary _____ on coral reefs, even though they are kept in check by _____ limitation and grazing.
9. Sessile reef organisms must often compete for _____. Among the most important competitors are corals, _____, _____, and _____.
10. The _____ sea star has undergone population explosions on many reefs in the _____. This may be due to natural or human factors.
11. Among the most important grazers on reefs are _____, _____, and microherbivores. Grazers are important because they prevent _____ from dominating slow-growing organisms.

THOUGHT QUESTIONS

1. What factors might account for the fact that the vast majority of atolls occur in the Indian and Pacific oceans and that atolls are rare in the Atlantic?
2. Scientists predict that the ocean will get warmer and the sea level will rise as a result of the greenhouse effect described in Box 17-1. How might this affect coral reefs?
3. There are only a few reefs off the northeast coast of Brazil (see map in Figure 13-13), even though it lies in the tropics. How would you explain this?

FOR FURTHER READING

Brower, K.: "State of the Reef." *Audubon*, March 1989, pp. 56-80. *A report on the status of coral reefs around the world.*

Bunkley-Williams, L. and E.H. WIlliams Jr.: "Global Assault on Coral Reefs." *Natural History*, April 1990, pp. 46-54. *There is growing concern about the bleaching of corals in many parts of the world.*

Goreau, T.F., N.I. Goreau, and T.J. Goreau: "Corals and Coral Reefs." *Scientific American*, vol. 241, no. 2, August 1979, pp. 124-136. *A classical account of several aspects of the biology of coral reefs.*

Levine, J.S.: "Coral Reef Fishes Use Riotous Colors to Communicate." *Smithsonian*, vol. 21, no. 8, November 1990, pp. 98-103. *The striking colors of coral reef fishes are not only beautiful, they act as signals.*

Life Near the Surface 14

"The ocean," for most of us, conjures up images of beaches and cliffs, breaking surf, or quiet bays. Such familiar inshore waters, however, make up only a small fraction of the World Ocean. The rest is the vast open sea, the **pelagic realm.** Though distant and unfamiliar, the open ocean affects us all. It regulates our climate, conditions our atmosphere, and provides food and many other resources. It is hard for most of us to really grasp how vast the pelagic is.

As we learned in Chapter 9, the pelagic environment is the water column itself, away from the bottom or the shore. Pelagic organisms live suspended in their liquid medium. With very few exceptions, the pelagic lacks the solid physical structure provided by either the bottom and other geologic features or by large organisms, like corals and kelps. There is no place for attachment, no bottom for burrowing, nothing to hide behind. Imagine what it would be like to spend your life floating weightless in the air, never touching the ground (Figure 14-1). This is what the pelagic realm is like for the organisms that live there. Pelagic organisms face very different problems from those which live near shore or on the bottom.

This chapter deals with the surface layers of the pelagic environment, the **epipelagic,** or upper pelagic realm. The epipelagic is often defined as the zone from the surface down to a given depth, commonly 200 m (650 feet). Being the shallowest part of the pelagic realm, the epipelagic is generally the warmest, and of course the best lighted. The epipelagic is therefore related to the **photic zone,** the zone from the surface to the depth where there is no longer enough light for plants to grow by photosynthesis. The depth of the photic zone varies, depending on water clarity and the amount of sunlight. In practice the epipelagic and photic zones are usually similar, and their differences will not be stressed in this chapter.

The epipelagic is divided into two main components. Epipelagic waters that lie over the continental shelf are referred to as **coastal,** or **neritic** (see Figure 9-22). The coastal environment is only a small part of the epipelagic, but it is important to humans because it lies relatively close to shore and is very productive. The surface waters beyond the continental shelf are known as the **oceanic** part of the epipelagic.

Figure 14-1 Jellyfish, inhabitants of the pelagic realm.

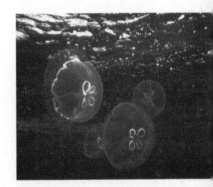

> The epipelagic realm is the layer of the ocean from the surface to a depth of 200 m. It is divided into coastal, or neritic, waters, which lie over the continental shelf, and oceanic waters, which lie beyond the shelf. The epipelagic is similar to the photic zone, the layer from the surface to the depth where light limits photosynthesis.

THE ORGANISMS OF THE EPIPELAGIC

primary production   The conversion of carbon dioxide into organic matter by autotrophs; that is, the production of food
Chapter 4, p. 92

Like nearly all ecosystems, the pelagic realm is fueled by energy from the sun that is captured in photosynthesis. Epipelagic ecosystems differ from shallow-water ones in that nearly all the **primary production** takes place within the epipelagic system itself. Coastal ecosystems often receive large amounts of food from elsewhere. The intertidal zone, for example, gets drifting seaweeds from offshore, and rivers carry organic material into estuaries. The pelagic realm, far from the shore and bottom, gets almost no outside input of organic matter.

On the other hand, the epipelagic supplies food to other communities. Large amounts of organic matter sink out of the epipelagic to feed the organisms that live below (see Chapter 15). Ocean currents carry epipelagic **plankton** into shallow water, where it is consumed by a profusion of filter feeders. Epipelagic fishes are also food for many seabirds, humans, and a few other land dwellers.

plankton   Small plants (*phytoplankton*) and animals (*zooplankton*) that drift with the currents
Chapter 9, p. 284; Figure 9-21

suspension feeders   Animals, including *filter feeders,* that eat particles suspended in the water column
Chapter 6, p. 147; Figure 6-20

deposit feeders   Animals that eat organic matter that settles to the bottom
Chapter 6, p. 158; Figure 6-20

Because there is no bottom where organic-rich sediments can build up, the epipelagic lacks deposit feeders. Suspension feeders, on the other hand, abound: After all, the food is suspended in the water column. There are also many large predators, like fishes, squids, and marine mammals.

The Phytoplankton

Because they have no place to attach, large plants like seaweeds and seagrasses are absent from most of the epipelagic. In a few places floating seaweeds are important, but by and large the only plants are single cells or simple chains of cells. These tiny algae and algae-like cells make up the **phytoplankton,** which occur over vast areas of the ocean, often in huge numbers. Phytoplankton perform nearly all the photosynthesis in the open ocean. They account for half of the world's primary production, producing half of the oxygen in our atmosphere in the process.

The standard way to collect plankton is to tow a net behind a boat (Figure 14-2). Diatoms and dinoflagellates are easily caught in such nets, so they are known as **net plankton.** Because they are relatively easy to catch,

Figure 14-2   A typical phytoplankton net **(A)** is relatively small and has very fine netting that prevents many microscopic plankton from passing through. This net **(B)** is called a bongo net because of the two rings that hold the mouth of the net open. It is used to sample zooplankton. Zooplankton nets are usually larger and have coarser mesh than do phytoplankton nets.

A

B

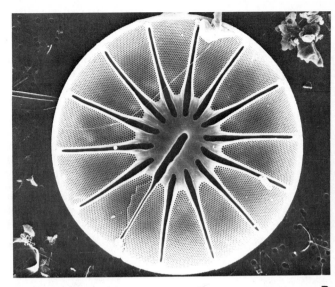

A

B

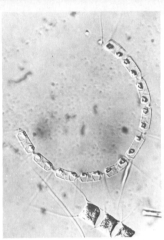

C

Figure 14-3 Diatoms have sculptured shells, or frustules, made of silica. *Thallassiosira allenii* (top and side view, **A**) and *Asteromphalus elegans* (**B**) consist of a single cell; the species of *Chaetoceros* (**C**) form chains. Also see Figures 4-11 and 5-5.

their importance has long been known. Biologists have learned more recently that a part of the phytoplankton known as the **nanoplankton** is a major primary producer in the epipelagic. The nanoplankton are defined as plankton that are too small to catch in standard nets.

Diatoms (Figure 14-3) are dominant members of the net phytoplankton. They are especially common in temperate and polar regions and other nutrient-rich waters. They are abundant both near the coast and in the open ocean.

Dinoflagellates (see Figure 5-7) are another major group of phytoplankton. Like diatoms, they are important in both coastal and oceanic waters. They are found throughout the oceans but tend to prefer warm areas. In the tropics they may replace diatoms as the most abundant members of the phytoplankton. Dinoflagellates may be better able to grow under low-nutrient conditions than diatoms. Given nutrients, on the other hand, dinoflagellates may **bloom**—that is, grow explosively into huge numbers—sometimes causing **red tides** (Box 14-1).

dinoflagellates Single-celled organisms with two flagella used to swim
Chapter 5, p. 124

BOX 14-1

Red Tides

At certain times, usually near the coast, the ocean loses its normal blue-green color and, practically overnight, becomes orange- or red-brown. This phenomenon, known as a **red tide** has been known for thousands of years. Indeed, the Old Testament may contain the earliest known reference to a red tide when it describes the waters of the Nile turning to blood (*Exodus* 7:20-21). Since then red tides have been reported many times and in many parts of the world.

Red tides have nothing to do with the tide. Instead, they are caused by massive blooms of phytoplankton in which the cells become so dense that they color the water. At the peak of a red tide there may be thousands of cells in a single drop! Nearly all red tides are caused by blooms of photosynthetic dinoflagellates, around 60 species of which are known to cause red tides. One organism that causes red tides in tropical areas and is not a dinoflagellate is *Trichodesmium,* a blue-green alga. *Trichodesmium* is not really blue-green, but reddish brown. The Red Sea got its name from the blooms of *Trichodesmium* that often occur there.

Red tides can have much more severe effects than just discoloring the water. A few species of red-tide dinoflagellates produce poisons that are among the most powerful natural toxins known. Normally there are too few of the dinoflagellates to cause much trouble, but during a red tide the dinoflagellates may poison the water, turning the surface into a sea of floating dead fish. Fish kills can also happen if the dinoflagellates suddenly die. As they decompose, the oxygen in the water is depleted and the fish suffocate.

Red tides can threaten human health as well. Several species of mussels, clams, and other bivalves tolerate the toxin by storing it away in the digestive gland and other tissues. People who eat the shellfish, however, can get **paralytic shellfish poisoning.** Victims suffer numbness and tingling, loss of balance and coordination, slurred speech, and, in severe cases, paralysis and death by asphyxiation.

Paralytic shellfish poisoning has been known on the west coast of North America for a long time. In California it is illegal to harvest some species of shellfish during the summer months, when red tides are likely to occur. Red tides were relatively uncommon on the east coast until the early 1970's, when they suddenly broke out in eastern Canada and New England. This forced the closure of valuable shellfish beds. Red tides also broke out in Florida and the Gulf of Mexico during the 1970's. Massive fish kills hurt the tourist industry when the fish rotted and began to smell. The toxins of *Ptychodiscus brevis,* the dinoflagellate causing the Florida red tides, remain active in seawater for a long time. When sea spray blew the poison ashore, many people suffered sore throats and irritation of the eyes and skin.

No one really knows what triggers the explosive dinoflagellate growth that causes red tides, but they are clearly different from ordinary phytoplankton blooms. Most phytoplankton blooms, like the spring blooms that occur in cold waters, include many different species, all growing in response to some improvement in conditions, such as an increase in nutrient levels. Red tides, on the other hand, invariably result from the rapid growth of only a single species. Furthermore, red tides are not obviously related to increased nutrients or other environmental factors. In some cases, red tides may be related to mass germination of cysts, which are the resting stages of the dinoflagellates. Red tides are still poorly understood and therefore unpredictable. They are just one more reminder that we have much to learn about the sea.

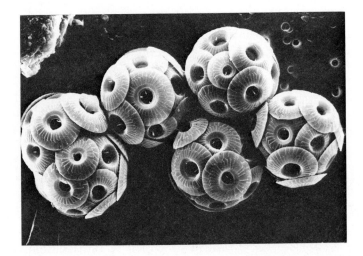

Figure 14-4 Coccolitho- phorids, like *Umbilicos- phaera sibogae* from Austra- lia, are tiny single-celled phytoplankton that can be very important primary producers in the epipe- lagic. The plates that cover the cell are made of calcium carbonate.

Because they are so small and hard to catch, nanoplankton are generally not as well known as the diatoms and dinoflagellates. They are extremely important, however, and contribute a major share of the epipelagic's photo- synthesis, up to 90% in some places. They are especially important in trop- ical areas.

Extremely small **blue-green algae** (or cyanobacteria) are major primary producers. They can account for up to 80% of the total production. **Cocco- lithophorids** (Figure 14-4) are common members of the nanoplankton and probably the best known. Though found in coastal waters, they do best in the open ocean. They sometimes far outnumber all other types of phyto- plankton.

In coastal zones another group of minute phytoplankton known as **cryp- tomonads** may be very plentiful. They could be very important in the econ- omy of the seas, especially in coastal waters, but not much is known about them. **Silicoflagellates** (see Figure 5-8) occasionally bloom and become im- portant primary producers.

> **Phytoplankton are the main primary producers in the epipelagic. The most important groups of phytoplankton are diatoms, dinoflagellates, and several types of very small plankton, or nanoplankton.**

A few other groups may form blooms, but only rarely, and such groups are not very important in the overall economy of the seas. They are not included in Table 14-1, which shows the major groups of marine phyto- plankton.

The Zooplankton

Phytoplankton form the base of the food web. The solar energy that they capture and store in organic matter is passed on to the other creatures of the epipelagic, from minute zooplankton to the gigantic whales. The first step in the flow of energy through the food web occurs when **herbivores** eat the

TABLE 14-1

The major groups of marine phytoplankton

SIZE CATEGORY	GROUP	COASTAL OR OCEANIC	LATITUDE/ TEMPERATURE	NOTES
Net plankton	Diatoms	Both	Everywhere, but most common in temperate and polar waters	Extremely important primary producers
	Dinoflagellates	Both	Everywhere, but most common in warm waters	Common red-tide organisms
	Blue-green algae (*Trichodesmium*)	Oceanic	Mainly tropical	Can fix atmospheric nitrogen; causes red tides in the Red Sea
Nanoplankton	Small blue-green algae (*Synechococcus*)	Both	Temperate and tropical	Extremely important primary producers
	Coccolithophorids	Oceanic	Everywhere, but most common in warm waters	Occasionally forms blooms
	Cryptomonads	Coastal	Everywhere	Poorly known, but perhaps very important
	Silicoflagellates	Coastal	Temperate and polar waters	Occasionally forms blooms

phytoplankton. Herbivores are the vital link between the primary producers, the phytoplankton, and the rest of the community. Most large epipelagic animals cannot feed directly on the tiny phytoplankton and rely instead on those herbivores who can. Of all the herbivores in the epipelagic, **zooplankton** are by far the most important. Thus a fundamental part of the epipelagic food web is the flow of energy from phytoplankton to herbivorous zooplankton. Surprisingly, relatively few types of zooplankton graze on phytoplankton.

A few types of zooplankton are the main grazers of phytoplankton in the epipelagic.

Very few zooplankton are strict vegetarians. The ones that do eat phytoplankton also eat other zooplankton occasionally. Most zooplankton species are primarily carnivorous and hardly eat phytoplankton at all. These carnivores may feed directly on herbivorous zooplankton and thus reap the energy produced by phytoplankton with only one intermediate step, or **trophic level,** in the food web. They may also eat other carnivores, adding links to the food web (Figure 14-5).

Copepods. Small crustaceans, especially **copepods** (Figure 14-6), dominate the zooplankton. Copepods are the most abundant members of the zooplankton practically everywhere in the ocean, typically numbering 70% or more of the community. This probably makes them the most abundant group of animals on earth.

copepods Tiny crustaceans that are often planktonic
Chapter 6, p. 172

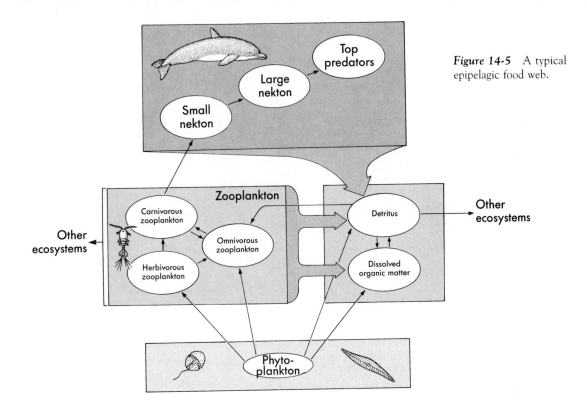

Figure 14-5 A typical epipelagic food web.

Nearly all epipelagic copepods eat at least some phytoplankton. It was once thought that they filter feed by continuously pumping a stream of water past their mouthparts. According to this view, water flows through a "sieve" formed by the bristles on the mouthparts and antennae (Figure 14-7). The copepod blindly catches whatever particles, phytoplankton or otherwise, happen to be the right size to snag in the bristles.

Feeding experiments and close-up photography, however, have shown that copepods can actively select the particles they capture. At least some copepods can sense individual phytoplankton cells, using both "smell" and sight. Then they use their limbs like paddles to draw in a stream of water

Figure 14-6 Some epipelagic copepods. Also see Figure 6-35.

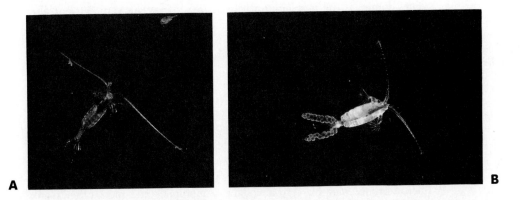

A   B

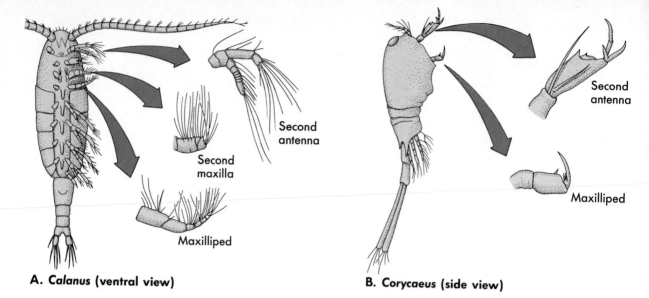

A. *Calanus* (ventral view)

B. *Corycaeus* (side view)

Second antenna

Second maxilla

Maxilliped

Second antenna

Maxilliped

Figure 14-7 The mouthparts and appendages of copepods that feed mainly on phytoplankton (**A**) have many long bristles. Copepods that prey on other zooplankton (**B**) have appendages with fewer and shorter bristles. Their appendages are better adapted for grasping.

that carries the cell closer. The cell is caught in the bristles like a butterfly in a net, and the phytoplankton cell is eaten.

Copepods are also major carnivores. Though most copepods eat at least some phytoplankton, many also eat other zooplankton, including other copepods, when they can. A few copepods are exclusively carnivorous. These generally catch their prey by seizing it with their clawlike appendages (Figure 14-7, *B*).

Other Crustaceans. Other important members of the zooplankton are crustaceans. One major group is the shrimplike **krill.** Though not as abundant around the world as copepods, krill may aggregate into huge, dense swarms (see Figure 16-16, *A*). Preferring cold oceanic waters, krill sometimes dominate the zooplankton in polar seas. They are very efficient filter feeders, capturing particles with their bristly appendages. Phytoplankton, especially diatoms, are a favorite food. Krill also eat detritus, especially **fecal pellets,** the solid waste excreted by other zooplankton. Small zooplankton may also be eaten.

Copepods are so small that most large animals cannot catch them, whereas krill are relatively big, up to 6 cm (2.5 inches) in the case of the Antarctic krill (*Euphausia superba*). They serve as food for fishes, seabirds, and even the great whales (see Figure 9-13). Krill are harvested by humans also (see the section in Chapter 16, "New Fisheries").

There are many other crustaceans—both herbivorous and carnivorous—in the zooplankton. At certain times and places, one of these crustacean groups may be very abundant. Amphipods and most other plank-

krill Planktonic, shrimp-like crustaceans
Chapter 6, p. 174

detritus Dead organic matter and the microscopic decomposers that live on it
Chapter 9, p. 277

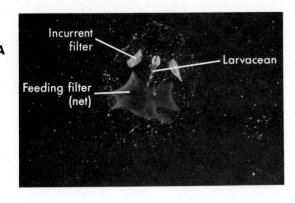

A

Incurrent filter

Larvacean

Feeding filter (net)

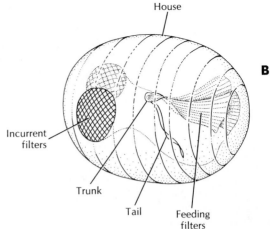

B

House

Incurrent filters

Trunk

Tail

Feeding filters

tonic crustaceans are small, like copepods. A few larger crustaceans, some similar to krill, are also found in the zooplankton. These include a number of planktonic species of crabs, shrimps, and their relatives—the **decapods** (see the section in Chapter 6, "Shrimps, Lobsters, and Crabs"). The decapods are almost exclusively carnivorous.

> **Copepods, krill, and other crustaceans dominate the zooplankton. Copepods are the main herbivores in the epipelagic and are by far the most abundant group of zooplankton.**

Non-Crustacean Zooplankton. Many groups of animals other than crustaceans are found in the zooplankton. Most of these are carnivorous. Among the most important of the non-crustacean herbivores are the transparent, planktonic **salps** (see Figure 6-55), relatives of the sea squirts, or tunicates, that live on the bottom. Salps filter out phytoplankton by pumping water through a sievelike sac or a fine mucous net.

Larvaceans (Figure 14-8) are also relatives of sea squirts, though they hardly look it. Unless, that is, you look at sea squirt larvae, which are also known as tadpole larvae because of their resemblance to tadpoles (see Figure 6-54, *B*). Larvaceans float inside a "house" they make of mucus. By beating its tail, the larvacean pumps water in through passages in the house (see Figure 14-8). Food particles are caught in a complicated mucous net that is secreted inside the house. This feeding mechanism enables larvaceans to capture extremely small food particles. Among the few animals that graze on nanoplankton, larvaceans are an important link between the nanoplankton and the rest of the epipelagic community. The same is true of some salps.

The openings through which water enters the house filter out particles too large for the larvacean to eat. As the larvacean feeds, these filters eventually get clogged. Some species can reverse the water flow, clearing the filters. If the filters get too clogged or a predator threatens, the larvacean simply abandons the house and swims away. It can produce a new house within minutes and resume feeding. When disturbed, some species can build and discard houses every 10 minutes! Even under normal circumstances, most species change houses every 4 hours or so. When larvaceans are abundant, their discarded houses can be an important source of detritus.

Figure 14-8 A larvacean spends most of its time inside a mucous "house." By beating its tail the larvacean draws water into the house through two intake, or incurrent, filters. Food particles are caught on the mucous feeding net; the filtered water flows outward. The photograph **(A)** shows *Stegosoma magnum* and its "house," which is about 1 cm (0.4 inch) long.

A

Figure 14-9 (A) Pteropods typically have a reduced shell and a foot that is expanded into winglike extensions. **(B)** This pteropod (*Gleba cordata*) features a proboscis that is greatly elongated when feeding. The mouth, which lies at the end of the proboscis, ingests food that is captured by mucous webs that reach 2 m (6.5 feet) in diameter!

A group of molluscs called **pteropods** are also phytoplankton grazers. Pteropods are small snails in which the foot has been modified to form a pair of "wings" that they flap to stay afloat (Figure 14-9, A). Some feed by capturing phytoplankton, including nanoplankton, in mucous nets or threads; others are carnivorous (Figure 14-9, B).

Larvaceans, salps, and some pteropods feed with mucous nets or threads. They are among the few zooplankton that eat nanoplankton. Discarded larvacean houses can be an important source of detritus in the epipelagic.

arrow worms or **chaetognaths** Small, wormlike predators with fins at the tail and spines on the head *Chapter 6, p. 163*

siphonophores Drifting colonial cnidarians in which different members of the colony are specialized for different tasks *Chapter 6, p. 151*

comb jellies or **ctenophores** Radially symmetric animals that resemble jellyfish but have eight rows of cilia *Chapter 6, p. 154*

Arrow worms, or **chaetognaths,** are extremely important predators in the zooplankton. They feed mostly on copepods. This may not be because they actually prefer copepods, but because there are so many copepods around. Arrow worms consume an assortment of other prey when it is available. They can be very abundant, and they have a significant role in epipelagic food webs.

Not all carnivorous zooplankton are tiny. Jellyfishes and **siphonophores,** for example, can be quite large but are weak swimmers and drift with the currents as part of the plankton. Jellyfishes (see Figure 14-1) are carnivorous, and many eat small fishes as well as zooplankton. The somewhat similar comb jellies (see Figure 14-23, B) are also carnivorous. The largest member of the plankton is the ocean sunfish (*Mola mola*; Figure 14-10).

Meroplankton. The zooplankton discussed to this point, with the exception of a few jellyfishes, spend their entire lives in the plankton and are called **holoplankton.** In addition to these permanent members of the zooplankton, there are a vast number of organisms that have planktonic larvae. These animals—whether from estuaries, the rocky intertidal, kelp beds, coral reefs, and even from the deep sea—release their eggs or young into

Figure 14-10 The ocean sunfish's (*Mola mola*) watery flesh gives it buoyancy, but this fish swims so weakly that it can be considered part of the plankton.

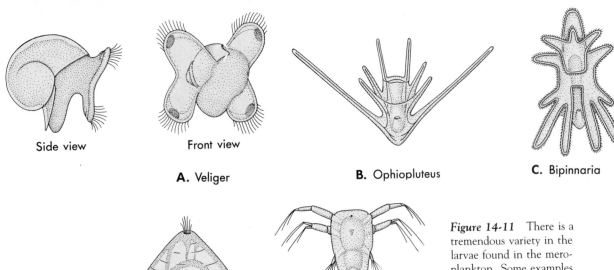

Side view   Front view

A. Veliger   **B.** Ophiopluteus   **C.** Bipinnaria

D. Trochophore   **E.** Nauplius

Figure 14-11 There is a tremendous variety in the larvae found in the meroplankton. Some examples are the veliger of molluscs (**A**), the ophiopluteus of brittle stars (**B**), the bipinnaria of sea stars (**C**), the trochophore of polychaete worms and some molluscs (**D**), and the nauplius of many crustaceans (**E**).

the water column, and the young spend the early part of their lives in the plankton. Such temporary members of the plankton are called **meroplankton.** Coastal waters are particularly rich in meroplankton.

Invertebrate animals often have a particular type of larva that is characteristic of the group (Figure 14-11). Some invertebrates have a whole series of different larval stages (see Figure 6-45). Nearly all marine fishes also have planktonic larvae (Figure 14-12).

> **The meroplankton are the larval stages of invertebrates and fishes that spend only part of their lives in the zooplankton.**

Small larvae tend to feed predominantly on phytoplankton; larger ones eat zooplankton. If there is a series of stages, as in crustaceans, the larvae may feed on phytoplankton initially and later switch to zooplankton. Fish larvae, too, often change from herbivores to carnivores as they grow.

Figure 14-12 The leptocephalus larva is the characteristic larva of freshwater eels (*Anguilla*; see Figure 7-26), as well as of other types of marine fishes.

The Nekton

Plankton are by far the most abundant organisms in the sea, and they form the foundation of the food chain in the epipelagic. Most of us, however, are much more familiar with the **nekton,** the large, strong swimmers. Fishes, marine mammals, and squids are the most abundant nekton. Turtles, sea snakes, and penguins are also included.

Practically all nekton are carnivorous. **Planktivorous** nekton, those that eat plankton, include small fishes like herrings, sardines, and anchovies, but also include the world's largest fishes, the whale shark (*Rhincodon typus*) and the basking sharks (*Cetorhinus maximus*)! The largest nekton of all, the ba-

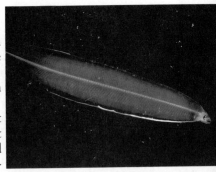

leen whales, eat plankton, mostly krill. Seals, penguins, squids, and an assortment of fishes including salmon, tuna, and flying fishes also eat krill.

Fishes, such as lanternfishes, that stay in deeper water during the day and swim up into the epipelagic at night (see the section in Chapter 15, "Vertical Migration and the Deep Scattering Layer") consume large amounts of zooplankton. These fishes are also important foods for some large epipelagic nekton, such as tuna and dolphins, that dive into deeper waters to feed.

Most nekton eat other nekton rather than plankton. Fishes, squids, and large crustaceans are the main foods. Epipelagic predators are usually not fussy and will eat many different prey, provided they are the right size. In general, the larger the predator, the larger the prey; however, there are exceptions like baleen whales and whale sharks. Thus small fishes like herrings have small prey: zooplankton. Bigger fishes eat these small fishes, and in turn are eaten by even bigger predators. At the top of the food chain are the largest predators of all—the **top predators,** or **top carnivores**—that eat the largest prey. The sperm whale (*Physeter catodon*), the largest of all nekton except the baleen whales, eats giant squid more than 10 m (33 feet) long. Other top predators also have massive prey. Among predators, killer whales (*Orcinus orca*) are second in size only to the sperm whale. They not only eat porpoises and seals, but may attack and kill baleen whales! Large sharks such as the great white (*Carcharodon carcharias*) and mako (*Isurus*) eat seals, billfishes (such as marlins and sailfish), and even other sharks.

Nearly all the nekton are predators. A few eat plankton, but most eat other—usually smaller—nekton.

LIVING IN THE EPIPELAGIC

Every environment places special demands on the organisms that live there, and the epipelagic is no exception. Compared with other marine environments, however, the physical features of the epipelagic are not very stressful and therefore do not require many special adaptations. The adaptations of epipelagic organisms largely center around two main needs: the need to stay in the epipelagic and the need to eat and avoid being eaten.

Staying Afloat

To live in the epipelagic, organisms must *stay* in the epipelagic. This simple fact presents all epipelagic organisms with a fundamental problem: Cells and tissue are denser than water. Shells and skeletons are even more dense. As a result, organisms sink unless they have some adaptation that prevents it. This is not much of a problem for organisms that live on the bottom or in deep water, since they don't need to float anyway. On the other hand, organisms that live near the surface must keep from sinking to stay in their habitat.

Phytoplankton have to stay in relatively shallow water to get enough light for photosynthesis. Once an algal cell sinks out of the photic zone it is

STRUCTURE AND FUNCTION OF MARINE ECOSYSTEMS

doomed, unless it can somehow get back into the sunlit layer. Animals, too, whether zooplankton or nekton, must be able to stay near the surface. This is not because these animals need sunlight but because their food is in the shallow water.

There are two basic ways that organisms can keep themselves from sinking out of the epipelagic. One of these is to increase their water resistance so that they sink slower. The other is to make themselves more buoyant so that they don't tend to sink in the first place. Buoyancy benefits not only drifters, but also organisms that swim, since it allows them to stay at a given depth without working as hard. This applies not only to the nekton, who by definition are strong swimmers, but to zooplankton and even some phytoplankton, like dinoflagellates, that swim up and down in the water column even though they cannot swim strongly enough to move against the currents.

Organisms must avoid sinking to stay in the epipelagic. This is achieved by increasing water resistance and increasing buoyancy.

Increased Resistance. How fast an organism of a given weight sinks depends on how much water resistance, or **drag**, it encounters. The water resistance depends mainly on the surface area of the organism: The higher the surface area, the higher the resistance and the slower the organism sinks. This is probably one reason that plankton are usually small: Small organisms have much more surface area for a given volume than large ones (see the section in Chapter 4, "Surface-to-Volume Ratio") and therefore relatively more drag. Other things being equal, small organisms sink slower than large ones.

An organism's shape can also increase its surface area, and therefore make it sink slower. Clearly, the parachute-like shape of many jellyfishes helps slow sinking (see Figure 14-1). Many planktonic organisms have extremely flat shapes (Figure 14-13). You can see how this helps slow sinking

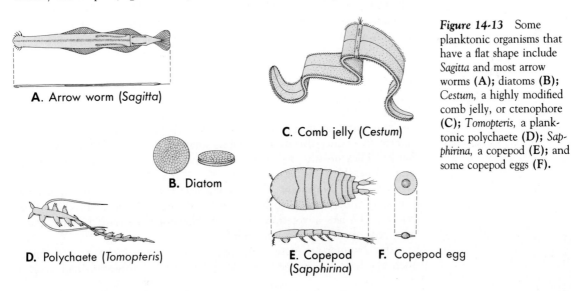

A. Arrow worm (*Sagitta*)

B. Diatom

C. Comb jelly (*Cestum*)

D. Polychaete (*Tomopteris*)

E. Copepod (*Sapphirina*)

F. Copepod egg

Figure 14-13 Some planktonic organisms that have a flat shape include *Sagitta* and most arrow worms (**A**); diatoms (**B**); *Cestum*, a highly modified comb jelly, or ctenophore (**C**); *Tomopteris*, a planktonic polychaete (**D**); *Sapphirina*, a copepod (**E**); and some copepod eggs (**F**).

Figure 14-14 Some planktonic organisms have long spines or projections. Others also form chains. Examples are diatoms **(A)**, copepods like *Augaptilus* **(B)**, zoea larvae of porcellanid crabs **(C)**, fish larvae (*Lophius*) **(D)**, and zoea larvae of *Sergestes*, a shrimp **(E)**.

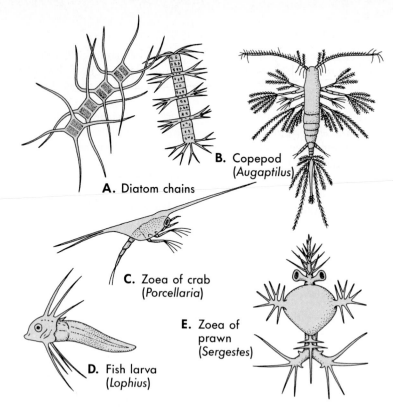

A. Diatom chains

B. Copepod (*Augaptilus*)

C. Zoea of crab (*Porcellaria*)

D. Fish larva (*Lophius*)

E. Zoea of prawn (*Sergestes*)

with a simple experiment. Take two sheets of paper, both of the same weight. Crumple one sheet into a tight ball. Drop both the flat sheet and the ball from the same height. The flat sheet will take much longer to hit the ground because it offers much more wind resistance. Exactly the same principle applies in the water. A flat shape has another advantage for phytoplankton. Note how the flat sheet of paper zigzagged from side to side as it fell. Flat phytoplankton cells do the same thing. The back-and-forth motion keeps seawater flowing over the cell's surface, renewing the layer of water with which the cell exchanges gases, nutrients, and waste products.

Long projections or spines are another common adaptation that increases surface area in planktonic organisms (Figure 14-14). Spines have the added advantage of making the organism harder to eat. Many phytoplankton form chains, another adaptation that helps slow sinking. Chains sink slower than the individual cells would if the chain were broken up.

Swimming organisms rarely have spines or other features that increase surface area, since this would increase water resistance and make swimming harder. They usually have adaptations that *reduce* drag, making it easier to move through the water.

Increased Buoyancy. The second way that epipelagic organisms can stay near the surface is to have special adaptations that make them more buoyant. Unlike adaptations that help the organism resist sinking, buoyancy reduces the tendency to sink.

One common adaptation that provides extra buoyancy is to store **lipids**, such as oils or fats, in the body. Being less dense than water, lipids tend to float and therefore provide buoyancy. Many plankton—notably diatoms, copepods, and fish eggs and larvae—contain droplets of oil. Many adult epipelagic fishes also gain buoyancy by storing large amounts of lipid. This is especially true in sharks, tunas and their relatives, and other species in which the swim bladder is poorly developed or missing altogether. Epipelagic sharks, for example, have enlarged livers with a very high oil content. Whales, seals, and other marine mammals also have a great deal of buoyant fat in a thick layer of blubber under the skin (see Figure 4-4).

Oils and fats are a good way to increase buoyancy, since they serve other functions at the same time. Lipids are the most efficient way to store excess food, for instance. In marine mammals the blubber provides insulation from the cold ocean water, as well as buoyancy. Insulation is especially important to warm-blooded animals, since they must burn up energy to maintain their body temperatures (see the section in Chapter 4, "Temperature").

Pockets of gas are another adaptation that provides buoyancy. Blue-green algae, for example, have tiny gas bubbles, or vacuoles, inside their cells. An advantage of this system is that the cell can change the size of the bubble and therefore how much floatation is provided. This allows the cell to regulate its depth: It can increase buoyancy to move up in the water column or decrease it to sink. Some other plankton, especially large ones, also have special gas-filled floats (Figure 14-15). Such floats are especially common in organisms that live right at the sea surface, as is described in the next section. Similarly, most epipelagic bony fishes have internal swim bladders that give buoyancy.

A gas-filled bladder or float provides much more buoyancy than, for example, lipid, but it has a major disadvantage. Gas expands and contracts as the animal moves up and down in the water column and the pressure changes. When the volume of gas changes, so does the amount of buoyancy. To control its buoyancy, a fish—like the blue-green algae just mentioned—must be able to regulate the amount of gas in the swim bladder. Most fishes can do this, but the actual mechanism varies. Some fishes simply pump gas in and out of the bladder through a special duct. These fishes can adjust fairly rapidly to changes in pressure. Other fishes regulate the amount of gas in their swim bladders much more slowly and cannot cope well with changes in depth. Most fishermen have pulled up fish with bulging eyes or with the stomach protruding out of the mouth (see Figure 3-18). This happens when the fish is brought to the surface; the swim bladder blows up like a balloon as the gas inside the bladder expands. In normal life the fish would never come to the surface, or it would surface much more slowly than it did on the end of the line.

The swim bladder is often poorly developed or absent in active fishes that frequently change depth, such as sharks and tunas. Such fishes must compensate for the loss of buoyancy. In addition to their large oily livers, sharks have large, stiff fins and asymmetric tails that provide lift as long as the shark is swimming (see Figure 7-13). Members of the tuna family (*Scomber*, *Thunnus*) also rely on stiff, almost winglike fins and constant swimming to provide lift.

Yet another way to gain buoyancy is to control the composition of the

swim bladder   A gas-filled sac that lies inside the body cavity of most bony fishes
Chapter 7, p. 198; Figure 7-13

Figure 14-15 This siphonophore consists of a colony of individuals specialized for swimming, feeding, or breeding. A small gas-filled float is located at the upper edge of the colony.

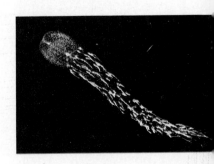

Figure 14-16 Sargasso weed (*Sargassum*) is a brown seaweed that has gas floats. Huge masses of it float out in the north-central Atlantic, or Sargasso Sea. This small branch was washed ashore in the Bahamas. A variety of animals lives among the floating seaweed (see Figure 9-7).

body fluids. The basic idea is that by excluding heavy ions—like sulfate (SO_4^{-2}) and magnesium (Mg^{+2})—and replacing them with lighter salts, especially one called ammonium chloride (NH_4Cl)—organisms can reduce their density and become more buoyant. Certain dinoflagellates do this. So do various zooplankton, including salps, comb jellies, and some squids.

Mechanisms that increase the buoyancy of epipelagic organisms include the storage of lipids, gas-filled floats, and the substitution of light ions for heavy ones in internal fluids.

The Floaters. The plankton include an unusual and highly specialized group of organisms that live right at the very surface of the ocean. Marine biologists usually refer to these surface-dwelling organisms as **neuston.**

All plankton need to keep from sinking, but the floaters have to go one step further and actually float. The most common method of doing this is to have some sort of gas-filled structure to provide buoyancy (Figure 14-16). There are many variations on this seemingly simple theme.

Organisms living at the surface constitute the neuston.

The upper surface of the by-the-wind sailor (*Velella*), a jellyfish-like cnidarian, is specialized as a float (Figure 14-17). The float protrudes into the air and acts like a sail, pushing the colony along in the wind. The Portuguese man-of-war (*Physalia*), a siphonophore, is famous for its powerful sting (see Box 6-1). Like the by-the-wind sailor, part of the colony acts as a sail (Figure 14-18). Several other species of cnidarians live at the sea surface. Some of these can even control the buoyancy of their floats and sink below the surface for shelter during storms.

Despite their powerful stinging cells, these relatives of jellyfishes have enemies. The violet shell (*Janthina*) makes a raft of mucus filled with bub-

Figure 14-17 The by-the-wind sailor (*Velella*).

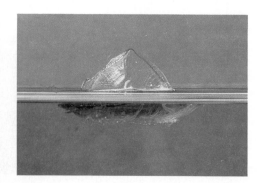

Figure 14-18 The man-of-war fish (*Nomeus gronovi*) lives unharmed and safe from predators among the tentacles of the Portuguese man-of-war (*Physalia physalis*).

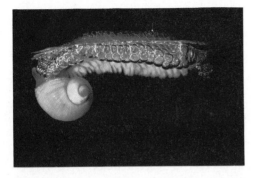

Figure 14-19 The violet shell (*Janthina janthina*) makes a bubble raft that floats it on the surface. Its deep blue shell is paper thin and therefore light, an adaptation to its unusual habitat.

Figure 14-20 A floating pelagic sea slug (*Glaucus*).

bles from which it hangs upside down (Figure 14-19). These inch-long snails drift about, eating Portuguese men-of-war and other delicacies when they find them. Another mollusc, a sea slug called *Glaucus*, stays afloat by swallowing a bubble of air (Figure 14-20). It feeds on the by-the-wind sailor and its relative, the small, disc-shaped *Porpita*. *Porpita*'s stinging cells, still capable of firing, end up in the flaps that protrude from the sea slug's back and will sting animals that disturb the sea slug. Thus *Porpita* provides not only food, but protection, for the sea slug.

One unusual inhabitant of the surface does not have a gas-filled float. In fact, it doesn't actually float at all. The water strider (*Halobates*), the only insect that lives in the open ocean, skims over the water surface (Figure 14-21). It is closely related to the water striders that are common in lakes and ponds. The water strider isn't marine in the strict sense, because it cannot swim and drowns if it falls through the surface. Still, it manages to live throughout the oceans.

Figure 14-21 *Halobates sericeus*, the marine water strider, relies on greatly elongated legs and on the water's surface tension to skate on the surface.

Predators and Their Prey

Many of the adaptations of epipelagic animals, or for that matter any animals, are related to the need to find food and at the same time avoid being eaten. In comparison with other marine environments, the need to adapt to predation, whether as predator or prey, is especially important in the epipelagic.

Sense Organs. Predators and their prey play a continual game of hide-and-seek. The predator tries to find its prey and attack before the victim has a chance to react, while the prey wants as much warning of approaching enemies as possible. Most animals in the epipelagic therefore have highly developed sense organs to help them detect their prey and enemies.

The epipelagic has plenty of light to see by, at least during the day, so it should come as no surprise that vision is important to many epipelagic ani-

mals. Many zooplankton have well-developed eyes. Copepods and other zooplankton may use vision to locate their prey. There is some evidence that vision also helps them avoid predators. Many zooplankton react to such changes in light as passing clouds and therefore perhaps to the shadow of a predator.

Squids, fishes, and marine mammals all have good eyesight. Vision is especially important to nekton in the epipelagic because there are no solid structures that may be used for concealment. In other environments, animals can hide in the sediments, under rocks, or behind seaweeds. There are no such hiding places in the epipelagic. The large, well-developed eyes of most fishes (Figure 14-22) and other nekton are used not only to find prey and watch for predators, but also to find mates and stay together in schools.

Fishes have another remote sensing system: the **lateral line.** Like all fishes, those in the epipelagic are extremely sensitive to vibrations in the water. The lateral line almost certainly plays a major role in schooling behavior (see the section in Chapter 7, "Schooling"). It also alerts the fish to predators: Fishes react instantly to the lunge of an enemy even if they don't see it. Fishes also use vibrations in the water to track down their food. Most predatory fishes—including sharks, tunas, and billfishes—are strongly attracted to splashes on the surface and irregular vibrations in the water, the kinds of vibrations made by an injured fish. Another remote sensing system, **echolocation,** is found in dolphins and some other cetaceans. This sophisticated built-in sonar allows them to locate their prey at a distance (see the section in Chapter 8, "Echolocation").

> **Epipelagic animals have well-developed sense organs, especially vision, the lateral line of fishes, and echolocation in cetaceans.**

Coloration and Camouflage. Just because there are no hiding places doesn't mean that epipelagic organisms cannot camouflage themselves. Indeed, **protective coloration** is nearly universal among epipelagic organisms.

One of the easiest ways to be inconspicuous in the vast watery world of the epipelagic is to be transparent. This adaptation is seen in a variety of zooplankton, some of which are almost perfectly clear (Figure 14-23). Gelatinous zooplankton like some jellyfishes, salps, larvaceans, and comb jel-

lateral line A system of canals on the sides of fishes that helps them detect vibrations in the water
Chapter 7, p. 212; Figure 7-20

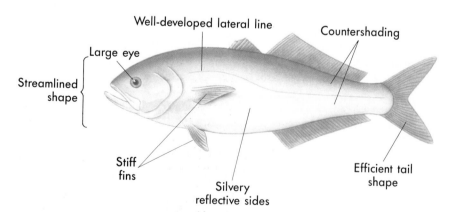

Figure 14-22 Typical adapatations of epipelagic fishes.

Well-developed lateral line

Countershading

Large eye

Streamlined shape

Stiff fins

Silvery reflective sides

Efficient tail shape

STRUCTURE AND FUNCTION OF MARINE ECOSYSTEMS

A

C

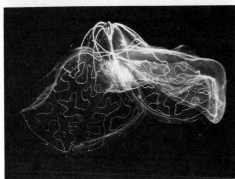

B

Figure 14-23 Many species of zooplankton are transparent or nearly so. Examples include hyperiid amphipods, which have two giant eyes that cover most of the head **(A),** comb jellies **(B),** and squids **(C).** Of course, the organisms that are best at being transparent cannot be shown in photos, since they are almost invisible!

lies are probably the best at being invisible. Many other groups of zooplankton are at least partially transparent, with perhaps only the eyes, a few spots of pigment, or the internal organs visible. Some nearly transparent zooplankton have a faint bluish tinge, perhaps helping them blend in with the surroundings.

A very common form of protective coloration, particularly in nekton, is **countershading,** where the back (or dorsal surface) is dark—usually green, blue, or black—and the belly (or ventral surface) is white or silver (see Figure 7-9). Countershading is particularly well suited to the epipelagic. To a predator looking down, the ocean depths are a dark blue. A dark object blends in against this dark background, and countershaded organisms are dark when viewed from above. Looking up, on the other hand, the predator sees the bright light filtering down from above and the silvery white sea surface. Against this bright background a dark object stands out like a sore thumb, but the silvery-white undersides of epipelagic nekton reflect light and match the bright surface. Laterally compressed bodies, also common, reduce the size of the silhouette whether viewed from above or below (Figure 14-24).

> **Countershading, in which the dorsal surface is dark and the ventral surface is white or silver, is a widespread adaptation among epipelagic nekton. Countershaded organisms blend in against the background whether they are viewed from above or below.**

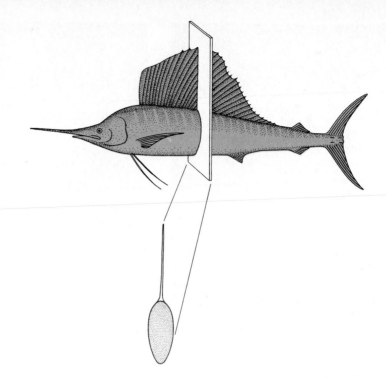

Figure 14-24 Like many epipelagic fishes, the sailfish *(Istiophorus platypterus)* has a laterally compressed body; that is, it is high and narrow when viewed in cross section. Also see Figure 1-30.

Most epipelagic fishes have silvery sides that reflect light. This helps them to blend in when viewed from the side. It is also common for them to have vertical bars or irregular patterns that help break up their outline (Figure 14-25).

Flying fishes *(Cypselurus)* have evolved a distinctive defense that is not exactly camouflage but does make them hard to see, at least temporarily. When threatened, flying fishes burst out of the water and glide through the air on their greatly enlarged pectoral fins (Figure 14-26). To a predator, the flying fish seems to have disappeared, because it is difficult to see past the surface into the air. This strategy is only partly effective, however, and flying fishes are a favorite food of many epipelagic predators.

Swimming: The Need for Speed.   Protective coloration or not, the keen senses of epipelagic predators enable them to find their prey. When this happens the prey's only hope is to flee, because there are no hiding places. Whether the prey gets away or the predator gets a meal depends on which swims faster. The emphasis is on sheer speed, as opposed to maneuverability, such as being able to burrow in the bottom or fit into crevices in the rocks. It is no coincidence that the epipelagic contains the world's most powerful swimmers.

Although many plankton have adaptations that increase water resistance and make them sink slower, practically all epipelagic nekton have streamlined bodies that *reduce* resistance and make swimming easier and more efficient (see Figure 8-15). Epipelagic nekton rarely have bulging eyes, long spines, or other projections that would increase resistance and slow them down. Instead, their bodies are sleek and compact. They often have smooth body surfaces to help them slide through the water. Epipelagic fishes

Figure 14-25 The vertical bars on this striped marlin *(Tetrapturus audax)* help break up its outline, making it harder to see in the dappled underwater light.

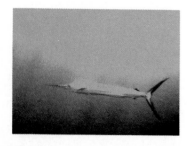

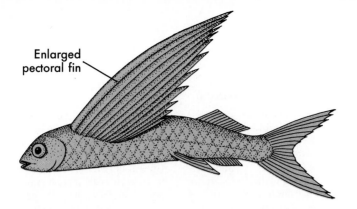

Enlarged
pectoral fin

Figure 14-26 A flying
fish *(Cypselurus).*

usually have small scales, or none at all, and dolphins and whales have lost nearly all of their hair. Fishes produce mucus that actually lubricates the body surface, allowing them to slip through the water even more easily.

Epipelagic nekton are also firm and muscular, another adaptation for swimming. The force produced by these muscles is delivered almost entirely by the tail, because epipelagic fishes rarely swim with their pectoral fins, which are used only for steering and lift (see Figure 7-12, B). Most epipelagic fishes have a tail that is narrow and high (see Figure 14-24). Studies have shown this shape to be the most efficient for high-speed swimming.

Typical bony fishes have fins that consist of movable spines connected by a thin membrane (see Figure 7-9, B). This makes them dexterous and flexible. By contrast, the fins of epipelagic bony fishes tend to be stiff, which gives them the strength to provide maneuverability and lift at high speed. These stiff fins are not much good for hovering, swimming backward, poking into holes and crevices, or other activities that require tight maneuvers at slow speed. Epipelagic fishes don't do such things often anyway.

Epipelagic nekton not only have a lot of muscle, but their muscles are strong and efficient, as you might expect. Fishes have two kinds of muscle, red and white—the equivalent of the light and dark meat on a turkey. Red muscle gets its color from a high concentration of **myoglobin** and therefore can store a lot of oxygen. Red muscle is best suited for long, sustained effort. White muscle, on the other hand, is better at providing short bursts of power. Typically, epipelagic fishes that swim continuously have a relatively high proportion of red muscle. The red muscle provides most of the power for sustained cruising. When the fish needs a quick burst of speed, the white muscle kicks in.

myoglobin   A red muscle protein in vertebrates that stores oxygen
Chapter 7, p. 210

> **Typical epipelagic nekton have sharp eyesight, countershading, streamlined bodies, well-developed and efficient muscles, and high, relatively narrow tails.**

Muscle tends to work more efficiently at warm temperatures. Epipelagic sharks, tunas, and billfishes have evolved a system to conserve the heat generated by their muscles and keep their internal temperature above that of the surrounding water. In fact, these fishes can almost be considered "warm-blooded." In most fishes, heat generated in the muscles is carried by the blood to the skin and then lost to the water. "Warm-blooded" fishes have a special arrangement of the blood vessels known as the **rete mirabile,** or

BOX 14-2

Swimming Machines

Tunas, mackerels, billfishes, and their relatives—the **scombroid fishes** or just plain **scombroids**—swim continuously. Feeding, courtship, reproduction, and even "rest" are done in constant motion. As a result, practically every aspect of the body form and function of these swimming machines is adapted to enhance their ability to swim.

The swimming feats of tunas and their kin are nothing short of amazing. Tunas routinely cover vast distances in their annual migrations. One northern bluefin tuna (*Thunnus thynnus*) was tagged southeast of Japan and recaptured off the Pacific coast of Baja California, Mexico. It swam a distance of 10,800 km (6,700 miles). Black marlin (*Makaira indica*) cover similar distances, and most other scombroids also make long-distance migrations (see Figure 7-24).

These lengthy journeys are made in an impressively short time. Tagged northern bluefin tuna have crossed the Atlantic in 119 days, averaging a straight-line distance of over 65 km (40 miles) per day. The fish actually cover a much greater distance because they constantly change direction, dashing here and there in search of food. Some tunas could cross the Atlantic in less than 2 months at their slowest cruising speed if they took a direct route.

Tunas and billfishes are primarily endurance swimmers, adapted for sustained high-speed cruising, but they are also accomplished sprinters and can make blistering high-speed bursts. Indeed, the scombroids include the fastest swimmers in the sea. The fastest of all is the sailfish (*Istiophorus platypterus;* see Figure 14-24), which can exceed 110 km/hour (70 mph) for short periods. Although no other fish can match this, most tunas and billfishes are fast swimmers. Several large species can reach 75 km/hour (50 mph); some smaller species are just as fast for their size. Most

scombroids are so fast that they can easily outdistance most of their prey. Their phenomenal speed has probably evolved not so much to catch prey (they could do that at slower speeds) as to compete with schoolmates. The first one to the prey, after all, gets the meal.

Many of the adaptations of scombroid fishes serve to reduce water resistance. Interestingly enough, several of these hydrodynamic adaptations resemble features designed to improve the aerodynamics of high-speed aircraft. Though human engineers are new to the game, tunas and their relatives evolved their "high tech" designs long ago. Perhaps we still have a trick or two to learn!

Scombroid fishes have made streamlining into an art form. Their bodies are sleek and compact. The body shapes of tunas, in fact, are nearly ideal from an engineering point of view. Most species lack scales over most of the body, making it smooth and slippery. The eyes do not protrude at all but lie flush with the body. The eyes are also covered with a slick, transparent lid that reduces drag. The fins are stiff, smooth, and narrow, which helps cut the amount of drag they create. When not in use, the fins are tucked into special grooves or depressions so that they lie flush with the body and do not break up its smooth contours. Airplanes retract their landing gear while in flight for the same reason.

Scombroid fishes have more complex adaptations to improve their hydrodynamics. The long bill of marlins, sailfish, and swordfish (*Xiphias gladius*) probably helps them slip through the water. Many supersonic aircraft have a similar needle at the nose. Most scombroids have a series of keels and finlets near the tail. Although most of their scales have been lost, tunas and mackerels retain a patch of coarse scales near the head called the **corselet.**

BOX 14-2 cont'd.

Swimming Machines

The keels, finlets, and corselet help direct the flow of water over the body surface in such a way as to reduce resistance. Again, supersonic jets have similar features.

Since they are always swimming, tunas simply have to open their mouths and water is forced in and over their gills. Accordingly, they have lost most of the muscles that other fishes use to suck in water and push it past the gills. In fact, tunas must swim to breathe. They also must keep swimming to keep from sinking, since most species have largely or completely lost the swim bladder.

One potential problem is that opening the mouth to breathe detracts from the streamlining of these fishes and tends to slow them down. Some species of tuna have specialized grooves in their tongue. It is thought that these grooves help to channel water through the mouth and out the gill slits, helping to reduce water resistance.

In addition to adaptations that reduce drag, scombroids have adaptations that increase the amount of forward thrust they can generate. Their high, narrow tails, with the tips swept back, are almost perfectly adapted to provide propulsion with the least possible effort. Their muscles are also highly efficient, and they have a high proportion of red muscle, as might be expected in continual cruisers. The mechanism that maintains a warm body temperature is also highly developed. A bluefin tuna in water of 7° C (45° F) can maintain a core temperature of over 25° C (77° F). This warm body temperature may help not only the muscles, but the brain and eyes, to work better.

Adaptations for high-speed swimming in tunas.

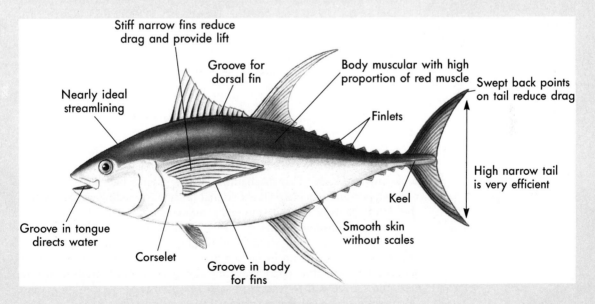

CHAPTER 14 LIFE NEAR THE SURFACE

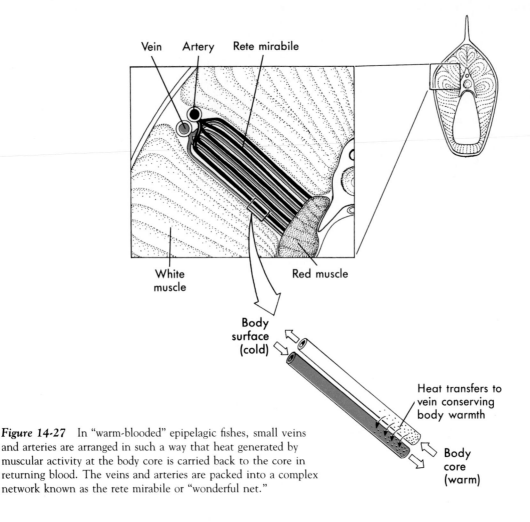

Vein   Artery   Rete mirabile

White
muscle

Red muscle

Body
surface
(cold)

Heat transfers to
vein conserving
body warmth

Body
core
(warm)

Figure 14-27 In "warm-blooded" epipelagic fishes, small veins and arteries are arranged in such a way that heat generated by muscular activity at the body core is carried back to the core in returning blood. The veins and arteries are packed into a complex network known as the rete mirabile or "wonderful net."

"wonderful net." In this arrangement (Figure 14-27), body heat being carried outward in the blood is transferred into inward-flowing blood and taken back into the body. This mechanism greatly reduces heat loss. The fish's core temperature remains well above that of the surrounding water, whereas the skin remains nearly the same temperature as the water.

Vertical Migration.   The epipelagic can be a dangerous place, with voracious predators zooming around and no place to hide. On the other hand, the surface layers contain by far the most food in the pelagic realm. Some zooplankton solve this apparent dilemma by spending only part of their time near the surface and retreating to safer territory when not feeding. These animals undertake **vertical migration.** During the day they live at considerable depth, usually at least 200 m (650 feet) or so (Figure 14-28). At these

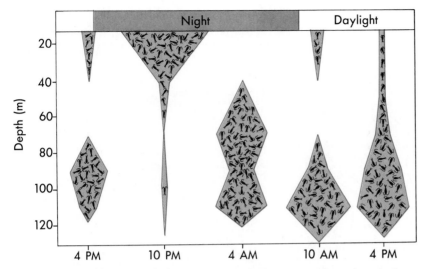

Figure 14-28 The depth distribution of a copepod that migrates vertically at different times of day.

depths there is not much light, so the zooplankton are relatively safe from the many predators that use vision. At night, the zooplankton swim up into the surface layers to feed on phytoplankton and other zooplankton.

Vertically migrating zooplankton stay below the photic zone during the day. At night they migrate to the surface to feed.

The reasons for vertical migrations in zooplankton are not entirely clear, and factors other than avoiding predation may be involved. These daily movements require a great deal of energy. For an organism that is 2 mm long, a 200-m migration is the rough equivalent of a 200-km (120-mile) journey for a human! This is a long way to go to avoid predators when many other zooplankton manage to withstand predation without migrating. Various explanations other than escaping predation have been proposed. Zooplankton may be able to slow down their metabolism and conserve energy by spending part of their time in the deep water, which is cold and reduces their body temperature. This is somewhat like sleep or hibernation. Vertical migration may also be related to the toxins produced by a number of phytoplankton, especially dinoflagellates. By avoiding the surface during the day, when photosynthetic organisms are productive, vertical migrators may be able to avoid the toxins. Many species of nekton, mainly fishes and some large shrimps, also make vertical migrations. These animals generally migrate to much greater depths than zooplankton (see the section in Chapter 15, "Vertical Migration and the Deep Scattering Layer").

THE NATURE OF EPIPELAGIC FOOD WEBS

Epipelagic food webs are of great interest, especially since epipelagic fisheries provide food and employment to millions (see the section in Chapter 16, "Food From the Sea"). Species that are caught commercially can be understood and managed only if we understand the rest of the food web on which they depend.

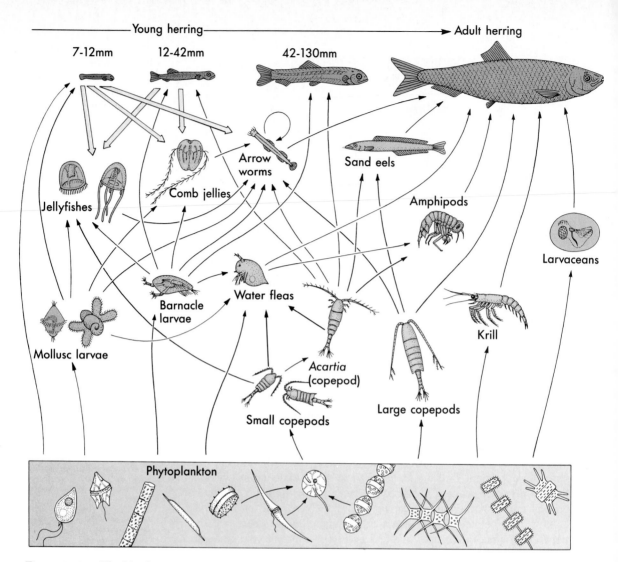

Young herring

Adult herring

7-12mm   12-42mm   42-130mm

Jellyfishes

Comb jellies

Arrow worms

Sand eels

Amphipods

Larvaceans

Mollusc larvae

Barnacle larvae

Water fleas

Acartia (copepod)

Krill

Small copepods

Large copepods

Phytoplankton

Trophic Levels and Energy Flow

Figure 14-29 The North Sea herring *(Clupea harengus)* eats different foods at different stages of its life *(black arrows)*. Its larval stages are eaten by predators that do not feed on adult herring *(open arrows)*. Many, if not most, epipelagic animals have similarly complex feeding relationships. Illustrations of food webs are usually greatly simplified and gloss over such complexities. For example, how would this partial food web fit into the scheme shown in Figure 14-5?

The trophic structure of the epipelagic is extremely complex and not very well understood. For one thing, the epipelagic contains a vast number of different species. The feeding habits of most of them are poorly known, if at all. It is hard to describe the trophic structure of a community if you do not even know what many of the animals eat!

Another difficulty in understanding epipelagic food webs is that most of the animals are **omnivores** and eat a diversity of foods. This means that they often eat prey from different trophic levels. Zooplankton such as copepods often eat both phytoplankton and zooplankton; thus they may function both as herbivores and as carnivores. Nekton, too, consume prey at different levels on the food web. A tuna, for example, may act as a secondary consumer when it eats krill, as a third-level consumer when it eats a sardine that ate zooplankton, and as a fourth-level consumer when it eats a mackerel that ate a flying fish that ate zooplankton. This makes it difficult to assign many animals to a particular trophic level.

Yet another complication is that most epipelagic animals consume different prey at different times in their lives. Many eat different food as larvae than they eat as adults, and even the larvae often switch from phytoplankton to zooplankton. Even after they leave the larval stage, many fishes eat different prey as they grow larger (Figure 14-29).

The basic flow of energy in the epipelagic can be depicted as phytoplankton → zooplankton → small nekton → large nekton → top predators. The top predators, of course, are also nekton. This scheme is greatly simplified, and each level contains a mini-web. Within the zooplankton, for example, are herbivores and several levels of carnivores, forming a food web among themselves.

Epipelagic food chains usually have many steps and generally are longer than in other ecosystems. The number of steps varies: Tropical food webs tend to have more levels than those in colder waters. Even in cold waters there may be five or six steps between the primary producers, the phytoplankton, and a top predator. There are exceptions, of course: The diatoms-to-krill-to-whales chain, for example, is among the shortest of food chains.

Epipelagic food chains tend to be long and complex, because they contain many species and many epipelagic animals feed at different trophic levels.

The epipelagic is an exception to the rule of thumb that only about 10 percent of the energy contained in one trophic level is passed on to the next (see the section in Chapter 9, "The Trophic Pyramid"). Epipelagic herbivores convert more than 20 percent of the energy derived from phytoplankton into growth. Epipelagic carnivores, too, are more than 10 percent efficient, though not as efficient as the herbivores.

Even with their high efficiency, the length of epipelagic food webs means that most of the energy fixed by phytoplankton is lost before it reaches the top carnivores. After all, even an efficiency of 20 percent means that 80 percent of the energy is lost in just one step. Large animals that feed directly on plankton have an advantage in that they have more food available than do predators who eat other high-level carnivores. By feeding directly on low trophic levels, they eliminate many intermediate steps. It is no accident that the largest epipelagic nekton, baleen whales and whale sharks, and the most abundant fishes, like anchovies and herrings, eat plankton.

Detritus and Bacteria

Detritus and bacteria (Figure 14-30) play an important part in epipelagic food webs, as they do in other marine ecosystems. Two important sources of detritus have already been mentioned: fecal pellets and abandoned larvacean houses. Larvacean houses and other accumulations of mucus can be very abundant and support rich populations of bacteria. They are often called **marine snow** because they look something like underwater snowflakes.

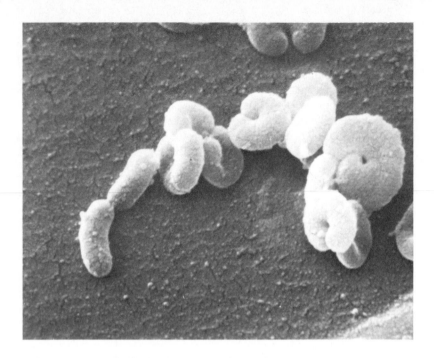

Figure 14-30 A scanning electron micrograph of several cells of *Cyclobacterium marinus,* a ring-forming pelagic bacterium. Also see Figure 5-2.

Many zooplankton and small fishes eat marine snow. Much of the detritus, however, sinks out of the epipelagic zone before organisms can use it. Furthermore, as previously noted, the epipelagic does not get much detritus from other ecosystems.

The role of bacteria in epipelagic food webs still is not understood very well. It has long been known that bacteria are important in breaking down detritus and in nourishing animals that eat detritus. In recent years, marine biologists nave learned that bacteria are of major significance in another way.

The ocean contains vast amounts of **dissolved organic matter (DOM),** organic compounds that exist not as particles but in dissolved form. Much of the DOM in the sea simply leaks out of phytoplankton cells and other organisms. Some DOM is spilled by grazers or excreted as waste. We may not think of such material as food in the ordinary sense, but it contains large amounts of energy. As dieters who like sugar in their coffee know, dissolved matter can contain calories.

Very few animals in the epipelagic can use DOM. Some phytoplankton may be able to use it, but it appears that the major consumers of DOM in the sea are bacteria. In fact, some estimates indicate that as much as half of the organic matter produced by phytoplankton photosynthesis may be used by bacteria in the form of DOM! The bacterial populations found on detritus particles may be using DOM at least as much as the detritus itself. Free-living bacteria, those not attached to particles, subsist almost entirely on DOM.

Bacteria that live on detritus provide nourishment to animals that eat the detritus. Free-living ones are preyed on by various protozoans, the smallest zooplankton. Protozoans are in turn preyed upon by minute zooplankton and so on up the food web. Bacteria are vital therefore in making the energy in DOM available to the rest of the food web again.

Epipelagic bacteria break down detritus and absorb dissolved organic matter (DOM) from the water, making the energy contained in the DOM available to the rest of the food chain.

Not much is known about pelagic bacteria; we do not know how efficiently they use DOM, the nature of their predators, and how their importance varies from place to place. The task of studying them is complicated because they are so small, which makes them hard to catch and to count, and because they reproduce so rapidly, which means that a sample of these bacteria can change dramatically before it is analyzed.

Patterns of Production

Epipelagic food webs are complex indeed, but they all share one simple feature: Primary production by phytoplankton is the base. All other organisms, from the smallest zooplankton to the largest predators, depend on this primary production. Some parts of the epipelagic rank among the most productive ecosystems on earth (Table 14-2). There are also large areas that are among the least productive, the "deserts" of the ocean. The abundances of animals, from zooplankton to whales, generally follow the pattern of primary production (Figure 14-31). It is therefore essential to understand the factors that control the amount of primary production by phytoplankton.

Like all plants, phytoplankton need two main things to perform photosynthesis. First, they need sunlight, the ultimate source of energy for the ecosystem. Second, they need a supply of essential nutrients. Without sunlight and nutrients, phytoplankton cannot grow and produce the food that fuels the food web.

Light Limitation. The epipelagic represents the sunlit layer of the ocean, but there still may not always be enough light for photosynthesis. In other words, primary production may be **light limited.** There is no light at night, of course, so phytoplankton must get enough light during the day to allow them to grow. At high latitudes, decreased light availability may limit phytoplankton growth during the winter, when the days are short and the sunlight weak. In winter, phytoplankton at these latitudes store less energy in photosynthesis than they burn up in respiration. In tropical and subtropical waters, on the other hand, there is enough light to support photosynthesis throughout the year, at least at the surface.

The total primary production in the water column depends not only on the intensity of light at the surface, but also on how far down the water column the light penetrates. If only a thin layer at the surface is sufficiently well lit for plant growth, there will not be as much production as if the photic zone were deep. The depth of the photic zone varies with the season, being deepest in summer. Photic zone depth also depends on the weather. On cloudy days, light does not penetrate as deeply as on bright, sunny days. Especially important is the amount of sediment and other material in the

photosynthesis $CO_2 + H_2O +$ SUN ENERGY $\longrightarrow$ organic matter $+ O_2$
Chapter 4, p. 88

respiration Organic matter $+ O_2 \longrightarrow$ ENERGY $+ CO_2$
Chapter 4, p. 90

Phytoplankton production (mg C/m²/day)

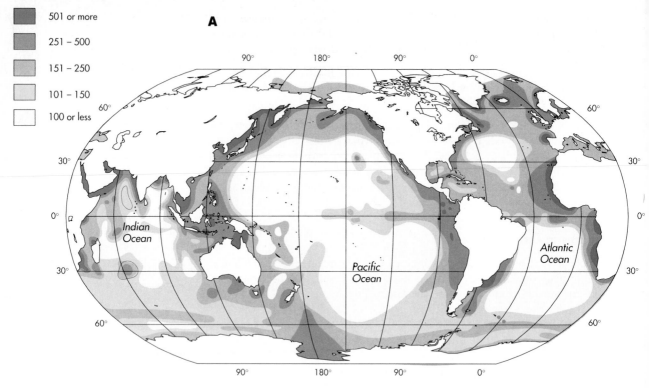

A

■	501 or more
▓	251 – 500
▒	151 – 250
░	101 – 150
□	100 or less

Indian Ocean

Pacific Ocean

Atlantic Ocean

Figure 14-31   The distribution of primary productivity in the oceans **(A).** Note the relationship of productivity to current and upwelling systems.

water. Light does not penetrate as deeply in dirty water, so the photic zone is not as deep as in clear water. The phytoplankton themselves affect the depth of the photic zone. Since they absorb light to perform photosynthesis, the phytoplankton cut down the amount of light available to deeper-living phytoplankton. This phenomenon, known as **self-shading,** is especially important in highly productive waters, which are rather murky because they contain so much plankton. Conversely, the barren waters of the **central gyres,** in the middle of the ocean basins (see Figure 3-23), are incredibly clear.

Figure 14-32 These flasks contain cultures of a microscopic plant. Extra phosphate and nitrate were added to the flask on the left, while the plants in the flask on the right were short of nutrients.

   Nutrients.   Nutrients, especially **nitrogen** and **phosphorus,** play a major part in controlling primary production. Even with plenty of light, plants cannot perform photosynthesis if there are not enough nutrients, that is, if primary production becomes **nutrient limited** (Figure 14-32). In the ocean, nitrogen is usually the **limiting nutrient,** the one in shortest supply. **Nitrate** (NO_3^{-2}) is the most important source of nitrogen. **Phosphate** (PO_4^{-3}), a source of phosphorus, and various other nutrients also limit phytoplankton growth occasionally. The supply of silicon sometimes limits diatom growth, for example, and vitamin B_{12} is limiting in the Sargasso Sea in the north-central Atlantic.

STRUCTURE AND FUNCTION OF MARINE ECOSYSTEMS

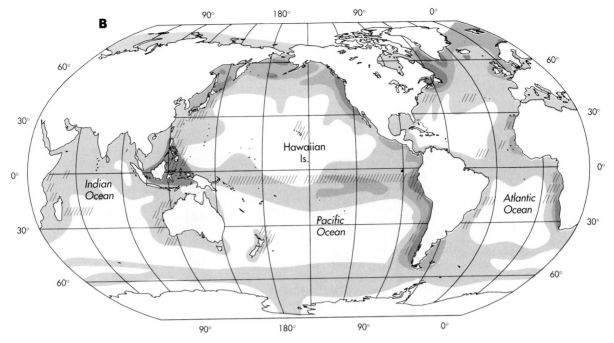

B

Distribution of zooplankton (mg/m³)

�(dark gray)	501 or more
▪(medium gray)	201 - 500
▫(light gray)	51 - 200
□(white)	50 or less

Figure 14-31, cont'd Since the primary production is the base of the food web, this pattern is reflected at higher levels—for example, the distribution of zooplankton abundance **(B)** and the abundance of sperm whales *(hatched areas)* as indicated by catches from 1760 to 1926. Some areas with low primary production, such as the Hawaiian Islands, show high catches because of whale migration and the proximity to land bases.

Some blue-green algae can fix free nitrogen (N₂) and are occasionally a significant source of nitrogen in tropical waters. *Trichodesmium*, one such blue-green alga, may form red-tide blooms (see Box 14-1). The vast majority of nutrients in the epipelagic, however, are recycled (see Figure 9-20). They are used again and again in the cycle of photosynthesis and decomposition. The cycle begins when dissolved nutrients are incorporated into living organic matter by phytoplankton. Later, when the living material dies, the nutrients are **regenerated,** or released, as bacteria decompose the material.

The dependence on recycled nutrients profoundly affects production in the epipelagic. Much of the organic material produced by phytoplankton eventually ends up as detritus: fecal pellets, dead bodies, and other organic particles. These organic particles tend to sink, and many sink out of the epipelagic into deeper waters before they break down and decay. The nutrients they contain are released below the photic zone (Figure 14-33). The net result of this process is that nutrients are removed from the surface and carried to the dark, cold waters below. Thus surface waters are usually depleted in nutrients, and phytoplankton growth in the epipelagic is often nutrient-limited. Deep water, on the other hand, is usually nutrient rich, but

nitrogen fixation The conversion of nitrogen gas (N₂) into nitrogen compounds that can be used by plants as nutrients
Chapter 9, p. 281

TABLE 14-2

Typical rates of primary production in various marine environments

Production rates can be much higher at certain times or in specific locations, especially at high latitudes. Values for some selected terrestrial environments are given for comparison.

ENVIRONMENT	RATE OF PRODUCTION (grams of carbon fixed/m^2/year)
Pelagic environments	
Arctic Ocean	0.7-1
Southern Ocean (Antarctica)	40-260
Subpolar seas	50-110
Temperate seas (oceanic)	70-180
Temperate seas (coastal)	110-220
Central gyres	4-40
Coastal upwelling areas	110-370
Equatorial upwelling areas	70-180
Benthic environments	
Salt marshes	260-700
Seagrass beds	550-1100
Kelp beds	640-1800
Mangrove forests	370-450
Coral reefs	1500-3700
Terrestrial environments	
Extreme deserts	0-4
Temperate farmlands	550-700
Tropical rain forests	460-1600

From several sources.

there is not enough light for photosynthesis and the nutrients cannot be used in primary production. The general nutrient profiles shown in Figure 14-34 are characteristic of most of the pelagic realm.

Because many organic particles sink out of the epipelagic before the nutrients they contain are regenerated, surface waters are poor in nutrients, and phytoplankton growth is nutrient limited. Deep waters are high in nutrients because of the rain of organic particles from the surface.

The phytoplankton that support epipelagic food webs, then, have a problem: At the surface there is sunlight but hardly any nutrients, whereas in deep water there are plenty of nutrients but not enough light. Coastal phytoplankton have it somewhat easier than oceanic forms. Since coastal areas are relatively shallow, the bottom tends to trap sinking organic particles, and some of the regenerated nutrients are returned to the water column. This is one reason why coastal waters are highly productive. Another reason is that rivers bring in fresh nutrients from land. Even so, coastal wa-

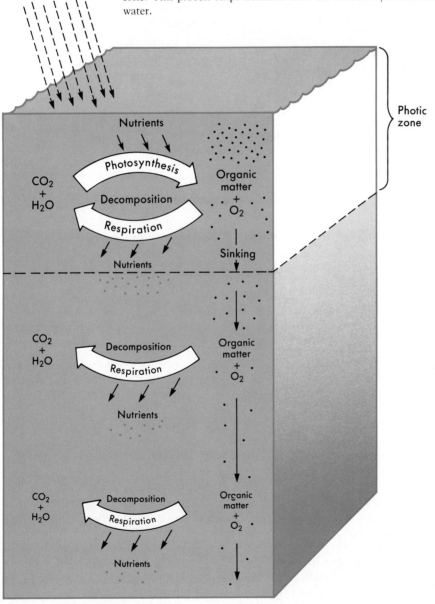

Figure 14-33 The amount of nutrients at different depths is controlled by photosynthesis, respiration, and the sinking of organic particles. During primary production, dissolved nutrients are removed from the water and incorporated into organic matter. Respiration, in this case mainly by bacterial decay, breaks down the organic matter and regenerates the nutrients. The uptake of nutrients can occur only at the surface, where there is light. Since organic particles sink, however, much of the regeneration occurs below the photic zone. This process strips nutrients from the surface layer and carries them into deeper water.

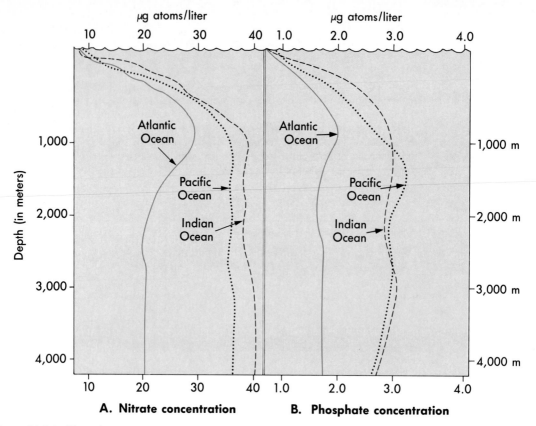

μg atoms/liter   μg atoms/liter

Atlantic Ocean

Pacific Ocean

Indian Ocean

Depth (in meters)

A. Nitrate concentration   **B. Phosphate concentration**

Figure 14-34 Typical profiles of nitrate **(A)** and phosphate **(B)** in the open ocean. Note that surface waters generally have very low levels of these nutrients.

ters may have very low surface concentrations of nitrate, phosphate, and other nutrients.

> Coastal waters are usually highly productive because the shallow bottom prevents organic particles and the nutrients they contain from sinking out of the photic zone and because rivers add fresh nutrients.

For primary production to occur in oceanic areas, the nutrients contained in the deep water must somehow get to the surface. The only effective way for this to happen is for the water itself to move to the surface, carrying the nutrients along with it. This sounds simple enough, but it occurs only at certain times and places. The ocean is usually **stratified;** that is, the warm surface layers "float" on the denser water below (see the section in Chapter 3, "The Three-Layer Ocean"). A **thermocline,** a zone of transition, lies between the warmer, less dense layer and the colder, denser layer below. Just as energy is needed to push a cork below the surface or lift a weight up from the bottom, it takes energy to push surface water down across the thermocline into the denser water below, mixing it with the nutrient-rich deep water, or to bring the dense deep water to the surface. Much of the time this simply does not happen, and surface production is often nutrient limited.

Seasonal Patterns. One way that deep-water nutrients get to the surface is for the water column to mix, just as shaking a bottle of salad dressing disperses the vinegar—which originally sat on the bottom—throughout the bottle. On the continental shelf, strong wind and waves may mix the water column all the way to the bottom. Often, however, wind and waves do not mix the water deeply enough to bring nutrients to the surface, especially in the open ocean.

If the surface water gets more dense, however, it loses its tendency to float. In the ocean, this happens at high latitudes when winter cold and winds cool the surface water, making it denser. If the surface water gets denser than the water below, **overturn** occurs (see Figure 3-36). The thermocline breaks down, the surface water sinks, and nutrient-rich deep water is mixed up to the surface. Overturn depends on very cold conditions and usually is not widespread except in polar waters. Even in the absence of overturn, however, surface cooling can promote mixing. Since cooler surface water is not as buoyant, less energy is needed to mix it with the deep water. The mixing of deep nutrients to the surface is assisted by winter storms, which bring strong winds and large waves. Because of overturn and mixing during the winter, polar and temperate waters are highly productive (Figure 14-31).

> Oceanic waters at high latitudes are highly productive because winter overturn and mixing bring nutrient-rich deep water to the surface.

The productivity figures in Figure 14-31 and Table 14-2 refer to total production during the entire year. This obscures the fact that primary production may vary over the course of the year. In warm temperate and tropical waters such seasonal variation is relatively minor. The water column remains stable throughout the year, restricting the transport of deep nutrients up into the photic zone. For this reason, most tropical waters have low but fairly constant levels of production throughout the year (Figure 14-35, A).

In temperate waters the effects of the seasons are profound. Overturn and mixing during the winter bring large amounts of nutrients to the surface. The phytoplankton are light limited in winter, however, and unable to use the nutrients in photosynthesis. Primary production during winter months, therefore, is low at temperate latitudes (Figure 14-35, B).

When spring comes, the days get longer and the sunshine becomes more intense. Light is no longer limiting, and there are plenty of nutrients in the surface water because of overturn and mixing during the preceding winter. This combination—sufficient amounts of both light and nutrients—provides ideal conditions for phytoplankton and produces a period of rapid growth called the **spring bloom** (Figure 14-35, B). In many areas the spring bloom represents most of the primary production for the entire year.

As the spring bloom develops, two things occur, neither beneficial for phytoplankton growth. First, the increase in sunlight not only stimulates photosynthesis, it also warms the surface water, making it less dense. This increases the stability of the water column, eliminating overturn and greatly reducing the mixing of nutrient-rich deep water to the surface. By late spring, the input of nutrients to the surface from below has practically stopped. Second, the phytoplankton are rapidly using up nutrients as they

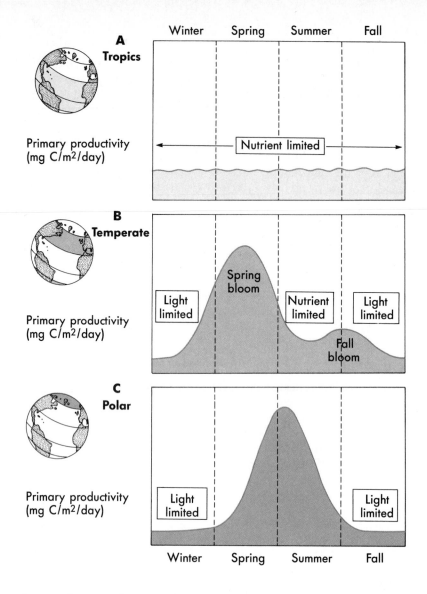

Figure 14-35 Generalized seasonal cycles of primary productivity in (**A**) tropical, (**B**) temperate, and (**C**) polar waters.

photosynthesize and grow. Thus nutrients are removed from the water at the same time that the supply is blocked by the stratification of the water column. The amount of available nutrients in the water falls, and the phytoplankton become nutrient limited. Nitrate is usually the nutrient that runs out first.

During most of the summer, primary production is nutrient limited. The level of production depends on nutrient recycling and the new input of nutrients. If the water column never becomes very stable, some mixing and therefore some production takes place. If the surface water warms considerably and the water column becomes stratified, the input of nutrients by mixing is almost nil and primary production falls. Temperate waters generally have low productivity during the summer months as a result of stratification.

Production in the autumn depends largely on which comes first: the short days and weak sunlight that bring light limitation or surface cooling and storms, which increase the mixing of nutrients. If the days are cool and sunny, for example, a strong wind can mix up nutrients while the phytoplankton still have enough light to perform photosynthesis. If so, there will be a burst of production known as the **fall bloom** (Figure 14-35, B). If warm, mild conditions persist late into the fall, the phytoplankton become light limited before overturn and mixing begin. Even though production at high latitudes is light limited during the winter and nutrient limited during the summer, there is enough production during blooms to give these waters a high total for the year (Figure 14-31).

> **At high latitudes, primary production shows a seasonal cycle. During winter, production is light limited, but overturn and wind mixing bring nutrients to the surface. In spring, increased sunlight allows the phytoplankton to use the nutrients and a spring bloom occurs. During summer, production is nutrient limited because the phytoplankton use up the nutrients and stratification prevents mixing. In autumn, stratification may break down while there is enough light for photosynthesis. If so, there will be a fall bloom.**

Cold polar waters may not become stratified at all and may continue to produce at very high levels throughout the short summer when the sun shines most of the day (Figure 14-35, C). Primary production falls as the days shorten, and it is much reduced during the long winter months. Polar waters, which are not nutrient limited during the summer, can be extremely productive, as is true around Antarctica (see Table 14-2). The Arctic Ocean, on the other hand, lies at such a high latitude and is so extensively covered by ice that it generally is light limited and unproductive in spite of overturn and mixing.

Upwelling and Productivity.   Overturn and mixing caused by the cooling of surface waters are not the only processes that transport nutrients up into the photic zone. At certain times and places large amounts of nutrient-rich deep water move up to the surface. This phenomenon, known as **upwelling,** is caused indirectly by the Coriolis effect.

The water column can be thought of as a column of thin layers, almost like a stack of paper. The wind pushes the top layer, which begins to move. Because of the Coriolis effect, this uppermost layer of water moves 45 degrees to the right of the wind direction in the Northern Hemisphere or to the left in the Southern Hemisphere (see the section in Chapter 3, "Surface Currents"). The top layer pushes on the layer below, and again the Coriolis effect comes into play: The second layer moves not in the same direction as the top one, but slightly to the right, and slightly slower. This process passes down through the water column, each layer being pushed by the sheet above and pushing the one below. The direction that the water moves changes down the water column in a spiral (Figure 14-36), called the **Ekman spiral** after the Swedish oceanographer who first discovered it. The effect of the wind decreases with depth, so that progressively deeper layers move slower and slower. Eventually, at a depth of at most a few hundred

Coriolis effect   As a result of the earth's rotation, anything that moves large distances on the earth's surface tends to bend to the right in the Northern Hemisphere and to the left in the Southern Hemisphere
Chapter 3, p. 65

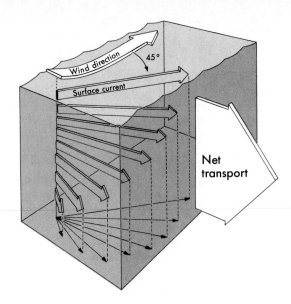

Figure 14-36 When a steady wind blows over the sea surface, the uppermost layer moves at 45 degrees from the wind direction. Each deeper layer moves farther to the right in the Northern Hemisphere, or to the left in the Southern Hemisphere. When the direction of the current at each depth is plotted, the result is a spiral, called the Ekman spiral. The net result of this process is that the affected layer of water, called the Ekman layer, is transported at right angles to the wind direction. Also see Figure 3-22.

meters, the wind is not felt at all. The upper part of the water column that is affected by the wind is called the **Ekman layer.** Though each "micro-layer" moves in a different direction, the Ekman layer as a whole moves at 90 degrees from the wind direction. This process is called **Ekman transport.**

> Winds blowing over the sea surface produce Ekman transport, in which the upper part of the water column moves perpendicular to the wind direction, to the right in the Northern Hemisphere and the left in the Southern Hemisphere.

Figure 14-37 When prevailing winds blow in the right direction along the coast, Ekman transport carries surface water offshore. Deep water rises into the photic zone, carrying nutrients with it. This is called coastal upwelling. Is the example in this figure from the Northern or Southern Hemisphere?

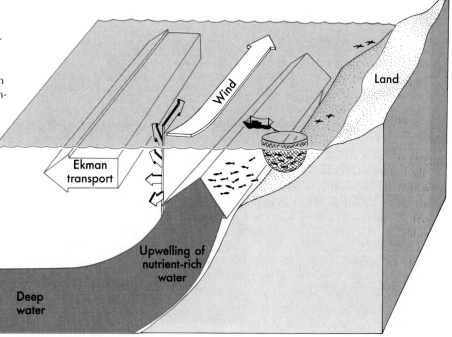

STRUCTURE AND FUNCTION OF MARINE ECOSYSTEMS

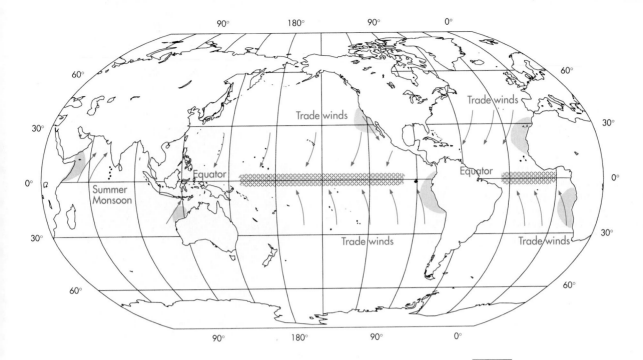

Coastal upwelling

Equatorial upwelling

Figure 14-38 The major coastal upwelling areas of the world and the prevailing winds. These areas are among the most productive parts of the ocean (see Figure 14-31).

In coastal waters, Ekman transport can produce extremely intense upwelling. In some places, mainly the eastern sides of the ocean basins, prevailing winds blow parallel to the coast so that Ekman transport carries the surface layer offshore. Cold deep water moves up to replace it (Figure 14-37), producing intense **coastal upwelling.** Coastal upwelling carries huge amounts of nutrients into the photic zone, and major coastal upwelling areas (Figure 14-38) are among the most productive waters of the epipelagic (see Figure 14-31 and Table 14-2). These coastal areas are therefore among the sea's richest fishing grounds (see Figure 16-6).

On some coasts, like the Pacific coast of South America, upwelling is fairly steady and takes place over a large geographic area. On other coasts, like that of California, upwelling tends to occur as localized, short-lived events (Figure 14-39). If the wind is strong for a few days, strong upwelling occurs, only to fade away when the wind dies and to spring up somewhere else. Even on the Pacific coast of South America, upwelling tends to be most intense in small, localized patches.

Upwelling also tends to be seasonal, occurring mainly during the time of year when winds are strong and blow in the right direction along the coast. Such seasonality is most pronounced on the coast of Somalia, in eastern Africa (Figure 14-38). This area is dominated by **monsoons,** strong winds that blow northward in summer and southward in winter. The summer monsoon causes very intense coastal upwelling, which dies when the winds change to the winter monsoon. When the upwelling stops, production falls dramatically.

Coastal winds are not the only source of upwelling. The Coriolis effect also produces **equatorial upwelling,** especially in the Pacific. Note that the

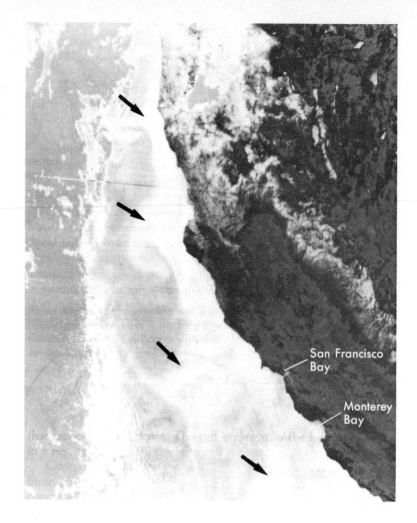

Figure 14-39 Upwelling events bring cold water to the surface, so they can be recognized in aerial and satellite photos by low surface temperature. This photo of the Pacific coast of North America from Oregon to central California shows sea surface temperature. Pockets of cold water—upwelling areas—can be seen as light areas along the coast (*arrows*).

San Francisco
Bay

Monterey
Bay

direction of the Coriolis effect changes from right to left at the Equator. Thus the north equatorial currents transport surface water to the right (north), whereas the south equatorial currents transport water to the left (south; Figure 14-40). In equatorial regions the sea surface is being pulled apart, or diverging, and deep water moves up to fill the void. Equatorial upwelling is not as intense as coastal upwelling, but it still raises primary production significantly.

> Upwelling brings nutrients to the surface and results in high primary production. Coastal upwelling is very intense. It occurs when winds cause the Ekman transport of surface water offshore. Equatorial upwelling is caused by the divergence of equatorial surface currents.

Geographic Patterns. Water depth and temperature, prevailing winds, and surface currents all affect the ocean's productivity, as we have seen in the previous sections. The geographic distribution of productivity (see Figure 14-31) reflects these factors. Coastal waters are highly productive be-

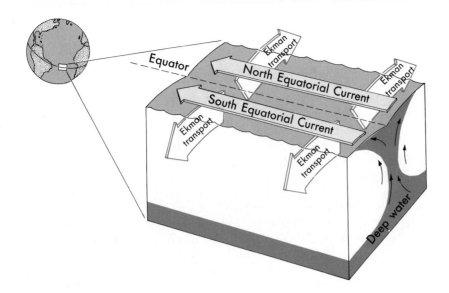

Figure 14-40 The Coriolis effect influences the equatorial currents differently on opposite sides of the equator. Surface water therefore moves away from the Equator. To replace it, deep water moves up into the sunlit zone. This is called equatorial upwelling. Also see Figure 14-31.

cause the shallow bottom prevents nutrients from sinking out of the photic zone and because wind and waves mix the water column. Coastal areas that experience upwelling are especially productive. Equatorial waters, too, are productive because of upwelling, but not as productive as coastal upwelling zones. Polar and cold temperate regions have high productivity because overturn and upwelling bring nutrients up into the sunlit layer.

None of the processes that bring nutrients to the surface operate in the vast central gyres. These lie in relatively warm latitudes, so the surface waters never cool enough for overturn or mixing. Far from both the coast and the equatorial currents, the central gyres never experience upwelling, either. The gyre regions are permanently limited by nutrients; with the exception of the light-limited Arctic Ocean, they have the lowest productivity in the epipelagic. In fact, the central gyres are among the least productive ecosystems on earth.

The El Niño–Southern Oscillation

The interplay of winds, currents, and upwelling affects not only the ocean's primary production, but fisheries, climate, and ultimately human welfare. This is brought home most strongly when normal patterns suddenly change, as in the phenomenon known as the **El Niño–Southern Oscillation (ENSO).** In recent years ENSO—often simply called El Niño—has received widespread attention not only from marine scientists, but also because of its impact on humans, from economists, politicians, and the general public.

The term "El Niño" originally referred to a change in the surface currents along the coasts of Perú and Chile. For much of the year the trade winds along this coast blow from south to north, producing strong upwelling (see Figure 14-38). The upwelling brings nutrients to the surface, making these waters one of the world's richest fishing grounds. Every year, usually in December, the trade winds slack off, upwelling decreases, and the water gets warmer. Local residents have been familiar with this event for centuries, since it signals the end of the peak fishing season. Every few years, however, the surface water gets much warmer than usual. Upwelling ceases completely, primary production drops to almost nothing, and the fishes that normally teem in these waters disappear. Fisheries along the Perú-Chile coast are devastated, and seabirds die in huge numbers (see Box 16-1). Because this change in currents comes around Christmas, it is called *El Niño*, or "The Child," after the baby Jesus.

The "Southern Oscillation" part of ENSO has also been known for a long time. Like El Niño, the Southern Oscillation was first noted because of its impact on humans. People in India depend on the summer monsoon winds to bring rain for their crops. Sometimes the monsoons fail to come, producing famine and hardship. Beginning in 1904, a British administrator named Gilbert Walker set out to predict when the monsoons would fail by examining and analyzing weather records from around the world. He never achieved his goal, but in studying weather patterns he discovered a major atmospheric phenomenon that he named the Southern Oscillation.

The Southern Oscillation refers to a long-distance linkage in atmospheric pressure, or barometric pressure as it is called in weather reports. When the pressure is high over the Pacific Ocean, it tends to be low over the Indian Ocean, and vice versa (Figure 14-41). Over a period of months or years, the air pressure swings back and forth like a giant seesaw that extends halfway around the world. The changes in atmospheric pressure bring dramatic changes in wind and rainfall, including the failure of the summer monsoon.

Figure 14-41 The Southern Oscillation consists of two gigantic pressure systems half a world apart. When the atmospheric pressure is unusually high in the western Pacific, it tends to be unusually low over the Indian Ocean, and *vice versa*. Shaded areas represent the centers of these two systems.

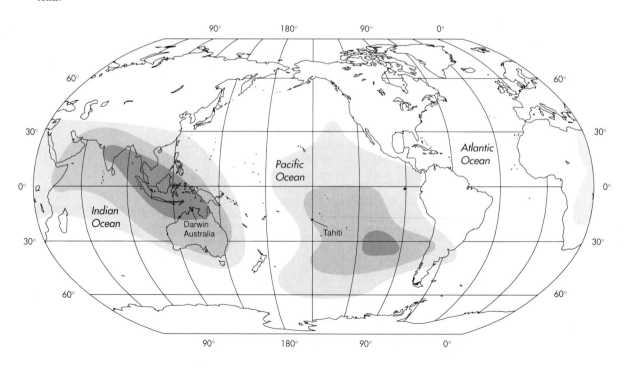

STRUCTURE AND FUNCTION OF MARINE ECOSYSTEMS

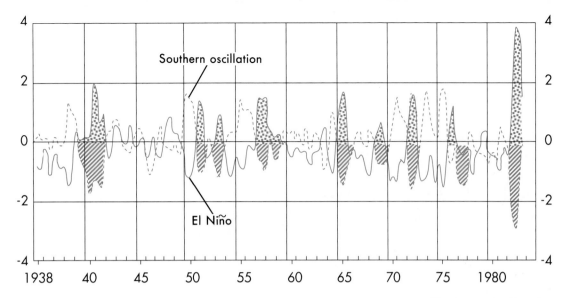

Southern oscillation

El Niño

Figure 14-42 This plot shows both an index used to measure El Niño and an index of the Southern Oscillation. The connection between El Niño and the Southern Oscillation is obvious. There were nine major ENSO events, indicated by shading, between 1935 and 1983. The 1982 to 1983 ENSO was by far the most intense. The index of El Niño is the departure from the average sea surface temperature at Puerto Chicama, Perú. The Southern Oscillation index is the atmospheric pressure at Tahiti minus the pressure at Darwin, Australia (see Figure 14-41).

In the late 1950's it became apparent that El Niño and the Southern Oscillation were two sides of the same coin: Instead of being isolated phenomena that take place in a particular region, they are part of a complex interaction of ocean and atmosphere that links the entire planet. El Niño corresponds closely to dramatic changes in the atmospheric Southern Oscillation (Figure 14-42). Because El Niño and the Southern Oscillation are both part of the same general phenomenon, they are now usually referred to by one name: ENSO.

The global effects of an ENSO event became painfully clear in 1982 and 1983, when the strongest ENSO in at least a hundred years occurred. In South America the 1982-1983 ENSO is blamed not only for the failure of fisheries, but also for droughts in southern Perú, Bolivia, and northeastern Brazil (Figure 14-43). In contrast, northern Perú, Ecuador, southern Brazil, northern Argentina, and Paraguay experienced severe flooding. The effects of the 1982-1983 ENSO were by no means restricted to South America. Australia, Indonesia, southern India, the Hawaiian Islands, and western and southern Africa experienced severe droughts. Floods, too, happened in many places, including some Pacific islands and the west coast of the United States. Tahiti and Hawaii were hammered by cyclones. Storms in California caused catastrophic waves and mudslides. Intense precipitation in the Colorado River basin resulted in an unusually deep snowpack, which caused flooding in the spring when the snow melted. ENSO is also blamed for a number of secondary effects. Droughts, for instance, brought famine, destructive bush fires and dust storms, and unbearable heat to many places. On the east coast of the United States, ENSO is thought to have brought warm, wet conditions that favored mosquitoes and resulted in outbreaks of encephalitis. In Montana, ENSO may have caused an increase in rattlesnake bites because hot, dry conditions forced mice to search for food at lower than normal altitudes and the snakes followed. An increase in shark

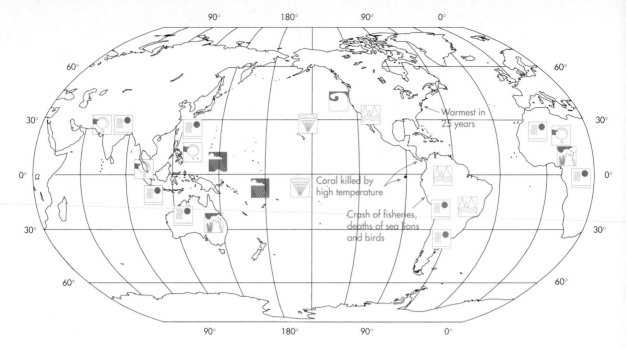

Labels on map: Warmest in 25 years · Coral killed by high temperature · Crash of fisheries, deaths of sea lions and birds

Figure 14-43 The effects of the 1982 to 1983 ENSO were varied and felt around the world. Some of them are shown on this map.

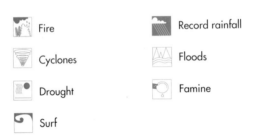

Fire

Record rainfall

Cyclones

Floods

Drought

Famine

Surf

attacks off the coast of Oregon was blamed on the unusually warm water. The unusually warm, wet conditions may have favored the rodents that carry fleas, triggering an outbreak of bubonic plague in New Mexico.

The effects of ENSO are particularly harmful because they are dramatic changes in the normal weather pattern. Heavy rains may cause floods in places where such rains are rare, and the lack of heavy rains can cause drought and famine in places, like India and much of Africa, that depend on such rains. Not all the effects of ENSO, however, are bad from the human point of view. The ENSO of 1982-1983 brought an unusually warm winter to much of the eastern United States, saving an estimated half of a billion dollars in fuel costs. Even in South America, where normal fishing patterns were severely disrupted, there were beneficial effects. Warm water species like dolphinfish, skipjack tuna, Spanish mackerel, and yellowfin tuna appeared in record numbers, and the fishery for scallops boomed.

Scientists still do not know what causes an ENSO event; they do not know whether El Niño causes the Southern Oscillation, whether the reverse is true, or whether both are caused by something else. Since the cause is unknown, ENSO events cannot be predicted. Nevertheless we have learned a great deal. Realizing that El Niño in South America and the Southern Oscillation in the Indo-Pacific are part of the same process was a major breakthrough: there is now no doubt that the climates of different parts of the world are linked. Furthermore, though ENSO events cannot be predicted, scientists have at least learned to recognize them in their early stages. This should provide at least some warning when future ENSO events arrive. There is, of course, still much to be learned, and ENSO may even hold a key to understanding patterns of climate around the globe.

Do It Yourself

SUMMARY

1. The _____ zone is the zone between the sea surface and the depth where plants can no longer grow because of light limitation.
2. The primary producers in the epipelagic are almost all _____. The most important are _____, _____, and _____.
3. _____ are extremely small photosynthetic organisms that live in the plankton.
4. _____ are by far the most abundant members of the zooplankton in the epipelagic. Shrimplike _____ and other crustaceans are also important.
5. Organisms that spend part of their lives as larvae in the plankton are called _____.
6. Organisms must avoid sinking to remain in the epipelagic. One way to do this is to increase water resistance by having a _____ body size, a _____ body shape, and long _____. Epipelagic organisms also have adaptations that increase _____, such as stored _____, gas-filled bladders, and changes in ionic composition.
7. Sensory organs, especially _____ and the _____ line, are well developed in epipelagic fishes.
8. _____, in which organisms have a dark back and silver or white belly, is common among epipelagic nekton. They also tend to have smooth, _____ bodies, well-developed and efficient _____, and high, relatively narrow _____.
9. Some zooplankton make vertical migrations. They stay below the photic zone, typically at depths of around _____ during the _____, and migrate to the surface at _____.
10. Organic compounds that are dissolved in the water rather than being in particulate form are called _____. _____ are the major consumers of _____ in the ocean, and make the energy it contains available to animals.

11. Nutrient levels are _____ near the surface and _____ in deep water. This is because _____ removes nutrients from surface waters and organic particles sink to deep water, where the nutrients are _____ by bacterial decomposition.

12. Coastal waters have high production because the shallow bottom prevents _____ from sinking out of the _____ and because _____ bring in nutrients from land.

13. At high latitudes, primary production is light _____ limited during the winter, but _____ and mixing brings nutrients to the surface. In the spring there is a spring _____, whereas in the summer phytoplankton become _____ limited.

14. The Coriolis effect produces _____ transport, in which the upper layer of water moves at right angles to the wind direction. When strong winds blow along a coast, this produces coastal _____, which results in very high production. _____ occurs because the north and south equatorial currents diverge at the Equator.

THOUGHT QUESTIONS

1. Plankton are unable to swim effectively, so they drift at the mercy of the currents. You might think that the currents would scatter planktonic organisms throughout the oceans, but many species are restricted to particular regions. What mechanisms might allow a species to maintain its characteristic distribution?

2. Spiny species of diatoms are found both in warm subtropical waters and in colder areas. Since warm water is less dense than cold water, would you predict any differences between the spines of warm water individuals and those of cold water individuals? Why?

FOR FURTHER READING

Hardy, J.T.: "Where the Sea Meets the Sky." *Natural History*, May 1991, pp. 58-65. *The ocean's uppermost layer, the home to unique forms of life, is also the place where many pollutants and other toxic chemicals concentrate.*

Kunzig, R.: "Invisible Garden." *Discover*, vol. 11, no. 4, April 1990, pp. 66-74. *Phytoplankton provide oxygen to the atmosphere, feed the ocean, and help counteract the greenhouse effect. Some of the most important ones are still being discovered.*

Ramage, C.S.: "El Niño." *Scientific American*, vol. 254, no. 6, June 1986, pp. 76-83. *A review of the causes and effects of El Niño and the Southern Oscillation.*

The Ocean Depths 15

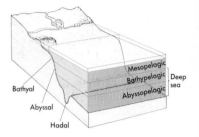

I t has been called "inner space." Dark and cold, inhabited by bizarre, fearsome-looking creatures, it *is* a little reminiscent of the outer space of science fiction movies (Figure 15-1). Like outer space, humans can venture into this mysterious realm only with the aid of elaborate, specially designed craft. Even then there is an element of risk. But "inner space" is here on earth: it consists of the waters of the sea that lie below the sunlit surface layer. The ocean depths are the least known of all our planet's environments.

The ocean depths include a number of distinct habitats. Immediately below the epipelagic lies the **mesopelagic,** or "middle pelagic," zone of the ocean (Figure 15-2). The epipelagic is roughly equivalent to the **photic zone,** the surface layer perhaps as deep as 150 to 200 m (500 to 650 feet) where there is enough light to support plant growth. In the mesopelagic there is still some dim light, but not enough for plants to grow. Below the mesopelagic there is no sunlight at all. This inky darkness is the world of the **deep sea.** The term "deep sea" is sometimes used to include the mesopelagic zone, but we use it only for the perpetually dark waters below the mesopelagic.

> **The waters below the epipelagic can be divided into the mesopelagic zone, where there is some light but not enough for primary production, and the deep sea, where there is no sunlight at all.**

Several different habitats are discussed in this chapter; each supports a distinct community of organisms. These communities share one important feature: there is not enough light for the primary production of food by photosynthesis. Without primary production to support the rest of the food web, the communities that lie beneath the photic zone depend on organic material produced in the surface layers of the ocean for food. Some of this surface production sinks into the dark waters below. Without a steady supply of food from above, there could be little life below the sunlit layer of the sea. The only known exception to this principle is discussed in the section "Deep-Sea Hot Springs" on p. 477.

Because of the dependence on surface production for food, life is much less abundant below the photic zone than in the sunlit surface layer. Most food particles get eaten before they sink into deeper water. With food in short supply, pelagic organisms become more and more scarce at greater depths. There are typically 5 or 10 times fewer organisms at 500 m (1,600 feet), for example, than there are at the surface, and perhaps 10 times fewer again at 4,000 m (13,000 feet).

Figure 15-1 A jellyfish, *Halicreas minimum,* photographed at a depth of 730 m (2,400 feet).

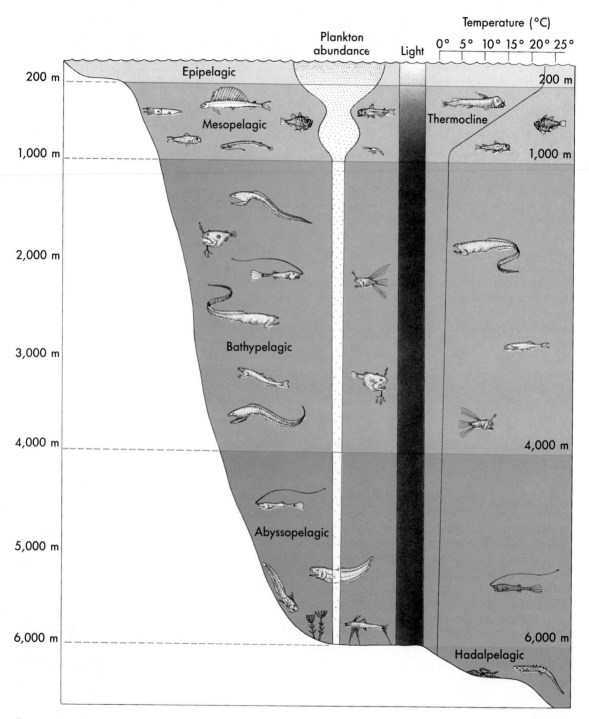

Figure 15-2 Life in the mesopelagic and deep sea is closely linked to the abundance of plankton and light intensity in the water column.

STRUCTURE AND FUNCTION OF MARINE ECOSYSTEMS

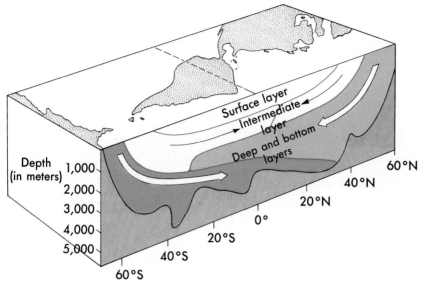

Figure 15-3 The deep water masses in the ocean originate at the surface in the extreme north and south Atlantic, then sink and spread out along the bottom. Water that originates in the Atlantic also spreads into the other ocean basins.

Deep-water organisms depend on the surface not only for food, but for oxygen (O_2). If the ocean were stagnant, the oxygen below the surface would be quickly used up and animal life would be impossible. Fortunately, however, there is gradual circulation to even the deepest parts of the sea, bringing life-giving oxygen. To sink all the way to the deep sea, oxygen-rich surface water must become very dense, that is, cold and relatively salty (see the section in Chapter 3, "Vertical Motion and the Three-Layer Ocean"). This takes place at only a few locations and only occasionally. The main places where this surface **overturn** reaches the bottom are in the Atlantic, south of Greenland, and just north of Antarctica (Figure 15-3). After sinking, the water spreads out through the ocean basins, carrying oxygen with it. Oxygen may become depleted in some places, but by and large this deep circulation replenishes the oxygen in the ocean depths.

overturn The sinking of surface water caused by an increase in its density, which results from a decrease in temperature, an increase in salinity, or both
Chapter 3, p. 81; Figure 3-36

THE TWILIGHT WORLD

The mesopelagic is a world of twilight. The dim light during the day is enough to see by, perhaps even enough to read a newspaper by in the upper part of the mesopelagic, but not enough to support phytoplankton or other plants. As the depth increases, of course, the sea gets darker. Eventually, typically at a depth of about 1,000 m (3,000 feet), there is no light at all. The absence of light marks the bottom of the mesopelagic zone, which thus stretches from the bottom of the epipelagic, around 200 m (660 feet) deep, to about 1,000 m deep.

The mesopelagic zone extends from about 200 m to about 1,000 m deep.

The temperature at a given depth in the mesopelagic generally varies much less than in the epipelagic. The mesopelagic, however, is the zone

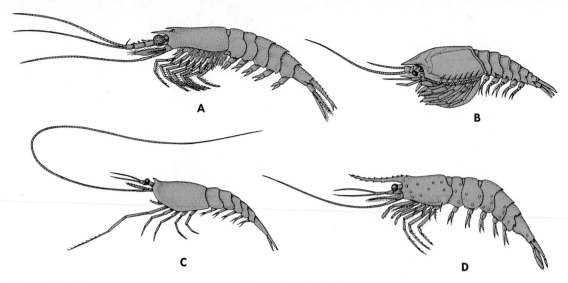

Figure 15-4 Some crustaceans representative of those that inhabit the mesopelagic zone include **(A)** krill (*Thysanopoda tricuspidata*); **(B)** mysid, or opossum, shrimps (*Gnathophausia ingens*), and "true," or decapod, shrimps like **(C)** *Sergestes similis* and **(D)** *Systellaspis debilis*.

main thermocline The zone between about 200 and 1000 m (650 and 3,300 feet) that is a transition between the warm surface layer of the ocean and the cold deep water
Chapter 3, p. 82

krill Planktonic, shrimplike crustaceans
Chapter 6, p. 174

copepods Tiny crustaceans that are often planktonic
Chapter 6, p. 172

where the **main thermocline** occurs (Figure 15-2), so organisms that move up and down in the water column encounter large changes in temperature. Mesopelagic organisms that stay at about the same depth experience more constant temperatures.

The Animals of the Mesopelagic

Though phytoplankton and other plants cannot live in the dim light, the mesopelagic supports a rich and varied community of animals, which are often called **midwater** animals.

Zooplankton. The major groups of animals in the mesopelagic zooplankton are much the same as in the epipelagic (see the section in Chapter 14, "The Organisms of the Epipelagic"). **Krill** (*Thysanopoda*, Figure 15-4, A; *Meganyctiphanes*, see Figure 16-16, A) and **copepods** (*Gaussia, Metridia*) are generally dominant, as they are in surface waters. Several different kinds of shrimps (*Sergestes, Gnathophausia*; Figure 15-4) are also common in the mesopelagic, relatively more so than in the epipelagic. Krill and most mesopelagic shrimps have a common adaptation of midwater animals: **photophores,** or **light organs,** which are specialized structures that produce light. The function of this "living light", called **bioluminescence,** is discussed in the section "Bioluminescence" on p. 461.

A group of crustaceans known as **ostracods** can be very abundant in the mesopelagic. Ostracods have a characteristic shell, or **carapace,** that makes them look like tiny clams with legs (Figure 15-5). They are crustaceans, however, and are unrelated to clams. Like copepods, most ostracods are small, usually only a few millimeters (⅛ inch) long. One group (*Giganto-*

Figure 15-5 A mesopelagic ostracod, *Giganto-cypris*. This one is about 1 cm (0.5 inch) long. Most ostracods, however, are much smaller.

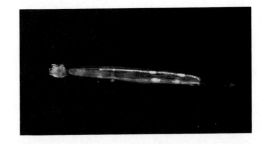

Figure 15-6 Arrow worms, or chaetognaths, are important predators in the midwater plankton, as they are in the epipelagic.

cypris), however, can reach 1 cm (½ inch). Amphipods (*Parathemisto*) and other crustacean groups are also part of the midwater plankton.

Arrow worms, or **chaetognaths** (*Sagitta, Eukrohnia*), are important midwater predators (Figure 15-6). At times they are among the most abundant components of the midwater zooplankton, especially in the upper parts of the mesopelagic. Jellyfishes (*Atolla, Solmissus;* Figure 15-1), siphonophores (*Lensia, Dimophyes*), comb jellies (*Aulacoctena*), larvaceans (*Bathychor-daeus*), and pteropods are common. Most of the other major groups that are part of the epipelagic zooplankton (see the section in Chapter 14, "The Zooplankton") are also represented in the mesopelagic, though the particular species are usually different.

Squids (*Abraliopsis, Histioteuthis;* Figure 15-7) are another prominent part of the midwater community. Some swim only weakly and therefore are considered to be planktonic, whereas strong-swimming squids are part of the nekton. Mesopelagic squids usually have photophores, which characteristically are arranged in a different pattern in each species. There are also a few mesopelagic octopuses (*Vitreledonella, Japatella*). The vampire squid (*Vampy-roteuthis*) looks something like an octopus but is actually neither a squid nor an octopus. It is in its own separate group (Figure 15-8).

Midwater Fishes. Nearly all mesopelagic fishes are quite small, about 2 to 10 cm (1 to 4 inches) long, though a very few species get considerably larger. Bristlemouths (*Cyclothone, Gonostoma;* Figure 15-10, A) and lanternfishes (*Myctophum, Diaphus;* Figure 15-10, B) are by far the most abundant fishes in the mesopelagic. These two groups may account for 90% or more of the fishes collected by midwater trawls (Figure 15-9). The bristlemouths are the most common of all. One species (*Cyclothone signata*) is thought to be the most abundant fish on earth, which is both surprising and impressive when you consider the huge schools of fishes like sardines and herrings that live in the epipelagic. Bristlemouths are named for their many sharp teeth. They have rows of photophores on their underside, or ventral surface.

arrow worms or chaeto-gnaths Small, wormlike predators with fins at the tail and spines on the head
Chapter 6, p. 163

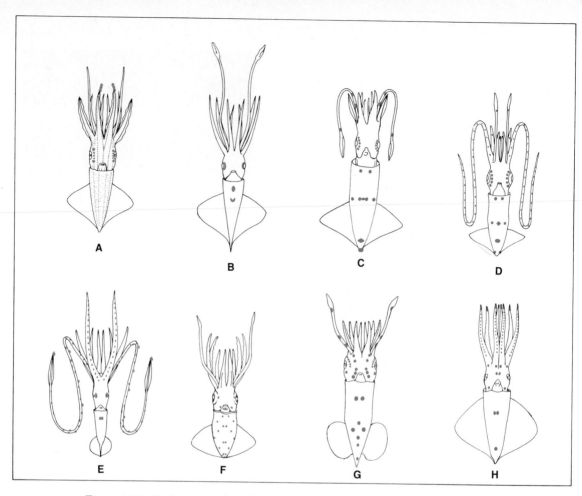

Figure 15-7 Each species of squid has a different pattern of photophores. The photophores are shown in color. **A,** *Abraliopsis;* **B,** *Cycloteuthis;* **C,** *Selenoteuthis;* **D,** *Nematolampas;* **E,** *Chiroteuthis;* **F,** *Thelidioteuthis;* **G,** *Pterygioteuthis;* **H,** *Octopoteuthis.*

Figure 15-8 The vampire squid, *Vampyroteuthis infernalis,* is not really a squid. It has ten arms like true squids, but two of the arms are modified into long, retractable "feelers" instead of being used to capture prey as in squids. Vampire squid only reach a length of 20 cm (8 inches).

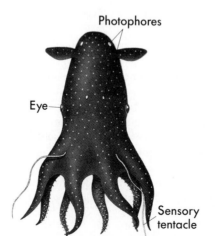

Photophores

Eye

Sensory tentacle

STRUCTURE AND FUNCTION OF MARINE ECOSYSTEMS

Figure 15-9 Nets like these, called rectangular midwater trawls, are commonly used to collect mesopelagic organisms. The net can be opened and closed by remote control while it is at the desired depth. This prevents surface organisms from being caught while the net is being lowered and retrieved. In these photos the mouth of the net is open.

Figure 15-10 Some typical mesopelagic fishes: **A,** bristlemouth *(Cyclothone braueri)*; **B,** lanternfish *(Myctophum affine)*; **C,** hatchetfish *(Polyipnus laternatus)*; **D,** Pacific viperfish *(Chauliodus macouni)*; **E,** dragonfish *(Leptostomias gladiator)*; **F,** barracudina *(Lestidium atlanticum)*; **G,** longnose lancetfish *(Alepisaurus ferox)*; **H,** sabertooth fish *(Coccorella atrata)*; and **I,** giganturid *(Gigantura vorax)*.

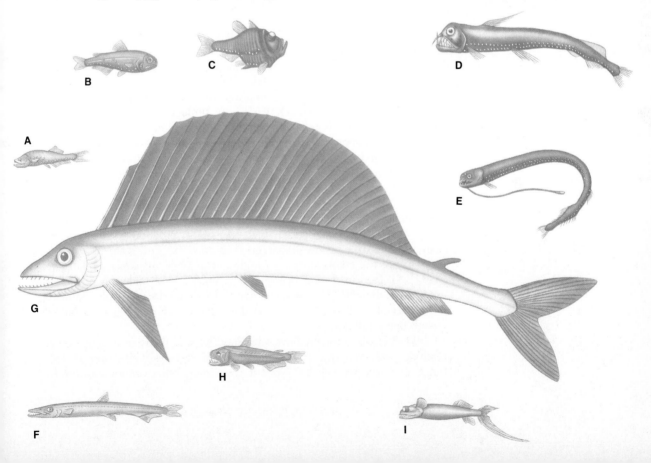

Figure 15-11 Black scabbard fish *(Aphanopus carbo)* being sold by Madeiran fishermen. The fish, known locally as *espada,* is a rare example of a midwater fish that is caught for food.

Lanternfishes get their name from the rows of photophores that adorn their heads and bodies. Like squids, each species typically has a distinctive photophore pattern. Lanternfishes have blunt heads, relatively large mouths, and large eyes, probably an adaptation to help them see in the dim light. Lanternfishes are very general in their food habits, eating just about anything they can get down.

The most common mesopelagic—or midwater—organisms are krill, copepods, shrimps, and small fishes like bristlemouths and lanternfishes.

Bristlemouths and lanternfishes are the most numerous, but many other fishes live in the mesopelagic (Figure 15-10). Marine hatchetfishes *(Argyropelecus, Sternoptyx)* look somewhat like the unrelated freshwater hatchetfishes sold in pet shops, except for their large eyes and mouth and ventral photophores. Viperfishes *(Chauliodus)*, dragonfishes *(Stomias, Idiacanthus)*, barracudinas *(Paralepis, Lestidium)*, sabertooth fishes *(Evermannella, Coccorella)*, lancetfishes *(Alepisaurus)*, snake mackerels *(Gempylus)*, and cutlassfishes *(Aphanopus)* are all long, eel-like fishes with large mouths and eyes. Many of these fishes have photophores, usually in rows along the ventral surface. Most of them are less than 30 cm (12 inches) long, but there are a few exceptions. One of the lancetfishes *(Alepisaurus ferox;* Figure 15-10, G), grows to about 2 m (6.5 feet) and may be the largest mesopelagic fish. The black scabbard fish *(Aphanopus carbo)*, a kind of cutlassfish, can get well over 1 m (3.3 feet) long. It is caught by fishermen from the Atlantic island of Madeira, off Portugal (Figure 15-11).

Adaptations of Midwater Animals

By the standards of the lighted world, many of the animals that live below the photic zone seem bizarre indeed. Though they look strange to us, they are well adapted to their unique environment. Species from the same depth often have very similar characteristics even though they are unrelated. On the other hand, closely related species that live at different depths may differ markedly in appearance and other characteristics.

Feeding and Food Webs. Most of the food produced in the epipelagic is used there, so only about 20% of the food makes it to the mesopelagic. This means that the mesopelagic is chronically short of food, which is why there are fewer organisms in the mesopelagic than in the epipelagic. The abundance of midwater organisms is related to the productivity of the waters above: there is more mesopelagic life under highly productive surface waters than under areas with low primary production.

Many of the characteristics of midwater animals are directly related to the lack of food in the mesopelagic. The small size of midwater fishes, for example, is thought to be an adaptation to the limited food supply. Animals need a lot of food to grow large, so there may be an advantage to limited growth and small size.

Midwater fishes usually have large mouths, and many have hinged, extendible jaws equipped with fearsome teeth (Figures 15-12 and 15-13). Since food is scarce below the photic zone, most fishes cannot afford to be

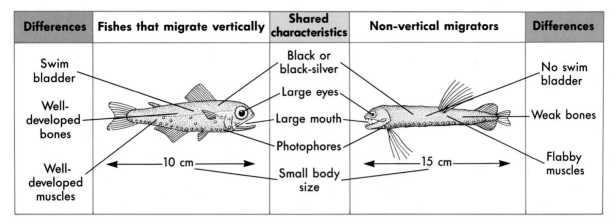

Differences	Fishes that migrate vertically	Shared characteristics	Non-vertical migrators	Differences
Swim bladder Well-developed bones Well-developed muscles	←10 cm→	Black or black-silver Large eyes Large mouth Photophores Small body size	←15 cm→	No swim bladder Weak bones Flabby muscles

Figure 15-12 Some adaptations of typical mesopelagic fishes, including some differences between vertical migrators like lanternfishes (left) and nonmigrators like dragonfishes (right). Compare these with the adaptations of epipelagic and deep-sea fishes shown in Figures 14-22 and 15-25.

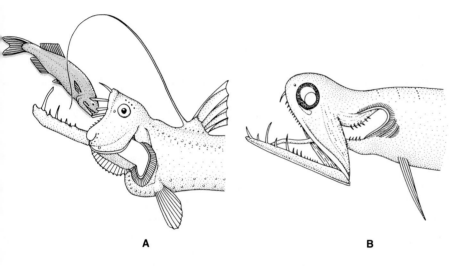

A

B

Figure 15-13 **A,** The viperfish (*Chauliodus*) has hinged jaws that can accomodate large prey. **B,** The rat-trap fish (*Malacosteus*) has a similar jaw arrangement.

picky. They usually have very broad diets and eat just about anything they can fit into their mouths. The large, protrusible jaws allow them to eat a wide range of prey; consequently, they don't have to pass up potential meals because they are too large to eat. Some mesopelagic fishes can even eat prey larger than themselves! The long, sharp teeth help keep prey from escaping.

> Common adaptations of midwater fishes include small size; large mouths; hinged, extendible jaws; needlelike teeth; and unspecialized diets. These adaptations result from the limited food supply in the mesopelagic.

Midwater animals fall into two major groups: those which stay in the mesopelagic and those which migrate to the surface at night. The non-migrators include a few species of small zooplankton, mainly copepods and

krill, that filter out **detritus** and the small amount of phytoplankton that
sinks out of the photic zone. The fecal pellets of epipelagic copepods and
other surface grazers form an important part of the detritus eaten by mesopelagic filter feeders. These pellets sink much faster than individual phytoplankton cells, so they stand a better chance of making it to the mesopelagic before being eaten.

Most midwater animals that do not migrate, however, are not zooplankton but fishes, shrimps, and squids. They are sit-and-wait predators that lurk
in the dim light, gulping down anything that comes within range. With
food so hard to come by, these organisms have a number of adaptations that
reduce their energy requirements. Instead of energy-consuming muscle,
non-migrating midwater fishes have flabby, watery flesh. Most of these fishes
have lost the swim bladder, which requires energy to fill and regulate. To
reduce weight and be more neutrally buoyant, they have developed soft,
weak bones and have lost defensive structures such as spines and scales.
This allows them to float at a constant depth without using up energy swimming. Since they don't swim much, they have no need of the streamlining
that is so characteristic of epipelagic fishes.

Vertical Migration and the Deep Scattering Layer.   Rather than staying put and waiting for food to descend from above, most mesopelagic organisms make **vertical migrations.** They swim up at night to feed in the rich
surface layers and during the day descend to depths of several hundred
meters or more, where in the dim light they probably are relatively safe
from predators. Some vertical migrators spend the day in a lethargic stupor,
conserving energy until their next foray to the surface. Vertical migration is
also seen in many zooplankton that live in the deeper parts of the epipelagic
(see the section in Chapter 14, "Vertical Migration").

Vertically migrating fishes differ in several important ways from those
which stay in the mesopelagic (see Figure 15-12). Well-developed muscles
and bones are needed to make the daily swim up and down the water column. These structures increase the weight of the fish, so these fishes have
retained the swim bladder for buoyancy. As they move up and down, they
experience dramatic changes in pressure. Vertically migrating fishes can rapidly adjust the volume of gas in the swim bladder to prevent it from collaps-
ing or exploding when they change depth. They are also able to tolerate the
temperature fluctuations they experience as they move up and down across
the thermocline.

> Midwater animals that make vertical migrations have well-developed bones
> and muscles, wide temperature tolerances, and—in fishes—swim bladders.
> Non-migratory midwater fishes have flabby muscles and lack swim bladders.

Vertical migration by mesopelagic animals was first discovered during
World War II, when sonar came into use. Sonar soundings regularly showed
a series of sound-reflecting layers, or "false bottoms." The real bottom gives
a sharp, clear echo, but these layers—collectively dubbed the **deep scattering layer (DSL)**—give a soft, diffuse echo that makes a shadowy trace on
sonar plots (Figure 15-14). During the day the DSL lies at depths of 300 to
500 m (1,000 to 1,600 feet), but at sunset it rises to the surface. The depth

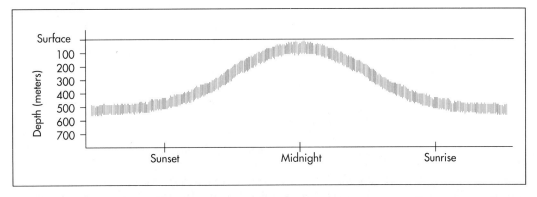

Figure 15-14 Diagrammatic representation of the deep scattering layer.

of the DSL is clearly related to light intensity: the DSL stays deeper when there is a full moon than on moonless nights and even moves up and down when clouds pass over the moon.

Net tows have shown that the DSL includes fishes—especially lanternfishes, krill, shrimps, copepods, jellyfishes, squids, and other midwater animals. Most of these organisms, though they are found within the DSL, do not contribute to its sound-reflecting properties. Since organisms are made mostly of water, sound passes through them much as it does through seawater, without bouncing off. Air pockets, however, reflect sound quite strongly. The echo of the DSL comes mostly from sound waves bouncing off of the swim bladders of the fishes.

The deep scattering layer (DSL) is a sound reflecting layer made up of vertically migrating midwater animals. Lanternfishes, krill, and shrimps, are the dominant organisms of the DSL.

Vertical migration is important in transporting food into deep water. When lanternfishes, krill, and other animals return to the mesopelagic after stuffing themselves near the surface, they carry the products of surface production down with them. This greatly increases the food supply in the mesopelagic. Many non-migratory midwater predators feed heavily on vertically migrating species. Since migrators are more muscular than non-migrators, they provide a much more nutritious meal.

Sense Organs. To help them see in the dim light, midwater fishes characteristically have eyes that are not only large but unusually sensitive. Large, light-sensitive eyes also occur in squids, shrimps, and other groups. Some midwater fishes have developed **tubular eyes** (Figure 15-15), a complex visual system that is almost like having two pairs of eyes. Tubular eyes give very acute vision in the direction that the eyes are pointed, either upward or forward, but are not good for lateral vision. To compensate, the **retina**—the light-sensitive part of the eye—extends part way up one side of the eye (Figure 15-16, A). This allows the fish to see to the side and below (Figure 15-17). In normal eyes the retina lies only at the back of the eye.

Some fishes with tubular eyes have special yellow filters in their eyes. The filters allow the fish to distinguish between natural light and bioluminescence. Thus the fish can at least partially defeat the defensive counterillumination of prospective prey.

Figure 15-15 Barreleyes, or spookfishes, (such as *Dolichopteryx longipes*) are midwater fishes with tubular eyes.

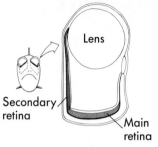

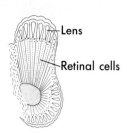

A. Fish

Lens

Secondary retina

Main retina

B. Deep-sea octopus

Lens

Retina

C. Krill (bilobed eye)

Lens

Retinal cells

Figure 15-16 **A,** The eye of a midwater fish *(Scopelarchus).* The blue layer is the retina, which is the part of the eye sensitive to light. The cross-hatched part of the retina at the bottom of the eye provides good upward vision. The part of the retina that extends up the side of the eye provides lateral vision (see Figure 15-17), which is less acute. Eyes with a similar structure have evolved in **(B)** octopuses *(Amphitretus pelagicus)* and **(C)** krill *(Stylocheiron suhmii).*

Superior visual field

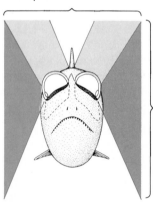
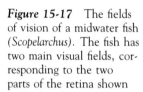

Lateral visual field

Figure 15-17 The fields of vision of a midwater fish *(Scopelarchus).* The fish has two main visual fields, corresponding to the two parts of the retina shown in Figure 15-16, A.

Figure 15-18 A midwater krill that features bilobed eyes. Also see Figure 15-4, A.

Figure 15-19 This squid *(Histioteuthis)* has different-sized eyes, visible here on opposite sides of the head.

Adaptations very similar to tubular eyes are found in at least one octopus (*Amphitretus*) and in the bilobed eyes of some krill (*Stylocheiron*; Figures 15-16, *B* and *C*, and 15-18). One squid, *Histioteuthis*, has taken a different path to the same apparent end. One eye in these squids is much larger than the other (Figure 15-19). The squid floats in the water column with the large eye pointed upward and the small eye downward.

Many midwater animals have evolved large eyes that provide good vision in dim light. Other adaptations include tubular eyes in fishes and bilobed eyes in krill.

Eyesight alone is probably not enough in the twilight of the mesopelagic. Midwater fishes also have very well-developed **lateral lines.** The elongated shape of many of these fishes may be related to the development of the lateral line: the longer the body, the longer—and therefore more sensitive—the lateral line.

Coloration and Body Shape. Like their epipelagic counterparts, mesopelagic predators rely heavily on vision. Since midwater prey cannot afford the energetic costs of fast swimming or heavy defensive spines and scales, camouflage is perhaps even more important than in the epipelagic. The basic strategies, however, are very similar: **countershading,** transparency, and reduction of the silhouette. There are many variations on these basic strategies, however, especially in relation to depth and light levels.

Transparency is particularly common in the shallower and more well-lit parts of the mesopelagic. Copepods, jellyfishes, shrimps, bristlemouth fishes, and other animals that live in the upper mesopelagic tend to be transparent, some almost completely so. Deeper in the mesopelagic, fishes tend to be more silvery, and in the deepest, darkest part, black. Zooplankton from the deeper parts of the mesopelagic are typically orange, red, or purple. These colors might be quite conspicuous at the surface, but colors change underwater (see the section in Chapter 3, "Transparency"). Since red light does not penetrate to mesopelagic depths, these organisms are an inconspicuous gray or black in their natural habitat.

Mesopelagic fishes often have black backs and silvery sides (Figure 15-20), reminiscent of the countershading of epipelagic animals (see the section in Chapter 14, "Coloration and Camouflage"). In the mesopelagic, however, there is not enough light to reflect off the white or silver belly typical of epipelagic countershading and mask the animal's outline. Even a white object produces a silhouette in twilight. A silhouette makes the animal conspicuous and vulnerable to all those sharp eyes peering up from the depths. To reduce their silhouettes, some mesopelagic fishes— like hatchetfishes—have laterally compressed bodies. This reduces the size of the body outline when the animal is viewed from above or below.

Bioluminescence. Most midwater animals have evolved a much more effective way to mask their silhouettes. Their bioluminescent photophores, found mostly on their underside, produce light that helps the animal blend in with the background light that filters down from the surface (Figure 15-21). This adaptation, which functions in a similar manner to countershading, is called **counterillumination.**

lateral line A system of canals on the sides of fishes that helps them detect vibrations in the water
Chapter 7, p. 212; Figure 7-20

countershading A pattern of coloration common in open water nekton in which the back is black or dark blue and the underside white or silver
Chapter 14, p. 421

Figure 15-20 To be as inconspicuous as possible to potential predators, hatchetfishes have black backs and silvery sides. Their laterally compressed bodies help reduce their silhouettes.

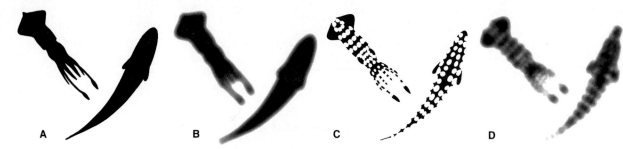

Figure 15-21 The value of having photophores on the lower, or ventral, surface of midwater animals is shown in this simple demonstration. **A** and **B** are the silhouettes of two midwater organisms as they would appear without photophores. **C** and **D** are the same animals, with the photophores in white to match the background. The photophores break up the silhouette, making the animal much less visible. This effect is more pronounced when the animals are a little out of focus (**B** and **D**), as they are when viewed through the water.

The light produced by midwater animals is closely matched to the background light. Like the natural light at these depths, nearly all midwater bioluminescence is blue-green. Furthermore, some—and perhaps most—mesopelagic animals can control the brightness of the light that they produce to match it to the brightness of the background. This has been shown experimentally by placing special "blinders" on shrimps and other animals.

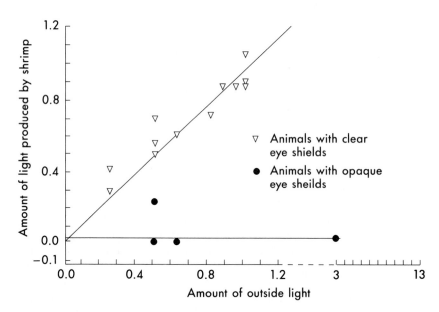

Figure 15-22 The following experiment showed that a midwater shrimp (*Sergestes similis*; see Figure 15-4, C) produces bioluminescence to match the background light intensity. Shrimp having both eyes covered with blinders (lower line) did not produce any light, no matter how much outside light there was. Those with clear blinders (upper line) produced more light as the outside level was increased.

STRUCTURE AND FUNCTION OF MARINE ECOSYSTEMS

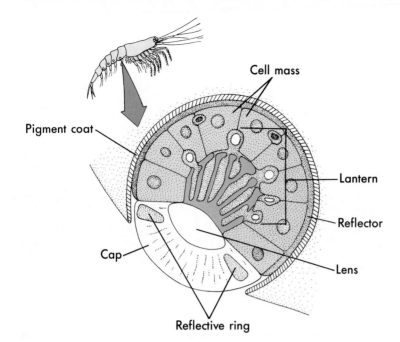

Cell mass

Pigment coat

Lantern

Reflector

Cap

Lens

Reflective ring

Figure 15-23 A cross section through a photophore of a krill.

These blinders allow experimenters to control the amount of light that the animal sees (Figure 15-22). When the animal is exposed to bright light, it produces bright bioluminescence; when the light is dim so is the bioluminescence. When opaque blinders are put on the shrimp so that it sees no light, it "turns off" its photophores completely. This control of the brightness of the bioluminescence is essential. It is easy to see an animal if it produces light at night or if the light is brighter than the background. On the other hand, the animal creates a silhouette if the light is not bright enough.

Most midwater animals are bioluminescent, and midwater organisms have evolved many different ways to produce light. Photophores, as we have seen, are common. In some species the light is produced by the animal's own specialized tissue. In other species, symbiotic bacteria live inside the light organ and produce the light. In either case, the photophore can be quite complex (Figure 15-23).

Bioluminescence is not always produced by specialized photophores. In many jellyfishes and other gelatinous animals, light is produced by cells that are scattered over the body surface. Some copepods, ostracods, shrimps, and other animals secrete bioluminescent fluids that they squirt out through special glands, either in addition to or instead of having photophores. Some squids and octopuses even produce bioluminescent ink.

Counterillumination is an important function of bioluminescence, but not the only one. The fact that the pattern of photophores is different among species, and even between sexes, may mean that bioluminescence is used to communicate and attract mates. Organisms that produce bioluminescent secretions may do so as a defense mechanism in the same way that shallow-water squids and octopuses use their ink. When disturbed, they often produce a burst of light and dart away. This presumably distracts preda-

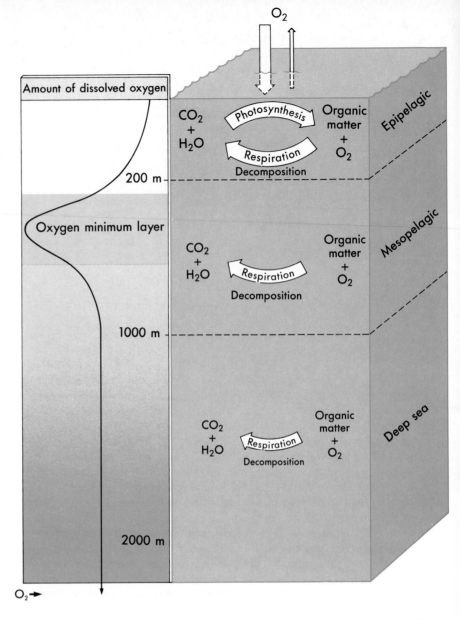

Figure 15-24 Surface waters are rich in oxygen, since oxygen enters from the atmosphere and is released by photosynthesis. In the mesopelagic zone neither the atmosphere nor photosynthesis can contribute oxygen to the water, but there is extensive bacterial decomposition of organic matter sinking from shallow water. This uses up oxygen and results in an oxygen minimum layer. Below the oxygen minimum layer, most of the organic matter has already decayed on its way down, and oxygen remains dissolved in the water.

tors and allows the prey to escape. Some animals have light organs around their eyes that may help them see. Some even use the light to lure prey.

Most midwater animals are bioluminescent. Bioluminescence is used in counterillumination to mask the silhouette, to escape from predators, to attract prey, and perhaps in communication and courtship.

The Oxygen Minimum Layer. In many places, midwater organisms have to deal with a shortage of oxygen in the water. Oxygen can enter the ocean in two ways: either by gas exchange with the atmosphere (see the section in Chapter 3, "Dissolved Gases") or as a by-product of photosynthesis. Once a water mass leaves the surface and descends to mesopelagic depths, there is no way for it to gain oxygen. It is removed from contact with the atmosphere, and there is not enough light for photosynthesis. Respiration and bacterial decay, however, continue to use up oxygen (Figure 15-24). In fact, a substantial proportion of the organic matter produced in the photic zone is decomposed in the mesopelagic. As a result the water becomes depleted in oxygen, often in a fairly well-defined layer around 500 m (1,600 feet) deep known as the **oxygen minimum layer.** The amount of oxygen in the oxygen minimum layer can drop to practically nothing. Below the oxygen minimum layer there is very little food and therefore very little respiration and decomposition, so the oxygen is not used up. Thus the water below the oxygen minimum layer retains most of the oxygen it had when it left the surface.

Animals live in the oxygen minimum layer despite the low oxygen concentration. Fishes, krill, and shrimps that live there usually have large, well-developed gills to help extract what little oxygen there is. They also tend to be inactive, which decreases their use of oxygen.

THE WORLD OF PERPETUAL DARKNESS

Below the mesopelagic lies the little-known world of the deep sea, where sunlight never penetrates. This alien environment is vast indeed. It is the largest habitat on earth and contains about 75% of our planet's liquid water. The deep sea can be divided into several depth zones. The **bathypelagic zone** includes depths between 1,000 and 4,000 m (3,000 to 13,000 feet), and the **abyssopelagic zone** lies from 4,000 to 6,000 m (13,000 to 20,000 feet). The **hadopelagic**—or **hadal pelagic zone**— consists of the waters of the ocean trenches, from below 6,000 m to just above the sea floor, as deep as 11,000 m (36,000 feet). Each of the depth zones supports a distinct community of animals, but they also have much in common. Here we stress the similarities, rather than the differences, among the depth zones of the deep sea. The deep sea also includes the ocean bottom beyond the continental shelf. Bottom-living organisms are covered separately (see the section "The Deep Ocean Floor" on p. 470).

trenches Deep depressions in the sea floor that are formed when two plates collide and one sinks below the other
Chapter 2, p. 29; Figures 2-14 and 2-15

> The deep sea includes the bathypelagic, from 1,000 to 4,000 m; the abyssopelagic, 4,000 to 6,000 m; and the hadopelagic, 6,000 m to the bottom of trenches. The deep sea also includes the deep-sea floor.

In the darkness of the deep sea there is no need for countershading. Many animals, especially zooplankton, are a drab gray or off-white. Deep-sea fishes tend to be black. Shrimps are often bright red, which in the deep sea has the same effect as being black. A few deep-sea fishes are also red.

Bioluminescence is as ubiquitous in the deep sea as in the mesopelagic. Deep-sea organisms do not use bioluminescence for counterillumination, however, since there is no light to create a silhouette. They have fewer photophores than midwater species, and the photophores tend to be on the head and sides rather than on the ventral surface. In the deep sea the primary uses of bioluminescence are probably to attract prey and in communication and courtship.

The large, sensitive eyes of midwater animals are superfluous in the darkness of the deep sea. Like animals that live in caves, many deep-sea animals are blind—especially in the deepest regions. Since they live in perpetual darkness, vision is not needed. Blindness is not universal, however, and many deep-sea animals retain functional eyes. This is probably because the deep sea is not *completely* dark. Sunlight may not reach the deep sea, but bioluminescence is common. Since bioluminescence is quite bright, the large eyes of midwater animals are not necessary. If deep-sea animals have eyes, they are almost always small (Figures 15-25 and 15-26).

The conditions of life in the deep sea change very little. Not only is it

Figure 15-25 Some typical characteristics of deep-sea pelagic fishes. Compare these with the adaptations shown in Figures 14-22 and 15-12.

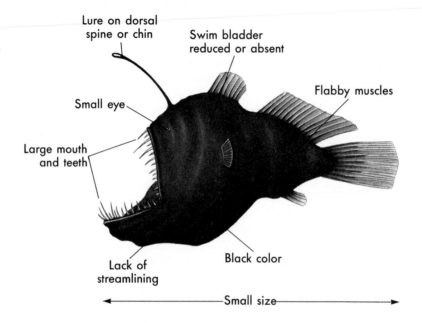

Lure on dorsal spine or chin

Swim bladder reduced or absent

Flabby muscles

Small eye

Large mouth and teeth

Lack of streamlining

Black color

Small size

Figure 15-26 Comparison of two closely related bristlemouths, one from the mesopelagic (*Gonostoma denudatum,* left) and one from the deep sea (*G. bathyphilum,* right). The deep-sea fish has smaller eyes, less muscle, and fewer light organs. It also has less developed nervous and circulatory systems, as indicated by the smaller brain and gill filaments.

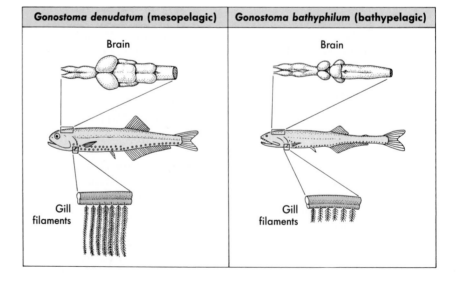

Gonostoma denudatum (mesopelagic)	*Gonostoma bathyphilum* (bathypelagic)
Brain	Brain
Gill filaments	Gill filaments

STRUCTURE AND FUNCTION OF MARINE ECOSYSTEMS

always dark, it is always cold: the temperature remains nearly constant, typically at 1° to 2° C (35° F). Salinity and other chemical properties of the water are also remarkably uniform.

The Lack of Food

Deep-sea organisms may not have to adapt to variations in the physical environment, but they face a continual shortage of food. Very little, about 5%, of the food produced in the photic zone makes it past all the hungry mouths in the waters above. Deep-sea animals do not make vertical migrations to the rich surface waters, probably because the surface is too far away and the change in pressure too great. With food critically scarce, deep-sea animals are few and far between.

The energy-saving adaptations to food shortage seen in midwater organisms are accentuated in the deep sea. Deep-sea fishes—the most common of which are bristlemouths and deep-sea anglerfishes (*Linophryne, Melanocetus*),—are very small, usually 10 cm (4 inches) or less (Figure 15-27). The

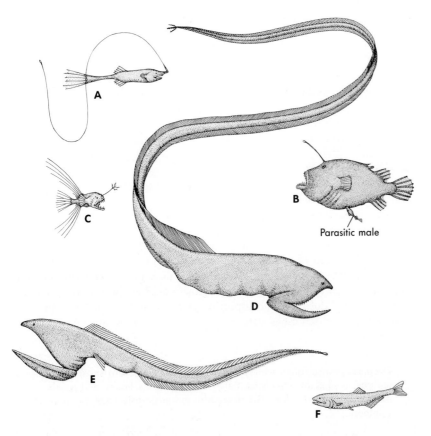

Parasitic male

Figure 15-27 Some examples of deep-sea fishes: **A,** deep-sea anglerfish (*Gigantactis macronema*); **B,** female deep-sea anglerfish with one attached male (*Cryptopsarus couesi*); **C,** deep-sea devilfish (*Caulophryne acinosa*); **D,** swallower (*Saccopharynx ampullaceus*); **E,** gulper (*Eurypharynx pelecanoides*); and **F,** deep-sea bristlemouth *Gonostoma bathyphilum*.

Figure 15-28 For unknown reasons, the deep-sea members of some groups are giants compared to their shallow-water relatives. This is a deep-sea amphipod (*Alicella gigantea*).

Figure 15-29 The gulper (*Eurypharynx pelecanoides*) has a huge mouth plus a very large, distensible stomach. See Figure 15-27, *E* for a view of the whole fish.

trend to small size, however, does not always hold in crustaceans and other invertebrates. In these groups, individuals are sometimes larger than in shallow-water environments. In some species, in fact, individuals are enormous by shallow-water standards (Figure 15-28), a phenomenon known as **deep-sea gigantism.** The reasons for gigantism in deep-sea invertebrates are unknown.

Deep-sea fishes are sluggish and sedentary, even more so than midwater fishes. They have flabby, watery muscles; weak skeletons; no scales; and poorly developed respiratory, circulatory, and nervous systems. Nearly all lack functional swim bladders. It appears that these fishes float in the water column, expending as little energy as possible, until a meal comes along. Most deep-sea fishes have huge mouths and can consume prey much larger than themselves. This trend reaches its peak in the gulper eels (*Eurypharynx*; Figures 15-27, *E* and 15-29) and swallowers (*Saccopharynx*; Figure 15-27, *D*), which look like swimming mouths. To go along with their large mouths, many species have stomachs that can expand to accommodate the prey once it has been engulfed (Figure 15-30).

Female anglerfishes have evolved an unusual method of catching food, from which they get their name. The first spine of their dorsal fin is modified into a long, movable "pole" that they wave in front of their mouths (Figures 15-25 and 15-27, *A* to *C*). Dangling from the end of the pole is the "bait," a fleshy bit of tissue that resembles a tasty meal. Symbiotic bioluminescent bacteria live in the bait, so it glows enticingly in the dark. The anglerfish gobbles down any unsuspecting victim that is attracted to the bait. Many other deep-sea fishes attract prey with lures, often on barbels attached to the chin.

> Deep-sea pelagic fishes are typically small and black, with small eyes, large mouths, expandable stomachs, flabby muscles, weak bones, and poorly developed swim bladders. Bristlemouths and anglerfishes are the most common.

Figure 15-30 A black deep-sea swallower (*Chiasmodon niger*) with distended stomach after swallowing a very large prey.

Sex in the Deep Sea

Food is not the only thing that is scarce in the deep sea. In such a vast, sparsely populated world, finding a mate can be a serious problem—even harder than finding food. After all, most deep-sea animals are adapted to eat just about anything they can get, but a mate must be both the right species and the opposite sex!

Many deep-sea fishes have solved the latter problem by becoming **hermaphrodites.** After all, it would accomplish nothing to finally get two members of the same species together only to find that they are both the same sex. If every individual can produce both eggs and sperm, breeding is ensured.

Deep-sea organisms probably have evolved ways to attract mates also. Bioluminescence, for example, may send a signal that draws other members of the same species. Many species, as we have seen, have a unique pattern of light organs. The lure of female anglerfishes also differs among species, so it may have a role in attracting mates. Chemical attraction can be important as well. Male anglerfishes have a very powerful sense of smell, which they use to locate females. The females apparently release a special chemical that the male can detect and follow. Such special chemicals are called **pheromones.**

Some anglerfishes (*Cryptopsaras, Ceratias*) have evolved a most unusual solution to the problem of finding mates. When a male finally locates a female, who is much larger, he bites into her side, where he remains attached for the rest of his life (Figure 15-27, B). In some species the male's modified jaws fuse with the female's tissue. Their circulatory systems join, and the female ends up nourishing the male. This arrangement, known as **male parasitism,** ensures that the male is always available to fertilize the female's eggs.

Neither hermaphroditism nor male parasitism seem to be common in deep-sea invertebrates. The mechanisms that bring males and females together, if any, are unknown. There is some evidence that they aggregate into breeding groups, perhaps attracted by bioluminescent signals.

hermaphrodites Individuals with both male and female gonads
Chapter 7, p. 220

Finding mates, a problem for deep-sea animals, is eased by the use of bioluminescent and chemical signals and by the development of hermaphroditism and male parasitism.

Living Under Pressure

Under the weight of the overlying water column, the pressure in the deep sea is tremendous. This is one reason that so little is known about the deep sea. Instruments, camera housings, and deep submersibles are very expensive because they must be carefully designed and built to withstand the pressure without being crushed. Only a very few submersibles can venture into the deepest trenches, where the pressure may exceed 1,000 atmospheres (14,700 psi; see the section in Chapter 3, "Pressure"). It is just as difficult to bring animals up from the deep sea as it is for us to go down to them. Unable to endure the enormous change in pressure, they invariably die when brought to the surface. A few scientists have succeeded in retrieving organisms from the deep sea in special pressurized chambers. Much has been learned from such work, but it is frustratingly difficult.

It is clear that pressure has important effects on deep-sea organisms. The absence of a functional swim bladder in deep-sea fishes, for example, probably is due to the high energetic cost of filling the bladder with gas under extreme pressure. Along with the availability of food, pressure seems to be the main factor causing zonation in deep-sea pelagic organisms—that is, dividing the deep sea into bathypelagic, abyssopelagic, and hadopelagic zones. Some organisms simply cannot live at great pressures. Fishes seem to be limited by depth: the deepest-living fish known does not go beyond about 8,400 m (27,500 feet).

The depth range of deep-sea organisms appears to be determined in part by the effects of pressure on the basic processes in living things. High pressure, for instance, profoundly affects the workings of the enzymes that control metabolism. It appears that deep-sea organisms have enzymes that are adapted to function under pressure. The enzymes of particular species are adapted to a specific range of pressure and therefore of depth. This is probably a crucial factor in determining the distribution of deep-sea organisms.

Hydrostatic pressure is great in the deep sea and partially controls the depth distribution of deep-sea organisms.

THE DEEP OCEAN FLOOR

The floor of the deep sea shares many of the characteristics of the pelagic waters immediately above: the absence of sunlight, constant low temperature, and great hydrostatic pressure. Nevertheless, the biological communities of the deep-sea floor are very different from pelagic communities because of one key factor: the presence of the bottom.

Marine biologists have learned a bit more about the **benthos**—or bottom-inhabiting organisms—of the deep sea than about deep-water pelagic communities, but they still have barely scratched the surface. What we do know has been learned using a variety of techniques. Devices called epibenthic sleds are dragged along the bottom, scooping up organisms, and corers actually bring a chunk of the bottom to the surface. Remote-controlled robots (Figure 15-31) are used to collect samples and perform experiments. Deep submersibles like *Alvin* (Figure 1-19) have been useful, even

enzymes Proteins that speed up and control chemical reactions in organisms
Chapter 4, p. 88

metabolism The vast set of chemical reactions that sustains life
Chapter 4, p. 86

Figure 15-31 The remote underwater manipulator (RUM) is used to sample the deep-water benthos. Operated by remote control, RUM can move along the bottom. The arm can be used to collect samples or manipulate equipment. The device can also hold cameras, water bottles, and other devices.

more so than in the water column. Deep-sea cameras are used to photograph fast-swimming animals such as fishes that we have not yet learned to catch in nets.

Feeding in the Deep-Sea Benthos

As you might expect, food shortage is of extreme importance on the floor of the deep sea. Very little of the surface production makes it all the way to the bottom. Benthic organisms, however, have a major advantage over pelagic ones. In the water column, food particles that are not immediately located and eaten sink away and are lost. Once food reaches the bottom, on the other hand, it stays put until it is found. Thus, although pelagic animals may get the "first crack" at food sinking out of the photic zone, benthic ones get a much longer opportunity. Food particles that reach the bottom tend to be those that sink fairly rapidly, minimizing the chance that they are eaten on the way down. Fecal pellets, for example, are an important source of organic matter for the deep-sea benthos.

Still, the "rain" of organic matter to the sea floor is actually more like a drizzle. Very little food is available to the benthic community. Furthermore, much of the material that reaches the sea floor, like the chitinous remains of crustacean zooplankton, is not immediately digestible. On the sea floor, however, bacteria decompose the chitin, which becomes food for other organisms.

Most of the deep-sea floor is covered in fine, muddy sediments. The **meiofauna,** tiny animals that live among the sediment particles (see Box 11-1), graze on bacteria and absorb nutrients from the water between the particles. The meiofauna probably play a major role in making the energy in bacteria and dissolved organic matter (DOM) available to larger benthic animals.

Suspension feeders are rare among the larger organisms in the deep-sea benthos. Instead, **deposit feeders** are the dominant benthic organisms, especially polychaete worms, bivalve molluscs, crustaceans, sea cucumbers, and brittle stars. Many of these are **infauna** and burrow in the sediments. Others, the **epifauna,** rest on the sediment surface.

> **The deep-sea benthos is dominated by deposit feeders. The dominant animal groups are the meiofauna, polychaetes, bivalves, crustaceans, sea cucumbers, and brittle stars.**

There are predators in the deep-sea benthos, but they seem to be fairly rare. The main predators on deposit-feeding animals are probably sea stars, brittle stars, and crabs. Members of the nekton, like fishes and squids, are also important predators. Sea spiders, or pycnogonids, prey on other invertebrates by sucking out their soft parts. Sea spiders are small in shallow water (see Figure 6-47), but some deep-sea species can be as large as 80 cm (30 inches) across (Figure 15-32). Another interesting group of predators are the tripod fishes (*Bathypterois, Benthosaurus*). Nearly blind, these fishes sit on the bottom on their elongated fins (Figures 15-33 and 15-36, B), facing into the current and snapping up passing plankton.

Occasional "baitfalls"—large pieces of food that sink rapidly, like the dead bodies of large fishes or whales—are a source of food to the bottom.

Figure 15-32 A deep-sea pycnogonid photographed at a depth of approximately 1,900 m (6,200 feet).

chitin Highly resistant material found in the skeleton of crustaceans and other structures
Chapter 6, p. 165

suspension feeders Animals, including *filter feeders*, that eat particles suspended in the water column
Chapter 6, p. 147; Figure 6-20

deposit feeders Animals that eat organic matter that settles to the bottom
Chapter 6, p. 158; Figure 6-20

BOX 15-1

The Chambered Nautilus

Ammonites like this fossil grew to great size and once dominated the oceans.

The chambered nautilus (*Nautilus*).

The chambered nautilus (*Nautilus*) is often called a living fossil. All other cephalopods—the squids, octopuses, and cuttlefishes—have only vestiges of the hard outer shell that characterizes most molluscs. Only *Nautilus* retains a large, heavy shell. It is the last surviving representative of a group of organisms that once ruled the seas.

Most of the early animals in the sea lived on the bottom, crawling and scratching for survival. Around 500 million years ago, a group of molluscs that looked something like today's limpets developed a new trick: the ability to partially fill their shells with gas. Buoyed by the gas, these ancient ancestors of modern cephalopods—called nautiloids—were able to float up off the bottom, away from predators. Before long they also developed the ability to move by squirting out water. Perhaps this was the the first form of jet propulsion.

The trick was immensely successful. For 200 million years or so, nautiloids and their descendants—especially a group with coiled shells called the ammonites—dominated the ocean. They had the water column more or less to themselves and could drop on their prey from above. Then fishes arrived on the scene. To be more accurate, fishes had been around for some time, but they finally had evolved the swim bladder and thus could maintain neutral buoyancy and compete with nautiloids. Fishes had a major advantage: they could swim faster and more efficiently than could the nautiloids, with their cumbersome shell. Out-competed by fishes, most of the nautiloids became extinct. Others, the squids, became more fishlike. They abandoned the shell and became streamlined, active swimmers. Of all the heavy-shelled cephalopods, only *Nautilus* remains. By studying living *Nautilus*, scientists hope to learn more about their extinct relatives, who left an abundant fossil record and are of great interest to geologists.

Nautilus lives in a coiled shell that is partitioned into a series of chambers. The chambers are walled off one by one as the animal grows, and only the last chamber is occupied. The other chambers are filled with gas and provide buoyancy, without which the heavy shell would drag the animal to the bot-

When a large piece of food such as a dead fish is placed as bait on the bottom (Figure 15-35), animals rapidly congregate. Among the most common of these are crustaceans, especially amphipods (*Eurythenes, Alicella;* Figure 15-28), who arrive soon after the bait touches down. Many deep-benthic amphipods are generalists that feed on detritus and perhaps prey on live organisms if nothing else is available. Some, however, seem to specialize as scavengers. They apparently have a well-developed sense of smell, which probably helps them find new baitfalls. When caught in traps, these amphipods often have nothing in their guts except the bait, which indicates that they had not fed for some time before. This, and the fact that they have an

tom. When the chamber is first partitioned, it is filled with seawater. Rather than pumping gas in, *Nautilus* removes all the ions from the water in the chamber. Being more dilute than the animal's blood, the water flows into the bloodstream by osmosis, and gas diffuses into the chamber to replace it. Thereafter the chamber is more or less sealed off and does not need repeated filling and emptying as does a fish's swim bladder. This gas-filled chamber allows the animal to move rapidly up and down in the water column.

Nautilus is found near coral reefs but is not really a coral reef animal because it generally lives much deeper than corals, typically between 100 to 500 m (330 to 1,650 feet). In some places it occasionally ventures into shallow water for short periods. The upper limit of its normal depth range is set by temperature: water temperatures above 25° C (77° F) are lethal, and temperatures below 20° C (68° F) are preferred. The lower depth limit is set by pressure. Since the shell chambers are sealed, gas cannot be pumped in to counteract the pressure at great depth. Instead, the animal relies on the strength of its shell,

just as a submarine relies on the strength of its hull. When the pressure gets too great, at about 800 m (2,600 feet), the shell is crushed. Actually, water begins to leak into the chambers at much shallower depths, and the animal probably doesn't spend much time as deep as even 500 m.

Not much is known about the biology of *Nautilus*. It appears to be primarily a scavenger, locating animal remains and the molted shells of lobsters by its sense of smell. When it senses food, the animal swims to the food by jet propulsion, squirting water out of a muscular funnel. It may also prey on live hermit crabs and other crustaceans. When not feeding, it spends most of its time "asleep" conserving energy. The animal can tolerate conditions of almost no oxygen, but the benefits of this ability are not known. The vast majority of *Nautilus* caught in traps are males. Perhaps females are hard to catch, or perhaps there is a shortage of females. Scientists will continue to study *Nautilus* not only to learn about the animal itself, but to try to get a glimpse into the ocean's past.

Longitudinal section of a shell of *Nautilus*.

expandable gut, may mean that they are adapted to capitalize on large but infrequent meals.

Various fishes also find freshly placed bait quickly (Figure 15-35). The most common of these are the grenadiers, or rattails (*Coryphaenoides, Lionurus*); brotulas and cusk eels (*Bassogigas, Abyssobrotula*); deep-sea spiny eels (*Notacanthus*); and hagfishes (*Eptatretus*; see Figure 7-3). These bottom scavengers tend to be large, relatively well muscled, and active (Figure 15-36), unlike bathypelagic fishes (Figure 15-37). They seem to be adapted for cruising along the bottom in search of the occasional bonanza. At depths above 2,000 m, sharks may also show up, quickly putting an end to the feed.

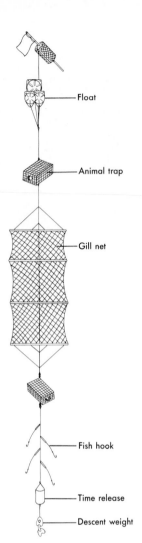

Float

Animal trap

Gill net

Fish hook

Time release

Descent weight

Figure 15-34 A free-vehicle midwater net and traps used to collect deep-water animals.

Figure 15-33 A tripod fish (*Bathypterois*). The long feelers are used to "taste" the bottom (see Figure 15-36, *B*).

Figure 15-35 Grenadiers, or rattails (see Figure 15-36, *D*), and other deep-sea fishes coming to bait.

STRUCTURE AND FUNCTION OF MARINE ECOSYSTEMS

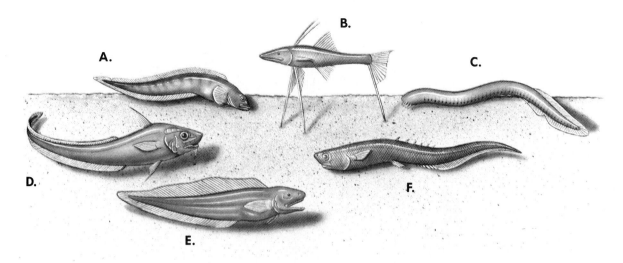

Figure 15-36 Some typical deep-sea bottom fishes: **A,** eelpout *(Zoarces);* **B,** tripod fish *(Bathypterois);* **C,** hagfish *(Eptatretus stouti);* **D,** grenadier *(Lionurus carapinus);* **E,** brotulid *(Bassogigas profundissimus);* **F,** deep-sea spiny eel *(Notacanthus bonapartei).*

The Nature of Life in the Deep Sea

There is a growing realization that life in the deep sea proceeds at a very different pace than it does at the surface. Most deep-sea animals seem to grow very slowly, probably because of the lack of food. On the other hand, they live for a long time. Deep-sea clams have been estimated to be 50, 60, or even 100 years old. Perhaps the low temperature and high pressure slow down the processes of life in the deep sea.

It may also be that deep-sea animals need to live a long time to store up enough energy to reproduce. The larvae of deep-sea forms do not spend time in the food-rich photic zone. The chances of making it all the way to the surface and then back to the deep-sea floor are simply too small. Instead, deep-sea animals tend to produce large eggs, with enough yolk to see the larva through its early stages without eating. It takes a lot more energy to produce a large egg than a small one, so deep-sea animals produce only a few eggs. In at least some animals, reproduction may be tied to feeding. In some species of amphipods, individuals caught in baited traps are all sexually immature. This has led biologists to speculate that they do not reproduce until they manage to find a good meal.

Bacteria in the Deep Sea

As mentioned previously, bacteria play an important role in deep-benthic food webs. They are eaten by meiofauna and deposit feeders, and they break

RANGE	EPIPELAGIC	MESOPELAGIC (vertical migrators)	MESOPELAGIC (non-migrators)	DEEP PELAGIC	DEEP-SEA BOTTOM
Appearance	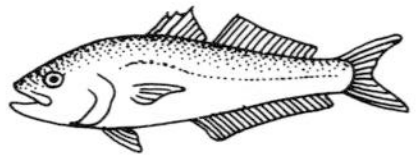	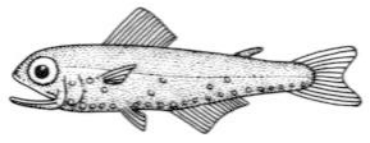	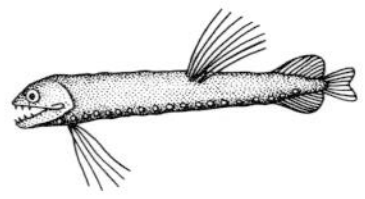	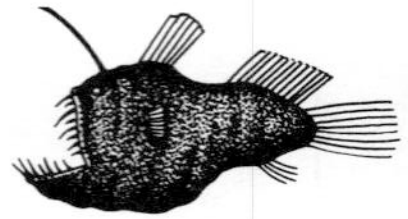	
Size	Wide size range, from tiny to huge	Small	Small	Small	Relatively large
Shape	Streamlined shape	Relatively elongated and/or laterally compressed	Relatively elongated and/or laterally compressed	No streamlining, often globular in shape	Very elongated
Musculature	Strong muscles, fast swimming	Moderately strong muscles	Weak, flabby muscles	Weak, flabby muscles	Strong muscles
Eye characteristics	Large eyes	Very large, sensitive eyes	Very large, sensitive eyes, sometimes tubular eyes	Eyes small or absent	Small eyes
Coloration	Typical countershading: dark back and white or silver belly	Black or black with silver sides and belly; counterillumination	Black or black with silver sides and belly; counterillumination	Black	Dark brown or black
Bioluminescence	Bioluminescence relatively uncommon	Bioluminescence common, often used for counterillumination	Bioluminescence common, often used for counterillumination	Bioluminescence common, often used to attract prey	Only a few groups bioluminescent

Figure 15-37 Some typical characteristics of fishes from different depth zones in the pelagic realm.

STRUCTURE AND FUNCTION OF MARINE ECOSYSTEMS

down indigestible organic matter. Scientific interest in deep-sea bacteria got a major boost as the result of an accidental "experiment." In late 1968 the crew of *Alvin* was preparing to make a dive. The cable holding the submersible to the side of the tender ship suddenly broke; *Alvin*, hatch open, dropped to the bottom of the sea about 1,540 m (5,000 feet) down. The crew escaped safely, but their lunchbox was left on board. It was to become the most famous lunch in the history of marine biology.

Alvin was not recovered for over 10 months. When it was finally raised, scientists discovered that the long-lost lunch, instead of having rotted away, was in amazingly good condition. Though soggy, the sandwiches still looked almost fresh; the bologna was a little grey but still pink inside. The rest of the lunch—apples and a thermos of soup—also looked good enough to eat. Once brought to the surface, the food soon spoiled, even though refrigerated.

Why was the food preserved in the deep sea? Though the deep sea is cold, it is no colder than a refrigerator. Are bacteria absent from the deep sea? Does the pressure somehow inhibit bacterial decay? Is there some other explanation? These questions sparked a flurry of research.

It is now known that bacteria do live in the deep sea, as they do in every other environment on earth. Pressure slows down bacterial growth, however, and most shallow-water bacteria cannot grow at the pressures of the deep sea. Therefore the bacteria that were already in the lunch probably died when *Alvin* sank. Deep-sea bacteria, however, are adapted to tolerate high pressure; some, in fact, cannot grow without it. Even these pressure-loving bacteria, however, grow slower than do surface bacteria.

Many deep-sea bacteria live inside amphipods and other animals; they probably help the animals digest chitin and other detritus. The animals provide transportation, carrying the bacteria to rich food sources. By keeping the animals out, the lunchbox probably prevented the bacteria from getting to the food. There are also free-living bacteria in the deep sea. It may be that these just grow too slowly to have decomposed the lunch. It has also been suggested that free-living bacteria in the deep sea are adapted to use nutrients in very low concentrations. If this is the case, they may be overloaded by such rich food as a bologna sandwich, which they would not normally encounter.

DEEP-SEA HOT SPRINGS

The year 1977 marked one of the most exciting discoveries in the history of biology, and it wasn't even made by biologists! A group of marine geologists and chemists were using *Alvin* to look for **hydrothermal vents** on a section of mid-ocean ridge near the Galápagos Islands in the eastern Pacific. Geologists had predicted the existence of the vents, and there were even photos made with deep-sea cameras, but no one had actually seen a hydrothermal vent. The photos also showed clumps of clams in the area of suspected vents, but this didn't attract much attention. As the *Alvin* scientists moved along the ridge, they found not only a hydrothermal vent but something completely unexpected: a rich, flourishing community unlike anything ever

hydrothermal vents
Undersea "hot springs" associated with mid-ocean ridges
Chapter 2, p. 45

Figure 15-38 The animals that inhabit hydrothermal vents in the mid-ocean ridge near the Galápagos Islands include giant tube worms (*Riftia pachyptila*; shown here), mussels, clams, fishes, and crabs. This photograph was taken from the *Alvin* at a depth of 2,800 m (9,200 feet).

seen (Figure 15-38). In the days that followed, more vents were found, each flourishing with animals. There were gigantic worms up to 1 m (3.3 feet) long; 30-cm (12-inch) clams; dense clusters of mussels, shrimps, crabs, and fishes; and a variety of other unexpected life. The vents were like oases of life on the barren deep-sea floor. The scientists collected a few specimens, gave the vents fanciful names like "Clambake," "Garden of Eden," and "Dandelions" and went back to give the news to a stunned world.

A series of expeditions was soon mounted to the Galápagos and other vent areas around the world. With every dive it seemed, *Alvin* brought up more exciting discoveries. Nearly all of the organisms in the rich communities around the vents were new to science. What is unique about these communities is that they do not depend on photosynthesis.

Around the mid-ocean ridge, seawater trickles down through cracks and fissures in the earth's crust, is heated to very high temperatures, and emerges at hydrothermal vents (see the section in Chapter 2, "The Mid-Ocean Ridge and Hydrothermal Vents"). When it emerges, the hot water contains large amounts of dissolved material—especially **hydrogen sulfide** (H_2S) and various sulfide minerals. When the hot vent water hits the cold ocean water, many of the minerals crystallize, creating "black smokers," "chimneys," and other mineral deposits.

Hydrogen sulfide is an energy-rich molecule, but it is toxic to most organisms. Some bacteria, however, can use the energy contained in hydrogen sulfide molecules to make organic matter in much the same way that green plants use light energy. This process is called **chemosynthesis.** The primary producers in hydrothermal vent communities are not green plants—for there is no sunlight—but **chemosynthetic bacteria.**

Deep-sea hydrothermal vents harbor rich communities. The primary production that supports these communities comes from bacterial chemosynthesis, not photosynthesis.

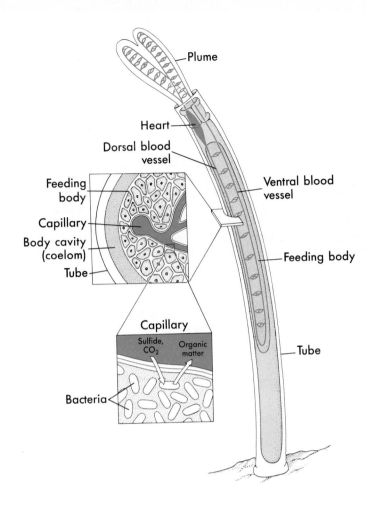

Labels on figure:
Plume
Heart
Dorsal blood vessel
Feeding body
Capillary
Body cavity (coelom)
Tube
Ventral blood vessel
Feeding body
Tube
Capillary
Sulfide, CO$_2$
Organic matter
Bacteria

Figure 15-39 The anatomy of the giant hydrothermal-vent tube-worm (*Riftia pachyptila*). The plume at the end acts like a gill, except that it exchanges hydrogen sulfide as well as carbon dioxide and oxygen. The carbon dioxide and sulfide are carried in the blood to the feeding body, where bacteria use them to make organic matter by chemosynthesis. *Riftia* is a vestimentiferan, a group of worms that is usually classified with the beard worms, or pogonophorans (see the section in Chapter 6, "Odds and Ends in the World of Worms").

The water near hydrothermal vents contains huge numbers of bacteria, so many that they cloud the water. Some vent animals feed by filtering bacteria from the water, but this does not seem to be the principal mode of feeding by vent animals. One of the dominant animals in many vent communities, the giant tube worm (*Riftia*), does not filter out bacteria. In fact, it doesn't even have a mouth or digestive tract! Instead, these worms have a highly specialized organ called a "feeding body" (Figure 15-39) that is packed with symbiotic bacteria. The bacteria perform chemosynthesis inside the worm's body and pass much of the organic matter they produce on to the worm. The worm, in turn, supplies the bacteria with raw materials. The bright-red plume acts like a gill, exchanging not only carbon dioxide and oxygen, but hydrogen sulfide. Though hydrogen sulfide kills most organisms, the worm's blood actually transports it to the bacteria in the feeding body.

A number of other vent animals, like mussels (*Bathymodiolis*) and large clams (*Calyptogena*), also contain symbiotic bacteria, though they can filter feed as well. After finding such symbiotic relationships at vents, biologists began to look elsewhere. Symbioses between invertebrates and sulfide bacteria have now been found in other environments that are rich in hydrogen sulfide, like mangrove sediments, oil seeps, sewage outfalls, and even the remains of dead whales.

Some vent communities in the Atlantic are dominated by a shrimp (*Rimicaris*). The shrimp cover the mineral chimneys that form at vents (see Figure 2-29). They scrape off bits of the mineral to eat. The mineral provides no nourishment, but it is coated with large numbers of bacteria. The bacteria are digested and the remaining mineral eliminated.

These shrimp are unusual in another way. They do not have recognizable eyes and were originally thought to be blind. On their upper surface, however, are two shiny patches that contain light-sensitive cells like those found in eyes. Though a lens and other features of eyes are absent, the patches apparently can detect much fainter light than can be seen by humans. Before the discovery that these shrimp could "see," no one suspected that there was any light at all at deep-sea vents. Using a special low-light camera similar to ones that are used to study distant stars, scientists discovered that there is a faint glow, invisible to the human eye, around the vents. This glow is apparently caused by the heat of the emerging water. The shrimp may use this dim light to locate active vents and to avoid coming too close to the hot water and getting cooked.

SUMMARY

1. The waters below the photic zone can be divided into the _____ zone and the deep sea. The deep sea is often divided into the _____, _____, and _____ zones. The distinction between the mesopelagic and deep sea is based on the penetration of _____, whereas the division of the deep sea into different zones is due mainly to _____ and/or _____, and the availability of food.

2. As ocean depth increases, both the amount of food and the abundance of life _____.

3. Most midwater fishes have adapted to the lack of food by developing _____ mouths and _____ jaws equipped with formidable teeth. They also have very _____ feeding habits.

4. Midwater fishes that make vertical migrations differ from those which do not in the degree of development of their _____, bones, and _____.

5. Sensory adaptations in midwater animals include _____ and, in fishes, a well-developed _____, which may explain the long bodies of many fishes.

6. The _____ refers to the concentration of vertically migrating midwater animals that show up on sonar. It consists mainly of _____, _____, and _____.

7. _____ refers to the use of bioluminescence to mask the silhouette. Other possible uses of bioluminescence include startling _____, _____, _____, and _____.
8. The oxygen minimum layer results from the effect of the _____ in the mesopelagic.
9. Physical characteristics of the deep sea include a lack of light, constant salinity, low _____, and great _____.
10. The most common deep-sea pelagic fishes are _____ and _____.
11. Common adaptations in the deep pelagic realm include _____ size, _____ mouths, _____ stomachs, _____ coloration, _____ muscles, and the lack of a _____.
12. Solutions to the problem of finding mates in the deep sea include the use of _____ and _____ as signals, _____, and _____, in which the male attaches to the female.
13. The two main factors that probably control the depth ranges of deep-sea organisms are _____ and _____.
14. _____ feeding is the main mode of feeding in the deep-sea benthos.
15. Deep-sea hydrothermal vents support rich biological communities. The primary producers in these communities are not plants but _____ bacteria that use _____ instead of light to produce organic matter.

THOUGHT QUESTIONS

1. The deep-sea floor has been considered as a potential site for the disposal of toxic and radioactive wastes. What questions about the biology, geology, and chemistry of the deep-sea environment do you think should be answered before such plans are approved?
2. How do you think that non-migratory midwater fishes, with their flabby muscles, are able to prey on vertical migrators, which have well-developed muscles?

FOR FURTHER READING

Alper, J.: "The Methane Eaters: Strange Life Blossoms at the Seep." *Sea Frontiers*, vol. 36, no. 6, December 1990, pp 22-29. *Mussels, tube worms, and chemosynthetic bacteria thrive in spots where methane and brine seep out from the Gulf of Mexico's floor.*

Childress, J.J., H. Felbeck, and G.N. Somero: "Symbiosis in the Deep Sea." *Scientific American*, vol. 256, no. 5, May 1987, pp. 114-120. *The amazing life at hydrothermal vents features invertebrates that harbor chemosynthetic bacteria in their tissues.*

Clark, E.: "Down the Cayman Wall." *National Geographic Magazine*, vol. 174, no. 5, November 1988, pp. 712-731. *A submersible surveys life from a shallow-water coral reef to a depth of 975 m (3,200 feet) in the Caribbean.*

"Deep-Sea Hot Springs and Cold Seeps." *Oceanus*, vol. 27, no. 3, Fall 1984. *A series of 13 articles on various aspects of hydrothermal vents and deep-sea seeps of brine and oil.*

Wu, N.: "Fangtooth, Viperfish, and Black Swallower: at 3,000 Feet, the Light Goes Out and Life Depends on Strange Adaptations." *Sea Frontiers*, vol. 36, no. 5, September-October 1990, pp. 32-39. *Excellent photos of deep-sea fishes.*

Part Four

Humans and the Sea

16 Resources from the Sea

Once upon a time, thousands of years ago, a group of humans was walking along the seashore for the first time and came upon a clump of oysters. Someone opened one up and ate it. It tasted good so they went back for more. Though purely fictional, this may have been the first use of the sea's resources by humans.

Times have changed and our exploitation of the sea's many resources is now commonplace and much more sophisticated. For example, the albacore tuna seen in Figure 16-1 was caught by several modern fishing boats in the Pacific. The catch was unloaded into a Dutch freezer ship, carried across the Panamá Canal to the Caribbean, and unloaded at a cannery in Puerto Rico. The valuable catch was canned for human consumption, and the fish by-products were processed into pet food, fertilizer, and chicken feed for sale around the world.

This chapter examines the resources of the sea, and their exploitation and use by humans. Our use of marine resources is a vast and fascinating subject. In discussing it, we will draw on our knowledge of the sea floor and the chemical and physical features of the ocean (discussed in Part I of the book) and of the types and distribution of marine life (discussed in Parts II and III). Also involved are a broad range of disciplines, ranging from the application of technology to the marketing of new products.

THE LIVING RESOURCES OF THE SEA

The oceans, which cover most of the surface of the earth, are the planet's largest factory of organic matter. Humans take advantage of this productivity and harvest many different kinds of marine organisms. Most are harvested for food, but they also provide numerous other products and materials. Millions of people also use the living resources of the sea in recreational fishing, sport diving, or even by keeping an aquarium at home.

Food from the Sea

The oceans have been a source of food since prehistoric times. They provide not only food, but employment for millions. Confucius said, "Give a man a fish, and you feed him for a day. Teach him how to fish, and you feed him for life."

Figure 16-1 The unloading of albacore tuna at a modern cannery.

Many different types of marine plants and animals are harvested. Seaweeds and creatures as diverse as jellyfishes, sea cucumbers, sea turtles, and even polychaete worms are part of the diet in many cultures. The vast majority of the harvest, however, consists of fishes, which are known in fisheries terminology as **finfish**. Also harvested are molluscs and crustaceans, which together are called **shellfish**. Freshwater animals also are eaten, but in much smaller amounts.

Most of the world's food is grown on land, and food from the sea represents only about 1% of all the food eaten. Still, seafood, especially finfish, is considered to be one of the world's most important foods. This is because it is rich in protein, which is essential for normal growth but which is lacking in the diet of millions. Finfish and shellfish provide close to 10% of the animal protein consumed by people around the world. The percentage is largest in nations like Korea (close to 70%) and Japan (45%). It is relatively low in the United States, a mere 3.3%.

> The groups of marine organisms that are most widely used as food are fishes, molluscs, and crustaceans. Seafood is important to the world's population because it is a good source of protein.

Marine fish is a valuable and potentially inexpensive source of protein for an expanding world population. The world's population reached 4 billion in 1975, 5 billion in 1987, and will reach 6 billion by the end of the 1990's (Figure 16-2). It is expected to reach 10 billion by 2050. The population will double much faster in the places where it hurts the most: the poor, already crowded countries of the so-called Third World. Protein deficiency is already a major cause of disease and death in these countries. Some people look hopefully to the ocean as a source of food for the several thousand new mouths that enter the world every day. As we shall see, however, most of the world's fisheries are already being exploited, and overfishing and pollution loom on the horizon. Some fisheries have already been exhausted.

Fishing is one of the oldest human professions. **Traditional fisheries**, which involve simple gear and methods, are still practiced in many countries, especially developing ones (Figure 16-3). Traditional fisheries, in fact, employ most of the world's fishermen. These fisheries provide animal protein to people whose diets are often protein-deficient. Because the catches from traditional fisheries usually go unrecorded, it is difficult to estimate their contribution to the total world harvest.

Figure 16-2 The curve representing the growth of the world human population over the centuries looks like a giant **J**: very slow, almost nonexistent population growth followed by a sudden explosion of very fast growth. This **J**-shaped curve means that the more people there are, the faster the population grows. Growth was slow for millennia due to food shortages and disease. The domestication of plants and animals helped provide more food, but periodic famines and plagues kept the population in check. Better sanitation, advances in medicine, and more efficient agriculture eventually helped to sharply reduce the death rate and within a few centuries the population began to rapidly increase.

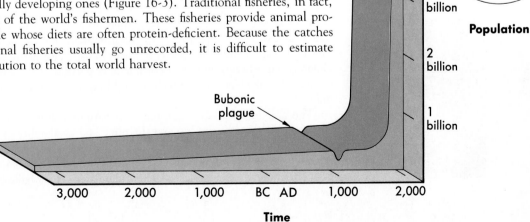

Figure 16-3 Traditional fishing with gill nets.

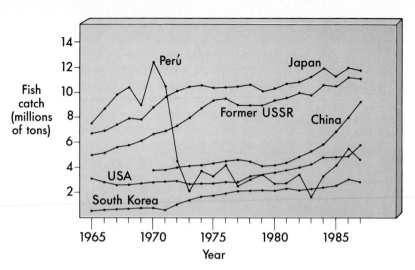

Figure 16-4 Fish catches by the major fishing nations. Notice that worldwide annual catches have remained relatively constant. Perú is an exception because of the collapse of its huge anchovy fishery (see Box 16-1). [Source: FAO]

Figure 16-5 A Japanese boat that uses light to fish squid at night.

In the last few decades the large-scale commercial operations of developed nations have dominated world fisheries (Figure 16-4). The continuous expansion of the world's population and a higher demand for protein in affluent countries have steadily increased the pressure on the food resources of the sea. The application of technology, from the development of more effective gear to the use of satellites to provide data to help find fish, has greatly increased the efficiency with which we harvest these resources. Nations like Japan and the former U.S.S.R. have developed high-tech fishing fleets that can remain in the fishing grounds away from the home port for long periods (Figure 16-5). Factory, or mother, ships process and store the fish taken by the smaller fishing boats. The catch is processed, marketed, and sold in a multitude of ways: fresh, frozen, canned, dried and salted, smoked, marinated, as fish sticks, and as fish meal. These operations have become multimillion-dollar efforts that include not only the detection and harvest of the catch, but the design and development of boats and fishing gear, the processing and marketing of the catch, and sometimes the management of the resources to prevent them from disappearing. This important business provides jobs not only to fishermen, but to technicians, marine biologists, marketing specialists, economists, and many other experts. Commercial fisheries provide a considerable portion of the export earnings of countries like Canada and Iceland and help sustain these nations' economies.

The amount of seafood consumed in various cultures varies widely because of both consumer tastes and the price and availability of seafood. The average Japanese eats the most seafood by far, a whopping 86 kg (190 pounds) a year! Scandinavians and Spaniards are also consummate eaters of fish. Inhabitants of poor, landlocked countries like Afghanistan, on the

HUMANS AND THE SEA

other hand, typically do not eat much seafood at all. It has been estimated that Americans consume an average of 10 kg (22 pounds) per person a year, a bit lower than a world average of 12 kg (27 pounds). Seafood consumption has been steadily increasing in the United States, Canada, and other developed countries as consumers look for healthful, low-fat sources of protein.

Major Fishing Areas. Most of the world's major fisheries are located near the coast (Figure 16-6). These coastal fisheries exploit the rich waters of the continental shelf. Since shelf waters are shallow compared with the open ocean, it is relatively easy to harvest bottom-dwelling—or **demersal**—species. Coastal fisheries also include **pelagic** catches: fishes, squids, and other animals that live in open water. Primary production is higher over the shelf than farther offshore, supporting much more profuse life (see Table 14-2 and maps in Figure 14-31). Some of the richest fishing areas of the world are located in the highly productive waters where upwelling takes place, such as the coasts of Perú and southwest and northwest Africa (Figure 16-6). Other rich coastal fisheries are found where the continental shelf is very wide, such as the Grand Banks of Newfoundland and the North Sea.

> Most of the world's fisheries are coastal. Coastal fisheries take both bottom, or demersal, and open-water, or pelagic, catches. They are mostly concentrated in waters where primary production is increased by upwelling.

Notice from the map in Figure 16-6 that the largest catches are taken from the northwest Pacific and the northeast Atlantic. Being close to the major industrial nations, these areas have been intensively exploited for the longest time. Many other areas, though richer in the number of species, are less heavily exploited and still have some untapped resources. This is the case in the Indian Ocean, the southern Pacific and Atlantic regions, and around Antarctica. These areas are more remote and therefore less economical to exploit. Fisheries in such remote areas, however, are rapidly increasing. One of the fastest growing and most prosperous fisheries today is centered around the Falkland Islands in the southwestern Atlantic. This new and well-managed fishery is being exploited by several nations.

Though most marine food resources are taken from the continental shelf, which constitutes only about 8% of the ocean, a second major type of fisheries exploits the remaining 92%, a vast area that supports fewer but still very valuable fisheries. Many of these are pelagic fisheries located in upwelling areas just outside the continental shelf. Fisheries for migratory species like tuna, on the other hand, may be concentrated in the open ocean far from the coast.

Major Food Species. Although there are thousands of species of fishes, crustaceans, and molluscs, relatively few are of commercial value. A summary of the major commercial catches for the world are given in Table 16-1.

The largest catches in the world are those of **clupeoid fishes,** small plankton-feeding fishes that form huge schools. They include herrings (often sold as "sardines"), anchovies, sardines (or pilchards, as they are known in many parts of the world), menhadens, and shads (Table 16-2; Figures 16-7 and 16-8).

continental shelf The shallow, gently sloping edge of the continents, extending from the shore to the point where the slope gets steeper
Chapter 2, p. 41; Figure 2-22

primary production The conversion by autotrophs of carbon dioxide into organic carbon, that is, the production of food
Chapter 4, p. 92

upwelling The upward flow of nutrient-rich deep water to the surface, resulting in greatly increased primary production
Chapter 14, p. 439

Plankton-feeding fishes use gill rakers, slender projections of the gills, to strain plankton from the water
Chapter 7, p. 203

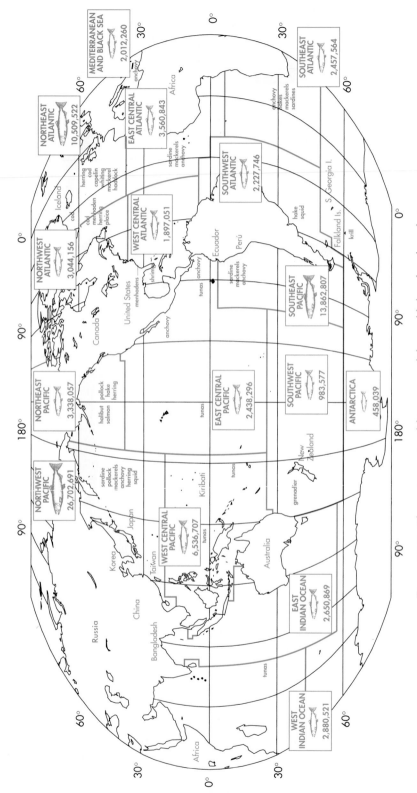

Figure 16-6 The major marine fishing areas of the world, based on boundaries established by the Food and Agriculture Organization (FAO) of the United Nations. Figures refer to the 1988 catches (in million tons) for each region. The figure for Antarctica incorporates catches for three separate FAO regions.

TABLE 16-1

World commercial catches of selected fishes, molluscs, and crustaceans in millions of metric tons

Catches tabulated in groups defined by the Food and Agriculture Organization of the United Nations; figures for catches are rounded to the nearest ten-thousandth and, when added, may not equal totals.

CATCH	1982	1983	1984	1985	1986	1987	1988
Herrings, sardines, etc.	17.87	17.45	19.61	21.10	23.96	22.30	24.10
Cods, haddocks, hakes	10.98	11.22	12.26	12.46	13.54	13.77	13.62
Miscellaneous marine fishes	7.82	7.97	8.15	8.41	9.02	9.00	9.58
Jacks, mullets, sauries	7.99	8.20	8.82	8.31	7.47	8.25	9.02
Rockfishes, basses, congers	5.36	4.95	5.43	5.21	5.97	5.68	5.65
Mackerels	3.84	3.67	4.27	3.83	4.01	3.61	3.87
Tunas, bonitos, billfishes	2.81	2.96	3.10	3.18	3.43	3.54	3.78
Flounders and other flatfishes	1.13	1.12	1.19	1.35	1.32	1.29	1.34
Salmon, smelts	0.85	0.97	0.96	1.17	1.09	1.08	1.15
Sharks, rays	0.61	0.56	0.60	0.62	0.63	0.66	0.67
Total marine fishes	58.39	58.07	63.35	64.40	69.27	68.02	71.56
Freshwater fishes	6.77	7.48	8.03	8.74	9.73	10.45	11.15
Anadromous and catadromous fishes (other than salmon)	2.09	2.21	2.37	2.57	2.50	2.62	2.60
Squids, octopuses	1.65	1.71	1.71	1.79	1.76	2.30	2.24
Clams, cockles	1.17	1.21	1.37	1.51	1.66	1.75	1.80
Mussels	0.91	0.90	0.90	0.97	1.00	1.07	1.17
Oysters	0.97	1.04	1.04	1.09	1.07	1.11	1.09
Scallops	0.53	0.56	0.84	0.60	0.53	0.74	0.87
Total marine molluscs	5.45	5.64	6.08	6.18	6.30	7.29	7.49
Shrimps, prawns	1.73	1.83	1.91	2.12	2.22	2.36	2.48
Crabs	0.79	0.79	0.87	0.89	0.90	0.96	1.05
Krill	0.53	0.23	0.13	0.19	0.46	0.38	0.37
Lobsters	0.16	0.17	0.18	0.20	0.20	0.20	0.21
Total marine crustaceans	3.27	3.06	3.12	3.42	3.79	3.89	4.09
WORLD TOTAL (marine and freshwater)	76.73	77.41	83.82	86.26	92.61	93.41	97.99

Source: FAO

They are usually concentrated over the continental shelf, but they can also be found offshore in upwelling areas. They are caught with large **purse seines** that surround and trap the schools (Figures 16-8, *B* and 16-9, *B*).

Herrings, anchovies, and sardines are eaten fresh, canned, or pickled. **Fish protein concentrate (FPC),** or **fish flour,** an odorless powder used as a protein supplement for human consumption, is also made from these and other fishes. Most of the clupeoid catch, however—particularly that of menhadens and herrings—is ground into **fish meal,** an inexpensive protein supplement used in feed for poultry, livestock, and even other species of fish that are farmed. The fish are also pressed to obtain **fish oil,** used in the manufacture of products like margarine, cosmetics, and paints. Some of the catch finds its way into fertilizers and pet food as well. Fisheries such as

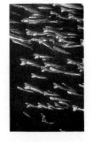

Figure 16-7 A feeding school of the northern anchovy *(Engraulis mordax).*

B

C

A

D

Figure 16-8 The European anchovy *(Engraulis encrasicholus)* is an important food fish. It is canned, salted, or cooked fresh. Basque fishermen from the port of Bermeo **(A)** in the Bay of Biscay, northern Spain, make bountiful catches using purse seines **(B).** Fish is unloaded **(C)** and sold in the fish market **(D).** Women workers, or *neskatillas,* play an important role in the fishing routine. They are in charge of carrying and selling the fish.

these, where the catch is used for purposes other than direct human consumption, are known as **industrial fisheries.** Industrial fisheries are now estimated to account for over a third of the world's total.

> **Clupeoid fishes, herrings and their kin, provide the largest catches of fish. Most are used not directly as food but in industrial fisheries for fish meal and other products.**

Cods and related fishes—the pollock, haddock, hakes, and whiting (Table 16-2)—constitute the second most important group worldwide. These fishes are all demersal, cold-water species. They are caught with **trawls** that are dragged along the bottom (Figure 16-9, C). The Atlantic cod *(Gadus morhua)* has supported a very important fishery for centuries. European fishermen were already fishing for it in the Grand Banks of New-

HUMANS AND THE SEA

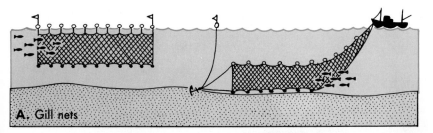

Figure 16-9 Common fishing methods.

A. Gill nets

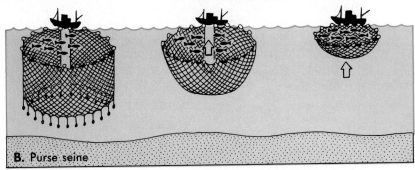

B. Purse seine

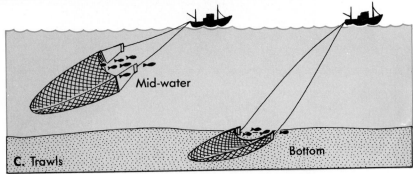

Mid-water

Bottom

C. Trawls

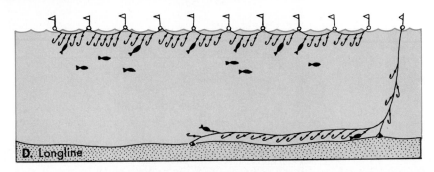

D. Longline

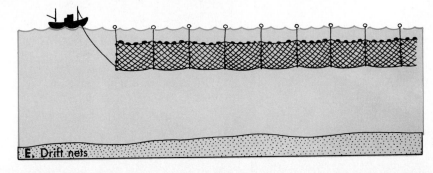

E. Drift nets

TABLE 16-2

Examples of commercially important fishes around the world

Fishes are not drawn to scale; the measurements given are the approximate maximum length.

SPECIES	DISTRIBUTION AND HABITS
Herrings (*Clupea*)	North Atlantic and Pacific; schooling, plankton feeders; 38 cm (15 inches)
Sardines or pilchards (*Sardinops, Sardinella*)	Mostly temperate worldwide; schooling, plankton feeders; 30 cm (12 inches)
Anchovies (*Engraulis*)	Worldwide; schooling, plankton feeders; 20 cm (8 inches)
Menhadens (*Brevoortia*)	Temperate and subtropical Atlantic; schooling, plankton feeders; 38 cm (15 inches)
Cods (*Gadus*)	North Atlantic and Pacific; demersal; feed on bottom invertebrates and fishes; 1.5 m (5 feet) in the Atlantic cod
Haddock (*Melanogrammus aeglefinus*)	North Atlantic; demersal, feed mostly on bottom invertebrates; 90 cm (35 inches)

TABLE 16-2

Examples of commercially important fishes around the world—cont'd

Fishes are not drawn to scale; the measurements given are the approximate maximum length.

SPECIES	DISTRIBUTION AND HABITS
 Hakes and whiting (*Merluccius*)	Temperate worldwide; demersal, feed on bottom invertebrates and fishes; 1 m (3 feet)
 Flatfishes: flounders, halibuts, soles, and others (*Platichthys, Hippoglossus,* etc.)	Mostly temperate worldwide; demersal, feed on bottom invertebrates and fishes; 2 m (6.5 feet) in some halibuts
 Tunas (*Thunnus, Katsuwonus,* etc.)	Tropical and temperate; schooling, carnivores; 4.3 m (14 feet) in the bluefin tuna
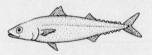 Mackerels (*Scomber, Scomberomorus*)	Tropical and temperate worldwide; schooling, carnivores; 235 cm (8 feet)
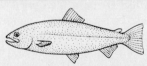 Salmon (*Oncorhynchus, Salmo*)	North Pacific and Atlantic; open ocean and rivers, carnivores; 1 m (3 feet)

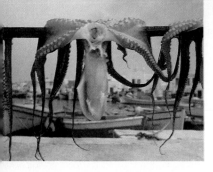

Figure 16-10 Octopuses have long been valued as a culinary delicacy. These freshly-caught individuals were photographed on the Greek island of Paros in the Aegean Sea.

cephalopods Squids, octopuses, and a few related molluscs that use sucker-bearing arms to capture prey and—with few exceptions—lack a shell
Chapter 6, p. 167

land a century before the arrival of the first colonists (and most probably before Columbus). It was a vital source of inexpensive protein in many parts of the world. In an age before refrigeration, the cod was salted and dried for preservation. The dried fish was soaked in water before cooking, just as it is done today in the Mediterranean and Caribbean. Now, however, most cods and related fishes are sold fresh or frozen.

Jacks, mullets, rockfishes (including the ocean perches, also known as redfishes), and mackerels (Table 16-2) follow in importance in terms of worldwide tonnage. Canned mackerel from Japan and South America has become a cheap source of protein in some parts of the world. Flounders, halibuts, and other flatfishes (Table 16-2) are important catches in the United States, Canada, and other countries. Salmon (Table 16-2), though now caught in smaller numbers than formerly, are a very valuable catch. The salmon fishery is of considerable importance in the north Pacific in terms of the dollar value of the catch. Salmon are caught both along the coast and in the open ocean.

The most important open-ocean fishery is that for several species of tunas (Table 16-2). The migration routes of tunas crisscross the oceans, mostly in tropical waters (see Figure 7-24). The skipjack, yellowfin, albacore, bigeye, and bluefin are large tunas that command high prices in world markets. Most are canned for consumption in the more affluent countries. They are caught by modern fleets that use sophisticated techniques to spot the schools in mid-ocean. The fishes are caught with large seines, surface **longlines,** and **gill nets** (Figure 16-9). The fishing boats are equipped with freezers so that the fishes can be brought to port long after they are caught. In addition to these high-tech fleets, some traditional fisheries continue to exploit bonito and other small tunas.

After fishes, molluscs are the most valuable group of marine food species. Several types of cephalopods provide the largest catches of molluscs (Table 16-1). Squids, cuttlefishes, and octopuses are especially popular in Far Eastern and Mediterranean countries (Figure 16-10). Japanese fishermen use lighted boats to catch squid at night (see Figure 16-5). Other important food molluscs are clams (*Macoma* and many others), oysters (*Ostrea*), mussels (*Mytilus*), scallops (*Pecten*), and abalones (*Haliotis*).

Crustaceans are prized as food the world over. Major fisheries exploit many types of shrimps and lobsters, all of which bring a high price. Blue crabs (*Callinectes sapidus*), king crabs (*Paralithodes camtschatica*), dungeness crabs (*Cancer magister*), and others are also harvested. Krill and other small but plentiful planktonic crustaceans are being increasingly exploited, partly for human food (see the section "New Fisheries" on p. 502).

Most of our food from the sea is finfish and shellfish, but many other groups of marine organisms are also eaten (Figure 16-11). Seaweeds, for instance, are eaten in many cultures, especially in the Far East (see Box 5-2). Sea urchins are much esteemed for their gonads, or roe, which command an astronomical price, particularly in Japan. In California there is an expanding fishery for the red sea urchin (*Strongylocentrotus franciscanus*). Most of the roe is exported to Japan or sold to Japanese restaurants where, known as *uni,* it is eaten raw. Some species of sea cucumber are another common seafood. Called *trepang* or *bêche-de-mer,* they are dried, boiled, smoked, or eaten raw in China and throughout the Pacific. It might not be long before they turn up in our corn flakes. Most other invertebrate groups, from jellyfish to worms to barnacles, are eaten somewhere in the world.

Sea turtles are still hunted and their eggs still gathered nearly everywhere they occur, even where they are officially protected. Seals and whales are also eaten in some places, and there are traditional fisheries for marine mammals in Alaska, Arctic Canada, and Siberia. These traditional fisheries are a source of considerable controversy, because the species involved are endangered. Should the fisheries be banned or left entirely unregulated? A compromise between these two extremes involves questions of the rights of indigenous cultures, the ethics of conservation, and international politics.

Optimal Yields and Overfishing. Like all living things, the organisms we harvest for food can reproduce to replace the individuals that are lost to disease and predators, including humans. Because these living resources can replace themselves, they are known as **renewable resources.** Resources that are not naturally replaced, like oil and minerals, are called **nonrenewable resources** (see the section "Resources from the Sea Floor" on p. 510). Experience has shown, however, that though the food resources of the ocean are renewable, they are not inexhaustible. Renewable, yes; unlimited, no. Let's see why.

Imagine that fishermen discover a new, untapped population of sardines. The new fishery is very successful, and the fishermen make handsome profits. The news spreads, and soon other fishermen join in. After several years, **overfishing** results: catches dwindle, and the fish that are caught get smaller. The fishery becomes uneconomical, and many—if not most—of the fishermen go out of business.

What happened? The reproductive rate of sardines—or salmon, scallops, or sharpnose surfperches—depends, at least in part, on the size of the population (**stock,** as it is known to fisheries biologists). Like dinoflagellates growing in a jar (see Figure 9-2), fish stocks grow fastest when there are neither too few nor too many individuals in the population. If the population size is very small, the number of young being born is also small because there are not many potential parents. If the population is too large, competition, overcrowding, and so forth slow down the growth rate (see the section in Chapter 9, "How Populations Grow"). The number of new young—that is, the population growth rate—is highest at some intermediate density (Figure 16-12).

Figure 16-11 Raw abalone and salmon or sea urchin roe are used in *sushi* and *sashimi*, Japanese delicacies that have become popular around the world.

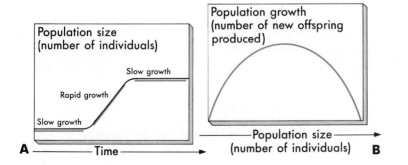

A Time →

B Population size (number of individuals)

Figure 16-12 The graph (A) shows the theoretical growth of a population that begins with just a few individuals. At first, with only a few adults to produce offspring, the population increases very slowly. As the population size gradually increases, so does the number of offspring, and the *rate* of growth begins to increase. Eventually the population reaches the point where limiting factors such as food shortage and crowding prevent further growth. Thus the rate of growth is directly related to population size (B) and is maximal at intermediate sizes.

To harvest the stock in the optimal way, one must consider this feature of population growth. Obviously, it would be foolish to harvest the entire stock at once, because no individuals would be left to reproduce, and the fishery would be destroyed forever. If the fishery is to last indefinitely, the number of fish caught can be no more than the number of new fish added through reproduction; if more are caught, the population will decline. The **sustainable yield** of a population is the amount that can be caught and just balance the growth of the population. In other words, the catch is just large enough to prevent the population from growing but not so large as to reduce it.

Since the growth rate depends on the population size, so does the sustainable yield. When the stock is large, it is held in check by natural mortality; even a small harvest causes a reduction in the population. Very small populations also have a low growth rate. In this case, fishing is more dangerous, since it can threaten the "seed" stock and drive the species to the brink of extinction.

The highest catch that can be continued year after year without threatening the stock, called the **maximum sustainable yield,** occurs at medium population size, when the natural growth rate is highest. From the point of view of profit and food production, this can be thought of as the optimal catch.

The size of the catch, and therefore of the stock, is directly related to **fishing effort:** how many boats and fishermen there are, how long they stay at sea, and so forth. When little effort is put into the fishery, of course, the

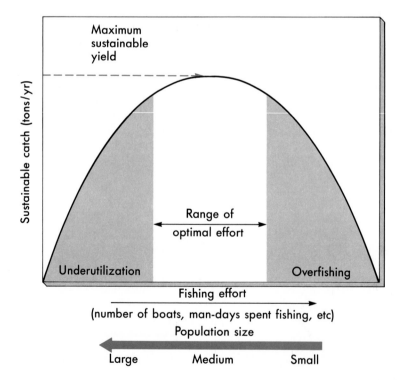

Figure 16-13 A generalized, theoretical curve showing that as fishing effort increases, so does the catch, but only up to a point, which is the optimal catch. After this point the overexploited fishery will yield smaller and smaller catches as the effort increases. The black curve is called a catch-effort curve.

catch is small. Only a small fraction of the stock is removed by fishing, and the population grows until limited by natural factors. The population remains secure, but the fishery yields less food and less profit than it could sustain safely.

With intense effort the catch exceeds the maximum sustainable yield. Before long the stock declines and catches fall no matter how much fishing effort is expended: overfishing has occurred.

A plot of the sustainable yield versus fishing effort, then, is very similar to a plot of the population growth rate versus the size of the stock (Figure 16-13). The maximum sustainable yield occurs at moderate levels of effort, which result in intermediate stock sizes.

Unfortunately, the biological properties of most fisheries' stocks do not mix well with the principles of economics. Most fisheries are profitable when harvested at the level of the optimum catch. As long as there is money to be made, more fishermen and more boats will be attracted to the fishery, and the catch will exceed the optimal level. In an open fishery—that is, one that is completely unregulated—market forces inevitably lead to overfishing. In some cases, economic forces can even cause a species to be fished to extinction.

Overfishing results when catches are higher than the maximum sustainable yield, or optimal catch, of a fishery. Free-market forces almost always result in overfishing if a profitable fishery is left unregulated.

Overfishing is not just a theoretical problem. It has already affected many fisheries. Most major commercial fisheries, in fact, are now being fully exploited or overexploited, if they are not already exhausted. This is especially true of fisheries that have been exploited the longest, such as the stocks of cod, haddock, herring, halibut, salmon, and other fishes in the North Sea and the north Atlantic and Pacific oceans. As in the case of the great whales (see the section in Chapter 17, "The Case of the Whales"), fishing effort in these fisheries has long surpassed the optimal catch. As expected, these fisheries are far less productive than they once were.

Most major commercial fisheries have been affected by overfishing, and some are already exhausted.

The danger of overfishing seems to be particularly serious in herrings, anchovies, and other clupeoid fishes. These species often undergo dramatic population cycles and therefore may be especially vulnerable to fishing pressure. The Pacific sardine (*Sardinops caerulea*), for example, supported a booming fishery in California that collapsed in the 1940's. The collapse is thought to have occured when heavy fishing amplified a natural low point in the cycle. Though there are signs of recovery and a limited catch is now allowed, the fishery may never fully recover. Another infamous collapse of a large clupeoid fishery is the case of the Peruvian anchovy, or *anchoveta* (*Engraulis ringens*; Figure 16-14 and Box 16-1).

There are other threats to fishery resources. Pollution from oil spills, sewage, toxic chemicals, and other sources is a growing menace. The destruction of habitats is another serious cause for concern. Estuaries, mangrove forests, seagrass beds, and other threatened environments are critical

Figure 16-14 The Peruvian fishing fleet, once the world's largest, was greatly affected by the collapse of the anchovy fishery (see Box 16-1).

BOX 16-1

Of Fish and Seabirds, Fishermen and Chickens

Islas Ballestas, guano islands off Pisco, Perú.

The ocean has been good to Perú. As the cold Perú Current flows northward, coastal upwelling (see Figure 14-38) pumps nutrients from the depths in huge amounts. The resulting productivity supports a rich harvest of plankton that in turn feeds immense schools of the Peruvian anchovy, or *anchoveta* (*Engraulis ringens*). This is a critical part of a complex cycle that has had tremendous consequences for the economy of Perú.

From time immemorial, clouds of cormorants, boobies, and other seabirds have gorged themselves on the anchovy schools. The birds roost and nest on small islands along the coast. With hardly any rain to wash them away, the birds' droppings slowly build up in the dry climate. Over thousands of years the droppings accumulated into layers as thick as 45 m (150 feet). This thick crust, called **guano,** is rich in nitrogen and phosphorus and is an excellent fertilizer.

The fertilizer starts out thousands of miles away, when organic matter sinks from the surface to the dark depths of the Pacific. When the organic material decomposes, nitrate and phosphate—the main nutrients for phytoplankton—are released into the water. The nutrients remain in the deep, unavailable to phytoplankton, until upwelling eventually brings them into shallow water along Perú's coast (see Figure 14-37). Here they are taken up in primary production by diatoms and other phytoplankton, which form the base of the food web. The plankton are eaten by the anchovies that feed the guano birds. Nutrients thus travel from the bottom of the Pacific to the coastal islands of Perú with the help of upwelling, plankton, anchovies, and seabird droppings!

And then people came along. Guano, a valuable agricultural fertilizer, was mined and exported. The guano industry brought considerable wealth to Perú, and by the end of the nineteenth century millions of tons had been mined. What took thousands of years to deposit was used up in a matter of decades.

The anchovies were next. A commercial anchovy fishery was established

breeding grounds and nurseries for fishes, lobsters, shrimps, scallops, and other valuable species. Heavy fishing, not surprisingly, can worsen the impact of pollution and habitat destruction. Intensive fishing techniques that are used to maximize efforts, such as **drift nets** and gill nets (see Figure 16-9), are a threat to many forms of marine life. The effects of pollution, habitat destruction, and drift nets will be examined further in Chapter 17.

Managing the Resources.   Given the danger of overfishing, most people agree that fishery resources should be harvested in a way that does not deplete them beyond recovery. In other words, fisheries must be managed to ensure their long-term value. The wise management of fishery stocks is much harder than it sounds. For one thing, the maximum sustainable yield can be very difficult to estimate. To do so, fishery biologists need detailed information about the size of fishery stocks, such as how fast they grow and reproduce, how long the organisms live, and what they eat (which often changes at different stages of the life cycle; see Figure 14-29). Such information is rarely easy to obtain and often unavailable. Biologists may have to

in the early 1950's and grew to unprecedented dimensions. It became the world's largest single fishery and made Perú the largest fishing nation (Figure 16-4) in history. Catches continued to increase, except in years when El Niño, the warm currents that occasionally arrive at Christmas time (see the section in Chapter 14, "The El Niño–Southern Oscillation"), interfered with upwelling.

Ironically, anchovies are so small that they are not often used for human consumption. Most of the catch was exported as fish meal for chicken feed. Fish meal from Perú became the world's single largest source of protein meal. Most of this protein was consumed by farm animals in other parts of the world, while many Peruvians continued to have protein-deficient diets.

By the late 1960's, the annual catch surpassed the estimated optimal, or maximum sustainable, yield of around 10 million tons a year. After the 12.3 million–ton catch of 1970, the fishery collapsed. The guano birds also suffered a dramatic decline. The El Niño of 1972 added to these woes. The an-

chovy fishery has never recovered, and annual catches over 3 million tons were a rarity during the 1980's. The collapse had widespread effects. The price of other protein meals, particularly soybean meal, increased. In Brazil alone, several million acres of tropical rain forest were cleared to grow soybeans.

The reasons for the collapse of the anchovy fishery are not entirely clear. Overfishing may not have been the only cause. Clupeoid fishes like the Peruvian anchovy naturally undergo population cycles. It may well have been that the *combination* of heavy fishing pressure, a dip in the natural cycle, and—a little later—an El Niño pushed the anchovy over the brink. Maybe the management of fisheries for species like the Peruvian anchovy needs to be tuned to the fishes' biology, with heavy fishing in some years and sharply curtailed harvests in others.

Will the Peruvian anchovy—and guano birds—eventually recover? Perhaps more important is whether any lessons have been learned: will the same mistakes be repeated elsewhere?

Immature boobies (*Sula*) on a Peruvian guano island.

rely on rough "guesstimates" or questionable assumptions, which makes their estimates of the optimal catch less reliable.

Furthermore, real fisheries are much more complex than we have indicated. The smooth **catch effort** curve shown in Figure 16-13 represents a greatly simplified model and does not take many natural features into account. For example, the harvested species may be competing with some other species. Fishing pressure might alter the competitive balance. Schools may get smaller as a result of fishing. This might not only make the fish harder to catch, it might adversely affect their behavior and reproduction. Simple models do not consider the size of the fish caught, but it might make a difference whether large, older individuals or small, younger ones are caught. Similarly, the time of year when fishing is done may be important, such as whether the catch is taken before or after the breeding season. These natural complexities can have dramatic consequences, perhaps causing a fishery to fail unexpectedly if not taken into account.

Given the uncertainties of determining optimal catches and the poten-

competition   When one organism uses a scarce resource to the detriment of another
Chapter 9, p. 267

tially disastrous effects of overfishing, it can be argued that the catch should be set somewhat lower than the estimated optimum just to be on the safe side. A fisherman, however, with a family to feed and boat payments to make, might have different ideas. The same can be said of cannery and dockyard workers, gear merchants, bankers, and all the other people who depend directly or indirectly on the fishery. Thus fishery management is a complex, often controversial, matter that is affected by economic and political factors, as well as biological ones. If the catch is harvested by more than one nation, international relations also come into play.

Once a desired catch level is agreed on, there are a number of ways to manage the fishery. Limits may be placed on the catch of each boat or, in an international fishery, each nation. Alternatively, the total catch may be limited, with the season being closed as soon as the target catch is reached. Restrictions can be placed on the number of boats or fishermen, on the length of the season, or on the areas open to fishing. Fishermen may be prohibited from taking individuals of certain sizes or, with species such as crabs and lobsters, from taking females.

Many forms of fisheries management involve controls on the type of gear used. The size and power of boats, for example, can be limited. Certain methods of fishing can be banned altogether. Longlining, for example, might be permitted but not bottom trawls. The mesh size of nets, which partially determines their efficiency and the size and species of the individuals caught, can be regulated (Figure 16-15).

> The management of fisheries to prevent overfishing includes the establishment of quotas, restrictions on the type of fishing gear, and other measures.

In addition to preventing overfishing, management measures may be undertaken to help restore stocks that have become depleted. These measures may involve the improvement and protection of essential habitats or the transplantation of artificially raised young.

Fishing regulations can be implemented and enforced fairly easily in a small bay or estuary, but waters offshore are a different story. It has long been agreed that the resources of the open ocean outside the territorial waters of any nation are the common property of all nations. This has led to contention over the boundaries of a nation's territorial waters, especially where the continental shelf was wide and rich in resources. Traditionally, a country's territorial waters extended 3 nautical miles (5.5 km) offshore, though this varied from country to country. Some nations then began to exclude fishing fleets from other nations by extending their borders farther offshore. Foreign vessels that failed to respect the new boundaries were confiscated, and fishing disputes almost led to war in some cases. In certain fisheries, nations have been able to agree on joint management schemes. The International Whaling Commission is one example; the shrimp fishery in the Gulf of Mexico is another. Unfortunately, such international regulatory efforts have been limited in scope and jurisdiction.

A more general compromise was finally reached in 1982 with the **United Nations Convention on the Law of the Sea** (see the section in

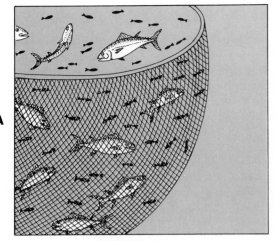

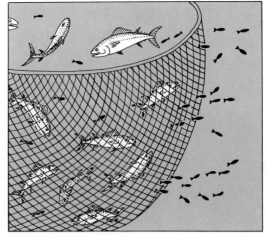

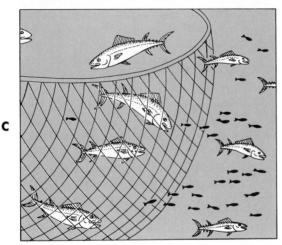

Figure 16-15 Both the size and species of fish caught by nets can be controlled by the net's mesh size. A fine net **(A),** for example, might catch both sardines and tuna. With a larger mesh size **(B)** the sardines escape and only tuna are caught. With an even larger mesh size **(C)** only large tuna are taken.

Chapter 18, "Prospects for the Future"). The United States became the 105th nation to sign the treaty in 1989. It allows nations to establish an **exclusive economic zone (EEZ)** 200 nautical miles (370 km) offshore. Within this zone, nations can completely control their fishing, oil, and mineral resources. They can keep foreign fishermen out if they wish or sell fishing licenses to foreign fleets. In 1985, for instance, the U.S.S.R. paid $1.5 million for a 1-year fishing agreement with the Republic of Kiribati. This Pacific nation consists of tiny islands with a total area of just 725 km^2 (280 square miles); but it has a 3.4 million km^2 (1.3 million square miles) EEZ! EEZ's cannot always be policed, of course, and foreign fishermen may sneak in and fish without permission or with illegal gear. When caught, these "pirate" vessels are usually impounded.

The establishment of an exclusive economic zone (EEZ) allows nations to protect fishing and other resources 200 nautical miles around their shores.

In the United States the **Fisheries Conservation and Management Act** that became effective in 1977 provides for the conservation and management of the fishing resources within the EEZ. Vessels of foreign nations that have signed fishing agreements with the United States are allowed to fish in the EEZ only after receiving a fishing permit. They are given an allocation for particular species and are not permitted to catch more than the allotted amount.

More than 90% of the ocean's present fisheries lie within the EEZ of one nation or another, but the high seas are still considered common property. A number of potential new fisheries lie outside territorial boundaries and thus are open to all nations. Fishing fleets can now reach every corner of the ocean and can take valuable resources that belong to us all. International concern about the unregulated exploitation of open-ocean fisheries is growing.

New Fisheries.   The specter of overfishing casts doubt on the ocean's future ability to provide food. More effective management, the control of pollution, and the conservation of threatened environments may help, but a growing world population will place increasing demands on the sea's resources. Fishery biologists have for decades been absorbed with the possibility of tapping new or underexploited fisheries.

One possibility is to increase the use of the **by-catch,** the so-called junk species that are caught while fishing for more valuable species. Most of the by-catch dies anyway in the process of being caught and thrown back, so keeping it has little added impact. The main problem with using the by-catch is not so much that there is anything wrong with it, but that people simply will not buy it. Consumer tastes are fickle and vary tremendously from place to place. What is considered a delicacy in one place is "junk" in another. Hake (see Table 16-2), for instance, is prized in Europe but not in the United States. Croakers, sea robins, and monkfish are common "throwaway" fishes in the United States that are eaten elsewhere. Squids are caught in large numbers in California, but most of the catch is exported, mostly to Japan and Europe. They are a delicious source of lean protein, but American consumers are suspicious of the sucker-armed molluscs. Marketing squid by its Italian name of *calamari* seems to increase its appeal. A name change also helped overcome consumer resistance against eating dolphinfish (*Coryphaena hippurus*), certainly a fish and not a dolphin. The more seductive Hawaiian name of *mahi-mahi* helped enhance its marketability.

A potential use of the by-catch is to process it into products that have more market appeal. In the *surimi* process, for example, fish is washed to remove fat, minced, and mixed with flavorings and preservatives. Appropriately shaped, it is sold as imitation crab, lobster, shrimp, or scallops. This process is rapidly gaining acceptance in the American market. Alaskan pollock is now the main component, but less desirable fish species could also be used. Processed fish products such as fish sticks could also open the market to less popular catches. Unwanted catches can also be converted into fish flour and used to protein-enrich all sorts of processed foods.

There are also some fisheries that appear to be underexploited. Most of these lie in areas outside the continental shelves, particularly in the tropics and in the Indian Ocean. These waters, however, are for the most part relatively unproductive, so their fisheries' potential is questionable. More scientific knowledge is needed to predict their likely success.

The highly productive waters around Antarctica and its associated island groups are mentioned often as potential sources of increased commercial catches. Even here, however, overfishing is taking its toll. The waters around South Georgia Island, for instance, are already seriously depleted. Antarctic catches include various fishes, spiny lobsters, the southern king crab, and, most significantly, **krill.**

The name krill, from the Norwegian *kril* meaning whale food, is generally used to refer to a group of about 85 species of crustaceans (Figure 16-16, A). In Antarctica the dominant species is *Euphausia superba.* Krill eat the planktonic diatoms that bloom during the long Antarctic summer. Feeding low on the trophic pyramid, they have an ample food supply and occur in immense swarms that can turn the sea red. Whaling has drastically reduced the number of large whales in Antarctic waters. This has left more krill for other animals that eat them, like the small minke whale (*Balaenoptera acutorostrata*) and the so-called crabeater seal (*Lobodon carcinophagus*). With more abundant food, these animals are on the increase, though the minke whale has been subjected to whaling.

Antarctic krill are of moderate size, up to 6 cm (little more than 2 inches). They are caught with large mid-water trawls. Russia, Japan, and a few other nations have developed fisheries for Antarctic krill and for a related species, *Euphausia pacifica,* in the north Pacific. People have been slow to develop a taste for krill, however. Though rich in protein and vitamins, krill are reputedly quite bland when cooked fresh, and are even less appetizing frozen, dried, or canned (Figure 16-16, B). Most of the catch is used instead to feed farm animals and farmed fish. Early predictions of annual catches in the millions of tons have not materialized. In fact, catches have actually decreased after a record harvest of a little over half a million metric tons in 1982 (Table 16-1). Some nations have given up fishing for krill altogether. The long trip to Antarctica, the maintenance of a fleet so far from

krill Planktonic, shrimp-like crustaceans that are most abundant in cold water and constitute an important food of whales, penguins and other marine animals
Chapter 6, p. 174; Figure 6-39

Only about 10% of the energy contained in one trophic level is passed to the next higher levels, so that about 90% is lost. There is therefore far more energy at lower levels of the trophic pyramid
Chapter 9, p. 275; Figure 9-15

A   **B**

Figure 16-16 **(A)** A swarm of krill (*Meganyctiphanes norvegica*) photographed at a depth of 700 m (2,300 feet). **(B)** Canned krill for sale in Prague, Czechoslovakia— not the most popular item in the store.

port, and the limited appeal of the catch has simply made the fishery uneconomical. Furthermore, there is some concern that harvesting krill, the food on which so many Antarctic forms of life depend (see Figure 9-14, A), could have disastrous consequences. This is particularly important now that the blue and other large baleen whales are protected and, it is hoped, on the road to recovery.

There are a number of other potential sources of food from the sea. Many of these are underutilized or unconventional species like squid and other cephalopods, sharks, flying fishes, and pelagic crabs (*Pleuroncodes*; Figure 16-17). Lanternfishes (see Figure 15-10, B) represent another potential fishery. They are found in deep water but migrate to the surface at night in the deep scattering layer. They form dense schools in most oceans. These and other species remain underexploited or completely unexploited mainly because of lack of consumer acceptance or the absence of efficient fishing methods. Some may eventually support successful fisheries, though probably for fish meal rather than food for human consumption.

Measures to increase the supply of food from the sea include an increase in the exploitation of underutilized species and the development of new fisheries.

Mariculture. Instead of continuing to overfish the dwindling natural food resources of the oceans, why not raise marine organisms as we do farm animals? This is not a new idea: the Chinese have been doing it with freshwater fishes for thousands of years! The Romans farmed oysters, and the early Hawaiians built walled fishponds along the coast to raise mullet and milkfish. The application of farming techniques to the growth and harvesting of marine organisms is known as **mariculture,** the agriculture of the sea. A more general term is **aquaculture,** which is applied to the farming of both freshwater and marine organisms. The culture of freshwater and marine organisms reached an estimated 13.2 million metric tons in 1987, with China alone accounting for 5.6 million metric tons. Japan was responsible for 1.23 million tons. Though mariculture is used primarily to grow food species, it is also being used to grow commodities such as cultured pearls (see the section in Chapter 6, "Bivalves") and fishes for the aquarium trade (see the next section).

Mariculture is the farming of the sea for food and other resources. The term aquaculture includes the culture of freshwater species as well.

Traditional mariculture is still successfully practiced in several parts of the world, particularly in the Orient. The milkfish (*Chanos chanos*) is farmed in ponds filled with brackish water, that is, partially diluted seawater (Figure 16-18). The immature fish—or fry—are captured at sea, transferred to the ponds, and fed with ground-up agricultural by-products. In this way, wastes are converted to fish flesh that is cheaper to produce than pork or beef. These fish farms provide not only high-protein food, but jobs as well.

With various modifications, similar operations are used to farm other fishes, molluscs, shrimps, and seaweeds in many parts of the world. The fry or spat—the immature bivalve molluscs—are transplanted to favorable

deep scattering layer A group of organisms, mostly fishes and shrimps, that migrate vertically between the mesopelagic and epipelagic zones
Chapter 15, p. 458

Figure 16-17 *Pleuroncodes*, also called the "lobster krill" or "tuna crab," is often suggested as a potential source of human food. Large swarms of this small pelagic crab are often washed ashore on the Pacific coast of California and Mexico, turning beaches bright red. A possible problem with this plan is that *Pleuroncodes* is an important food for tunas.

Figure 16-18 Milkfish (*Chanos chanos*) being harvested from a brackish water pond in central Philippines (**A**). Not all fish will be used as food. Some of the 1-year-old fish are transferred into floating cages on the coast (**B**). Here they are allowed to mature and spawn to provide fry to replenish ponds.

sites in coves (Figure 16-19, A), natural ponds, and estuaries, including fjords and mangrove forests. These may be enclosed with wooden fences, nylon netting, concrete, or other materials. Salmon and other fishes are raised in large floating pens or cages. Oysters, mussels, and other molluscs are grown on racks, in baskets, hanging from rafts (Figure 16-19, B), and in other ingenious ways. This type of mariculture, in which farming takes place under more or less natural conditions with relatively little manipulation by humans, is known as **open mariculture,** or **semi-culture.**

In some industrialized countries, however, the farming of high-value food species is often a different story. In what can best be described as the domestication of marine organisms, growth is maximized by almost totally controlling the organisms and their environment. This type of mariculture is called **closed** or **intensive mariculture.**

Figure 16-19 (**A**) The farming of oysters in Whangaroa, New Zealand. (**B**) The blue mussel (*Mytilus edulis*) is cultured on ropes that hang from barges that lie anchored in the deep, narrow bays, or *rías,* typical of northwestern Spain.

TABLE 16-3

Examples of marine and brackish-water species used in commercial mariculture

SPECIES CULTURED	MAJOR AREAS WHERE CULTURED
Fishes	
Milkfish (*Chanos chanos*)	Southeast Asia
Pacific salmon (*Oncorhynchus*)	U.S., Canada, Japan
Atlantic salmon (*Salmo*)	Canada, Europe
Mullets (*Mugil*)	Southeast Asia, Mediterranean
Flatfishes (*Solea* and others)	Europe
Yellowtail (*Seriola*)	Japan
Red seabream (*Pagrus major*)	Japan
Horse Mackerel (*Trachurus*)	Japan
Puffer (*Fugu*)	Japan
Molluscs	
Oysters (*Crassostrea, Ostrea*)	U.S., Europe, Japan, Taiwan, Australia, New Zealand
Abalone (*Haliotis*)	U.S., Japan
Mussels (*Mytilus, Perna*)	U.S., Europe, New Zealand
Scallops (*Pecten, Patinopecten*)	Europe, Japan
Clams, cockles (*Anadara* and others)	Southeast Asia
Crustaceans	
Shrimps (*Penaeus* and others)	U.S., Mexico, Ecuador, Japan, China, Southeast Asia, Bangladesh
Seaweeds	
Reds (*Porphyra, Eucheuma, Laminaria*)	Japan, China, Southeast Asia

Intensive mariculture poses many problems, and only a small percentage of marine food species can be farmed under such artificial conditions (Table 16-3). Obviously, species like clupeoid fishes and tunas that require large, open spaces are not good candidates for intensive mariculture. To be farmed successfully, most species need specialized sites and equipment (Figure 16-20). Water pumped into tanks, pools, or other holding facilities must be free of pollutants and carefully checked for temperature, salinity, and other chemical factors. Toxic wastes from the organisms themselves must also be removed. Food can be a problem because the various stages of a species usually have different food requirements. This is a particular problem when trying to culture planktonic larval stages, since they require specially cultured food. Predators and disease-causing organisms must be periodically removed or killed. Parasites and disease can be devastating under the crowded conditions typical of culture operations. Even sneaky seabirds have learned to help themselves to the precious fish that are raised so painstakingly! Crowded conditions tend to cause cannibalism among fishes and crustaceans. Juvenile American lobsters, for instance, must be kept apart,

A

B

C

Figure 16-20 Research in mariculture has been responsible for remarkable progress. Ocean Farms of Hawaii, for instance, uses cold, nutrient-rich water that is pumped from a depth of 600 m (2,000 feet) **(A).** The water is used to grow giant kelp (*Macrocystis*), which lives only in cold water. The kelp is transplanted from nurseries to large ponds **(B),** where it provides a habitat for valuable salmon and oysters. The kelp is also fed to sea urchins (*Strongylocentrotus franciscanus*), harvested for its roe, and abalone (*Haliotis*) **(C).**

or they eat each other. Because of all these problems, the culture of some species, like lobsters, is still in the experimental stage. Commercial mariculture, however, is a challenge that many enterprising, business-oriented fishery scientists are willing to accept. It is now possible, for example, to freeze shrimp sperm, ensuring year-round reproduction. We have learned to induce the settlement of the planktonic larvae of abalone, a valuable mollusc. Scientists can even create oysters with three sets of chromosomes, instead of the normal two. This makes them sterile: since ripe oysters are strong tasting, we can enjoy tasty, sterile ones anytime!

Rather than raising them all the way to harvest, some species are grown for only a short time, then released as fry or juveniles to enrich natural populations, a process known as **seeding.** Salmon are an interesting example of this. In some places, especially Europe, salmon are held in captivity their entire lives, from hatching to harvest. In other places the salmon are released as fry to feed and grow at sea. Several years later they return as full-grown adults to the stream where they were released. Not only can they be caught by commercial and sport fishermen in the usual way, but they swim right back to the hatchery, where the ones not used for breeding can be caught and harvested. This practice, known as **salmon ranching,** originated in the Pacific Northwest of North America but has been transplanted to locations as far flung as South America and New Zealand.

Most mariculture operations require expensive equipment and specialized personnel. These demands further restrict the number of different species that are cultured: some species that could be farmed are not because it is not profitable. Species that are cultured are usually those that command high prices. Thus, instead of mass producing cheap protein for a hungry

Figure 16-21 Saltwater crocodiles (*Crocodylus porosus*), here grown together with freshwater crocodiles (*C. novae-guineae*), are farmed in Papua New Guinea for their valuable skin. This farm was set up as a way to dispose of chicken offal from a chicken farm and earn some money in the process. Saltwater crocodiles are worth more than their freshwater cousins, because they have smaller scales and therefore more of the skin can be used. They are harvested when they reach approximately 1 m (3 feet) in length.

world, mariculture tends to provide variety in the diet of the affluent.

Another drawback of mariculture is pollution. Salmon farming—in which thousands of fish are concentrated in extensive floating pens—release huge amounts of feces, urine, and uneaten food that ultimately pollute the water. The decomposition of these wastes deplete oxygen in the water and release nutrients that trigger harmful algal blooms.

Still, some fisheries scientists and economists are optimistic. They predict that someday intensive mariculture will provide a significant amount of relatively cheap food. They pin these hopes on the development of technological improvements and fast-growing varieties of food species. Already, shrimp mariculture is a significant source of foreign exchange earnings for several developing countries. The Organization for Economic Cooperation and Development has estimated that the world's aquaculture production could double by the end of the 1990's. The application of **genetic engineering** and other forms of **biotechnology** to fish farming is an exciting possibility. Genetic engineering may soon allow scientists to alter the genetic information contained in a species' DNA to produce faster growing, more disease-resistant, or better tasting strains. Another possibility is to pump nutrient-rich water from the deep sea or from sewage discharges and use it to grow the planktonic algae that are fed to fish and other species. The use of warm-water discharges from power plants, producing faster growth rates, is another prospect.

In the long run, however, mariculture will probably not be able to keep up with the increase in world population. For the masses, the prospects of mariculture are not that promising, at least for now.

Marine Life as Items of Commerce and Recreation

The living resources of the sea are used in many ways other than food. Timber and charcoal are obtained from mangroves; jewelry from pearls, shells, and black and precious corals; and leather from the saltwater crocodile (Figure 16-21), sea snakes, sharks, and other fishes. Seaweeds provide chemicals that are widely used in food processing, cosmetics, plastics, and other products (see the section in Chapter 5, "Economic Importance"). Some endan-

DNA A complex molecule that contains the cell's genetic information *Chapter 4, p. 88*

gered species, unfortunately, are still taken. Fur seals, for example, are killed for their pelts, and hawksbill turtles are taken for tortoiseshell.

Drugs from the Sea. Chemical compounds, or **marine natural products,** that are obtained from marine organisms have an exciting potential for use in medicine. The medicinal use of marine natural products by the Chinese probably goes back a few thousand years, but in general such use has been very limited. In the West the discovery of such drugs has been slow. Recent years, however, have seen a dramatic upsurge in the systematic collection, analysis, and testing of new material (Figure 16-22). The search for drugs from the sea has become one of the most exciting branches of marine science, combining the fields of biology, chemistry, and pharmacology.

Figure 16-22 This bryozoan (*Bugula neritina*) is the source of bryostatin 1, a most promising drug in the fight against cancer.

Some red seaweeds provide laxatives or compounds that kill infectious viruses. Sponges and soft corals are a rich and varied source of marine natural products. Some contain compounds that act as antibiotics, whereas others produce antiinflammatory agents or painkillers. The toxins, or poisonous substances, found in many marine organisms are also of potential value in medicine. The powerful toxin from certain puffers and porcupine fishes, for example, is used in Japan as a local anesthetic and as a painkiller in terminal cancer patients. The oils from several cold-water fishes show promise in preventing heart attacks by retarding the clogging of arteries. Marine natural products offer hope in the fight against such deadly diseases as cancer and acquired immunodeficiency syndrome (AIDS). Scientists have combed remote corners of the oceans, collecting organisms as varied as blue-green algae, soft corals, sea cucumbers, sea squirts, and sharks, in the search for cures for these killers.

> Several types of chemical compounds, called marine natural products, have been isolated from marine organisms and are used or are of potential use as medical drugs.

Fisheries for Fun. For millions, the living resources of the sea provide not a livelihood but relaxation, an ideal way to spend leisure time (see the section in Chapter 18, "Oceans and Recreation"). For the avid angler, the struggle with a powerful fish offers more than just a meal (Figure 16-23). In fact, a growing number of recreational fishing enthusiasts do not keep their catch at all, choosing instead to release it. Many assist fishery research by marking the fish with special tags before releasing them. In many places sport fishermen are a powerful voice for conservation. Marine recreational fisheries are the basis of a multi-billion dollar business worldwide. Fishing enthusiasts purchase all types of gear, charter boats, and take trips to exotic places to pursue their hobby.

Figure 16-23 Recreational fishing can be an enjoyable way to spend leisure time.

In the United States, the marine recreational catch is estimated to be around 30% of the commercial catch of food fishes. The variety of the catch is awesome, including everything from lobsters to big-game fishes like swordfish (*Xiphias gladius*), marlins (*Makaira, Tetrapturus*), and the Mako shark (*Isurus oxyrinchus*). Some of the most frequently caught species in the United States are the black sea bass (*Centropristis striatus*), bluefish (*Pomatomus saltatrix*), Atlantic croaker (*Micropogon undulatus*), several species of rockfishes (*Sebastes*), and the Pacific mackerel (*Scomber japonicus*).

Business and fun are also combined in the thriving **aquarium trade.** Although the aquarium hobby is dominated by freshwater fishes, marine aquaria are growing quickly in popularity. Special packaged salts that turn ordinary tap water into seawater and bottled seawater allow anyone to duplicate a corner of the ocean right at home. Studies have shown that watching aquarium fish reduces stress and may even help lower blood pressure! The export of colorful marine tropical fishes such as lionfishes, butterflyfishes, and damselfishes is a significant component of trade in countries like the Philippines. Unfortunately, the unrestricted and indiscriminate collection of marine life for the aquarium trade has been very destructive to the natural environment. In the Philippines and other places, aquarium fish are often collected by using poisons or explosives that kill hundreds or thousands of fish for every one that makes it to the neighborhood pet shop. The farming of aquarium fishes, as is done in the Bahamas and a few other places, is a recent development that perhaps will reduce the impact of fish collecting on the environment.

The fishery resources of the ocean are used for recreation by millions around the world.

RESOURCES FROM THE SEA FLOOR

In addition to its living resources, the sea contains many **nonrenewable resources,** resources that are not naturally replaced, such as oil, natural gas, and minerals. These resources are much harder and more expensive to locate and exploit under the sea than they are on land. Not surprisingly, the sea's nonrenewable resources went largely unused while supplies on land were plentiful. More and more, however, land reserves are running out, and we are turning to the sea floor for new resources.

Oil and Gas

Oil and natural gas were among the first of the sea's nonrenewable resources to be used commercially. In the late 1800's it was discovered that some California oil fields extended well offshore beneath the sea bed. The oil was extracted using oil rigs mounted on wooden piers that extended out from shore.

The offshore oil industry underwent tremendous expansion during the 1970's, when the high price of oil and gas made offshore drilling immensely profitable. Oil and gas production on the continental shelf now occurs in the Persian Gulf, the Gulf of Mexico, the North Sea, and in other areas on all continents except Antarctica. In fact, a large percentage of the continental shelves around the world is considered a potential source of oil and gas. The enormous economic potential of these reserves was one of the factors that prompted the establishment of EEZs to protect national interests.

The continental shelves are potentially a major source of oil and natural gas.

Exploiting offshore oil and gas can be a very complex, difficult, and costly operation, especially when the deposits lie many miles offshore as they do in the North Sea. Powerful currents, huge waves, and miserable weather compound the problems.

Overcoming these difficulties has been a major accomplishment of ocean engineering. Exploratory drilling is done from drill ships or partly submerged or jack-up platforms that can be towed from place to place and anchored in position on the sea floor. Once oil or gas is found, huge steel or concrete platforms are erected and secured to the bottom to extract it (Figure 16-24). Undersea pipelines transport the oil or gas to terminals on land.

The offshore production of crude oil has been increasing steadily, even though total world production declined slightly during the 1980's because of slow growth in demand. The 6 million barrels a day taken offshore in 1969 made up around 13% of the total world production. By 1985 offshore production was 15.3 million barrels a day, or 28% of the total.

The potential threat of oil pollution, however, is a factor of great concern (see the section in Chapter 17, "Oil"). Drilling operations can affect coastal fisheries, tourism, and recreation. As a consequence, oil and gas drilling have been banned or strictly regulated along many coastal regions.

Ocean Mining

The seabed is a potentially rich source of many types of minerals. Some people believe the bottom of the sea to be the planet's richest source of minerals. Although most marine deposits are not exploited presently, ocean mining will become more viable as we develop new technologies and as high-grade ores on land become exhausted.

Sand and gravel for the construction industry are mined offshore in several parts of the world. The galleries of some coal mines on land extend offshore underground. Large deposits of coal have been discovered in deep water but remain unexploited. Some tin, iron, and even diamonds are mined from offshore deposits brought to the seabed by rivers. Other metals, including gold, are found in offshore sand and gravel (Figure 16-25); in some places, beach sands are mined.

Among the most promising sources of minerals from the sea are **manganese nodules,** lumps of minerals that are scattered on the sea floor beyond the shelf (Figure 16-26). Manganese nodules contain not only manganese but nickel, copper, and cobalt, all of which are important to industry. The nodules are most common in the deep-ocean basins, but only a few areas have nodules with a high enough mineral content to make mining econom-

Figure 16-24 Oil-drilling rigs, like this one in the Gulf of Mexico off the Louisiana coast, are artificial islands built to cope with inhospitable conditions. They are serviced from land and incorporate the latest developments in undersea technology.

Figure 16-26 Manganese nodules 5 to 10 cm (2 to 4 inches) in diameter taken at a depth of 5,000 m (16,400 feet) south of Australia. Most nodules are pebblelike in shape, 1 to 15 cm (0.4 to 6 inches) in diameter, but some occur as crusts or thick slabs. Most appear to have been slowly deposited in layers, but the exact mechanism remains unknown.

Figure 16-25 This gold dredge operated off Nome, Alaska, until 1990.

ically feasible. Some of the richest deposits are found at a depth of about 5000 m (16,000 feet) in the equatorial Pacific south of the Hawaiian Islands.

Other mineral-rich sediments are found along the mid-ocean ridges and hydrothermal vents. The minerals are extracted from deep in the earth's crust as hot water percolates through volcanic fissures and vents; they are deposited when the hot water emerges and hits the cold ocean water. These deposits are rich in iron, copper, and zinc.

Manganese nodules and other mineral-rich deep-sea deposits are potential sources of valuable minerals.

There are many technical problems associated with mining the deep-sea floor. For now it is too expensive, mainly because of the tremendous depths. The question of possible detrimental effects of mining on the surrounding marine environment has also been raised. The ownership of mineral resources that lie outside the EEZ is another unresolved issue. As metal prices rise, however, deep-sea mining may become more attractive. Several agencies and private companies continue to investigate the prospects of deep-water mining. Future prospects include the use of specially designed dredges or suction pipes suspended from ships.

RESOURCES FROM SEAWATER

Ordinary seawater, which of course contains a combination of many different ions, is a potential source of almost incalculable resources. It is certainly plentiful and, to most coastal nations, easily accessible.

Fresh Water

Desalination plants that convert seawater into fresh water have been built in coastal regions that lack a sufficient supply of fresh water, mainly in desert and semidesert regions such as the Arabian Peninsula. Desalination is also used to supplement the water supply in populous areas like Hong Kong.

Several desalination systems are currently employed. The most widely used techniques are based on the principle of distillation. Seawater is boiled; the resulting water vapor, when cooled, condenses into fresh water. Distillation, however, requires a great deal of energy and is therefore expensive. It also produces a highly saline residue that can cause environmental problems. Several promising new methods may one day make desalination more practical and economical.

Minerals

Every element on earth is found in seawater, but most in extremely small quantities. The chief product presently obtained from seawater is **table salt,** or sodium chloride (NaCl), which of course is composed of the two main ions in seawater, sodium and chloride. Salt, a precious commodity to ancient civilizations, has been produced for centuries by evaporating seawater (Figure 16-27). Most of the other constituents of seawater are difficult to

---

hydrothermal vents Undersea "hot springs" associated with the mid-ocean ridges encircling the earth, where new material is added to the earth's crust *Chapter 2, p. 45*

ions Charged atoms or groups of atoms that are formed when salts are dissolved in water *Chapter 3, p. 55*

Figure 16-27 Aerial view of salt-evaporating ponds on the southwestern tip of Puerto Rico. The shallow ponds are flooded with seawater, and the sun and wind evaporate the water. Ponds like these have also become an important habitat for seabirds.

extract because they are present in such small amounts (see Table 3-1). Magnesium and bromine are the only other materials obtained from seawater in significant amounts, but valuable trace metals like uranium and gold may one day be commercially extracted.

Seawater is a potentially rich but currently limited source of fresh water and minerals.

Energy

Most people don't realize that the ocean is a vast reservoir of energy that might be put to human use. The energy is contained not in oil or gas, but in the seawater itself. Harnessing the sea's energy is the objective of several bold new concepts that may help meet the energy needs of the twenty-first century.

Mill wheels have been used since ancient times to harness **tidal energy**, the tremendous energy contained in the normal ebb and flow of the tides. Modern schemes call for the construction of large barriers across narrow bays and river estuaries in areas where the tidal range is high, at least 3 m (10 feet). Water moving in with the high tide is caught behind the barrier, and locks are opened to release the water at low tide. The flowing water drives turbines that generate electricity, just as in the hydroelectric plants in river dams. The mechanical energy contained in the tide is thus used to obtain electricity. One such electrical generation plant has been operational in Brittany, northwestern France, since 1966. A few other facilities have been built on an experimental basis. Large projects are envisioned on the River Severn in western England, the Bay of Fundy in eastern Canada, and other suitable areas.

The use of tidal energy is relatively efficient and pollution-free, but the resulting changes in the tidal patterns can be highly destructive to the nearby environment. The rich marshes and mudflats in estuaries may be damaged or destroyed. Pollutants from other sources tend to accumulate upstream since normal tidal flushing is restricted. River flows can also be altered, increasing the risk of floods inland.

Wind-generated waves and strong ocean currents are other potential sources of energy, particularly in areas subject to powerful, regular waves. As with tidal energy, water flow can be converted into electricity by turbine generators. One such scheme is operational in Norway. Wave energy is used

tidal range The difference in height between successive high and low tides
Chapter 3, p. 75

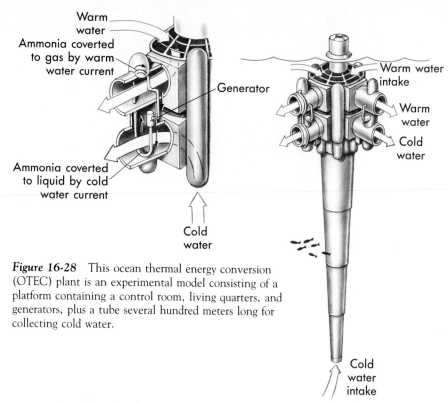

Warm
water
Ammonia coverted
to gas by warm
water current

Generator

Ammonia coverted
to liquid by cold
water current

Cold
water

Warm water
intake

Warm
water

Cold
water

Cold
water
intake

Figure 16-28 This ocean thermal energy conversion
(OTEC) plant is an experimental model consisting of a
platform containing a control room, living quarters, and
generators, plus a tube several hundred meters long for
collecting cold water.

to provide power for the operation of distant navigation buoys. Several
projects plan to use waves to produce energy for seawater desalination or to pump nutrient-rich deep water to the surface for use in mariculture.

Another possible source of energy from the ocean involves taking advantage of the temperature difference between surface and deep water. Pro-
ponents of **ocean thermal energy conversion (OTEC)** envision the development of electrical generating plants that float like giant buoys or are anchored to the bottom (Figure 16-28). Temperature differences of at least 20°
C (36° F) between the surface and below are essential, conditions that are
met in the tropics. Ammonia, propane, or another liquid that boils at low
temperature is circulated through pipes that are bathed in the warm surface
water. The liquid evaporates and is forced through turbine generators to
produce electricity. The pipes then flow through cold water pumped up
from the deep, condensing the gas back into a liquid. The cycle is repeated
over and over. This process, in which a difference in temperature is used to
produce electricity, is sort of the reverse of a refrigerator, which uses electricity to produce a difference in temperature between the inside and outside. Electricity generated by OTEC could be sent ashore by power lines, or
used at sea for various industrial operations. Cold water can also be used in
mariculture (see Figure 16-20).

**Potential sources of electricity from seawater include tides, waves, currents,
and the temperature difference between surface and deep water.**

Do It Yourself

SUMMARY

1. Most of the world's fisheries are concentrated along the coast, especially in areas enriched by _____. Bottom catches are known as _____; open-water ones, as _____.

2. _____ fisheries are those where the catch is not directly consumed by humans but processed into _____ and other products.

3. Catches above the optimal, or _____, yield lead to _____. Catches below this level are in tune with the natural growth of the catch.

4. To prevent overfishing and protect other offshore resources, many nations ascribe to an international agreement allowing them to establish an _____, which extends _____ nautical miles offshore.

5. The development of new fisheries, like that of the Antarctic _____, and an increase in the use of underutilized species are two ways of increasing the harvest of food from the sea.

6. _____, the farming of the sea, is another way of increasing our supply of food. It is referred to as _____ when dealing with both freshwater and marine organisms.

7. Marine organisms are also used as a source of marine natural products. Many of these are used as _____.

8. Resources from the sea floor include oil, gas, and minerals. _____, lumps of minerals found in the deep-ocean basin, are another promising resource.

9. Through _____, seawater can provide fresh water, as well as table salt and other minerals.

10. Various schemes have been designed to harness energy from tides, waves, and currents. _____ is one such scheme that takes advantage of the temperature difference between the surface and bottom.

THOUGHT QUESTIONS

1. It is discovered that for the last 3 years the annual catches of a commercially important fish have been above the maximum sustainable yield. One option is to decrease the fishing effort by decreasing the number of fishermen. This, however, would cause unemployment in a region where unemployment is already high. What other options might ensure a lower fishing effort? How would they be carried out?
2. Mariculture of many food species is expensive, and often only high-priced species are raised. This is of little help to the poor nations where food is needed the most. What measures and new developments might help increase the value of mariculture to these poor nations?
3. It has been suggested that cheap electricity generated from tides, waves, or currents could be used to pump nutrient-rich water from the deep. The water could then be used to grow algae to feed farmed fish. How else could this energy be used to decrease the costs of mariculture?

FOR FURTHER READING

Baldwin, R.S.: "Fundy Farming: World's Highest Tides Give Salmon Aquaculture an Edge." *Sea Frontiers*, vol. 36, no. 5, September-October 1990, pp. 40-45. *The Atlantic salmon, farmed in net pens and harvested after 18 months, generates a multimillion dollar market in eastern Canada.*

Conniff, R.: "Flashy New 'Items' Make a Big Splash in the Aquarium World." *Smithsonian*, vol. 20, no. 2, May 1989, pp. 90-101. *From featherduster worms to psychedelic fish, many fascinating marine creatures are being introduced to the many devotees of home aquaria.*

El Sayed, S.Z.: "Living Resources: The BIOMASS Program." *Oceanus*, vol. 31, no. 2, Summer 1988, pp. 75-79. *The Antarctic krill is a potentially bountiful fishery, but can it be fully exploited without hurting other marine species?*

Hapgood, F.: "The Quest for Oil." *National Geographic Magazine*, vol. 176, no. 2, August 1989, pp. 226-259. *Discusses the search for oil and natural gas offshore and elsewhere.*

Krakauer, J.: "A Fishing Frenzy Strikes on Sitka When Herring Run," *Smithsonian*, vol. 17, no. 7, October 1986, pp. 96-109. *A glimpse into the Alaskan herring fishery, a short but profitable affair that banks on our taste for the fishes' roe.*

Leigh, J.H.: "The Slug and the Damselfish." *Sea Frontiers*, vol. 36, no. 4, July-August 1990, pp 8-15. *Several groups of marine organisms are being used in biomedical research to help control diseases such as cancer and Alzheimer's disease and even to learn how our nervous system works.*

Lewis, R.: "Twenty Thousand Drugs Under the Sea." *Discover*, vol. 9, no. 5, May 1988, pp. 62-69. *A review of the potential commercial uses of marine plants and animals, from wonder drugs extracted from sponges to glue obtained from mussels.*

Mac Leish, W.H.: "New England Fishermen Battle the Winter Ocean on Georges Bank." *Smithsonian*, vol. 16, no. 2, May 1985, pp. 40-51. *The account of a 1-week trip for cod and pollock profiles the tough and precarious life of modern-day fishermen.*

Sitwell, N.: "The 'Queen of Gems'—Always Stunning, and Now More Cultivated than Ever." *Smithsonian*, vol. 15, no. 10, January 1985, pp. 40-51. *Oysters are appreciated not only as food, but as a source of pearls, which can be natural or cultured.*

Van Dyk, J.: "Long Journey of the Pacific Salmon," *National Geographic Magazine*, vol. 178, no. 1, July 1990, pp. 2-37. *A peek at the important salmon fishery across the north Pacific, its difficulties, and the implications of salmon farming.*

The Impact of Humans on the Marine Environment

17

$\mathcal{I}$t certainly doesn't look very good: the deep, blue sea is becoming messy, and the troubled waters are not that blue anymore. Rusting drums containing deadly chemicals rest on the bottom of the North Sea. A smelly brown substance that the Japanese call *hedoro*, a combination of the words for "vomit" and "muck," floats on the once-placid waters. Coral reefs are poisoned by fishermen in Southeast Asia. Oil spills threaten rare marine mammals from the Persian Gulf to Alaska (Figure 17-1) and oil globs are found even on the once-pristine ice shelf of Antarctica. Sick, dying dolphins and seals are washed up on beaches. Slimy algae bloom along the seaside resorts of the Adriatic.

The list is long and very alarming. It is, however, only a small sample of the impact we humans are making on the marine environment, a consequence of the fact that well over half of the more than 5 billion people inhabiting our planet live along seashores, estuaries, and river deltas. Everywhere, not just along industrial areas, the pressures of civilization are modifying the marine world. Water quality has decreased, fisheries and mariculture schemes are imperiled, recreational areas are at risk, and new health hazards develop. All of these consequences result from the mistaken belief that a bottomless ocean is a convenient, unlimited receptacle for our wastes and that we have the right to exploit and change anything within its limits.

Figure 17-1 Dead sea otter—an aftermath of the 1989 *Exxon Valdez* oil spill in Alaska.

POLLUTION

Pollution—visible or invisible; on land, air, or water—is by now an unwanted but familiar part of our lives. Pollution can be described as the introduction by humans of substances or materials that decrease the quality of the environment. These substances, or **pollutants,** are added to the environment by humans and not as a result of events such as natural oil seeps and volcanic eruptions, which may be called natural pollutants. Many pollutants are artificial substances that are foreign, and therefore harmful, to us and other organisms.

The role of humans in decreasing the quality of life in the marine envi-

517

ronment has been enormous. The potentially detrimental effects of pollu-tion can directly or indirectly affect all forms of life from beaches and estu-aries to the deepest corners of the ocean. There is in fact only *one* ocean: all seas and oceans are connected so that what we throw, empty, or dump in one place may inevitably affect a shore or seabed elsewhere. Pollution can similarly present a hazard to human health when marine organisms are eaten or when we go swimming, diving, or surfing.

Pollution of the ocean by humans involves adding substances that decrease the quality of the marine environment. Most of these pollutants are toxic or harmful to marine life.

The diversity in the types, distribution, and effects of marine pollutants is almost unlimited. The sources of these pollutants also are widespread and alarming. Pollutants can be traced to cities, oil drilling, agriculture, ship-ping, land mining, and nuclear tests, to name but a few. Here we will briefly discuss only the most important types of marine pollutants and their significance to the well-being of marine life.

Oil

Crude oil, or **petroleum,** is a sticky, dark greenish brown mixture of many chemical compounds. Most of these chemicals are **hydrocarbons,** long chains of carbon and hydrogen. Crude oil is a valuable commodity that is refined to yield not only fuels, but raw materials for making plastics, syn-thetic fibers, rubber, fertilizers, and countless other products.

Sources. Oil is at the same time one of the most widespread pollutants in the ocean. Though in several coastal regions oil comes from natural

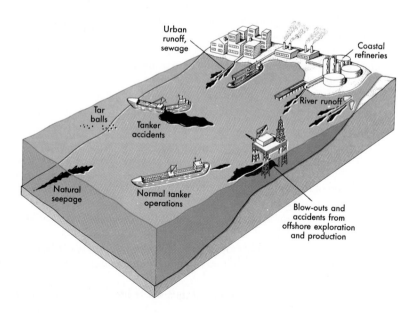

Figure 17-2 Sources of oil in the marine environ-ment.

HUMANS AND THE SEA

seeps, most enters the marine environment as a pollutant released and distributed by humans (Figure 17-2). It is extracted from the seabed (see the section in Chapter 16, "Resources from the Sea Floor"), transported across the oceans, and refined in plants along the coast. All of these processes are potential contributors to pollution. Large amounts of crude oil and derivatives like fuel, heating, and lubricating oil also are present in runoff from coastal cities and rivers. Marine pollution by ships is strictly controlled by the International Convention for the Prevention of Pollution from Ships (MARPOL). Yet oil tankers occasionally discharge oil while unloading, illegally cleaning their tanks at sea, and emptying the water they use as ballast on their return trips to load oil. Tanker operations and wastes from cities are considered the two most important sources of oil pollution. Infrequent but potentially disastrous are the blowouts of offshore rigs used in the extraction of oil (see Figure 16-25). The underwater blowout of an exploration well in the Gulf of Mexico in 1979 spilled at least half of a million tons (about 3.6 million barrels) of oil over a period of 9 months! The magnitude of this spill was not much higher than another blowout in the Persian Gulf in 1983.

1 metric ton of crude oil
= 7.33 barrels (bbl)
= 308 U.S. gallons

Of all mishaps, however, the massive oil spills that result from the sinking or collision of supertankers are the most dramatic and by far the most devastating to the marine environment. The *Torrey Canyon*, a tanker fully loaded with 118,000 tons of crude oil, ran aground and broke up near the southwestern tip of England in 1967. The gigantic spill affected both British and French coasts. The 1978 grounding of a supertanker, the *Amoco Cadiz*, poured 230,000 tons of crude oil along the coasts of Brittany, in northwestern France. In 1989 an estimated 35,000 tons of crude oil were spilled by the *Exxon Valdez* along the unspoiled coasts of southern Alaska, the home of whales, sea otters, salmon, fish-eating bald eagles, and other precious wildlife. Accidents have prompted the imposition of tighter restrictions on the construction (such as having double hulls) and operation of tankers. Altough the annual estimates of accidental spills have decreased, the amounts of polluting oil are still high. The industrialized world's thirst for energy demands the use of supertankers to transport huge amounts of crude oil along congested sea routes. Most accidents continue to occur near the coast, where the effects on the environment are potentially the most harmful.

Most of the components of oil are insoluble in water and float on the surface. They can be seen in most harbors as thin, iridescent slicks on the surface or as black deposits on sandy and rocky beaches. You would expect large areas of the ocean to be covered with the oil that has accumulated over the years. Fortunately some of its lighter components evaporate, and bacteria ultimately break it down. Oil is said to be almost completely **biodegradable** because, though very slowly, it is broken down, or decomposed, by bacteria.

Some of the components of crude oil sink and accumulate in sediments, especially after the lighter ones evaporate. Floating oil residues, or tar balls, have become common on the water surface along the shipping lanes that crisscross the oceans. They may persist for many years in the water and have been observed in remote areas far from shipping traffic. Some even have barnacles living on them! Extensive oil spills take the form of huge layers that coat everything in their path as they are carried by wind and currents.

Oil is a widespread pollutant that enters the sea as waste from land, from accidents during its extraction from the seabed, and as a result of its transportation across water.

Effects on Marine Life.   Even in small amounts, oil has been shown to cause a wide variety of effects in marine plants and animals. Organisms are able to accumulate oil from the water, sediments, and their food. Some of the substances contained in crude oil are known to be toxic. There is evidence, for instance, that oil interferes with the reproduction, development, growth, and behavior of many organisms, particularly the eggs, larvae, and other immature stages. It increases susceptibility to diseases in fishes and inhibits growth in phytoplankton.

Major oil spills can have disastrous effects on marine life, especially in coastal environments. Seabirds and marine mammals like the sea otter are particularly susceptible. Many die of exposure when feathers or hair are coated with oil. They lose the ability to maintain the thin layer of warm air that is needed for insulation from cold water (Figures 17-1 and 17-3). Birds that rely on flying to catch food are unable to do so and die of starvation. It is difficult to determine the number of birds killed by oil spills because many sink without reaching the coast, where the corpses are counted. About 3,200 dead birds, some belonging to rare species, were counted immediately after the *Amoco Cadiz* spill. The *Exxon Valdez* spill is believed to have killed between 350,000 and 390,000 seabirds and 3,500 to 5,500 sea otters. Estimates are that it will take up to 70 years for the wildlife in the area of the spill to recover.

The effect of oil spills on exposed rocky shores is less devastating than it may appear at first sight. Though initially there is mortality among many of the attached inhabitants, wave action and tides help clean away the oil. Rocky shore communities do recover, though recovery is dependent on factors such as the amount of oil, wave action, and temperature. Degradation, or breakdown, by bacteria takes place, but it is very slow, especially in cold water. Spills are degraded more quickly by bacteria if an oil-soluble fertilizer is added to the water or sprayed on rocks and sediment. Experience has shown that recovery begins to be seen within months and that an apparent near-normal condition may occur as early as 1 or 2 years after the spill. Heavier oil concentrations in sediments and isolated pockets, however, have been found to remain for 15 years or longer.

Large spills, on the other hand, can have catastrophic effects when they drift to salt marshes and mangrove forests. These communities are characteristic of sheltered coasts and estuaries where oil cannot be dispersed by wave action. Massive mortality of the dominant plants takes place, and recovery is very slow. Oil is absorbed by the fine sediment characteristic of these communities and may persist here for decades. Shallow-water coral reefs and seagrass beds also are greatly affected by oil spills. Coral colonies show swollen tissues, excessive production of mucus, and areas simply devoid of tissue. Reproduction and feeding in surviving coral polyps are known to be affected by oil.

phytoplankton The minute, drifting plants that are a critical part of open-water communities since they perform practically all of the photosynthesis in the open ocean *Chapter 14, p. 404*

Figure 17-3 An oil-coated loon (*Gavia*) during the *Exxon Valdez* spill in Alaska. Detergents, were widely used to disperse the oil during the *Torrey Canyon* spill on the English coast in 1967 but proved to be toxic and actually caused more damage than the oil itself.

Oil is harmful to most marine organisms. It is especially destructive to communities typical of shallow, sheltered waters.

Containing an oil spill and cleaning the mess still remains a major headache. Fencing off the spill with floating, fire-resistant booms prevents the spill from moving into shore. "Skimmers," boats equipped with U-shaped booms, are used to skim off and recover some of the oil. These methods, however, cannot be used in heavy seas. Chemical **dispersants** are added to the spill to break the surface oil into small droplets that can then disperse in the water. Unfortunately, dispersants also are harmful to marine life, and the dispersed oil still remains toxic underwater. The use of powerful streams of hot water to wash off the oil spill on beaches by the *Exxon Valdez* was also found to be harmful to many forms of life.

The cost of oil spills to the local economy can be enormous. Commercial fishing is one of the first areas to be affected. Fish and filter-feeding shellfish like oysters and clams that are tainted with oil are simply unmarketable. The size of successive fish catches, particularly of bottom-dwelling species, may decrease because of an initial mortality of adults and juveniles or a drop in the abundance of their food. Furthermore, oil-soaked beaches will prove to be disastrous to resort areas dependent on tourism. Claims by those affected plus the cleanup bill can run well past the billion dollar mark in major spills like that of the *Exxon Valdez*. Costs to Exxon were estimated to be over $2 billion and more than $100 million to the state of Alaska.

Figure 17-4 Sewage pollution of the seacoast is a serious health hazard on coasts around the world. In this particular area, shellfish became contaminated by filter feeding in sewage-polluted water.

Sewage

Disposing of ever-increasing amounts of wastewater, or **sewage,** is a major problem in cities around the world. **Domestic sewage** carries all kinds of wastewater from homes and city buildings. It may also carry storm runoff water. **Industrial sewage** contains a variety of wastes from factories and the like. For ages the discharge of sewage in the ocean has been a cheap alternative to waste disposal on land. Increasing amounts of sewage at sea ultimately has proved to be a major menace to the health of the marine environment and of humans as well.

Raw Sewage. Raw, or **untreated, sewage** discharged directly into the water is a serious health hazard to people. These sewage effluents contain many viruses and bacteria that cause diseases. Infectious hepatitis, for example, is carried by viruses found in human feces. Numerous cases of hepatitis result from people eating oysters, clams, and other shellfish that concentrate the viruses by filtering the water for food (Figure 17-4). Swimming in sewage-polluted water can also be hazardous. People can get sick just by swallowing water. The closing of beaches because of spills of raw sewage, common when sewers overflow after a rain storm, has now become routine in many areas. Pipes do not always dump sewage in deep water. Our coasts are full of embarrassing surprises that are easily seen floating on the surface.

Sewage is discharged directly into the sea by many communities around the world. Raw sewage is a major health hazard to humans because it spreads disease.

Figure 17-5 Treated sewage effluents from the Hyperion Treatment Plant being released at a depth of 60 m (180 feet) at the end of an 8-km (5-mile) outfall in Santa Monica Bay, southern California.

Sewage Treatment and Sludge. To prevent diseases and unwanted sights and odors, some but not all countries require by law the treatment of sewage dumped at sea. **Treated sewage** is partially purified in sewage-treatment plants by using chlorine to kill bacteria and some of the viruses. Most suspended solids, including feces and decaying food, are removed and allowed to be partially broken down by decay bacteria. The resulting semiliquid material, **sludge,** is often disposed of by discharging it into the ocean by way of pipes. Unseen by human eyes, sewage outfalls may line the coast. New York City uses barges to unload sewage far out at sea. Unfortunately, currents are known to occasionally bring some of it back! Furthermore, not all sewage is treated. In the United States only about 70% of the population is served by sewage-treatment plants. The percentage is abysmally low in most of the poorer countries bordering the sea.

Discharging huge amounts of sludge at sea has actually created a new set of problems for the marine environment. The millions of gallons of sludge dumped every day by municipal sewage systems accumulate around the outfalls. Sludge simply overcomes the natural communities on the bottom, creating black deserts around outfalls (Figure 17-5). It is clearly impossible for most detritus-feeding animals to handle the massive amounts of organic matter contained in the sludge. The organic matter is therefore decomposed by marine bacteria that thrive under these conditions. The decay bacteria use oxygen to such an extent that **anaerobic** conditions develop. The total number of species decreases as many of the natural inhabitants disappear. They are replaced by hardier forms of life, such as some species of polychaete worms. Bottom fishes collected around sewage outfalls tend to show skin tumors, erosion of fins, and other abnormalities, apparently a result of the high concentration of toxic substances and bacteria.

Sludge also contains large amounts of nutrients. Although these nutrients—like nitrate and phosphate—are needed by marine plants, excessive amounts overfertilize coastal waters, a phenomenon known as **eutrophication.** Fertilizers from agriculture runoff or wastes from fish farms complicate things in some areas. In shallow, enclosed areas eutrophication takes its toll, and a few species of hardy seaweeds may overtake the bottom (see the section in Chapter 13, "The Kaneohe Bay Story"). Eutrophication triggers phytoplankton blooms, explosive increases in the number of individuals. Blooms are responsible for raising turbidity and cutting down the penetration of light. Others are toxic. Coastal pollution may be responsible for frequent red tides and other phytoplankton blooms (see Box 14-1). They have become common in Japan, where they endanger valuable mariculture operations. Blooms of phytoplankton and other algae have become a recurrent problem in the Baltic and Adriatic seas.

High volumes of sludge discharged at sea greatly modify or destroy bottom communities and overfertilize the water.

Alternatives. It is possible to implement alternatives to the worldwide discharge of sewage at sea. Additional and improved treatment, mandated in the United States by the Clean Water Act of 1972, reduces the amount of suspended solids in the discharged sewage. But why not consider sewage a resource and turn sludge into something useful? Ideally we should be able to recycle the large amounts of precious water and fertilizers that go to waste. Some small communities are taking advantage of the ability of marshes to recycle large amounts of nutrients by using the marshes for natural sewage treatment (Figure 17-6). Sludge is also being recycled into landfill, building blocks used in construction, and compost. It is also spread on farmlands as a fertilizer, burned to generate electricity, and converted into fuel oil. Sewage unfortunately also includes large amounts of pesticides, heavy metals, and other toxic chemicals. These must be first removed by pretreatment, especially in the case of industrial wastes.

Synthetic Chemicals

Other pollutants also reach the sea from land. A most important group comprise **synthetic** chemicals, that is, those manufactured by humans. Though organic, and thus made up of at least carbon, they are nevertheless foreign to all forms of life as a consequence of being artificial. They are invisible to our eyes but nevertheless represent a serious threat to the marine environment.

Chlorinated Hydrocarbons. One major group of synthetic chemical pollutants are the **chlorinated hydrocarbons,** or organochlorides. They comprise a huge family of **pesticides,** chemicals that are used to kill insects and to control weeds. Chlorinated hydrocarbons include widely used pesticides such as DDT, aldrin, dieldrin, heptachlor, and chlordane. Millions of tons of these and many thousands of other chlorinated hydrocarbons have been used since the 1940's when DDT, the first to be manufactured, made its debut. They have been used to control insects that carry diseases to humans and to save plant crops from hungry insect pests. Pesticides have saved millions from disease and starvation, but unchecked use of chlorinated hydrocarbons has now brought to our attention a more sinister side, the fact that they are harmful to many forms of life.

These pesticides have not been used directly in the ocean. Nevertheless, chlorinated hydrocarbons are highly mobile and large amounts end up contaminating the marine environment. They are easily carried by wind from land, particularly when crops are sprayed from planes (Figure 17-7). They are not soluble in water but are brought in by rivers, runoff from land, and domestic and industrial sewage. They are carried everywhere, even way out at sea far from where originally used. Pesticides are absorbed by phytoplankton and particles suspended in the water. It is in this way that they find their way into other marine organisms.

Chlorinated hydrocarbons are synthetic, and organisms have not yet evolved ways of breaking them down. They are therefore **nonbiodegradable,** and getting rid of them is not easy. Since they hardly dissolve in water and are not excreted, they accumulate in fats, remaining there almost indefinitely. For these reasons, chlorinated hydrocarbons and other nonbiodegradable chemicals are said to be **persistent.** They circulate in the environment for many years, even decades. Sooner or later we find them in the tissues of living organisms. By being persistent, the concentration of chlorinated hydrocarbons in a particular animal is higher than in its food supply, and it increases as we move higher in the food chain. Among marine ani-

Figure 17-6 This lush, sweet-smelling artificial marsh is part of a network of marshes where domestic sewage from the town of Arcata in northern California is naturally purified before it is piped into the ocean. Sewage, which is mostly pure water, is first treated with chlorine to kill harmful bacteria. It is then pumped into the marshes where mud bacteria break down the organic matter. The released nutrients fertilize the marsh plants. The marshes prevent the pollution of the ocean by sewage and attract numerous species of birds and other wildlife.

food chain The steps of transfer of energy from *producers*, the plants, through *consumers*, the animals
Chapter 9, p. 273; Figure 9-13

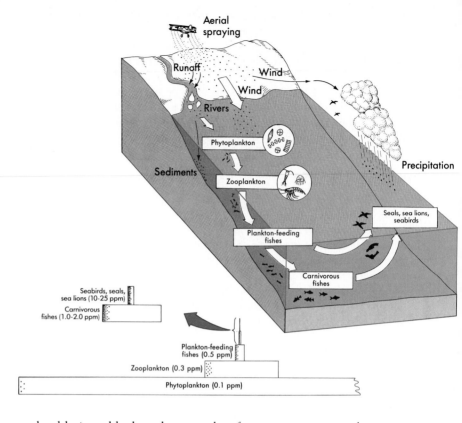

Figure 17-7 The concentration of chlorinated hydrocarbons increases with the relative position organisms have in the food chain, thus showing biological magnification. In the trophic pyramid that summarizes this generalized food chain, the concentration of pesticides is expressed in ppm, or parts per million.

Aerial spraying

Runoff

Wind

Wind

Rivers

Phytoplankton

Sediments

Zooplankton

Precipitation

Seals, sea lions, seabirds

Plankton-feeding fishes

Carnivorous fishes

Seabirds, seals, sea lions (10-25 ppm)

Carnivorous fishes (1.0-2.0 ppm)

Plankton-feeding fishes (0.5 ppm)

Zooplankton (0.3 ppm)

Phytoplankton (0.1 ppm)

trophic pyramid The pyramid-like relationship of energy, number of individuals, or biomass of the organisms found in a food chain
Chapter 9, p. 273; Figure 9-15

mals, chlorinated hydrocarbons are therefore more concentrated in carnivorous fishes and especially in the fish-eating birds and mammals that feed on them (Figure 17-7). These carnivores are said to be at the top of the trophic pyramid and are known as top carnivores. Since most of the chlorinated hydrocarbons in the trophic pyramid are not broken down, the carnivores end up concentrating the pesticides that had accumulated along the food chain leading to them. This phenomenon is known as **biological magnification.**

> Chlorinated hydrocarbons, which include many widely used pesticides, are nonbiodegradable and persistent. They accumulate in the food chain and thus show biological magnification.

The effects of the worldwide use of chlorinated hydrocarbons began to turn up at alarming rates during the 1960's. Fish caught for human consumption in the United States had to be destroyed since it contained too much pesticide—concentrations higher than the 5-parts-per-million level allowed by the Food and Drug Administration. Pesticides began to accumulate in top carnivores in concentrations that were thousands and even millions of times above those found in seawater.

The effects on birds, on land as well as at sea, were particularly dramatic. Birds were not actually poisoned, but the high concentrations of chlorinated hydrocarbons in their body fat interfered with reproduction, specifically with the deposition of calcium in eggshells. Eggshells became so thin that they broke during incubation before chicks became fully developed.

Such was the case of the brown pelican (*Pelecanus occidentalis*; Figure 17-8). The once-abundant bird became a rare sight in most of the United States as the result of a near-disastrous failure in its reproduction during the late 1960's and early 1970's. In Louisiana, where the brown pelican is the official state bird, whole nesting colonies vanished. The same situation was reported along most of the Gulf of Mexico and southern California. Females sat on broken eggs, or adults simply did not attempt to nest. The Channel Islands off southern California, the only nesting colony in the Pacific coast north of Mexico, recorded one chick in 1970 and seven in 1971. High amounts of DDT and some related chemicals were found in the tissues of birds taken from these and other nesting colonies. In the case of southern California, the most important source of DDT was traced to a sewage outfall where a chemical plant—the largest producer of DDT in the world at that time—was discharging its wastes. High amounts of DDT also were found in other marine animals in the area, from filter feeders like sand, or mole, crabs (*Emerita analoga*) to top carnivores like sea lions (*Zalophus californianus*).

By 1972 most uses of DDT and several other chlorinated hydrocarbons were banned in the United States and many other industrialized nations. DDT residues in marine animals and sediments then began to decrease. The brown pelican—once nearly extinct in the United States—has recovered, and its reproduction appears to have returned to normal. DDT and related residues, however, can still be found in bottom-dwelling animals such as the white croaker (*Genyonemus lineatus*) in southern California. Chlorinated hydrocarbons provide inexpensive control of insects and are still used in poorer nations. Because of their persistence and high mobility, the menace of these pesticides to the marine environment is not over yet.

PCBs.   Another group of synthetic chemical pollutants are the **PCBs,** the polychlorinated biphenyls. Like chlorinated hydrocarbons, PCBs are nonbiodegradable and persistent and therefore show biological magnification. Their uses proliferated, and after the 1930's they were employed in electric transformers and capacitors, in the manufacturing of plastics and paints, and in many other products. They were eventually found to be highly toxic, causing cancer and birth defects in humans. Other organisms also are affected. They have been detected, together with chlorinated hydrocarbons, in the blubber of whales and other marine mammals. The production and uses of PCBs were gradually regulated or banned by many nations but not until 1979 in the United States.

> **PCBs are pollutants notable for their toxicity. They are persistent and show biological magnification.**

This ban unfortunately came after PCBs had been spread throughout the oceans. PCBs continue to be used in some parts of the world, and electrical equipment containing them is still around us. Equipment is required to remain sealed, and the PCBs must be carefully disposed of when the equipment wears out. Invariably, PCBs are a main ingredient of the large

Figure 17-8   The brown pelican *(Pelecanus occidentalis)* is once again a common sight along the southeastern, gulf, and western coasts of the United States.

Figure 17-9 Hazardous chemicals such as PCBs and pesticides, and cancer-causing agents such as vinyl chloride and dioxin can be disposed of by incinerating them at high sea rather than by using dumps on land. Incinerator ships, like this one, burn hazardous organic chemicals at high temperatures, resulting in safe chemicals like carbon dioxide and water.

fouling organisms Those organisms, such as barnacles and shipworms, that bore or encrust on boats and underwater structures *Chapter 6, p. 167*

amounts of hazardous wastes that accumulate and that have to be disposed somehow somewhere (Figure 17-9). PCBs have become widespread in landfills and waste dumps. Like chlorinated hydrocarbons, decades of use and dumping have also left significant amounts of these persistent pollutants, especially around sewage outfalls and in the sediment of harbors of industrial cities. Some marine bacteria, however, are known to degrade PCBs that have accumulated in sediments. They have been found mostly in areas with high concentrations of PCBs such as the Hudson River estuary.

Heavy Metals

An additional category of *dissolved* chemical contaminants of the world's oceans are metals, particularly those classified as **heavy metals.** Although very minute amounts of some metals are actually needed by most if not all organisms, an excess of some dissolved metals in seawater can be very toxic.

Mercury.   One particularly troublesome heavy metal is **mercury.** Mercury reaches the ocean through several natural processes: the weathering of rocks, volcanic activity, rivers, and dust particles from the atmosphere. Even so, human activities appear to have played an increasingly important role, especially along the coast. Mercury is an active ingredient in chemicals used to kill bacteria and molds and in antifouling paints. It also is used in the production of chlorine and plastics, in other chemical processes, in fluorescent lamps, in drugs, and even in tooth fillings. Discharges from industries and cities and the burning of coal, which contains traces of mercury, has inevitably increased the concentration of mercury in the marine environment.

What is the problem with mercury? Pure, liquid mercury, like that in your thermometer, is harmless unless its vapors are inhaled. But it is a different story when mercury combines with organic chemicals, as when transformed by some bacteria and other microorganisms in the water or sediment. These organic compounds, especially **methyl mercury,** are persistent and accumulate in the food chain. Levels of mercury too high for human consumption have been found in large fishes such as tuna and swordfish. The older the fish, the higher the mercury content. Methyl mercury also can be found in coastal sediments, particularly around areas where wastes are dumped. Mercury compounds are very difficult to eliminate and are highly toxic to practically all forms of life. In humans they cause brain, kidney, and liver damage and are responsible for birth defects.

> **Mercury compounds are highly toxic. They are persistent and accumulate in the food chain.**

The dangers of the presence of mercury in seafood was tragically demonstrated by the mass appearance in the 1950's and 1960's of a crippling, permanent neurological disorder among the inhabitants of a town in southern Japan (Figure 17-10). The victims were poisoned by eating fish and shellfish that had concentrated mercury discharged at sea by a chemical plant.

Other Heavy Metals.   There are several other heavy metals that are carried to the ocean as toxic pollutants. **Lead** is one of the most widely dis-

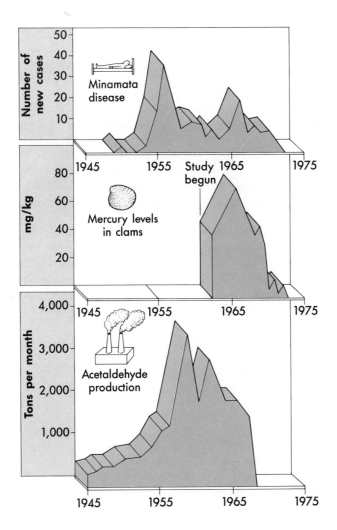

Figure 17-10 A very serious neurological disorder that often ends in severe brain damage, paralysis, or death appeared among the inhabitants of Minamata, Japan. The number of new patients (*upper graph*) was directly related to the production of acetaldehyde (*lower graph*) by a chemical plant in this traditional fishing town. A mercury compound was used in the production of acetaldehyde and vinyl chloride, which are used in making plastics. As a result, an estimated 200 to 600 tons of waste mercury were discharged into Minamata Bay between 1952 and 1968, when it was finally stopped. The disease, now called Minamata disease, was related to the ingestion of mercury in seafood, here indicated by its concentration (*middle graph*) in an edible clam (*Venus*).

tributed. As with mercury, organic lead compounds are persistent and concentrate in the tissues of organisms. Lead is very toxic to humans, causing serious nervous disorders and death. The principal source of lead pollution in the marine environment appears to be the exhaust of vehicles run with leaded fuels. Lead eventually reaches the water by way of rain and windblown dust. And, as you may have already suspected, lead has found its way into all sorts of products, such as paints and ceramics, that sooner or later find their way to the ocean.

Cadmium and **copper** are among other toxic heavy metals that are slowly concentrated in marine life. Levels of these and other heavy metals in seawater are especially elevated where wastes from mining and dredging operations reach the sea. This has become a problem in several areas in the tropics, where copper mining threatens coastal fisheries and highly-vulnerable coral reefs.

Radioactive Wastes

Also invisible and potentially deadly, **radioactive** pollutants have been contaminating the marine world since the first atomic bomb explosions in the early 1940's. **Radioactivity** is a property exhibited by certain unstable atoms that emit **radiation** in the form of energy or particular types of particles. Exposure to some types of radiation can be very harmful to all forms of life. In humans it causes cancer, leukemia, and other disorders. It does not need to be ingested since it can penetrate through living matter. Radioactive material also may continue to emit radiation for thousands of years. Some radioactive atoms occur in nature but in minute amounts; some harmful radiation reaches us from outer space. A most important source results as a by-product, or waste, of the use of the atom as a source of energy, in medical applications, and nuclear weapons.

Radioactive waste is extremely dangerous and must be disposed of somewhere. Some is stored in containers and dumped into designated areas of the ocean, with the obvious risk of leakage. Some radioactivity reaches the ocean as **fallout** from the test of nuclear weapons. Sunken nuclear-powered submarines and ships, fallen satellites, and crashed planes carrying nuclear weapons have been additional sources. Nuclear reactors along the coast and industrial effluents add to the risk.

> There is a potential danger of contamination of the ocean with radioactive materials, a serious health hazard.

Thermal Pollution

Seawater is often used as a coolant in power plants, oil refineries, and other industries; that is why so many of these industries are built along the coast. The heated water that results from the cooling process is pumped back to sea, causing alterations to the environment known as **thermal pollution.** Local warm pockets can be created in poorly mixed bays. Even if some fishes are attracted to the site, sudden temperature increases are known to affect plants and animals in the immediate vicinity. Higher temperatures also decrease the ability of water to dissolve oxygen. Since cold, dense water that is rich in nutrients sinks under warm water, thermal pollution induces stratification and depletion of nutrients in the upper layer of water.

> Thermal pollution results when heated water is pumped into the sea.

The effects of thermal pollution on marine life are especially pronounced in the tropics. In contrast to organisms living in temperate and polar regions, tropical species normally live just below the highest tolerable temperatures. Reef-building corals are particularly sensitive (Figure 17-11).

Solid Waste

A look at the upper reaches of any beach will reveal an amazing assortment of trash brought in by high water. Most of it is plastic: bottles, bags, styro-

atoms   Particles that make up molecules; in the case of water (H_2O), each molecule is made of two hydrogen atoms and one oxygen atom
Chapter 3, p. 52

stratification   The separation of the water column into layers, with the densest, coldest water at the bottom, thus preventing the mixing of the nutrient-rich deep water with the less dense, warmer upper layer
Chapter 3, p. 78; Figure 3-35

Figure 17-11 These corals were affected by the heated effluents of a power plant on the island of Oahu, Hawaii. Almost all coral colonies in water 4° to 5° C (7° to 9° F) above ambient temperature died. Some, like the colonies of *Porites lobata* in the center, became paler after losing their symbiotic algae, or zooxanthellae, a phenomenon known as bleaching.

foam cups and packaging, disposable diapers, nets, and thousands of other items (Figure 17-12). Add rubber, glass, and metal and you have quite a heap of trash. It is amazing how much of our throwaway society's mess finds its way to sea. Although ocean dumping by ships is now banned in about half of the world's merchant fleets, solid waste from land still flows to sea at an alarming rate. Landfill sites are filling fast and burning trash pollutes the air, so ocean dumping is the easy way out.

Plastic is especially troublesome because it is strong and durable. It is nonbiodegradable, and all of those millions of tons of plastic will keep polluting for hundreds and hundreds of years. Styrofoam and other plastics eventually break down into tiny particles that are by now found in every remote corner of the ocean. They have been found in the guts of many animals that ingest them by mistake. Larger plastic debris is a major threat to marine life. Sea turtles, seabirds, seals, and others are maimed or killed after getting entangled in plastic fishing line. Many die with their digestive tract clogged with bags and brightly colored plastic junk.

MODIFICATION AND DESTRUCTION OF HABITATS

Pollution is unfortunately not the only way we affect the marine environment. This section briefly summarizes cases where human activities like dredging, dumping silt or mud, landfilling, or even using explosives causes problems. In these cases we modify or destroy **habitats,** the places where organisms live. The environment is affected *directly* rather than indirectly, as when pollutants are released somewhere else.

Figure 17-12 Plastic debris can be more than an eyesore. It has been estimated that it kills as many as 2 million seabirds and 100,000 marine mammals every year. Plastic bags kill sea turtles that swallow them thinking they have caught a jellyfish. Six-pack rings are particularly risky to seabirds ensnared with them around the neck. Unable to feed or fly, they face a certain slow death.

Figure 17-13 Salt marshes in many locations in California and elsewhere have been obliterated by landfilling.

Estuaries

Dredging navigation channels can be very destructive to coastal environments, particularly estuaries. An increase in exposure to wave action or in water circulation may result in the destruction of the fragile salt marshes and other coastal wetlands that border estuaries in temperate areas (see the section in Chapter 11, "Types of Estuarine Communities"). Not only are channels dredged for pleasure boats and ships, but artificial marinas and harbors have made unspoiled estuaries an uncommon sight. Estuaries are highly productive environments where many species of commercial and sport-fishing importance reproduce and grow. About 65% of the shellfish and fish catches in the United States depend on estuaries to complete their life cycles. Estuaries also serve as feeding and resting grounds for many migratory birds. Numerous estuaries have been abused and even completely wiped out by having them landfilled to build everything from oil refineries to cities (Figure 17-13). About one third of all estuaries in the United States have disappeared altogether; California alone has lost about 67% of them.

Mangrove Forests

The same factors that threaten salt marshes and other estuarine communities are menacing the mangrove forests that develop along estuaries and other protected coasts in the tropics. Mangrove forests also are very productive, providing food and shelter to many species and helping to reduce coastal erosion, but like salt marshes, they are considered by many as "waste lands." They are being cleared at an alarming rate to provide space for crops, houses, mariculture ponds, and garbage dumps. The growing use of mangrove wood as fuel and timber in some areas is another cause for concern.

Figure 17-14 The Persian Gulf shoreline—and coral reefs—were originally where this sign is. The site became a landfill site so now the shoreline is 1 mile away!

Coral Reefs

Rich but fragile, coral reefs are threatened by human intervention all over the tropics (Figure 17-14). Though they support luxuriant life that provides much needed protein and potentially life-saving drugs, coral reefs are being subject to much abuse (see the section in Chapter 13, "The Kaneohe Bay Story"). As are their also-threatened terrestrial counterparts, the tropical rain forests, coral reefs are among the oldest and richest environments on earth. Ironically, the rapid disappearance of tropical rain forests threatens coral reefs too. The clearing, or deforestation, of the rain forests for agriculture and urban expansion increases the amount of soil that is washed out to sea by the plentiful tropical rain. Soil from land and sediment from dredging at sea are serious problems (Figure 17-15). The accumulation of silt and other particles over live coral colonies can be damaging, even if the corals themselves are usually able to keep their outer surfaces free from moderate amounts of particles. Besides, coral larvae do not settle on sediment-covered bottoms. Fine sediments suspended in the water also affect corals by decreasing the amount of light they get.

Figure 17-15 Dredging in or near coral reefs releases large amounts of sediment that are harmful to reef-building corals in many ways. Dredging is common throughout the Republic of the Marshall Islands and the Federated States of Micronesia, where the dredged material is used for landfill or as construction material.

Coral reefs are extensively damaged or destroyed by explosives used to kill fishes. Even if technically illegal, dynamite fishing is practiced in many areas. It may take at least several decades for damaged coral reefs to recover their former splendor. Explosives also are used to open channels for navigation and in military testing, including nuclear bombs.

Fishing with poisons such as bleach and cyanide kills coral reefs as well. Although generally banned as a fishing method, cyanide is used widely in the Philippines to stun reef fishes that are then sold for the lucrative aquarium trade. Another related threat is the indiscriminate collection of corals for sale as souvenirs and decoration. Shell collectors can be very destructive to coral reefs by turning over or breaking corals to get their specimens. Damage by anchors, fish traps, reef walking, and unrestricted diving have also taken its toll.

Another indication of stress to coral reefs is a worldwide outbreak of bleaching that began during the late 1980's. Bleaching occurs when corals expel their symbiotic zooxanthellae, a phenomenon that results in the formation of white patches on the colonies. Recovery may take place, but bleached coral does not grow and is vulnerable to disintegration. Some biologists have suggested that bleaching may be the result of increased temperatures brought about by global warming (see Box 17-1).

zooxanthellae Dinoflagellates adapted for living within animals
Chapter 5, p. 124

Estuaries, mangrove forests, and coral reefs are being adversely modified and even destroyed in many parts of the world as a result of direct human interference.

THREATENED AND ENDANGERED SPECIES

The alteration or destruction of habitats by humans may have another disastrous effect: driving species to their eventual disappearance, or **extinction,** from the face of the earth. Recall that individuals become adapted to changes in the environment as a result of natural selection. If they cannot adapt, extinction will be the eventual outcome. Extinction is therefore a natural consequence of the process of evolution. To make a distinction,

natural selection The process in which some members of a population show better chances of surviving and reproducing than others because they are more successful in facing conditions in the environment
Chapter 4, p. 109

BOX 17-1

Living in a Greenhouse: Our Warming Earth

Evidence has been accumulating over the years, and some scientists had even issued early warnings. Most people, however, had never heard about it until recently.

Carbon dioxide, a normal component of the atmosphere, provides a warming effect on the earth's surface, a phenomenon known as the **greenhouse effect.** As sunlight strikes the planet, most of the solar energy—about 70%—is absorbed by the earth. Of the absorbed energy, some is reflected back as warming infrared radiation. Carbon dioxide traps part of this heat energy, warming the earth like the glass of a greenhouse. It has been estimated that without the greenhouse effect the earth would be about 10° C (50° F) colder.

Living organisms, both at sea and on land, contribute to the carbon dioxide in the atmosphere. Plants remove carbon dioxide from the atmosphere through photosynthesis; both plants and animals return it through respiration. Life on the planet removes about as much carbon dioxide from the atmosphere as is added. Humans, however, have been increasing the amount of carbon dioxide in the atmosphere by burning enormous amounts of fossil fuels. These fuels, oil and coal, are nothing but the fossilized remains of ancient forests. We release their energy and turn them into carbon dioxide when we drive our cars. We also cut down the tropical rain forests that consume a great deal of carbon dioxide, burn them, and release even more carbon dioxide in the process!

Carbon dioxide has increased by 25% since 1850, and the planet may be warming up, an effect known as **global warming.** How warm it will get and what the consequences will be is highly debatable. A rise of about 0.5° C has occurred this century, but some predict a rise of 2° to 4° C in the next century. Warming will cause more water to evaporate from the oceans, increasing precipitation, hurricanes, and other storms. Some areas will be wetter, others drier. Scientists fear that the polar ice caps will begin to melt and that sea levels will rise and flood coastal lands. How far up and how fast they would rise is something scientists can't agree on. Projections vary from a rise of 0.3 to 1.5 m (1 to 5 feet) by 2030. Though this may look like a small change, some nations have started planning for the consequences of rising waters. Large portions of Florida and the Netherlands will be flooded, and island-group nations like the Maldives and Kiribati may disappear altogether!

Other gases increase the greenhouse effect. One is **methane,** which is normally released from rice paddies and swamps. A major culprit is the **chlorofluorocarbons (CFCs)** used in sprays and air conditioners. CFCs also are involved in another potential disaster: the gradual destruction of the **ozone layer** in the atmosphere. This natural layer of ozone gas (O_3) protects life from the sun's ultraviolet radiation, harmful because it causes genetic mutations and cancer. The destruction of the ozone layer will not, by the way, allow more heat to leave the atmosphere and cancel out global warming. Already an "ozone hole" appears every year in the layer over Antarctica and the Arctic. Disaster looms high over the planet unless we stop increasing the amount of carbon dioxide in the atmosphere.

The role of carbon dioxide (CO_2) and greenhouse gases in the retention of heat by the earth's atmosphere.

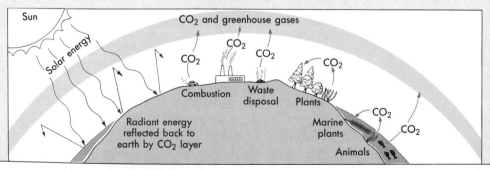

some biologists refer to human-induced extinction as **extermination.**

Species that face extermination are classified as **rare** when they are not in immediate danger but are at risk, **threatened** when their numbers have become low, and **endangered** when in immediate danger of disappearing forever.

Extermination of marine species can be brought about by overfishing or overhunting for food, hides, and other products; by habitat destruction; by pollution; and by overcollecting for the aquarium or shell trade. The Steller's sea cow, a sirenian, is a shocking example of a marine animal that was rapidly exterminated as a consequence of unregulated hunting for food (Figure 17-16).

> Species are categorized as rare, threatened, or endangered when they face the possibility of extinction or extermination.

sirenians A group of marine mammals collectively known as the sea cows and characterized by a pair of front flippers, no rear limbs, and a paddle-shaped tail
Chapter 8, p. 239

The Case of the Whales

Of the many examples of marine species in danger of extermination, perhaps the most widely publicized and tragic is that of the whales.

Whale hunting, or **whaling,** is an old tradition with a rich history. Na-

Figure 17-16 The Steller's sea cow (*Hydrodamalis gigas*) was a giant kelp-eating creature. It became known to science in 1741, when it inhabited the kelp beds of Commander Islands in the western Bering Sea. It has been estimated they weighed 10 tons or more! Its meat was described "as good as the best cuts of beef," the reason for its extermination at the hands of whalers. The species was slaughtered to extinction soon after its discovery: the last known live individual was taken in 1768.

great whales The large-size whales, comprising the plankton-feeding, or *baleen* whales, and the sperm whale, a carnivorous—or *toothed*—whale
Chapter 8, p. 244

baleen Flexible plates that hang from the upper jaws of filter-feeding whales, Figure 8-18
Chapter 8, p. 242; Figure 8-18.

tive Americans hunted gray whales in prehistoric times. Basques may have hunted them off Newfoundland before Columbus. It was not until the 1600's, however, that Europeans started a substantial exploitation of the great whales in the north Atlantic and Arctic oceans (Figure 17-17). Americans, who eventually dominated worldwide whaling, began hunting off New England by the late 1600's. Whales were harpooned from small open boats, a technique whalers learned from the natives. It was a rewarding fishery, though not one exploited primarily for food. Blubber (see Figure 4-4) provided oil, or "train oil," that was used to make soap and light cities. Baleen was used to make stays for corsets and other goods. Meat and other valuable products also were obtained from the huge animals. Fishing efforts rapidly increased, and during the 1800's fast steamships and the devastating explosive harpoon were introduced. The largest and fastest whales, like the blue whale (*Balaenoptera musculus*) and the fin whale (*Balaenoptera physalus*), were now at the mercy of whalers.

Whales are long-lived mammals with a very low reproductive rate. The great whales invariably give birth to one well-developed calf that has been carried by the mother for a year or more (see the section in Chapter 8, "Biology of Marine Mammals"). Females do not give birth to another calf until after another 1 or 2 years. As a result of this low reproductive potential, whale stocks could not stand the intense fishing pressure, and many of them collapsed. Almost all great whales are now classified as endangered (Table 17-1).

The first to be seriously depleted was the slower-swimming north Atlantic right whale (*Eubalaena glacialis*), "right" to be killed because it floated after being harpooned. By the early 1900's whaling had moved to the rich feeding grounds around Antarctica. This location proved to be a real bonanza. Whaling nations developed factory ships able to haul whole carcasses

Figure 17-17 Sperm whales being harpooned in the south Pacific by the crew of the *Acushnet,* an American whale ship from Fairhaven, Massachusetts. This water-color painting is part of the sea journal of a 1845 to 1847 voyage.

TABLE 17-1

Estimated numbers of great whales before exploitation and during the late 1980's

WHALE SPECIES	STATUS	ESTIMATED PRE-EXPLOITATION NUMBER	ESTIMATED NUMBER IN EARLY 1990's
Blue	Endangered	+200,000	5-10,000
Bowhead	Endangered	30,000	7,800
Bryde's	Protected	100,000	90,000
Fin	Endangered	+500,000	120,000
Gray	Endangered	+20,000	21,000
Humpback	Endangered	115,000	10,000
Minke	Hunted	140,000	+700,000
Northern right	Endangered	unknown	1,000
Southern right	Endangered	100,000	3,000
Sei	Endangered	+250,000	+50,000
Sperm	Endangered	+2,000,000	+1,000,000

Sources: IWC and National Geographic Society.

for processing. The Antarctic fishery reached its peak in the 1930's. The whales received a reprieve during World War II, but it was too late. It is estimated that more than a million whales were taken from Antarctica alone.

Blue whales, the largest of them all, were especially sought. A large specimen yielded more than 9,000 gallons of oil. It has been estimated that over 200,000 blue whales were taken worldwide between 1924 and 1971, close to 30,000 during the 1930-1931 whaling season alone. Soon catches pushed way above the optimal yield level. Catch per whaler's day's effort declined every year since 1936. As many as 80% of all blue whales caught by 1963 were sexually immature, which means that there were even less individuals in the ocean able to perpetuate the species.

Fin whales, the second largest of all whales, became the next major target as blue whales became more and more scarce. The 1950's and early 1960's saw annual catches of 20,000 to 32,000 fin whales per year, mostly from Antarctica. As their stocks dwindled, whalers shifted once again in the mid-1960's, this time to the smaller sei whale (*Balaenoptera borealis*). The sei whale averages a length of around 13 m (44 feet), whereas the fin whale averages around 20 m (65 feet).

Intense whaling has lead to the near-extermination of most species of great whales. These species are now endangered.

The abrupt disappearance of the more commercially valuable whales, one after the other, meant lower profits for the whaling industry. In 1946, twenty whaling nations set the **International Whaling Commission (IWC)** in an attempt to regulate whale hunting to stop overfishing. It collected data on the number of whales, though the numbers came mostly from the whalers themselves. It set annual quotas for the number of whales to be

Figure 17-18 Beginning in 1988 Japan opted to continue hunting whales in Antarctica, referring to it as "scientific whaling." This photo, taken in 1989, shows the carcass of a minke whale *(Balaenoptera acutorostrata)* being processed by the crew of a whaling ship. The Japanese were allowed to take up to 300 minkes during the 1988 to 1989 hunting season.

Figure 17-19 California gray whales *(Eschrichtius robustus)* are once again a common sight along their long migration routes from Alaska to Mexico (see the map in Figure 8-29).

killed each year, quotas that unfortunately were nonbinding and could not be enforced. Furthermore, some whaling nations did not belong to the IWC. Saving the whaling industry became more important than saving the whales. The blue whale was not completely protected by the IWC until the 1965-1966 season, long after its numbers had been drastically reduced; by then then they were so hard to find that it was no longer a profitable fishery. Even under the protection of the IWC, blue whales were hunted at least until 1971 by the fleets of countries that did not belong to the IWC.

Under mounting pressure from conservationists, the IWC gradually banned the hunting of other whales. Demand for whale products, mostly oil used in the manufacture of margarine and lubricants, was reduced since substitutes had been found for most of them. Whale meat, however, remained used as pet food and valued as human food, mostly in the *kujiraya*—or whale-meat bars—of Japan. The lower quotas of the IWC were unfortunately not always accepted by all nations. The United States Congress separately passed the **Marine Mammal Protection Act of 1972,** which bans the hunting of all marine mammals (except in the traditional fisheries of Alaskan natives) and the importation of their products. By 1974 the IWC had protected the blue, gray *(Eschrichtius robustus)*, humpback *(Megaptera novaeangliae)*, and right whales around the world, but only after their stocks were no longer economically viable. Sperm *(Physeter catadon)*, minke *(Balaenoptera acutorostrata)*, fin, and sei whales were still hunted in large numbers, but worldwide catches began to dwindle. Catches fell from 64,418 in 1965, to 38,892 in 1975, to 6,623 in 1985. A moratorium on all commercial whaling was finally declared by the IWC in 1985, a move long-sought by conservationist groups. The U.S.S.R. halted all whaling in 1987. Japan, Iceland, and Norway, however, opted in 1988 to continue hunting minke, fin, and sei whales, a fishery permitted by the IWC under the controversial guise of "scientific whaling" (Figure 17-18). Iceland abandoned it in 1990. The IWC moratorium was extended for 5 years in 1991. In the meantime, hunting of smaller cetaceans not protected by IWC, like the Dall porpoise *(Phocoenoides dalli)*, has increased, apparently as a substitute for whale meat.

Nobody knows when the great whales will again roam the oceans in numbers equal to those before the start of large-scale whaling. Some experts are afraid that a few critically endangered species will never recover completely. Small-scale whaling still remains part of the traditional fisheries of the native inhabitants of the Arctic region from Greenland to Siberia. One of the whales hunted there, the bowhead *(Balaena mysticetus)*, is endangered. Other smaller whales are hunted there: the killer whale *(Orcinus orca)*, narwhal *(Monodon monoceros)*, and beluga *(Delphinapterus leucas)*.

Slow recovery is under way in other species. The California gray whale, protected since 1947, has made a phenomenal comeback (Figure 17-19; also see Table 17-1). Even the blue whale, whose reproduction is severely limited by its restriction to small populations scattered around the world, is making a comeback of sorts. They have returned to the southern reaches of the Arctic Ocean north of Norway, a region where they flourished before their near-extermination by whalers. Their numbers in Antarctica, however, are even lower than initially predicted: around 500 animals, only 0.2% of those feeding there before whaling began!

Dolphins, not protected by the IWC, are also at great risk. They have replaced the larger whales as the most threatened of all cetaceans. As many as 28 species of small cetaceans are in immediate danger of extinction. Only 200 to 500 *vaquitas*, or "little cows" *(Phocoena sinus)*, are left. This shy, shovel-nosed porpoise, known only from the northern Gulf of California, remained unknown to science until 1958. Everywhere, fishermen are depleting stocks of fish and squid on which dolphins feed. Dolphins themselves are being hunted for human food. It is becoming popular in countries like Perú, where dolphin meat is much cheaper than beef or chicken. Tuna fishermen using giant purse-seine nets (see Figure 16-9, B) trap and drown the many dolphins that often swim above schools of tuna. Fishermen often find their catch after spotting dolphins, which is known as "setting on tuna." During the early 1970's an estimated 200,000 dolphins died annually, mostly in the hands of American fishing fleets. The slaughter induced such a public outrage that the United States, through the newly enacted Marine Mammal Protection Act of 1972, called for a drastic reduction in the accidental deaths of dolphins. It imposed a quota of 20,500 for the number of dolphins that could be killed by American fleets. The use of special nets was enforced, and observers were placed on board vessels to verify compliance with the ruling. By 1990 it was estimated that the number of dolphins killed by the United States tuna fleet, by then operating in "dolphin-safe" western Pacific waters, had reached zero. Environmentalists won a major victory when in 1990 the three biggest tuna packers in the United States pledged not to buy or sell fish that was caught using methods that injure or kill dolphins. Tuna cans began to display "dolphin-safe" labels.

Dolphins, however, continue to drown in fishing nets. By the late 1980's annual deaths had begun to rise to more than 100,000, mostly in the eastern tropical Pacific and at the hands of fishing fleets not regulated by American laws. The number of dolphins in the wild has noticeably decreased, particularly in the coastal spotted *(Stenella attenuata)* and eastern spinner *(Stenella longirostris;* see Figure 8-20, B) dolphins. Most of the tuna caught in the eastern Pacific, the yellowfin tuna *(Thunnus albacares)*, is taken by herding dolphins that swim over schools of tuna.

Dolphins also are tangled and killed by the thousands in huge **drift nets** (see Figure 16-9, E) that also threaten sea turtles, seals, seabirds, and other marine life (Figure 17-20). The nets, in some cases as large as 60 km (37 miles) long and 15 m (50 feet) deep, are used to catch fish, squid, and practically anything that tries to swim by. They are deployed in international waters at high sea and allowed to drift overnight. Not only do they deplete valuable commercial fisheries like albacore tuna and salmon, but they trap a large number of noncommercial species. These "walls of death" also are very wasteful since a large percent of the catch drops out during hauling. Their use in the north Pacific salmon fishery has been particularly deadly to the Dall porpoise. Hundreds of fishing boats outfitted for drift netting have been used to catch tuna in the south Pacific with potentially disastrous results. Since they operate in international waters, there is little that individual governments can do.

Figure 17-20 This Pacific white-sided dolphin *(Lagenorhynchus obliquidens)* drowned after getting caught in a drift net in the north Pacific.

Other Marine Species Risking Extermination

Whales are joined by other marine species whose future is at stake. Giant clams (*Tridacna*; see Figure 13-42) are taken for food and for their shells in such numbers that they have become exterminated in many parts of the tropical Pacific. Marine snails like cowries (*Cypraea*) and cones (*Conus*), with shells eagerly sought after by collectors (Figure 17-21), have similarly disappeared. The exploding shark-meat market, plus the pressure of recreational catches, threatens many species of sharks. Sharks, like whales, are slow-reproducing animals. It is feared that many species may be pushed to the brink of extinction in a decade or two. Sea snakes have been exterminated in some places because of hunting for their skins. All seven species of sea turtles are endangered. They have been shamelessly exploited for food, adults as well as eggs, and tortoiseshell (Figure 17-22; also see the section in Chapter 8, "Sea Turtles"). Their nesting sites have been overrun by development, and they drown in fishing nets. Nesting females may be confused by lights along the coast. Seabirds have not fared too well either. The brown pelican was almost exterminated by pesticides. The great auk (*Pinguinus impennis*), a penguinlike seabird that inhabited the north Atlantic, is gone forever as a result of our appetite for its eggs.

Many of the other marine organisms that face extermination are marine mammals with low reproductive rates. Some species of seals, sea lions, and walrus were brought to the brink of extinction for their skins, meat, blubber, or the precious ivory of their tusks. Monk seals (*Monachus*; see Figure 8-10, *B*) are endangered. One species, the Caribbean monk seal (*M. tropicalis*), is most probably extinct by now. The number of Steller sea lions (*Eumetopias jubatus*) has sharply declined in the north Pacific. The sea otter (*Enhydra lutris*; see Figure 8-12) has made a comeback. It is still threatened, however (see the section in Chapter 12, "Kelp Communities"). Manatees (*Trichechus*; see Figure 8-14) and the dugong (*Dugong*), the smaller cousins of the extinct Steller's sea cow, are all in danger of extinction.

Figure 17-21 Shells and reef-building corals from the tropical Pacific for sale in Venice, Italy.

Figure 17-22 One of the many types of tortoiseshell products that are regularly confiscated by the U.S. Fish and Wildlife Service. They are part of the multi-million dollar world trade in illegal wildlife.

Many marine species are at a risk of being exterminated as a consequence of unregulated exploitation and other direct human intervention. Species of whales, sea turtles, manatees, and other marine mammals are especially imperiled.

ALIEN SPECIES

The deliberate or accidental introduction of **alien,** or **exotic, species** into areas where they are naturally absent can cause unsuspected harm (Figure 17-23). Alien species can have devastating effects, including the elimination of **native**—or local—species. Estuaries are particularly vulnerable. Alien species can also bring with them parasites that may eventually infect native species. Seaweeds and invertebrates like barnacles, sea squirts, and shipworms that grow as fouling organisms on ships have become established around the world.

Many other species have been introduced in ballast water and with commercial fishery products. One good example is the Asian clam (*Potamocorbula amurensis*) that was accidentally introduced in San Francisco Bay in

 A

 B

Figure 17-23   *Penaeus japonicus* (A) and *Charybdis longicollis* (B) are two of over thirty alien species of crustaceans that have migrated from the Red Sea to the Mediterranean. The shrimp is now of commercial importance in the western Mediterranean. The crab is a pest: it has taken over sandy bottoms at depths of 35 to 50 m (115 to 165 feet). These alien species migrated by way of the Suez Canal, which opened in 1869. Most migrants have moved from the richer Red Sea into the Mediterranean, since high tide carries larvae farther up the canal from the Red Sea end. The migration of these alien species has been called Lessepsian migration after the builder of the canal, Ferdinand Lesseps. Migration between the Atlantic and Pacific oceans across the Panamá Canal has not taken place, because only freshwater flows through the canal, killing any marine life that is brought in by the tide (see Figure 8-4).

California, apparently as a direct result of the opening of trade with the People's Republic of China. These clams did not live in the bay in 1985, but by 1990 they literally covered whole sections of the bay's muddy bottom.

Introduction of alien species is another threat that results from human intervention.

The intentional introduction, or **transplantation,** of species for commercial purposes can also bring in alien species. The Japanese oyster (*Crassostrea gigas*) was introduced as spat, or young individuals, into the Pacific coast of North America. It was a successful transplantation in the sense that the oyster is now of significant commercial value. Unfortunately, many species living on the spat shells were also introduced. One nasty alien species is an oyster drill (*Ceratostoma inornatum*), a marine snail that preys on oysters and other native bivalves. Also introduced was *Sargassum muticum*, a Japanese brown seaweed that is now established along the coast from British Columbia to southern California (Figure 17-24). It grows fast and takes over space otherwise occupied by native seaweeds and, most importantly, eelgrass (*Zostera*), a source of food and shelter for many species (see the section in Chapter 12, "Seagrass Beds"). The same species of *Sargassum* has also been introduced into England, most probably with transplanted Japanese oysters.

A similar situation has taken place in the northeastern coast of the United States. A green seaweed (*Codium fragile tomentosoides*), apparently brought over on oysters transplanted from Europe or the west coast of North America, has become a pest by growing in masses on rocks and oysters.

Figure 17-24   *Sargassum muticum* is a brown seaweed that has been accidentally introduced into the western coast of North America and northeastern Europe, most probably from Japan.

CONSERVING AND ENHANCING THE ENVIRONMENT

The threats of pollution, destruction of habitats, and overexploitation that the marine environment has to defy does not suggest a very optimistic future. A growing human population guarantees further hazards. At stake is the survival of life, not only at sea but on the whole planet as well. Is it too late? Can something be done?

BOX 17-2

Sand on the Run, or What to do with Our Shrinking Beaches

Barrier island on the Cape Lookout National Seashore, North Carolina, in the eastern United States.

Sandy beaches constitute one of our most valuable coastal resources. Millions use them for recreation, and their importance as key tourist attractions is evident from Atlantic City to Waikiki. Sandy shores also happen to be among the most restless of all marine environments. Sand shifts, so disruptions like storms, hurricanes, wind, and currents periodically modify the shore. Beaches have been shrinking and disappearing everywhere, and overdevelopment is not the only culprit.

The Atlantic and Gulf coasts of the United States are protected by **barrier islands,** the long, low, sandy islands that run parallel to the coast. This long stretch of islands constitutes one of the world's most splendid sandy beaches. Barrier islands are characteristic of shores bordering wide continental shelves. They were formed as sea levels began to rise between 12,000 and 14,000 years ago (see the section in Chapter 2, "Climates and Changes in Sea Level"). Waves and wind began pushing bottom sediments and formed bars on the flat shelf. Sand bars eventually formed barrier islands, which migrated toward the shore as the sea level continued to rise.

The value of barrier islands goes beyond protecting coastline by absorbing the stress of storms and currents. Their **sand dunes** are inhabited by salt-resistant plants and land animals. Seabirds use them as nesting sites. Barrier islands may include salt marshes, seagrass meadows, freshwater sloughs, and even forests. Florida's barrier islands highlight mangrove forests. Untouched barrier islands, however, are becoming rare. Bridges and roads have made some more accessible, and as a result they have been overrun by residential and commercial development.

Development is not the only concern. Their origin may give us a hint that barrier islands were not planned to be with us for a long time. Their size and shape is always changing. Wind and **longshore currents** erode their seaward side by continually shifting sand from one tip to the other, north to south in the case of the eastern United States. Channels between islands may fill up, and islands may be sliced in half.

Not everybody realizes that sandy beaches and barrier islands will not tolerate permanent structures for too long. Seawalls, breakwaters, jetties, and groins may be built as a safeguard. Erosion may be controlled by encouraging the natural development of dunes or by planting vegetation that helps stabilize the sand. Replenishing sand by bringing it periodically from offshore is one expensive, but temporary, way to save a beach. This is what is being done in highly urbanized southern California, where rivers, which once brought in sand to the shore, now end behind dams or have been transformed into concrete-bottomed channels. Managing beaches and barrier islands, however, may simply be a waste of time and money. In many cases, efforts serve only to disrupt the natural coastal processes that sooner or later will triumph over our ingenuity.

Conservation

The preservation, or **conservation,** of marine life and its protection from the abuse of human intrusion is one solution. Conservation efforts include many local, national, and international projects dedicated to the protection of species and environments that face extermination (Figure 17-25). Some countries regulate the types and amounts of pollutants that can be dumped at sea. Oil drilling has been banned along some coasts. Commercial fisheries are being regulated by national governments, especially as a result of the establishment of exclusive economic zones, or EEZs, and by international commissions (see the section in Chapter 16, "Managing the Resources"). Governments also have established wildlife sanctuaries, marine parks, and reserves for the preservation of areas of ecological significance. One example is the National Estuarine Reserve Research System of the United States. It designated 17 reserves in 14 states and Puerto Rico protecting almost 1,200 km^2 (3,100 square miles) of estuaries, salt marshes, and mangrove forests. Groups and organizations like the Cousteau Society, Greenpeace, Sierra Club, the World Wildlife Fund, and the United Nations Environment Programme play a very active role in conservation through education, the sponsoring of projects, and even lobbying for legislation.

Protection of coastal regions against the advancing wave of development is of crucial importance. The conflict between economic development and the preservation of coastal resources, though especially intense in developing countries, is found around the world. Although laws have been passed by many governments, there is still a lot of work to be done. **Coastal management** aims at promoting the intelligent use of our coasts and at the same time helping to preserve them for future generations. The multiple use of coastal resources and the need to preserve them calls for judicious planning to accommodate the often-conflicting interests of developers, fishermen, surfers, power-plant builders, beach goers, and nesting seabirds. Coastal management deals with issues as varied as historical preservation, beach access, military uses, tourism, and water quality. In the long run, it is crucial to realize that some areas must be retained in their natural state and that conservation should be the goal behind the preservation of our coasts.

> Efforts to stop the deterioration of the marine environment include conservation and effective coastal management.

Antarctica is of special significance in the agenda of conservationists. Its coasts support unique forms of life found nowhere else, and its ice cap holds about 70% of the earth's fresh water supply. But nobody owns Antarctica. Potentially rich oil, coal, and mineral deposits may attract development, with disastrous consequences.

Restoration of Habitats

Another strategy for improving the quality of the environment is to help habitats recover from modifications caused by pollution and habitat destruction. **Habitat restoration** helps recovery from stress by transplanting, or re-

exclusive economic zone (EEZ) A zone 200 miles (370 km) wide along the coast, in which nations have exclusive rights to fishing and other resources *Chapter 16, p. 501*

Figure 17-25 Biologists tagging a hawksbill turtle (*Eretmochelys imbricata*) in Papua New Guinea in an effort to know more about the migrations and habits of the sea turtle. Papua New Guinea is the only place with breeding grounds for six species of sea turtles. This photograph was taken at Wuvulu Island, where sea turtles are common since its inhabitants cannot eat turtle meat and eggs because of religious restrictions.

Figure 17-26 Restoration of marine habitats has included the transplantation of young giant kelp (Macrocystis) plants into areas where kelp forests have diminished or vanished.

stocking, key species from healthier areas. The loss of priceless salt marshes and mangrove forests through landfilling or the building of boat marinas can be compensated for by creating or improving a similar habitat nearby. These efforts of course assume that the new location meets the physical requirements (such as tides, salinity, and the type of substrate) that are needed for the development of the community. Successful examples of habitat restoration include the transplant of cord grass (Spartina), one of the dominant plants of salt marshes, and mangrove seedlings. Tides slowly bring in the larvae of other components of the community. Young giant kelp (Macrocystis) plants, with their rootlike holdfasts tied to concrete blocks or to submerged nylon lines, have been used to help restore kelp forests in California (Figure 17-26).

Artificial Reefs

Fishing can be greatly enhanced by building **artificial reefs.** The irregular surface and the hiding places provided by the reefs attract fishes, lobsters, and other forms of life, as well as anglers and divers. Shellfish and seaweeds like kelp can thrive on their surface. Everything from concrete blocks, discarded tires, and toilets to scuttled ships and complex frames and structures have been used to build artificial reefs around the world (Figure 17-27). Hundreds have been built in Japan, where *tsukiiso*, or reef-construction, has become a successful way of increasing commercial yields of fish, abalone, sea urchins, and seaweeds.

> Habitat restoration by transplanting key species and the building of artificial reefs are techniques employed to enhance the marine environment.

Figure 17-27 Artificial reefs built of various types of prefabricted concrete blocks.

HUMANS AND THE SEA

BOX 17-3

Ten Simple Things We Can Do to Save the Oceans

Our impact on the health of the oceans is much more serious than we often assume. We are all guilty, no matter how far we live from the seashore. You can do a lot to help save the oceans. Here are ten simple things inspired by *50 Simple Things You Can Do To Save The Earth* by the Earthworks Group.

1. *Use less plastic!* Tons of plastic bags, styrofoam cups and packaging, plastic bottles, disposable diapers, and similar products are used each minute around the world. Practically all of it is nonbiodegradable. A large share of this garbage finds its way into the ocean, killing the marine life that eats it. Six-pack rings and plastic fishing line are particularly deadly. Don't forget to pick up any plastic you may leave around the beach. Cut open those six-pack rings anytime: you do not know where they may end up.

2. *Don't use pesticides!* Almost all pesticides are nonbiodegradable and sooner or later they may reach the ocean. They are toxic not only to pests but to other forms of life.

3. *Dispose of hazardous materials properly!* Pesticides aside, the list of toxic chemicals we use is endless. Paints, disposable batteries, permanent-ink markers, household cleaners, used crankcase oil, and many other products contain toxic chemicals, everything from heavy metals to PCBs. Dumping them down the drain simply adds them to sewage, which may ultimately drain right into the ocean.

4. *Don't buy products made from endangered or threatened species!* Do not buy tortoiseshell or ivory, which may come from walruses. Corals and shells should be left alone, and alive, in their natural home. Be sure to tell merchants you object to their shells, corals, sand dollars, and other marine life, which were most probably collected alive and killed for sale. Don't buy the yellowfin tuna that is caught in nets that trap dolphins.

5. *Save energy!* Saving gasoline, heating oil, coal, and electricity means saving the marine environment from more offshore oil drilling and oil spills. We would also need fewer nuclear power plants, which means less radioactive wastes and less thermal pollution. The saving of energy also keeps carbon dioxide, a greenhouse gas that promotes global warming, out of the atmosphere.

6. *Recycle plastics!* Reduce the risk of plastics being washed to sea by recycling them. Your local recycling center can tell you how.

7. *Recycle motor oil!* Oil dumped onto the ground or into the sewer or trash easily pollutes rivers, which ultimately empty into the ocean.

8. *Recycle other forms of trash!* Reduce the amount of trash that may wind up at sea—from bottles to aluminum cans to tires—by recycling. Rubber or metallic balloons may land in the water and cause harm to marine animals that swallow them.

9. *Don't use sprays that contain CFCs!* Chlorofluorocarbons (CFCs), which are still present in some sprays, help destroy the ozone layer that protects the earth from ultraviolet radiation harmful to life. It is also a greenhouse gas. CFCs can even be released from air conditioners, styrofoam, and fire extinguishers.

10. *Get involved! Keep informed!* Be aware of and keep up with environmental issues. Read, ask questions, listen carefully. Numerous organizations, societies, and agencies are committed to help save our planet. Sponsor them and hear what they have to say.

Do It Yourself

SUMMARY

1. Pollution may be defined as a decrease in _____ caused by the addition of substances or materials by humans.
2. Oil is a common marine pollutant with widespread effects. Of all marine communities, those found in _____ and _____ water are the most affected.
3. Pollution by sewage is a major problem along the coast. Raw sewage is particularly hazardous because it _____. Partial treatment usually results in large amounts of _____ that overcome bottom communities and overfertilize the water. Overfertilization of the water is known as _____.
4. _____ is the phenomenon shown by chlorinated hydrocarbons and other synthetic chemicals when they accumulate in the food chain. It is a consequence of the fact that these chemicals are _____; that is, they cannot be broken down by organisms.
5. Other marine pollutants prominent for their toxicity are PCBs and heavy metals like _____, _____, _____, and _____ .
6. The pumping of heated water from power plants and other operations into the ocean is known as _____.
7. Modification or destruction of habitats in rich but fragile communities like _____, _____, and _____ is a serious threat to the marine environment.
8. Species become _____ when they disappear forever. We say they become _____, however, when their disappearance is brought about by humans. Species like the blue whale that are in immediate danger of disappearing are known as _____.
9. _____ introduced on purpose or by accident may be very harmful since they can eliminate _____, that is, those species characteristic to a particular location.
10. Efforts toward the enhancement of the marine environment include _____ by transplanting individuals of depleted species and building artificial _____ to improve fishing.

THOUGHT QUESTIONS

1. Wastes from duck farms used to wash into two shallow-water bays on Long Island, New York. The wastes, rich in nutrients such as nitrate and phosphate, polluted the water. What do you suppose was the immediate effect of the pollutants? Can you speculate on the likely effects to the commercially valuable shellfish of the area?
2. It is found that a chemical present in effluents coming from a factory is being stored in the tissues of herring, a plankton-feeding fish. What type of observations and possible experiments would you suggest to find out if the chemical is biodegradable? What is the significance of finding out if the chemical is biodegradable or not?
3. Tourism and its effects (for example, pollution from hotels and the impact of boats and tourists on fragile habitats) often clash with conservation efforts. Sometimes, however, tourism may help. The economic impact of banning the hunting of harp-seals in eastern Canada has been compensated in part by the influx of tourists that now come to see the seals. Can you think of other examples? What recommendations can you make to minimize the impact of tourism on unspoiled marine environments?

FOR FURTHER READING

Brower, K.: "State of the Reef." *Audubon*, vol. 91, no. 2, March 1989, pp. 56-81. *A discussion of how the richest environment on earth is endangered by humankind, with an emphasis on the efforts of dedicated marine scientists to reverse the tide.*

Dolan, R., and H. Lins: "Beaches and Barrier Islands." *Scientific American*, vol. 257, no. 1, July 1987, pp. 68-77. *The current erosion problems along the Atlantic and Gulf coasts of the United States and the prospects for their future.*

Duedall, I.W. and M.A. Champ: "Artificial Reefs: Emerging Science and Technology." *Oceanus*, vol. 34, no. 1, Spring 1991, pp. 94-101. *The development of artificial reefs has emerged as a viable technique to enhance the marine environments.*

Hodgson, B.: "Alaska's Big Spill: Can the Wilderness Heal." *National Geographic Magazine*, vol. 177, no. 1, January 1990, pp. 4-43. *A superbly illustrated account of the Exxon Valdez oil spill, its effects on the environment, and the massive cleanup operation.*

Hodgson, B.: "Antarctica: A Land of Isolation No More." *National Geographic Magazine*, vol. 177, no. 4, April 1990, pp. 2-51. *The impact of tourism and pollution, including the potential consequences of the ozone hole, on the once pristine Antarctic environment.*

"Managing Planet Earth." *Scientific American*, vol. 261, no. 3, September 1989. *Eleven articles dealing with the dilemma of how to harmonize economic development with our planet's fragile natural environment, including articles on changes in the atmosphere and climate and on threats to the aquatic environment.*

Matthews, S.W.: "Under the Sun—Is Our World Warming?" *National Geographic Magazine*, vol. 178, no. 4, October 1990, pp. 66-99. *A comprehensive look at the greenhouse effect and global warming, with a preview of what may happen to our planet.*

Stewart, D.: "Nothing Goes to Waste in Arcata's Teeming Marshes." *Smithsonian*, vol. 21, no. 1, April 1990, pp. 174-180. *How a community recycles sewage into a lush wildlife sanctuary.*

"The Oceans and Global Warming." *Oceanus*, vol. 32, no. 2, Summer 1989. *Fourteen articles on global warming, including the role of the oceans and the possible consequences.*

Ward, F.: "Florida's Coral Reefs are Imperiled." *National Geographic Magazine*, vol. 178, no. 1, July 1990, pp. 114-132. *Pollution, boats, tourists, coral bleaching, and other types of stress threaten coral reefs in southern Florida and the Caribbean.*

Weisskopf, M.: "Plastic Reaps a Grim Harvest in the Oceans of the World." *Smithsonian*, vol. 18, no. 12, March 1988, pp. 58-67. *Plastic debris of every imaginable type pollutes beaches and harms marine life in the Gulf of Mexico.*

The Oceans and Human Affairs

18

Nowhere is the impact of the oceans on human experience more dramatic than in Venice, Italy (Figure 18-1). The city has been literally sculpted by the sea. It stretches, rather precariously at times, over a few islands that sit in a lagoon in the northern Adriatic Sea. Canals, gondolas, white and pink palaces situated just barely above high-tide level, and *seppie con polenta* (cuttlefish with corn mush) all have the flavor and fragrance of the sea. The city owes its initial survival and eventual prominence to the sea. It became a maritime republic that exerted economic and political dominance over the eastern Mediterranean for over 1,000 years. With its powerful navy and sea-based commerce, Venice was the vital link between the East and Europe. The sea even exerted profound influence on Venetian art. The celebrated colors of the city, bright but muted by the sea fog, are conspicuous in those wonderful paintings of Giorgione, Titian, and Tintoretto.

We would like to close the book by briefly outlining the many ways the ocean, as in Venice, has molded, affected, or influenced human cultures. We have already seen in Chapter 16 how we use the resources of the ocean, and in Chapter 17 how we are affecting the health of the ocean. Now we turn to the other side of the coin to see how the ocean has influenced us.

Figure 18-1 Isola di San Giorgio Maggiore seen from San Marco, Venice, Italy.

OCEANS AS BARRIERS AND AVENUES

Once upon a time, people believed that our planet was flat (Figure 18-2) and that sailing beyond the horizon meant falling into the gaping mouths of sea monsters. Sirens and evil spirits lured sailors to perdition. The peoples of the world were separated from each other by unknown continental masses or by the great expanses of uncharted waters. The oceans were nothing but barriers between cultures. Even the few kilometers of the channel that separates England from mainland Europe have had a lasting effect on what it means to be English. And, of course, the same can be said about Ireland, Madagascar, Japan, and other island nations.

Few people risked venturing beyond the horizon. There were, however, exceptions. For centuries, Arab sailors regularly sailed across the western Indian Ocean. The Vikings crossed the north Atlantic during the ninth and tenth centuries; so did Basque whalers not long after. Polynesian double canoes sailed the great distances of the Pacific, and it has been

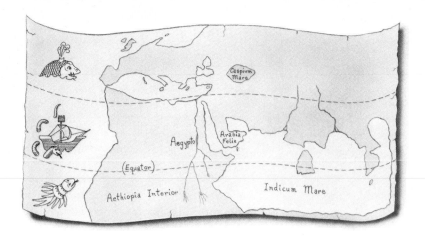

Figure 18-2 Fifteenth-century cartographers, greatly influenced by the geography of Ptolemaeus (second century AD), thought the earth was flat. Notice that, according to this map, circling Africa to reach India was impossible since the Indian Ocean was landlocked!

suggested that Chinese sailors may have reached the easternmost shores of the Pacific Ocean. It was not until the fifteenth century, however, that the European voyages of discovery began to change the ancient and medieval view that the world was flat.

The quest to discover new lands beyond the sea was pioneered by the Portuguese, who sailed around the southern tip of Africa, the Cape of Good Hope, and on to India by the late fifteenth century (see Box 3-1). Their intention was to take a share of the profitable spice trade. Also looking for a shorter way to the Orient was Christopher Columbus, who first crossed the Atlantic in 1492. Unlike the Vikings, who landed in America centuries before, Columbus' "discovery" was soon known by everyone. Many other seaborne explorers followed, not all looking for spices (see the section in Chapter 1, "The History of Marine Biology"). Between 1480 and 1780, these explorers had opened all oceans and—with the exception of most polar regions—few coastlines remained unexplored. The oceans then became very powerful avenues of culture and commerce, war and disease. Colonialism and imperialism sailed across the oceans, and so did immigrants, slaves, religions, languages, traders, scientific discoveries, products, and ideas.

The oceans, which before the age of discovery and exploration helped to effectively isolate the peoples of the earth, eventually became an avenue for change.

Today the oceans have become vital freeways that link world economies by transporting raw materials and manufactured goods to the four corners of the globe. Shipping remains the cheapest way of moving large quantities of goods over long distances, so most international trade moves by sea. The volume and type of goods moved by sea vary according to the geographic distribution of resources, location of industries and markets, population, and economic growth. Total seaborne trade, which began to grow dramatically after World War II, decreased by around 10% during the 1980's but only as a result of a drop in crude oil shipments. Crude oil, nevertheless, still accounts for the largest volume of seaborne trade (Figure 18-3). Those

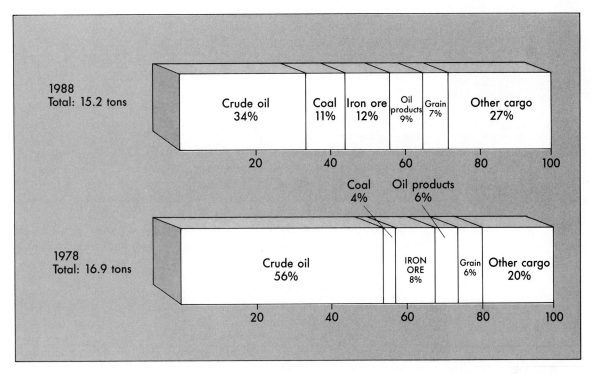

1988
Total: 15.2 tons

| Crude oil 34% | Coal 11% | Iron ore 12% | Oil products 9% | Grain 7% | Other cargo 27% |

20   40   60   80   100

Coal 4%   Oil products 6%

1978
Total: 16.9 tons

| Crude oil 56% | IRON ORE 8% | Grain 6% | Other cargo 20% |

20   40   60   80   100

Figure 18-3 World seaborne trade actually decreased between 1978 and 1988, a drop due to the fact that we have been using less oil. Crude oil, however, continues to be the number one commodity. Goods like cars, bananas, coffee, and tin compose the category of "other cargo." [Source: *The Economist*]

who deal in commodities such as iron ore, coal, and grain rely heavily on shipping. The enormous variety of goods being shipped has resulted in the development of vessels specialized in the transport of such exotic cargoes as liquified gas, livestock, wood-chip, and wine. Today a large percentage of maritime cargo moves by large steel containers that are carried to the docks by trucks and lifted by cranes into giant containerships (Figure 18-4). Cargo is then unloaded at the final destination without ever having been touched by human hands.

As far as carrying people is concerned, the oceans are no longer the crowded roads of yesterday. Except for short distances along the coast, jets have captured the business of transporting people between the continents. Large passenger ships, however, have found a new life carrying tourists to sunny islands.

OCEANS AND CULTURES

Our association with the ocean probably goes back to our early beginnings. It has even been suggested that some of the early stages of human evolution were spent at sea and that we were coastal inhabitants that frequently waded in the water searching for food. Evidence for this, so goes the argument, is our scanty hair, a relatively streamlined body, and the presence of a layer of fat for insulation; interestingly enough, these traits are also found in cetaceans! Most scientists, however, dismiss this hypothesis. There is

Figure 18-4 This giant containership, seemingly making its way across land, is actually crossing from the Pacific to the Atlantic Ocean by way of the Panamá Canal.

red tides Blooms of some dinoflagellates and blue-green algae (cyanobacteria) that discolor the water and produce toxins that, if accumulated in shellfish, cause *paralytic shellfish poisoning* in humans *Chapter 14, p. 405, Box 14-1; p. 406*

nevertheless no doubt of the strong influence of the ocean on the human cultures that began developing along coastal regions from prehistoric times.

Culture, which refers to the components of the environment created by humans, is reflected in a multitude of ways. It embodies objects like tools, ornaments, and dwellings, and immaterial things like customs, institutions, and beliefs. Intimate relationships with the ocean have molded many cultures around the world. Some anthropologists refer to these cultures as **maritime cultures.**

Fishhooks were one of the first tools made by humans. Excavations of prehistoric waste dumps give evidence of the importance of marine life as food and of the advantages of living near the seashore. Prehistoric shell middens, the accumulation of leftover shells, have been found in coastal areas around the world (see Figure 1-2). Remains of marine fishes, sea turtles, seabirds, sea urchins, and pinnipeds are often found in prehistoric deposits located near the coast. They have allowed archeologists to describe the food habits, diet changes, and sometimes the overexploitation of food resources by ancient cultures.

Humans then began to learn to fish using nets, traps, and other more sophisticated techniques. The extraction and trading of salt was of tremendous importance, influencing the development of cities and states. Boats were subsequently improved, allowing fishermen and salt traders to get farther from shore. The trade of food and manufactured goods was a small step away.

Fishing was a most important activity for many coastal American Indians. Nowhere is this more evident than among the Northwest Coastal Indians that inhabited the Pacific shores from southern Alaska to northern California. Their livelihood depended to a great extent on salmon, sea mammals, shellfish, and other marine life. These Indians knew the seasonal changes in abundance and the migration patterns of the species that fed them. Some tribes were even aware that red tides made mussels poisonous. Marine life also provided a basis for religious beliefs. Salmon, for instance, was for some tribes a supernatural being that took the shape of a fish and sacrificed itself every year to help humans. The fish spirits returned to their homes at sea to be transformed back into fish if their bones were thrown back into the water. Rituals and observances were carried out to prevent salmon from being offended and thus refusing to return and swim up the rivers to be caught for food. Other rituals welcomed salmon back before their migrations to spawn upstream. Marine life was also the subject of inspiration for superb wood carvings, such as totem poles, carved by some of the tribes (Figure 18-5). Shells from abalone and other marine snails were used in making masks and ornaments.

The Inuit and Inupiaq Eskimos and other peoples native to the Arctic carved a precarious existence from the icy waters. They fished and hunted whales and other marine mammals (Figure 18-6) from *kayaks*, one- or two-person boats made of seal or walrus skins, or *umiaks*, larger boats also made from skins. Their folklore is rich in stories of hunting incidents in which the hero is helped by spirits and other supernatural forces arising from the sea. Things have changed since then, but some continue to follow tradi-

Figure 18-5 Totem poles carved by the native inhabitants of the Pacific Northwest often include marine motifs such as halibut (shown here), killer whales, and sea otters.

Figure 18-6 A polar bear carved from a walrus tusk. The artist was a native to the Arctic region of eastern Siberia.

tional ways of fishing and hunting. The closely related Aleuts inhabited the Aleutian Islands farther south. They hunted seals, sea otters, whales, and other marine animals. The Aleuts are now greatly reduced in number, and their once splendid association with the ocean is just a memory.

The Seri Indians of northwestern Mexico regularly used the seeds of eelgrass (*Zostera*) from the Gulf of California as a traditional food source, the only known case of a grain being harvested right from the ocean! Eelgrass is now harvested only occasionally by the Seri. Its flour is cooked to make a gruel that is flavored with honey, cactus seeds, or sea turtle oil. In the old days, toasted grains were used to cure children's diarrhea. The dry eelgrass was used as roofing and to make dolls. A deer or bighorn sheep scrotum filled with dry eelgrass made a ball for children to play with.

The early Polynesians, Micronesians, and other peoples of Oceania had the whole Pacific Ocean as their backyard. As a result, they developed extraordinary skills as navigators (see Figure 1-4). Unlike Europeans, they were not afraid of the ocean that stretched beyond the horizon. They settled the farthest reaches of the Pacific, from New Zealand and the Mariana Islands to Hawaii and Easter Island. For this remarkable feat, they relied on double canoes carved from trees and secured with coconut fiber (Figure 18-7).

Figure 18-7 *Wa'a kaulua*, a Hawaiian double canoe.

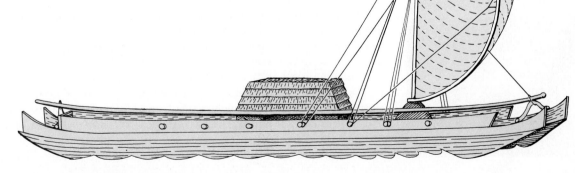

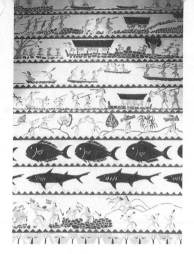

Figure 18-8 The Micronesian inhabitants of Palau, or Belau, have traditionally carved story boards on wood to illustrate their traditional legends, many of which tell about encounters with fantastic marine animals.

Figure 18-9 The Romans often used marine motifs in the mosaics used in the decoration of floors and walls.

The ocean and its creatures also provide images and symbols to the folklore and art of the peoples of Oceania (Figure 18-8). According to legend, New Zealand was discovered when Polynesian fishermen pursued an octopus that stole their bait. The ocean enjoyed a chief role in their creation myths and other aspects of their mythology. In some cultures a giant clam is said to have been used in the making of heaven and earth. Others have a great spotted octopus holding heaven and earth together. There are also countless stories involving marine life: tales of porpoise girls, pet whales, and shark gods falling in love with local girls.

The sea has helped mold many of the maritime cultures that have developed in coastal regions around the world.

Ancient sea-faring cultures also emerged in the Mediterranean and the Middle East. Egyptians, Phoenicians, Minoans, Greeks, Persians, Romans, Arabs, and other ancient mariners fished, traded, and made war with each other at sea. They were often inspired by the sea (Figure 18-9). Many of their shipwrecks have yet to be discovered (see Box 18-1).

Fishing and commerce remained the basis of coastal economies in Europe during the Middle Ages. The Baltic Sea herring fishery, for example, was the livelihood of the Hanseatic League, a thriving federation of Baltic and North Sea ports. When the fishery collapsed in the fifteenth century, so did the League.

England, the Netherlands, and Portugal owed much to seaborne trade and sea power for their emergence as leading nations during the late Middle Ages and early modern times. In many ways they simply followed the example of Venice and Genoa, the great rival seaports in the Mediterranean trade. Fishing, trade, and naval power in particular began to play other important roles. By stimulating shipbuilding, they strongly encouraged exploration, science, and technology (Figure 8-10).

"Those who rule the sea, rule the land" soon emerged as a powerful strategy in the emerging nation-states of the world. Echoes of Lepanto, the Invincible Armada, Trafalgar, Navarino, and other decisive naval actions still ring in our ears as heroic sagas or trivial defeats, depending on where our ancestors came from (Figure 18-11). *The Influence of Sea Power upon History*, a book written in 1890 by the American naval officer and historian Alfred T. Mahan, emphasized the importance of sea power. This book was extremely influential in persuading major powers to build modern fleets in the years before World War I. Mahan's predictions were quickly confirmed. In 1905 the Russian Empire was lethally crippled by the sinking of their Pacific fleet by the Japanese. Jutland, Midway, Coral Sea, and other World War I and II naval battles further proved the strategic importance of the oceans in modern warfare.

Since the birth of "gunboat diplomacy" nations have continued to employ sea power to exert their influence by defending their local or global interests. Missile-carrying nuclear submarines, long-range bomber planes, and nuclear missiles have replaced battleships as a first-strike force for the superpowers. Nevertheless, incidents and actual combat at sea over maritime boundaries and fishing rights have been making news headlines year

HUMANS AND THE SEA

BOX 18-1

Marine Archeology: A Science is Born

The discovery, salvage, and interpretation of that part of humankind's cultural heritage that remains undersea is the aim of a relatively new and exciting field, **marine archeology.** New tools, like scuba diving and remotely operated undersea vehicles, have allowed marine archeologists to reveal secrets that had remained concealed by sediment at depths beyond our reach.

Marine, or underwater, archeology as a modern science was born in 1960 when archeologists excavated a 3,000-year-old ship off the Mediterranean coast of Turkey. Before that time, the chance recovery of sunken artifacts was occasionally carried out by sponge fishermen, not by trained archeologist-divers. Underwater digs have since been investigated around the world. The oldest so far seems to be that of a Bronze Age ship loaded with fascinating artifacts that sank off the southwestern coast of Turkey 3,500 years ago.

These underwater digs have provided invaluable information about different aspects of ancient cultures: ship construction, life aboard ship, trade and discovery routes, and naval warfare. Tools, utensils, footwear, weapons, coins, and jewelry found on sunken ships have given us details of the products that were traded, food habits, and other details of everyday life.

Underwater sites may also yield art treasures, gold, and other valuables. Already a few priceless bronze Greek sculptures have been recovered from the Mediterranean. Wrecks of Spanish galleons, as well as of more modern ships, can hide a fortune in gold, coins, and jewelry. The plundering of wrecks for valuables is a major problem, since archeologically significant sites may be damaged or dispersed.

Shipwrecks may be located by analyzing old documents and charts. Shallow-water wrecks have been discovered by alert divers attracted by half-buried hull fragments, an anchor, or a pottery amphora that stored olive oil or wine in better days. Underwater surveying of wrecks involves dives, underwater photography and television, air photography, and charting the site using a grid system. More sophisticated techniques are also being used. Side-scan sonars use reflected sound waves to survey large sections of the ocean floor, whereas subbottom sonars can profile sites covered with sediment. A computer-controlled system, sonic high-accuracy ranging and positioning system, or SHARPS, provides archeologists with tridimensional maps of sites. A remotely controlled underwater vehicle sends live pictures to a surface vessel that beams them to a satellite in charge of relaying them to museums and institutions to be studied by experts; the image moves from the bottom to the surface to space to the surface thousands of kilometers away in a matter of seconds!

Marine archeologists are also involved in excavating towns and harbors that became submerged. They can provide an awesome variety of data on the economy and the use of technology of past cultures. Port Royal in Jamaica, which sank as a result of an earthquake in 1692, is an example of such a submerged dig. So is the now-submerged port of Cesarea Maritima, a Roman city on the coast of Palestine.

Sometimes sunken ships remain intact, as in the case of the *Vasa,* a Swedish warship that sank during its maiden voyage in 1628. It rested quietly in Stockholm harbor until 1961, when it was lifted whole from a depth of 33 m (108 feet)! The wreck had been preserved by cold water and thick mud. Luckily, wood-eating shipworms were absent because of the low salinities of the Baltic. Another case is that of the *Mary Rose,* an English warship that sank in 1536. Its starboard half, the only surviving part, was raised in 1982 after yielding valuable artifacts. The *Monitor,* a Civil War ironclad, lies 67 m (220 feet) below the surface off Cape Hatteras. Its turn may be next, or perhaps that of the *Titanic,* some 4,000 m (13,000 feet) under!

Site to be excavated

Photography

Mapping

Lift bag

Air lift

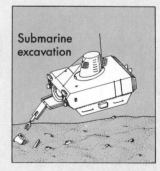

Submarine excavation

Examples of techniques used by marine archeologists.

Figure 18-10 Shanghai became China's leading port and the base for commercial penetration by the Western powers after the city was opened to unrestricted foreign trade in 1842. This 1860 drawing shows the great variety of ships and vessels used in trans-oceanic and local trade.

Figure 18-11 Sea power was a crucial ingredient in the building of European overseas empires that began in the sixteenth century.

after year. The control of sea lanes, critical to world economies, as well as to the defense of oil, fisheries, and other coastal resources, will ensure the continuous importance of naval power to all nations.

In the meantime, many of the maritime cultures that were able to survive until historic times were greatly modified and transformed. Only traces may still remain of times when life depended on the cycles of the ocean and on the food caught from it. Many of the maritime cultures were simply obliterated from the face of the earth.

Fortunately for us, some unique maritime cultures still survive. Fishing villages populated by maritime peoples who are still faithful to a unique lifestyle and who depend on the ocean for their survival have managed to survive in isolated spots such as the coast of Labrador, the Faeroe Islands, and the island of Tristan da Cunha. Some have managed to survive along coasts no longer classified as isolated (Figure 18-12). The Bajaus, or "sea gypsies," of the southern Philippines are boat dwellers who, like their ancestors, have followed a nomadic life at sea, fishing and diving for pearls. Korea's women divers, the *henyo,* continue a grueling livelihood of diving for sea urchins, octopus, and abalone. The Kuna Indians of the Caribbean coast of Panamá literally cling to several hundred tiny offshore islands, some built out of coral and filled in by the Kuna themselves (Figure 18-13). And then there are the Dutch, a maritime people by tradition who still manage to inhabit land they have reclaimed from the bottom of the North Sea by using dikes and canals. A large portion of their densely populated country actually lies below sea level! Many modern societies still include fishermen, fish and shellfish farmers, sailors, commercial divers, and other occupational groups that depend on the ocean (Figure 18-14). As such, they are regarded as maritime subcultures.

The imprint of the oceans on humankind fortunately still survives in our daily lives in things that transcend food, commerce, politics, and war. The

Figure 18-12 Fishermen and traders have used the traditional Arab dhows for centuries. These fishermen in Tanzania, East Africa, are using a small one-masted version. Dhows still sail between India and ports in the Arabian Peninsula, Red Sea, and East Africa in what is considered the world's oldest commercial sailing route.

Figure 18-13 A small and overcrowded village of Kuna Indians in one of the tiny islands of the San Blas group on the Caribbean coast of Panamá.

Figure 18-14 Sailors like this man in the Greek island of Crete in the eastern Mediterranean belong to a maritime subculture that remains tightly linked to the rhythms of the ocean.

Figure 18-15 *The Gulf Stream* (1899), an oil painting by Winslow Homer, an American painter much influenced by marine subjects.

sea and its lore endures from the work of writers, painters (Figure 18-15), and musicians who left their work for all of us to enjoy: Conrad, Melville, Hemingway, García Márquez, Turner, Monet, Debussy, Mendelssohn, and—of course—the Beatles.

OCEANS AND RECREATION

Another legacy of pure enjoyment from the ocean is that of **recreation.** Rising standards of living and increased leisure time, particularly in the developed countries, have opened up the opportunities people have for the recreational enjoyment of the ocean.

Practically anybody who can afford it, even those living far from the coast, can now fly to a faraway beach or take a cruise among once-remote paradise islands. It is an opportunity welcomed by the millions who save their pennies and fly to the beaches of the Caribbean, Florida, Tahiti, or faraway Bali, to name just a few vacation spots. Europeans flock to Palma, the islands of the Aegean, and Sardinia every summer. Russians prefer the Black Sea resorts; the Japanese, Hawaii and the Mariana Islands. There are even tours to spots like the Galápagos and Antarctica especially organized for divers and nature lovers. Catering to those seeking the sun and fun of the sea has developed into a giant tourist industry, one that keeps afloat many national and local economies. Seaside resorts sprang up along European coasts since Roman times, but by now they have been built practically anywhere there is sand, sun, and reasonably warm water. In many areas, however, tourism has displaced fishing and other traditional practices. Water pollution is often a legacy of those extra million gallons of sewage. The destruction of habitats resulting from the building of hotels and other facilities is another harmful side effect.

Recreation for many revolves around a multitude of water sports (Figure 18-16). The list is endless: snorkeling, scuba diving, sailing, windsurfing, water skiing, and surfing. Surfing alone is for many a way of life, a subculture of its own immortalized by Beach Boys' songs and echoes of "hanging ten" at Pipeline. The significance of these sports in the life of devotees is enormous. Together with marine sports fishing (see the section in Chapter 16, "Fisheries for Fun"), they provide relaxation and support a thriving recreation industry.

> Recreational opportunities offered by the ocean influence the way people spend their leisure time. It has also affected those who depend on the tourist industry for a living.

Figure 18-16 Sailing, once a principal means of transportation, is now a popular form of recreation around the world.

PROSPECTS FOR THE FUTURE

It is difficult, if not impossible, to predict accurately what lies in the future of our world ocean. It is not difficult, however, to reflect on what may lie ahead.

For some, the future of the ocean is bright indeed. Optimists have visions of cities built underwater or on floating islands, plenty of food provided by mariculture, unlimited energy from the water and seabed, and no pollution to speak of. Others are less confident. First of all, our impact on the marine environment is expected to further escalate as the world's population continues to expand. There is little doubt that our use of ocean resources will continue to increase. Today's high-energy technology will demand new means of exploitation, even for resources never imagined before. The prospects are therefore not very promising. Pollution, the destruction of habitats, and the disappearance of unique species will certainly escalate if no drastic measures are taken. Perhaps even more alarming is the possibility of a rise in sea levels as a result of global warming brought about by the greenhouse effect.

global warming Rise in the temperature of the earth as a result of an increase in the intensity of the greenhouse effect as the levels of carbon dioxide (CO_2) and other gases in the atmosphere increase *Chapter 17, p. 532, Box 17-1*

The oceans, a common heritage to all humankind, must be safeguarded by all to save them from overexploitation by a few and pollution by everyone. This has been one of the aims of the **Third United Nations Conference on the Law of the Sea (UNCLOS-3),** a 1982 treaty that culminated many years of complicated deliberations by the numerous nations that were represented. The treaty needs to be ratified by at least 60 of the signatory nations to take effect world wide. Only relatively few nations however, have already done so. Nevertheless, many nations have already implemented some of the provisions that resulted from the treaty.

The treaty, for example, lead to the establishment of the **exclusive economic zones (EEZs)** that extend a nation's economic interests 200 nautical miles (370 km) from the coast. This agreement has already had significant consequences in the present and future management of marine resources, particularly fisheries (see the section in Chapter 16, "Managing the Resources"). Nations owning tiny islands and reefs barely above high water can now claim large portions of the surrounding seabed, which may hold untold riches in fish, oil, or minerals! Such is the case of the Paracel and Spratly islands in the middle of the South China Sea, now claimed by several nations in the region. The treaty also allows free passage beyond a 12–nautical mile (22-km) **territorial sea** around a nation's shores. Unimpeded

Figure 18-17 The Bosporus, shown here from the European side at Istanbul, Turkey, is one of the most strategic straits in the world. It separates Europe from Asia and connects the Black Sea with the Aegean and Mediterranean seas. Its narrowest point is only 750 m (2,450 feet).

surface and underwater transit is guaranteed through straits that fall within the territorial sea of one or more nations. This includes the straits of Hormuz (in the Persian Gulf), Bering, Gibraltar, and the Bosporus (Figure 18-17), all of great strategic and economic importance.

The Third United Nations Conference on the Law of the Sea initiated a series of international agreements in an effort to define the use of the oceans by all nations. They included the establishment of exclusive economic zones.

More significantly, perhaps, the treaty calls for individual nations to enact laws and regulations to prevent and control pollution. We can say that it is on paper at least. It also guarantees the freedom to undertake scientific research in the high seas. Some important issues, however, still remain unresolved. There is, for instance, the case of Antarctica. It is encircled by highly productive waters crucial to the survival of many species. The continent itself, potentially rich in oil and mineral resources, is nevertheless claimed by many nations.

Other questions, disturbing but decisive, remain on the horizon. Will the world ocean be eventually overtaken by pollution, offshore oil platforms, and mere memories of whale songs and edible fish? Or will technology be ultimately employed to clean up the ocean and ensure that future generations may continue to enjoy its many wonders? One essential concern should always prevail: the ocean belongs to all of us and it is our responsibility to help preserve it.

Do It Yourself

SUMMARY

1. The oceans may be referred to as _____, because until the age of European discoveries they tended to keep many peoples isolated from each other.
2. We now call the oceans "avenues" because they help link the world mostly through _____.
3. The oceans have molded many cultures, particularly those which developed in coastal regions. They are known as _____ cultures.
4. Modern societies are directly influenced not only by the recreational opportunities the ocean offers millions of people, but by the significance of _____ as a major industry in both developing and developed countries.
5. The _____ has provided some important agreements related to the use of the oceans by all nations.

THOUGHT QUESTIONS

1. Most maritime cultures are either long gone or have been radically modified by others. Which elements of a rapidly changing maritime culture do you predict would be the first to disappear? Which would tend to linger unchanged the longest?
2. The Third United Nations Conference on the Law of the Sea made no provisions for Antarctica. Some of the land is probably rich in resources like oil, so the eventual exploitation of the land is inevitable. Several nations have already established claims, sometimes overlapping each other, to sections of the continent. How would you deal with these claims? Would you give first preference to nations, like Argentina or New Zealand, that claim geographic proximity to the continent, or to those, like Norway or Britain, that arrived to the claimed land first? How can resources be exploited if it is decided that the land does not belong to any particular nations?

FOR FURTHER READING

Bass, G.F.: "Oldest Shipwreck Reveals Splendors of the Bronze Age." *National Geographic Magazine*, vol. 172, no. 6, December 1987, pp 692-733. *A look at the salvage of a ship that sank off Turkey 3,400 years ago. The recovered artifacts provide fascinating evidence of seaborne trade in the ancient Mediterranean.*

Graham, F.: "Neptune's Bookshelf." *Audubon*, vol. 91, no. 2, March 1989, pp. 34-45. *A personal selection of the best books about the sea, including Conrad, Darwin, Heyerdahl, and a few pleasant surprises.*

Krakauer, J.: "Gidget has Grown Up, But Surfing is Still a 'Totally Happening' Sport." *Smithsonian*, vol. 20, no. 3, June 1989, pp. 106-119. *Surfing is more than an international sport; it is a way of life that has even influenced the language.*

Mathers, W.M.: *"Nuestra Señora de la Concepción."* National Geographic Magazine, vol. 178, no. 3, September 1990, pp. 38-53. *After more than 350 years underwater, the cargo of a trans-Pacific galleon carrying precious spices and jewels is recovered.*

Milovsky, A.: "Sea Hunters of Sireniki." *Natural History*, January 1991, pp. 30-37. *The hunting of whales, walruses, and seals from skin boats is still practiced by Inuits that live on the Bearing Sea coast of the Soviet Union.*

Weaver, M.A.: "Great Ships go to the Boneyard on a Lonely Beach in Pakistan." *Smithsonian*, vol. 21, no. 3, June 1990, pp. 30-41. *The sea provides an unusual livelihood to hundreds of Pakistanis, who break up old ships that are beached for scrap metal.*

Answers to the Do It Yourself Summaries

A

Chapter 1
1. observed
2. induction, deduction
3. hypothesis, testable, false
4. experiments
5. variables
6. theory

Chapter 2
1. Pacific, Atlantic, Indian, Arctic, Southern Ocean
2. density
3. core, mantle
4. mid-ocean ridge
5. mid-ocean ridges, trenches or subduction zones
6. subduction, mantle
7. Pangaea, 180 million
8. lithogenous, biogenous
9. continental shelf, continental rise
10. abyssal plain, 3,000, 5,000
11. hydrothermal vents, hot water

Chapter 3
1. hydrogen bonds
2. solid, liquid, gas or vapor, water
3. heat capacity
4. universal solvent, solutes
5. ions
6. relative composition of seawater
7. density
8. absorbed, blue
9. stratified
10. Coriolis effect
11. gyres
12. energy
13. moon, location, depth

Chapter 4
1. carbohydrates, proteins, lipids, nucleic acids
2. sunlight, simple sugars, carbon dioxide, oxygen
3. organic matter or glucose, respiration, ATP
4. nitrate
5. membrane, organelles
6. diffusion, osmosis, selectively permeable membrane
7. osmoregulation
8. ectotherms, endotherms
9. mitosis, daughter cells
10. gametes, meiosis
11. species
12. kingdoms, Monera, Protista, Fungi

Chapter 5
1. bacteria, blue-green algae or cyanobacteria
2. decomposers or decay bacteria
3. leaves, stems, roots
4. dinoflagellates, cold water or temperate and polar regions
5. multicellular
6. chlorophyll, brown, red
7. seagrasses, mangroves

Chapter 6
1. Protista
2. tissues, organs
3. radial, cnidarians
4. dorsal, anterior
5. flatworms, nematodes
6. coelom
7. shell, radula
8. jointed, Arthropoda, crustaceans
9. radial, water vascular system
10. gill or pharyngeal slits, notochord, dorsal

Chapter 7

1. Chordata
2. hagfishes or slime eels, lampreys
3. gill slits, ventral
4. suspension or filter
5. gills, swim bladder
6. body, fins, buoyancy
7. diffusion, opposite
8. to leave, urea, gills
9. lateral line
10. protection against predators, increased swimming efficiency
11. catadromous, anadromous
12. oviparous, ovoviviparous, viviparous

Chapter 8

1. marine turtles, sea snakes
2. maintain a more-or-less constant body temperature
3. viviparous, placenta
4. pinnipeds, seals, external ears and/or rear flippers that can move forward
5. blubber
6. manatees, posterior
7. baleen, krill
8. toothed
9. lungs, blood supply
10. echolocation, echo
11. sounds or vocalizations, visual, assistance
12. feeding, tropical or warm
13. delayed implantation

Chapter 9

1. ecology
2. resources, limiting resource
3. competition
4. competitive exclusion
5. resource partitioning
6. ecological niche
7. predation, animals, plants
8. symbiosis, commensalism, the symbiont harms the host, mutualism
9. trophic, primary producers, eat the food produced by primary producers, food chain or food web
10. 10%, organic matter or biomass
11. decomposers, detritus
12. nitrogen fixation
13. benthic, open water, plankton, against

Chapter 10

1. substrate
2. Cape Cod, west coast, Gulf coast

3. epifauna, infauna
4. emersion, physical factors, biological factors or competition and predation
5. refraction, headlands
6. byssal threads; limpets, chitons, etc.
7. space
8. lichens, blue-green algae, periwinkles
9. middle intertidal, emersion or drying out, snails or dog whelks
10. keystone predators
11. ecological succession, bacterial/algal film, barnacles, mussels or mussels/barnacles/gooseneck barnacles, climax community
12. seaweeds, light
13. sand, turbulent, clay, mud, calm
14. detritus, deposit

Chapter 11

1. partially enclosed, fresh water
2. sand bars or barrier islands, sinking
3. salinity, muddy or soft
4. oxygen, bacteria, circulation
5. osmoregulators, osmocomformers
6. deposit, detritus
7. immersion in seawater
8. salt marshes
9. primary producers or plants
10. detritus, plant-feeders or herbivores

Chapter 12

1. shelf break or edge of the continental shelf
2. stratification or accumulation in the bottom, larger
3. deposit feeders, detritus
4. soft or muddy or sandy, shallow, unprotected, detritus
5. hard or rocky, grazers or herbivores
6. brown, cold, nutrients
7. beds, forests, canopy
8. sea urchins, warm

Chapter 13

1. hermatypic or reef-building, calcium carbonate, polyps
2. zooxanthellae, nourishment, skeleton
3. coralline, calcium carbonate, waves, sediments
4. light, warm, tropics, sediment, pollution
5. fringing reefs, barrier reefs, atolls
6. volcanic, atoll
7. nutrients, primary, recycled, fixed
8. producers, nutrient
9. space, seaweeds, soft corals, sponges
10. crown-of-thorns, Pacific
11. fishes, sea urchins, seaweeds

Chapter 14
1. photic
2. phytoplankton, diatoms, dinoflagellates, nano-plankton
3. nanoplankton
4. copepods, krill
5. meroplankton
6. small, flat, projections or spines, buoyancy, lipids
7. vision, lateral
8. countershading, streamlined, muscles, tails
9. 200 m or 650 feet, day, night
10. dissolved organic matter (DOM), bacteria, DOM
11. low, high, photosynthesis or primary production, regenerated or released
12. nutrients, photic zone, rivers
13. light, overturn, bloom, nutrient
14. Ekman, upwelling, equatorial upwelling

Chapter 15
1. mesopelagic, bathypelagic, abyssopelagic, hadope-lagic or hadal pelagic, sunlight, pressure and/or depth
2. decreases
3. large, hinged extendible, general or broad
4. muscles, swim bladder
5. sensitive low-light vision, lateral line
6. deep scattering layer, lanternfishes, shrimps, krill
7. counterillumination, predators, attraction of prey, communication, courtship.
8. decomposition of organic matter
9. temperature, pressure
10. bristlemouths, deep-sea anglerfishes
11. small, large expandable, black, flabby, swim bladder
12. bioluminescence, chemicals, hermaphroditism, male parasitism

13. food, pressure
14. deposit
15. chemosynthetic, hydrogen sulfide

Chapter 16
1. upwelling, demersal, pelagic
2. industrial, fish meal
3. maximum sustainable, overfishing
4. exclusive economic zone (EEZ), 200
5. krill
6. mariculture, aquaculture
7. drugs
8. maganese nodules
9. desalination
10. ocean thermal energy conversion (OTEC)

Chapter 17
1. the quality of the environment
2. shallow, sheltered
3. spreads diseases, sludge, eutrophication
4. biological magnification, nonbiodegradable
5. mercury, lead, cadmium, copper
6. thermal pollution
7. estuaries, mangrove forests, coral reefs
8. extinct, exterminated, endangered
9. alien species, native species
10. habitat restoration, reefs

Chapter 18
1. cultural barriers
2. commerce and trade
3. maritime
4. tourism
5. Third United Nations Conference on the Law of the Sea (UNCLOS-3)

B   *Units of Measurement*

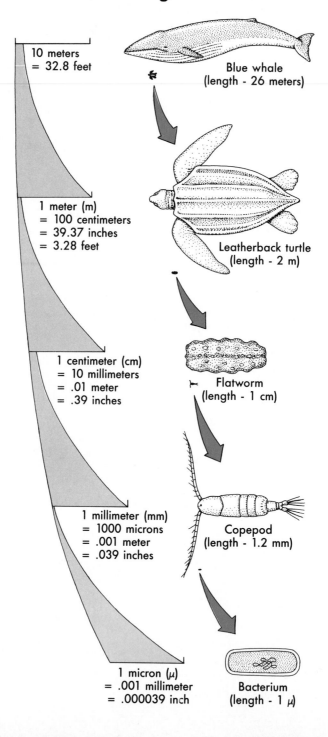

10 meters
= 32.8 feet

Blue whale
(length - 26 meters)

1 meter (m)
= 100 centimeters
= 39.37 inches
= 3.28 feet

Leatherback turtle
(length - 2 m)

1 centimeter (cm)
= 10 millimeters
= .01 meter
= .39 inches

Flatworm
(length - 1 cm)

1 millimeter (mm)
= 1000 microns
= .001 meter
= .039 inches

Copepod
(length - 1.2 mm)

1 micron (μ)
= .001 millimeter
= .000039 inch

Bacterium
(length - 1 μ)

Selected Maps

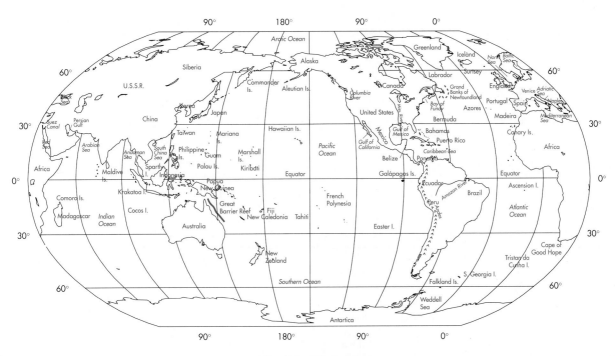

90°   180°   90°   0°

Arctic Ocean

Greenland   Iceland   North Sea   Baltic Sea

Siberia   Alaska

60°   U.S.S.R.   Commander Is.   Aleutian Is.   Columbia River   Canada   Labrador   Surtsey   England   Venice   Adriatic Sea   60°

China   Korea   Japan   Hawaiian Is.   United States   Grand Banks of Newfoundland   Bay of Fundy   Azores   Portugal   Spain   Naples

30°   Suez Canal   Persian Gulf   Taiwan   Mariana Is.   Gulf of Mexico   Bermuda   Mediterranean Sea   Madeira   30°

Red Sea   Arabian Sea   Andaman Sea   South China Sea   Philippine Is.   Guam   Marshall Is.   Pacific Ocean   Gulf of California   Bahamas   Puerto Rico   Caribbean Sea   Canary Is.   Africa

Africa   Maldive Is.   Spratly Is.   Palau Is.   Indonesia   Kiribati   Belize   Panama

0°   Equator   Galápagos Is.   Ecuador   Equator   0°

Comora Is.   Krakatoa I.   Papua New Guinea   French Polynesia   Peru   Amazon River   Brazil   Ascension I.

Madagascar   Indian Ocean   Cocos I.   Great Barrier Reef   Fiji   New Caledonia   Tahiti   Andes   Atlantic Ocean

30°   Australia   Easter I.   30°

New Zealand   Cape of Good Hope

Tristan da Cunha I.   S. Georgia I.

60°   Southern Ocean   Falkland Is.   60°

Weddell Sea

Antartica

90°   180°   90°   0°

The World Ocean

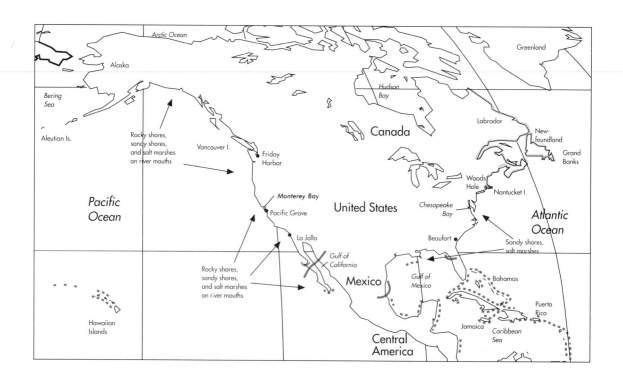

Rocky shores	————
Kelp forests approximate southern limits	————
Mangrove forests approximate northern limit	————
Coral reefs	··········

The major coastal communities of North America.

Glossary

Words in **boldface** are defined separately in the glossary. Other key terms are given in *italics*.

A

abiotic A nonliving (physical or chemical) component of the environment. *See* **biotic.**

aboral surface The surface opposite the mouth (or *oral surface*) in **echinoderms.**

abyssal plain The nearly flat region of the sea floor.

abyssal zone The bottom between approximately 4,000m (13,000 feet) to 6,000 m (13,100 to 20,000 feet) (Figure 9-22).

abyssopelagic zone The **pelagic** environment from a depth of 4,000 m (13,000 feet) to 6,000 m (20,000 feet) (Figure 9-22).

acorn worms (or **enteropneusts**) *See* **hemichordates.**

active continental margin A **continental margin** that is colliding with another plate and as a result is geologically active (Figure 2-24). *See* **passive continental margin.**

ADP (adenosine diphosphate) A two-phosphate molecule that converts itself into **ATP** whenever **energy** is released by a reaction (Figure 4-9).

agar A commercially-important **phycocolloid** extracted from **red algae.**

ahermatypic coral A coral that does not build reefs.

alga (plural **algae**) Any of several groups of **autotrophs** that lack the structural features (true leaves, roots, etc.) of the higher plants.

algal ridge A ridge of **coralline algae** that is found on the outer edge of some **coral reefs** (Figure 13-10).

algal turf (or **turf algae**) A dense growth of often filamentous **algae.**

algin A **phycocolloid** extracted from **brown algae** that is widely used in food processing.

alien species (exotic species) A species introduced by humans.

alternation of generations A reproductive cycle in which a **sexual** stage alternates with an **asexual** one, as in the case of a **gametophyte** alternating with a **sporophyte.**

ambergris The undigested material that accumulates in the intestine of the sperm whale.

ambulacral groove Each of the radiating channels of **echinoderms** through which **tube feet** protrude.

amino acid One of the 20 nitrogen-containing molecules that make up **proteins.**

amphibians (class *Amphibia*) The **vertebrates** that lay their eggs in fresh water: frogs, salamanders, and allies.

amphipods A group of small, laterally-compressed **crustaceans** that includes *beach hoppers* (Figure 6-37) and others.

ampulla (plural **ampullae**) Each of the muscular sacs that extend inside the body opposite the **tube feet** of **echinoderms.**

ampulla (plural **ampullae**) **of Lorenzini** One of several sensory structures in the head of sharks that detect weak electric fields.

anadromous The marine fishes that migrate to fresh water to breed. *See* **catadromous.**

anaerobic bacteria **Bacteria** that do not need oxygen.

anaerobic respiration The breaking down of organic matter in the absence of oxygen. *See* **respiration.**

anal fin Each of the last pair of **ventral** fins of fishes (Figure 7-9).

animals Members of the kingdom *Animalia*, which consist of **heterotrophic, eukaryotic,** and multicellular organisms.

angiosperms *See* **flowering plants.**

annelids *See* **segmented worms.**

anoxic Lacking oxygen.

antenna A sensory appendage in the head of **arthropods.**

anthozoans (class *Anthozoa*) The **cnidarians** that only include a complex **polyp** in their life cycle.

appendicularians *See* **larvaceans.**

aquaculture The farming of marine and freshwater organisms. *See* **mariculture.**

Aristotle's lantern The set of jaws and associated muscles used by **sea urchins** to bite off food.

arrow worms (or **chaetognaths;** phylum *Chaetognatha*) The planktonic **invertebrates** characterized by a streamlined and transparent body (Figure 6-24).

arthropods (phylum *Arthropoda*)   The **invertebrates** that have jointed appendages and a **chitinous,** segmented **exoskeleton.**

asexual (or **vegetative**) **reproduction**   The type of reproduction that takes place without the formation of **gametes.** *See* **sexual reproduction.**

atoll   A **coral reef** that develops as a ring around a central **lagoon** (Figure 13-28).

atom   The smallest unit into which an **element** can be divided and still retain its properties.

ATP (adenosine triphosphate)   A three-phosphate molecule that stores **energy** and releases it by converting itself into **ADP** to power chemical reactions in the organism (Figure 4-9).

autotroph   An organism that manufactures its own organic matter by using **energy** from the sun or from other sources. *See* **heterotroph.**

auxospore   The resistant stage of **diatoms** that restores the maximum size characteristic to the species.

B

back reef   The inner part of a **barrier reef** (Figure 13-21) or **atoll** (Figure 13-28).

bacteria (singular **bacterium**)   **Prokaryotic** and unicellular microorganisms included in the kingdom *Monera* (Figure 4-12).

baleen   The filtering plates that hang from the upper jaws of **baleen whales.**

baleen whales   The filter-feeding whales.

bar-built estuary   An **estuary** that is formed when a **barrier island** or *sand bar* separates a section of the coast where fresh water enters (Figure 11-3).

barnacles   A group of **crustaceans** that live attached to surfaces and are typically enclosed by heavy **calcareous** plates (Figure 6-36).

barrier island   A long and narrow island that is built by waves along the coast.

barrier reef   The type of **coral reef** that develops at some distance from the coast (Figure 13-21).

basalt   The dark-colored rock that forms the sea floor, or oceanic **crust.**

bathyal zone   The bottom between the **shelf break** and a depth of approximately 4,000 m (13,000 feet) (Figure 9-22).

bathypelagic zone   The **pelagic** environment from a depth of 1,000 m (3,000 feet) to 4,000 m (13,000 feet) (Figure 9-22).

beard worms (or **pogonophorans;** phylum *Pogonophora*)   The tube-dwelling **invertebrates** that lack a digestive system (Figure 6-23).

benthos   The organisms that live on the bottom (Figure 9-21).

big bang theory   A hypothetical cosmic explosion that resulted in clouds of dust and gas from which the earth and solar system are thought to have originated.

bilateral symmetry   The arrangement of body parts in such a way that there are only two identical halves, with different *anterior* and *posterior* ends and **dorsal** and **ventral** surfaces (Figure 6-16). *See* **radial symmetry.**

binomial nomenclature   A system of naming species using two names, the first of which refers to the **genus.**

biodegradable   A chemical that can be broken down by bacteria or other organisms.

biogenous sediment   The type of sediment that is made up of the skeletons and shells of marine organisms. *See* **calcareous** and **siliceous ooze.**

biological clock   A repeated rhythm that is synchronized with time.

biological magnification   The increased concentration of **nonbiodegradable** chemicals in the higher levels of the **food chain.**

bioluminescence   The production of light by living organisms.

biomass   The total weight of living organisms.

biotic   A living component of the environment. *See* **abiotic.**

birds (class *Aves*)   The **vertebrates** with feathers and eggs with calcified shells that are laid on land.

bivalves (class *Bivalvia*)   The clams, mussels, and other **molluscs** that possess a two-valved shell, filtering gills, and a shovel-like **foot.**

black corals (order *Antipatharia*)   The colonial **anthozoans** that secrete a black protein skeleton (Figure 6-14).

black smoker   A chimneylike accumulation of mineral deposits that is found at **hydrothermal vents** (Figure 2-29).

blade   The leaflike portion of the **thallus** of a seaweed.

bleaching   The expulsion of **zooxanthellae** by reef corals in response to stress.

bloom   The sudden increase in the numbers of an **alga** or **phytoplankton.**

blowhole   The nostrils, or nasal openings, of cetaceans.

blubber   A thick layer of fat under the skin of many marine **mammals.**

568

blue-green algae (or **cyanobacteria**) The **prokaryotic** and generally **photosynthetic** organisms included in the kingdom *Monera.*

bony fishes (class *Osteichthyes*) The fishes with a skeleton made mostly of bone and the presence of **gill covers.**

boring sponges The **sponges** that bore through **calcareous** skeletons and shells.

bottom layer *See* **deep layer.**

brachiopods *See* **lamp shells.**

breaching A behavior of whales that consists of leaping up in the air (Figure 8-26).

brittle stars (class *Ophiuroidea*) The **echinoderms** with five flexible arms that radiate from a conspicuous central disk and **tube feet** that are used in feeding.

brown algae (division *Phaeophyta*) The **seaweeds** with a predominance of yellow and brown pigments.

bryozoans (phylum *Bryozoa* or *Ectoprocta*) The colonial **invertebrates** where the anus is located outside the edge of a **lophophore** (Figure 12-18, *E*).

budding A type of **asexual reproduction** by which a separate new individual is produced from a small outgrowth, or *bud.*

by-catch The less valuable catch that is captured while fishing for a more valuable catch.

byssal threads Strong fibers secreted by mussels for attachment.

C

calcareous Made of **calcium carbonate.**

calcareous ooze A type of **biogenous sediment** that is made of the **calcium carbonate** shell and skeleton of marine organisms.

calcium carbonate (CaCO₃) A mineral that is the major component of the shell, skeleton, and other parts of many organisms.

carapace 1) The shieldlike structure that covers the anterior portion of some **crustaceans.** 2) The shell of **sea turtles.**

carbohydrate An **organic compound** that consists of chains or rings of **carbon** with **hydrogen** and **oxygen** attached to them.

carbon (C) An **element** that is an essential constituent of all **organic compounds.**

carbon cycle The cyclic reconversion of **carbon dioxide** into **organic carbon** (Figure 9-19).

carbon dating A procedure used to determine the age of fossils.

carbon dioxide (CO₂) A colorless gas that is required in the process of **photosynthesis.**

carbon fixation The conversion of inorganic **carbon** into energy-rich **organic carbon** in the process of **photosynthesis.**

carnivore An animal that eats other animals. A *top carnivore* is one that feeds at the top of the **food chain.** *See* **predation.**

carnivores (order *Carnivora*) The **mammals** with teeth that are adapted to eat other animals. The marine members of this group are the *sea otter* and the *polar bear.*

carotenoid One of a group of yellow, orange, and red plant pigments.

carposporophyte A **diploid** generation found in the **red algae.** It produces non-motile *carpospores.*

carrageenan A **phycocolloid** extracted from **red algae** that is widely used in food processing.

cartilaginous fishes (class *Chondrichthyes*) The fishes with a skeleton made of **cartilage:** the *sharks, rays, skates,* and *ratfishes* (or *chimaeras*).

catadromous The freshwater fishes that migrate to sea to breed. *See* **anadromous.**

caudal fin The posterior or tail fin of fishes.

cellulose A complex **carbohydrate** that is the main component of fibers and other structures of support in plants.

central nervous system The *brain* (or a similar aggregation of **nerve cells**) and one or more **nerve cords.**

central rift valley A depression in the **mid-ocean ridge** (Figure 2-27).

centrifugal force The force that tends to push a body away from the center.

cephalopods (class *Cephalopoda*) The *octopuses, squids,* and other **molluscs** that possess a **foot** modified into arms that surround the head.

cephalothorax The anterior portion of the body of many **arthropods,** which consists of the head fused with other body segments.

cetaceans (order *Cetacea*) The marine **mammals** with anterior flippers, no posterior limbs, and a **dorsal** fin: the *whales, dolphins,* and *porpoises.*

CFCs *See* **chlorofluorocarbons.**

chaetognaths *See* **arrow worms.**

chemosynthetic (or **chemoautotrophic**) **bacteria** **Autotrophic** bacteria (such as the *sulphur bacteria*) that use **energy** by releasing it from particular chemical compounds.

chitin A complex derivate of **carbohydrates** that is the main component of the skeleton of many animals.

chitons (class *Polyplacophora*) The **molluscs** that have a shell that is divided into eight overlapping plates (Figure 6-33).

chloride cells Cells in the **gills** of fishes that are involved in the excretion of excess salts.

chlorinated hydrocarbons (or **organochlorides**) **Nonbiodegradable** and synthetic chemical pollutants used as *pesticides.*

chlorofluorocarbons (CFCs) Chemicals used in sprays and other products that affect the **ozone layer.**

chlorophyll A green **photosynthetic pigment.**

chloroplast The **organelle** where **photosynthesis** takes place (Figure 4-13, B).

choanocyte *See* **collar cell.**

chordates (phylum *Chordata*) The animals that display a hollow, **dorsal nerve cord; gill slits;** and a **notochord.** Includes the **protochordates** and the **vertebrates.**

chromatophore A skin cell that contains pigment.

chromosome The cell structure where **DNA** is located.

ciguatera A type of poisoning found in tropical fishes. It may result from a toxin produced by a **dinoflagellate.**

ciliary comb One of eight bands of **cilia** fused at the base that is found in **comb jellies.**

ciliates A group of **protozoans** that have **cilia.**

cilium (plural **cilia**) A short, hairlike **flagellum** that is found in large numbers and used in movement, for pushing food particles, and in other functions.

clasper A copulatory organ along the inner edge of each **pelvic fin** in male sharks and other **cartilaginous fishes** (Figure 7-28).

cleaning symbiosis The type of **symbiosis** in which the smaller partner regularly removes parasites from fishes.

climax community The final stage in a **ecological succession.**

cloaca The common opening for the intestine and the excretory and reproductive systems of **cartilaginous fishes** and other animals.

clone A series of identical cells or individuals that have developed from a single cell or individual.

cnidarians (phylum *Cnidaria* or *Coelenterata*) **Invertebrates** with **nematocysts** and **radial symmetry.**

coastal management The use of coastal resources with the intention of preserving them.

coastal plain estuary *See* **drowned river valley estuary.**

coastal zone *See* **neritic zone.**

coccolithophorids Unicellular, **eukaryotic** members of the **phytoplankton** that have calcareous, buttonlike structures, or *coccoliths* (Figure 14-4).

coelacanths A group of fossil lobed-fin fishes. *Latimeria* was discovered alive in 1952 (figure in Box 7-2).

coelom The body cavity found in the more complex animals.

coelomic fluid The fluid that fills the **coelom** of **echinoderms** and other **invertebrates.**

coevolution When a **species** evolves in response to another.

collar cell (or **choanocyte**) The flagellated, food-trapping cells of **sponges.**

colloblast Sticky cell used for the capture of small prey in **comb jellies.**

comb jellies (phylum *Ctenophora*) The **invertebrates** with a gelatinous body, **radial symmetry,** and eight rows of **ciliary combs** (Figure 6-15).

commensalism The type of **symbiosis** in which one species obtains shelter, food, or other benefits without affecting the other, or *host.*

community The **populations** that live and interact in the same area.

competition The interaction that results when a resource is in short supply and one organism uses the resource at the expense of another.

competitive exclusion The elimination of one species by another as a result of **competition.**

compound eye An eye typical of **arthropods** that consists of numerous light-sensitive units.

conditioning A form of learning in which a behavior is associated with a reward such as food.

constant proportions, rule of A principle that states that the relative amounts of ions in seawater are always the same.

consumer A **heterotroph.** A *primary consumer* feeds directly on a **primary producer,** whereas a *secondary consumer* feeds on a primary consumer.

continental drift The movement of continental masses on the surface of the earth.

continental margin The edge of a continent; the zone between a continent and the deep sea floor (Figure 2-22). *See* **active** and **passive continental margins.**

continental rise The gently-sloping area at the base of the **continental slope** (Figure 2-22).

continental shelf The shallow, gently-sloping section of the **continental margin** that extends from the shore to the point where the slope gets steeper (Figure 2-22).

continental slope The steeper, seaward section of the **continental margin** (Figure 2-22).

convection The movement that results when heat is transferred in a fluid, such as the heat-driven motion observed in the earth's mantle.

convergent evolution When two different species evolve similar structures because of similar lifestyles.

copepods Small, mostly planktonic **crustaceans** (Figure 6-35).

copulation The sexual union undertaken to transfer **gametes.**

coral knoll (or **pinnacle**) A column of coral within the **lagoon** of an **atoll** (Figure 13-28).

coral reef The massive deposition of **calcium carbonate** by the action of colonial **stony corals** and other organisms.

coral rubble Coral fragments.

coralline algae The **green** and **red algae** that deposit **calcium carbonate** in their **thallus.**

coralline sponges *See* **sclerosponges.**

cord grasses Salt-tolerant grasses, species of *Zostera*, that inhabit **salt marshes** (Figure 11-9).

core The innermost layer of the earth (Figure 2-4).

Coriolis effect The tendency of particles moving large distances on the earth's surface to bend to the right in the Northern Hemisphere and to the left in the Southern Hemisphere (Figure 3-19).

counterillumination When the emission of light by **midwater** animals is controlled to match the background light (Figure 15-21).

countershading A color pattern that results in a dark back and a light belly (Figure 7-9).

courtship behavior Behavioral patterns involved in the attraction of the opposite sex.

crinoids (class *Crinoidea*) The **echinoderms** with a small, cup-shaped body and feathery arms: the *sea lilies* and *feather stars* (Figure 6-53).

crown-of-thorns sea star (*Acanthaster planci*) A predator of reef corals (Figure 13-37).

crust The outermost layer of the earth (Figure 2-4).

crustaceans (subphylum *Crustacea*) The **arthropods** that have two pairs of **antennae** and an **exoskeleton** hardened by **calcium carbonate.**

cryptic coloration A color pattern that allows an organism to blend with the surroundings.

cryptomonads (division *Cryptophyta*) Unicellular, **eukaryotic** members of the **phytoplankton** that have two flagella and no skeleton.

crystal A solid that consists of a regular pattern of molecules.

crystalline style An **enzyme**-releasing rod in the stomach of **bivalves** (Figure 6-29, *B*).

ctenophores *See* **comb jellies.**

current A horizontal movements of water.

cyanobacteria *See* **blue-green algae.**

D

decapods A group of **crustaceans** with five pairs of walking legs and a well-developed **carapace.** It includes the *shrimps, lobsters, hermit crabs,* and *crabs.*

decomposer An organism, such as the *decay bacteria*, that breaks down dead organic matter into smaller molecules.

deduction Reasoning from general principles to specific conclusions.

deep (or **bottom**) **layer** The deepest and coldest of the three layers of the ocean (Figure 3-37).

deep scattering layer (DSL) A sound-reflecting layer of many types of organisms that migrates daily from the **mesopelagic zone** to the **epipelagic zone** (Figure 15-14).

deep sea The dark waters below the **mesopelagic zone** (Figure 9-22).

deep-sea fan A fanlike accumulations of sediment at the base of a **submarine canyon.**

deep-sea hot spring *See* **hydrothermal vent.**

delayed implantation The delay in the attachment of the early **embryo** to the womb in **pinnipeds** and some other **mammals** in order to time birth with favorable conditions.

demersal fish A fish that lives on the bottom.

density The weight (or more correctly the *mass*) of a given volume of a substance.

deoxyribonucleic acid (DNA) The **nucleic acid** that contains the inherited code that specifies how each organism is made up and how it functions.

deposit feeder An animal that feeds on organic matter that settles on the bottom (Figure 6-20). *See* **suspension feeder.**

desalination The conversion of seawater into fresh water.

detritus The dead organic matter and the **decomposers** that live on it.

diatomaceous ooze A **biogenous sediment** that consists mostly of the **siliceous frustules** of **diatoms.** It is known as *diatomaceous earth* when found inland.

diatoms (division *Chrysophyta*) Unicellular and **eukaryotic autotrophs** with a **siliceous frustule.** Mostly **planktonic** (Figure 14-3).

diffusion   The movement of molecules from high to low concentration.

digestive gland   An **enzyme**-producing gland in several groups of **invertebrates** where digestion and absorption takes place.

dinoflagellates (division *Pyrrhophyta*)   Unicellular, **eukaryotic,** mostly **autotrophic** organisms with two unequal **flagella** (Figure 5-7).

diploid (or **2n**) **cell**   A cell, such as a body cell, that contains two similar sets of **chromosomes,** one from each parent. *See* **haploid.**

dispersal   The manner in which organisms get from place to place.

disruptive coloration   A color pattern that helps break the outline of an organism.

dissociation   The breaking up of a molecule into **ions** when placed in water or other solvents (Figure 3-6).

dissolved organic matter (DOM)   Organic matter that is dissolved in water and not in particles.

diurnal tide   A tidal pattern with a high and a low **tide** each day (Figure 3-33, C).

diversity   The total number of species inhabiting a particular environment.

DNA   *See* **deoxyribonucleic acid.**

domestic sewage   Wastewater from homes and non-industrial buildings. *See* **industrial sewage.**

dorsal   The upper or back surface of an animal with **bilateral symmetry** (Figure 6-16).

drag   Resistance to movement through water or any other medium.

drift net   A very long fishing net that is allowed to drift for a long time before it is pulled on board (Figure 16-9).

drowned river valley (or **coastal plain**) **estuary**   An **estuary** that is formed when sea levels rose at the end of the last glacial age (Figure 11-2).

E

ear stone (or **otolith**)   A calcareous body in the **inner ear** of fishes and other **vertebrates** that is used in the maintenance of equilibrium. *See* **statocyst.**

echinoderms (phylum *Echinodermata*)   The **invertebrates** with **radial symmetry** and a **water vascular system.**

echiurans (phylum *Echiura*)   The burrowing **invertebrates** that have an unsegmented body and a non-retractable **proboscis** (Figure 6-22, B).

echolocation   The ability of some animals to sense their surroundings by analyzing the reflection of sound waves, or *clicks,* they emit.

ecological niche   The role of a species in a **community,** that is, feeding habits, **habitat,** and other requirements.

ecological succession   The regular replacement of **populations** by others in a given area.

ecology   The study of the interactions among organisms and their environment.

ecosystem   A **community** or communities plus all the **abiotic,** or nonliving, components of the environment.

ectoprocts   *See* **bryozoans.**

ectotherm (or **poikilotherm**)   An organism whose internal temperature varies with that of the environment. *See* **endotherm.**

Ekman spiral   The spiral change in the movement of water in the **water column** when the water is pushed by wind (Figure 14-36). *Ekman layer* is that part of the water column affected by wind; *Ekman transport* is the net water movement 90 degrees from the wind direction.

element   A substance consisting of **atoms** of the same kind and one which cannot be decomposed by ordinary chemical means.

El Niño–Southern Oscillation (ENSO)   Large-scale changes in the normal weather patterns of the Pacific basin and adjacent regions. *El Niño* is a warming of the surface currents in the eastern Pacific, only one of the many consequences of ENSO.

embryo   The early developmental stage that, through *embryological development,* ultimately becomes an adult individual.

encrusting   An organism that grows as a thin or thick crust over rocks and other hard surfaces.

endangered species   A species that is in immediate danger of **extermination.**

endolithic alga   An alga that burrows into **calcareous** rocks or corals.

endophyte   A plant that lives within the tissues of another plant.

endoplasmic reticulum   The extensive system of folded membranes present in most **eukaryotic cells** (Figure 4-13).

endoskeleton   A skeleton under the external surface of an animal. *See* **exoskeleton.**

endotherm (or **homeotherm**)   An organism that can regulate its internal temperature. *See* **ectotherm.**

energy   The ability to do work.

entoprocts (phylum *Entoprocta*)   The colonial **invertebrates** in which the anus lies within a crown of tentacles and the **coelom** is absent.

enzyme   A **protein** that speeds up a specific chemical reaction.

epifauna   The animals that live on the surface of the **substrate**. *See* **infauna.**

epipelagic zone   The **pelagic** environment from the surface to a depth of 100 to 200 m (350 to 650 feet) (Figure 9-22).

epiphyte   A plant that lives on another plant.

equatorial current   A current that moves parallel to the Equator (Figure 3-23).

estuary   Enclosed areas where fresh and sea water meet and mix.

eukaryote   An organism that consists of one or more **eukaryotic cells.**

eukaryotic cell   A cell that contains **organelles** (Figure 4-13). *See* **prokaryotic cell.**

euryhaline   An organism that can tolerate a wide range of salinities.

eutrophication   Pollution by excessive nutrient enrichment.

evaporation   The escape of molecules from the liquid phase into the gaseous, phase, or vapor.

evaporative cooling   The lower speed and hence the lower temperature of molecules remaining in the liquid phase after **evaporation** of the fastest molecules.

evisceration   The expulsion of internal organs in **sea cucumbers.**

evolution   The gradual change in the genetic makeup of a species as a result of **natural selection.**

exclusive economic zone (EEZ)   A zone 200 nautical miles (370 km) wide along the coast where nations have exclusive rights to any resource. It was initiated by the *United Nations Convention on the Law of the Sea.*

exoskeleton   A skeleton that forms the external surface of an animal, as in **arthropods.** *See* **endoskeleton.**

experiment   An artificially-created situation that is used to test a **hypothesis.** In a *controlled experiment* a variable that may affect the hypothesis is prevented from affecting the experiment.

extermination   The **extinction** of a species that is caused by humans.

extinction   The disappearance of a species.

extracellular digestion   The digestion that takes place outside cells, usually in a gut, or digestive cavity. *See* **intracellular digestion.**

F

fault   The friction zone along the **shear boundary** between two plates (Figures 2-19 and 2-20).

feather stars   *See* **crinoids.**

fertilization   The union of **gametes.** It can be *external* and take place in the water, or *internal* and take place within the body.

fetch   The span of the sea surface over which the wind blows to form wind-driven **waves** (Figure 3-27).

filter feeder   A **suspension feeder** that actively filters food particles (Figure 6-20).

fin ray   Each of the bony spines in the fins of **bony fishes.**

fish meal   A fish-protein supplement used in animal feeds.

fish protein concentrate (FPC or fish flour)   A fish-protein supplement for human consumption.

fission   A type of **asexual reproduction** that results when one individual splits into two individuals.

fjord   An **estuary** that is formed in a deep valley created by a retreating glacier (Figure 11-4).

flagellum   A long, whiplike **organelle** that is usually involved in locomotion.

flatworms (phylum *Platyhelminthes*)   The **invertebrates** that are dorsoventrally flattened and have an incomplete digestive tract, true **organs,** and **organ systems.**

flowering plants (or *angiosperms;* division *Anthophyta*)   The plants that have flowers, seeds, and true leaves, stems, and roots.

fluke   The finlike tail of **cetaceans.**

flukes   A group of parasitic **flatworms.**

food chain   A series of organisms along different **trophic levels** that feed on one another.

food web   All of the interconnecting feeding relationships in a **community.**

foot   The muscular locomotory structure of **molluscs.**

foraminiferan ooze   A **biogenous sediment** that consists mostly of the **calcareous** shells of **foraminiferans.**

foraminiferans   A group of **protozoans** with a **calcareous** shell, or *test,* and **pseudopodia** (Figure 6-2).

fore reef   The outer part of a **barrier reef** (Figure 13-21) or **atoll** (Figure 13-28).

fouling organisms   The organisms that live attached to submerged surfaces such as boats and pilings.

fringing reef   A **coral reef** that develops as a narrow band close to a shore (Figure 13-18).

frustule   The **siliceous,** boxlike cell wall of **diatoms.**

fucoxanthin   The yellow to golden brown **photosynthetic pigment** of **brown algae.**

fungi (singular **fungus**)   The members of the kingdom *Fungi,* which consists of plantlike but non-photosynthetic organisms.

funnel   The **siphon** of **cephalopods.**

G

gamete   A **haploid** reproductive cell that develops into a new individual after its union with another gamete.

gametophyte   The **haploid, gamete**-producing generation in many **seaweeds.** *See* **sporophyte.**

gas (or **respiratory**) **exchange**   The movement of **oxygen** and other gases between the atmosphere and the ocean or between the water or atmosphere and living organisms.

gastropods (class *Gastropoda*)   The snails and other **molluscs** that typically possess a coiled **dorsal** shell and a **ventral** creeping **foot.**

genetic engineering   The artificial alteration of the genetic information of a species.

genital slit   The genital opening of **cetaceans.**

genus   A group of similar **species.**

gestation   The length of time between **fertilization** and birth in **mammals.**

gill   Thin-walled extensions of the body that are used in **gas exchange.**

gill arch   A supporting structure of fish **gills.**

gill cover (or **operculum**)   The flap of bony plates that covers the **gills** of **bony fishes.**

gill filament   The thin projection of a fish **gill** where **gas exchange** takes place (Figure 7-18, B).

gill raker   Each of the projections along the inner surface of fish **gills** (Figure 7-18, B).

gill slit (or **pharyngeal slit**)   One of several pairs of openings along the **pharynx** in **chordates** (Figure 6-56).

glass sponges   The deep-water **sponges** with a skeleton of fused **silica spicules.**

global warming   An increase in the **greenhouse effect** brought about by the increase of **carbon dioxide** in the atmosphere.

glucose   A **simple sugar** that plays an important role in the energy-releasing reactions of most organisms.

Golgi complex   The sacs and membranes in many **eukaryotic cells** that is involved in collecting and transporting molecules (Figure 4-13).

gonad   An organ that contains the *germ tissue* that produces the **gametes:** *ovary* and *testis.*

Gondwana   One of two large continents, the southern one, that formed when the "supercontinent" **Pangaea** broke up about 180 million years ago (Figure 2-21, B). *See* **Laurasia.**

gorgonians (order *Gorgonacea*)   The colonial **anthozoans** that secrete a skeleton made of protein (Figure 6-13).

granite   The light-colored rock that forms most of the continental **crust.**

grazer   An organism that feeds primarily on plants.

great whales   The large whales: the sperm and the **baleen whales.**

green algae (division *Chlorophyta*)   The **seaweeds** in which **chlorophyll** is not masked by other pigments.

greenhouse effect   The increase in the earth's temperatures that results from the presence of **carbon dioxide** in the atmosphere.

guano   The accumulation of excrements of seabirds.

guyot   A flat-topped **seamount.**

gyre   A large, nearly circular system of wind-driven surface currents that center around latitude 30 degrees in both hemispheres (Figure 3-23).

H

habitat   The place where an organism lives.

habitat restoration   The recovery of stressed or destroyed habitats by **transplantation.**

hadal zone   The bottom below 6,000 m (20,000 feet) (Figure 9-22).

hadopelagic (or **hadal pelagic**) **zone**   The **pelagic** environment below 6,000 m (20,000 feet) (Figure 9-22).

halophyte   A salt-tolerant terrestrial plant.

haploid (or **n,** or **1n**)   A cell, such as a **gamete,** that contains only half the normal number of chromosomes that is found in **diploid,** or body, **cells.**

harem   The **reproductive strategy,** often observed in **pinnipeds,** that involves a male herding a large group of females.

heart urchins   A group of burrowing **sea urchins** with a flattened test and short spines (Figure 12-8).

heat capacity   The ability of a substance to hold heat **energy.**

heavy metals   A group of toxic metals: *mercury, lead,* and others.

hemichordates (phylum *Hemichordata*)   The **invertebrates** with a **dorsal nerve cord** and **gill slits.** Includes the *acorn worms,* or *enteropneusts.*

hemoglobin   A blood **protein** that transports **oxygen** in many animals.

herbivory   When an animal, or *herbivore,* eats a plant.

heredity The transmission of characteristics from one generation to the next.

hermaphrodite An organism that has both male and female **gonads.**

hermatypic coral A reef-building coral.

heterotroph An organism that obtains **energy** from organic matter taken from another organism. *See* **autotroph.**

holdfast The rootlike portion of the **thallus** of a seaweed.

holoplankton The **plankton** that spends their entire life in the plankton. *See* **meroplankton.**

homeotherm *See* **endotherm.**

homing The ability of an animal to find its home area.

hormone A substance that acts as a chemical messenger in an organism.

horseshoe crabs (class *Merostomata*) A group of **arthropods** with a large horseshoe-shaped **carapace** (Figure 6-46).

hydrogen (H_2) A colorless and odorless gas that consists of two atoms of the **element** hydrogen, a constituent of **organic compounds.**

hydrogen bond A weak bond between the **hydrogen** atoms of adjacent **molecules,** as in the case of water (Figure 3-2).

hydrogen sulfide (H_2S) The gas that is produced in **anoxic** sediments.

hydrostatic skeleton A system that uses water pressure against the body wall to maintain body shape and aid in locomotion.

hydrothermal vent (or **deep-sea hot spring**) A spot in the **mid-ocean ridge** where heated water forces its way up through the **crust** (Figure 2-29).

hydrozoans (class *Hydrozoa*) The **cnidarians** that typically include a **polyp,** which is often colonial, and a **medusa** stage in their life cycle.

hypothesis A statement that might be true.

I

ice age A period of time when significant amounts of ice formed on the continents, inducing a fall in the sea level.

Indo-west Pacific region The tropical Indian and west and central Pacific oceans.

induction Reasoning from specific observations to a general conclusion.

industrial fishery A fishery in which the catch is used for purposes other than direct human consumption.

industrial sewage Wastewater from industries. *See* **domestic sewage.**

infauna The animals that burrow in the **substrate.** *See* **epifauna.**

ink sac A gland found in some **cephalopods** that secretes a dark fluid that is used to discourage predators.

inner ear A paired, sound-sensitive organ in **vertebrates.**

insects (class *Insecta*) **Arthropods** with three pairs of legs and one pair of **antennae.** Few are marine, an exception being the *water strider* (Figure 14-21).

intermediate layer Of the three main layers of the ocean, the one below the **surface,** or **mixed,** layer. It includes the main **thermocline** (Figure 3-37).

International Whaling Commission (IWC) An agency that regulates *whaling* around the world.

interstitial fauna Animals living between sediment particles. *See* **meiofauna.**

interstitial water The water contained between sediment particles.

intertidal (or **littoral**) **zone** The area between the highest and lowest tide (Figure 9-22).

intracellular digestion Digestion that takes place within cells, usually those lining the gut or digestive tract. *See* **extracellular digestion.**

invertebrates The animals that lack a backbone.

ion An atom or group of atoms that is electrically charged.

iridophore A **chromatophore** with light-reflecting **crystals.**

island arc The curved chain of volcanic islands that form along **trenches.**

isopods A group of small, dorsoventrally flattened **crustaceans** such as the *sea louse* (Figure 10-24).

J

jawless fishes (class *Agnatha*) The fishes that lack jaws and paired fins: the *hagfishes* (or *slime eels*) and *lampreys* (Figure 7-3).

K

kelp A group of **brown algae** characterized by their large size and complexity. Some, like the giant kelp, form dense *kelp beds* or *kelp forests.*

keystone predator A predator that is very important in the maintenance of the structure of a community.

krill A group of planktonic **crustaceans** that are an important food of whales and other animals (Figure 6-39).

L

lagoon   The body of water that separates a **barrier reef** from land or one that is located in the center of an **atoll** (Figures 13-21 and 13-28).

lamella (plural **lamellae**)   Each of the many thin plates that make up the **gill filaments** of fish **gills** (Figure 7-18, C).

lamp shells (or **brachiopods;** phylum *Brachiopoda*)   The **invertebrates** that have a **lophophore** and a shell that consists of two **valves.**

lancelets (subphylum *Cephalochordata*)   The **chordates** with the three basic chordate characteristics but that lack a backbone (Figure 6-56).

larva   Immature stage of an animal that looks different from the adult.

larvaceans (or **appendicularians;** class *Larvacea*)   The **tunicates** that retain the body of a **tadpole larva** throughout life (Figure 14-8).

latent heat of evaporation   The amount of heat **energy** that is needed to evaporate a substance, that is, to change it from liquid to gas phase.

latent heat of melting   The amount of heat **energy** needed to melt a substance, that is, to change it from solid to liquid phase.

lateral line   A system of canals and sensory cells on the sides of fishes that helps them detect vibrations in the water (Figure 7-20).

Laurasia   One of the two large continents, the northern one, that formed when the "supercontinent" **Pangaea** broke up about 180 million years ago (Figure 2-21, B). *See* **Gondwana.**

leeches (class *Hirudinea*)   **Segmented worms** that are specialized predators and parasites.

leptocephalous larva   The leaf-shaped larva of freshwater eels and other fishes (Figure 14-12).

lichen   The organism that results from the **symbiosis** of a **fungus** and an **autotroph** such as a **green alga.**

limiting resource   An essential factor whose short supply prevents the growth of a **population.**

lipid   A group of **organic molecules** that are often used by organisms in the long-term storage of **energy** and in water-proofing.

lithogenous sediment   A marine sediment that is derived from the breakdown, or **weathering,** of rocks. *See* **red clay.**

lithosphere   The **crust** and the top part of the mantle that covers the earths's surface. It is broken into separate **lithospheric plates** (Figure 2-13).

littoral zone   *See* **intertidal zone.**

lophophore   A feeding structure that consists of ciliated tentacles.

loriciferans (phylum *Loricifera*)   The tiny **invertebrates** that live among sediment particles and have a body enclosed by six plates (Figure in Box 6-2).

lottery hypothesis   The assumption that unpredictability plays a role in the development of **communities.**

M

macrophytes   *See* **seaweeds.**

madreporite   A porous plate that connects the **water vascular system** of **echinoderms** to the exterior.

magnetic anomalies   The magnetic bands in the sea floor that run parallel to the **mid-ocean ridge** (Figure 2-11).

magnetite   A magnetic, iron-containing material.

main thermocline   *See* **thermocline.**

male parasitism   The permanent attachment of a male to a female in some deep-sea fishes (Figure 15-27, B).

mammals (class *Mammalia*)   The **vertebrates** that have hair and **mammary glands.**

mammary glands   The milk-secreting glands of **mammals.**

manganese nodules   Lumps of minerals (including manganese and other valuable minerals) that are found on the sea floor beyond the **continental shelf** (Figure 16-26).

mangroves   Shrubs and trees that live along the seashore in tropical and subtropical regions (Figure 5-27).

mantle   1) The semi-liquid region between the **crust** and **core** of the earth (Figure 2-4). 2) The outer layer of tissue that secretes the shell of **molluscs.**

mantle cavity   The space lined by the **mantle** of **molluscs.**

mariculture   The culture of marine organisms. In *open mariculture* (or *semi-culture*), organisms are cultured in natural environments; in *close mariculture* (or *intensive mariculture*), organisms are cultured in a controlled environment.

marine archeology   The discovery, salvage, and interpretation of those material remains of humankind's past that have been preserved at sea.

marine natural products   Chemical compounds that are obtained from marine organisms.

marine snow   **Detritus** and other particulate organic matter that is found in the **water column.**

maritime culture   A culture with a very close relationship with the sea.

maxillipeds   The food-sorting appendages of some **crustaceans.**

maximum sustainable yield   The maximum catch of a **stock** that can be harvested year after year without diminishing the stock.

medusa   The bell-shaped, free-swimming stage of **cnidarians.**

meiofauna   Microscopic animals that live on the bottom, often used as a synonym of **interstitial fauna.**

meiosis   The cell division that results in the formation of **gametes.**

melon   A fatty structure on the forehead of some **cetaceans** that is used to direct sound waves emitted during **echolocation** (Figure 8-23).

meroplankton   The planktonic organisms that spend only part of their life in the **plankton.** *See* **holoplankton.**

mesenterial filament   Any of the long, thin tubes attached to the gut of corals and other **cnidarians** that are involved in digestion and absorption.

mesopelagic zone   The **pelagic** environment from a depth of approximately 100 to 200 m (350 to 650 feet) to 1,000 m (3,000 feet) (Figure 9-22).

metabolism   All the chemical reactions that take place in an organism.

metamorphosis   A marked change in form during embryological development.

microfossils   The microscopic shells and other remains of marine organisms that make up **biogenous sediments.**

mid-ocean ridge   The continuous chain of volcanic submarine mountains that extends around the earth. It includes the *Mid-Atlantic* and *Mid-Pacific ridges* (Figures 2-6 and 2-7).

midwater   Pertaining to the **mesopelagic zone.**

migration   The regular movement of one organism from one place to another.

mitochondrion   The **organelle** in which **respiration** takes place (Figure 4-13).

mitosis   The cell division in which a cell divides into two daughter cells that are identical to the original cell.

mixed semidiurnal tide   A tidal pattern with two successive high **tides** of different heights each day (Figure 3-33, B).

molecule   A combination of two or more **atoms.**

molluscs (phylum *Mollusca*)   The **invertebrates** with a soft, unsegmented body; a muscular **foot;** and, with some exceptions, a **calcareous** shell.

molt   The **exoskeleton** that is shedded during the *molting* process.

monerans   The members of the kingdom *Monera*, which consists of **prokaryotic** organisms.

monoplacophorans (class *Monoplacophora*)   A small group of **molluscs** that are thought by some to represent a link with **invertebrates** that show **segmentation.**

monsoon   Winds in the northern Indian Ocean that blow from the southwest in summer and from the northeast in winter.

mud flat   A muddy bottom that is exposed at low tide.

mutualism   The type of **symbiosis** in which both partners benefit from the relationship.

myoglobin   A type of muscle **protein** that stores **oxygen** in **vertebrates.**

myomere   Each of the bands of muscles along the sides of fishes.

N

nanoplankton   The **plankton** that is too small to catch in a standard plankton net. *See* **net plankton.**

native species   A local species that has not been introduced.

natural selection   A mechanism of evolutionary change that results when those individuals who are better *adapted* than others in meeting the challenges of the environment survive longer and produce more offspring.

nauplius   The planktonic **larva** of many **crustaceans** (Figure 6-45).

neap tides   The **tides** with a small **tidal range.** They occur around the times when the moon is in quarter. *See* **spring tides.**

nekton   The organisms that swim (Figure 9-21).

nematocyst   The stinging structure of **cnidarians.**

nematodes (or **roundworms;** phylum *Nematoda*)   The **invertebrates** with a cylindric body, a conspicuous body cavity, and a complete digestive tract.

nemerteans   *See* **ribbon worms.**

neritic (or **coastal**) **zone**   The **pelagic** environment above the **continental shelf** (Figure 9-22).

nerve cell   A cell specialized to originate or transmit nerve impulses.

nerve cord   A long, compact bundle of **nerve cells** that is part of the **central nervous system.**

nerve net   The network of interconnecting **nerve cells** in **cnidarians** and other **invertebrates.**

net plankton   The **plankton** that is caught in a standard plankton net.

neuston The organisms that live on the surface of the sea.

niche *See* **ecological niche.**

nictitating membrane A thin layer of tissue that can be drawn across the eye in sharks and some other **vertebrates.**

nitrate (NO_3^{-2}) An important **nutrient** in the ocean.

nitrogen (N_2) A colorless and tasteless gas that is an essential constituent of **proteins.**

nitrogen cycle The cyclic reconversion of atmospheric **nitrogen** into **organic** nitrogen compounds (Figure 9-20).

nitrogen fixation The conversion of gaseous **nitrogen** (N_2) into nitrogen compounds that can be utilized by plants and other **autotrophs.** It is performed by *nitrogen fixers.*

nonbiodegradable A chemical that cannot be broken down by bacteria or other organisms. They are said to be *persistent.*

nonrenewable resource A resource that is not naturally replaced.

notochord A flexible rod that lies below the **nerve cord** of **chordates.**

nucleic acid Organic molecules that store and transmit genetic information. *See* **deoxyribonucleic acid.**

nucleus The **organelle** of **eukaryotic cells** that contains the **chromosomes** (Figure 4-13).

nudibranchs (or sea slugs) **Gastropods** that lack a shell and have exposed **gills.**

nutrient A raw material other than **carbon dioxide** and water that is needed by an **autotroph** to produce **organic matter.** Examples are **nitrate** and **phosphate.**

nutrient regeneration The release of nutrients from organic matter by **decomposers.**

O

ocean thermal energy conversion (OTEC) A process that envisions obtaining **energy** by exploiting depth differences in temperature (Figure 16-28).

oceanic zone The **pelagic** environment beyond the **shelf break** (Figure 9-22).

olfactory sacs Structures on both sides of the head of fishes that are sensitive to chemical stimuli.

omnivore An animal that feeds on a variety of foods, typically from different **trophic levels.**

operculum 1) The tough lid that closes the shell opening of many **gastropods** when the body is withdrawn. 2) The **gill cover** of **bony fishes.**

organ A group of **tissues** specialized in one function.

organelle A membrane-bound, specialized structure within a cell (Figure 4-13).

organic compound A **molecule** that contains **carbon, hydrogen,** and often **oxygen.**

organization, level of The extent of specialization and organization among the cells of an organism: *cellular, tissue,* or *organ* levels.

osculum A large opening in many **sponges.**

osmoconformer An organism that allows its internal salt concentration to change with the **salinity** of the surrounding water.

osmoregulator An organism that is able to control its internal salt concentration.

osmosis The movement of water across a *selectively permeable* membrane such as the cell membrane, which allows only certain molecules to pass through.

outwelling The export of **detritus** and other organic matter from estuaries to other ecosystems.

overturn The sinking of surface water that has temporarily become more dense than the water below (Figure 3-36).

oviparous An animal that releases eggs.

ovoviviparous An animal that produces eggs that hatch inside the female immediately before birth.

ovulation The release of an egg from the ovary.

oxygen (O_2) A colorless and tasteless gas that is one of the essential **elements** to life. It is used in **respiration.**

oxygen minimum layer A layer at approximately 500 m (1,600 feet) where **oxygen** becomes depleted (Figure 15-24).

oyster reef A dense oyster bed present in some **estuaries.**

ozone layer The ozone (O_3) in the atmosphere that deflects ultraviolet radiation, which is harmful to life.

P

Pangaea The "supercontinent" from which the other continents have drifted (Figure 2-21, A).

Panthalassa The large ocean that surrounded the "supercontinent" **Pangaea** and which was the ancestor of the modern Pacific Ocean (Figure 2-21, A).

paralytic shellfish poisoning A condition caused when humans eat shellfish that have become contaminated with the toxin present in the **dinoflagellates** that cause **red tides.**

parapodium A usually flat lateral extension present in each of the body segments of **polychaetes** (Figure 6-19).

parasitism The type of **symbiosis** in which one partner, the *parasite*, derives benefit from the other, the *host*.

passive continental margin A **continental margin** that is located at the "trailing edge" of a continent and as a result shows little geologic activity (Figure 2-24). *See* **active continental margin.**

patchiness The grouping of organisms in clumps or patches.

PCBs *See* **polychlorinated biphenyls.**

peanut worms (or **sipunculans;** *phylum Sipunculida*) The burrowing **invertebrates** with an unsegmented body and an anterior end that can be pulled into the body (Figure 6-22, A).

pectoral fin Each of the pair of fins just behind the head of fishes (Figure 7-9).

pedicellaria One of the minute pincerlike organs of some **echinoderms** that help keep the surface clean.

pelagic Those organisms that live in the **water column** away from the bottom (Figure 9-21).

pelvic fin Each of the second pair of **ventral** fins of fishes (Figure 7-9).

pen The reduced, thin shell of squids.

perennial A plant that lives more than 2 successive years.

pharynx The anterior portion of the digestive tract of many animals. It is located directly behind the mouth cavity.

pheromone Chemicals that organisms use to communicate with other members of their species.

phoronids (phylum *Phoronida*) The tube-dwelling, unsegmented **invertebrates** that possess a horseshoe-shaped or circular **lophophore.**

phosphate $[(PO_4{}^{-2}) - (PO{}^{-2}{}_4)]$ An important **nutrient** in the ocean. An **ion** that forms part of ATP and ADP.

photic zone The surface layer where there is enough light for **photosynthesis** to occur. *See* **epipelagic zone.**

photophore (light organ) An organ that produces **bioluminescence.**

photosynthesis The chemical process involved in the transformation of solar **energy** into organic matter (Figure 4-6):
$$CO_2 + H_2O + sun\ energy \rightarrow organic\ matter + O_2$$

photosynthetic pigment A molecule such as **chlorophyll** that is responsible for capturing solar **energy** in **photosynthesis.**

phycobilins A group of **photosynthetic pigments** that includes *phycocyanin*, a bluish pigment in **blue-green** and **red** algae, and *phycoerythrin*, a red pigment in **red algae.**

phycocyanin *See* **phycobilins.**

phycoerythrin *See* **phycobilins.**

phycocolloid One of several **starch**like chemicals found in some **seaweeds.** They are of significant commercial importance. *See* **agar, algin,** and **carrageenan.**

phylogeny The evolutionary history of a **species.**

phylum The **taxon** that represents the main divisions of a kingdom. An equivalent, the *division*, is used in the non-animal kingdoms.

phytoplankton The **autotrophic,** plantlike component of **plankton.** *See* **zooplankton.**

pinnipeds (order *Pinnipedia*) The **mammals** with paddle-shaped flippers: the *seals*, *eared seals* (*sea lions* and *fur seals*), and the *walrus*.

placenta A membrane that connects the mammalian **embryo** with the mother's womb to provide nourishment.

planktivore An animal that feeds on **plankton.**

plankton The organisms that are found drifting in the water (Figure 9-21).

plants The members of the kingdom *Plantae*, which consists of photosynthetic, **eukaryotic,** and mostly multicellular organisms.

planula The ciliated **larva** of **cnidarians.**

plate tectonics The process involved in the movement of large plates on the earth's crust.

Pleistocene A geologic period, which began about 2 million years ago, characterized by a series of **ice ages.**

pneumatocyst A gas-filled bladder in **seaweeds.**

pod A **school** of cetaceans.

pogonophorans *See* **beard worms.**

poikilotherm *See* **ectotherm.**

polar easterlies The variable winds that blow at high latitudes (Figure 3-21).

pollen The structure that produces the male **gamete** in **flowering plants.**

pollution A decrease in the quality of the environment as a result of the introduction by humans of substances (*pollutants*) or by means of other human activities.

polychaetes (class *Polychaeta*) A group of **segmented worms** that have **parapodia** (Figure 6-19).

polychlorinated biphenyls (PCBs). A group of **nonbiodegradable** pollutants.

polyp The cylindric, typically attached stage of **cnidarians.**

population A group of organisms belonging to the same **species** and living in the same place.

pore cell   The tubelike cell of **sponges** that forms a pore.

precious corals   The **gorgonians** that secrete a red or pink skeleton that consists of fused **calcareous spicules.**

predation   When an animal, or *predator*, eats another organism, or *prey*. A *top predator* is one that feeds at the top of the **food chain.**

pressure   The weight exerted over a unit area of surface: 1 *atmosphere* (14.7 pounds per square inch) at the sea surface and 1 atmosphere plus the pressure exerted by the **water column,** which equals 1 atmosphere per 10 m (33 feet) of depth.

primary producer (or **producer**)   An **autotroph.** An organism that carries out **primary production.**

primary production   The conversion of the inorganic carbon that is contained in **carbon dioxide** into organic carbon by **autotrophs.**

primary productivity   The rate of **primary production,** that is, the amount of carbon fixed under a square meter of sea surface in a day or in a year (Figure 9-16).

proboscis   An extension near the mouth that in some **invertebrates** helps in the capture or collection of food.

profile   A graph that shows changes in temperature, salinity, or any other parameter with depth (Figure 3-11).

prokaryote   An organism, such as bacteria, that consists of one or more **prokaryotic cells.**

prokaryotic cell   The simplest type of cell, one that lacks **organelles** (Figure 4-12). *See* **eukaryotic cell.**

protective coloration   Coloration that benefits the individual by providing concealment from predators.

protein   A large group of complex **nitrogen-**containing **organic molecules** that play many crucial roles in organisms.

protists   The members of the kingdom *Protista*, which consists of unicellular and eukaryotic organisms. Many combine characteristics of both **animals** and **plants.**

protochordates   The **chordates** that lack a backbone.

protozoans   The animal-like **protists,** that is, the various groups of unicellular and **eukaryotic** organisms that are mostly **heterotrophic.**

pseudopodium   A thin or blunt extension of the cytoplasm.

pteropods   **Pelagic gastropods** that have a **foot** modified for swimming and a shell that is reduced or absent (Figure 14-9).

pyloric caecum (plural **caeca**)   Each of the slender blind tubes found in the intestine of many **bony fishes.**

pyramid of biomass   The decrease in **biomass** that is observed in each succeeding level of a **food chain.**

pyramid of energy   The decrease in **energy** that is observed in each succeeding level of a **food chain.**

pyramid of numbers   The decrease in the number of individuals that is observed in each succeeding level of a **food chain.**

R

radial symmetry   The regular arrangement of similar body parts around a central axis (Figures 6-9 and 6-16). *See* **bilateral symmetry.**

radioactivity   The emission of *radiation* by unstable **atoms** in the form of particles and rays.

radiolarian ooze   A type of **biogenous sediment** that consists mostly of the **silica** shells of **radiolarians.**

radiolarians   A group of **protozoans** with a **silica** shell and **pseudopodia** (Figure 6-3).

radula   The ribbonlike band of teeth of **molluscs** (Figure 6-27).

rare species   A species that is in danger of **extinction.**

recombination   The formation of new genetic combinations, as the one that takes place during **fertilization.**

red algae (division *Rhodophyta*)   The **seaweeds** that have a predominance of red pigments.

red blood cell   A specialized type of blood cell that carries **hemoglobin** in **vertebrates.**

red clay   A fine sediment that is the most common type of **lithogenous sediment** in the open ocean floor.

red tide   An extensive **bloom** of **algae** that discolors the water.

reef crest   The shallow outer edge of the **reef slope** of a **coral reef** (Figures 13-18).

reef flat   The wide and shallow upper surface of a **coral reef** (Figures 13-18).

reef slope   The outer, steep margin of a **coral reef** (Figure 13-18). Also *see* **fore reef.**

refraction   The changing in the direction of a wave as it moves in shallow water (Figures 10-10 and 10-11).

regeneration   The ability of an organism to grow a body part that has been lost.

renewable resource   A resource that is naturally replaced.

replication   The copying of a cell's **DNA** before the cell divides.

reproductive isolation   Inability of separate **populations** to interbreed.

reproductive strategy   The reproductive patterns that are followed by a particular species.

reptiles (class *Reptilia*)   The **vertebrates** with scales on their skin and leathery eggs that are laid on land. Marine reptiles include *sea turtles, sea snakes,* the *marine iguana,* and the *saltwater crocodile.*

resource partitioning   The sharing of resources by specialization.

respiration   The chemical process involved in the release of **energy** from organic matter (Figure 4-8): organic matter + $O_2 \rightarrow CO_2 + H_2O$ + energy

respiratory exchange   *See* **gas exchange.**

respiratory tree   The branched extensions of the posterior end of the digestive tract of **sea cucumbers** that are involved in **gas exchange.**

rete mirabile   A network of blood vessels; in some fishes it functions as a heat exchange system that helps keep internal body temperature higher than that of the water (Figure 14-27).

reversing thermometer   A thermometer that permits the measurement of temperature at depth, since the mercury column breaks when the thermometer is turned upside down.

rhizoid   A simple, rootlike portion of the **thallus** of some **seaweeds** and **fungi.** *See* **holdfast.**

ribbon (or **nemertean**) **worms** (phylum *Nemertea*)   The **invertebrates** with a complete digestive tract, a true circulatory system, and a long **proboscis** to capture prey (Figure 6-18).

rift   A crack in the earth's crust formed as pieces of the crust separate.

rockweeds (or **wracks**)   The brown algae such as *Fucus* that are inhabitants of rocky shores in temperate zones.

rorquals   The blue, fin, and the other **baleen whales** that display long grooves on the underside.

roundworms   *See* **nematodes.**

S

salinity   The total amount of salts dissolved in seawater. It is generally expressed in parts per thousand (‰).

salivary gland   Any of the glands in **molluscs, vertebrates,** and other animals that release digestive **enzymes** into the mouth.

salmon ranching   When cultured juvenile salmon are released and allowed to migrate to sea, later to return as adults.

salps (class *Thaliacea*)   Pelagic **tunicates** with a transparent, cylindric body, sometimes forming long colonies (Figure 6-55).

salt   A substance that consists of **ions** that have opposite electrical charges.

salt gland   A gland that secretes excess salts in seabirds and sea turtles.

salt marsh   The grassy areas that extend along the shores of **estuaries** and sheltered coasts in temperate and subpolar regions.

salt wedge   A layer of denser, saltier seawater that flows along the bottom in **estuaries** (Figure 11-5).

sand dollars   A group of **sea urchins** with a flat, round test and short spines that live partly buried in soft sediments (Figure 6-51, C).

Sargasso Sea   The region of the Atlantic Ocean north of the West Indies that is characterized by floating masses of *Sargasso weed,* a **brown alga.**

scaphopods   *See* **tusk shells.**

scavenger   An animal that feeds on dead organic matter.

school   A well-defined group of fishes or **cetaceans** of the same species. *See* **pod.**

scientific method   The set of procedures by which scientists learn about the world (Figure 1-34).

sclerosponges (or **coralline sponges**)   The **sponges** with a massive **calcareous** skeleton (Figure 6-8, *D*).

scuba (*self-contained underwater breathing apparatus*)   The use of tanks of compressed air for breathing underwater.

scyphozoans (class *Scyphozoa*)   The **cnidarians** that include a conspicuous **medusa** and a much reduced or absent **polyp** in their life cycle.

sea   A **wave** that has a sharp peak and a relatively flat **trough.** Seas are found in the area where waves are generated by the wind (Figure 3-28).

sea anemones   The **anthozoans** that consist of one large **polyp.**

sea cows   *See* **sirenians.**

sea cucumbers (class *Holothuroidea*)   The **echinoderms** with a soft and elongate body that lacks spines.

sea floor spreading   The process by which new sea floor is formed as it moves away from *spreading centers* in the mid-ocean ridges.

seagrasses   The grasslike **flowering plants** such as *eelgrass* that are adapted to live at sea (Figure 5-26).

sea lilies   *See* **crinoids.**

seamount   A submarine volcano in the **abyssal plain.**

sea slugs   *See* **nudibranchs.**

sea spiders (class *Pycnogonida*)   A group of **arthropods** that have a reduced body and four pair of legs (Figure 6-47).

sea squirts (or **ascidians**; class *Ascidiacea*) The **tunicates** with a saclike, attached body as adults (Figure 6-54).

sea stars (or **starfishes**; class *Asteroidea*) The **echinoderms** with five or more radiating arms and **tube feet** that are used in locomotion.

sea urchins (class *Echinoidea*) The **echinoderms** with a round or flattened test and movable spines.

seaweed (or **macrophyte**) A large, multicellular **alga.**

sediment The loose material such as sand and mud that settles on the bottom. *See* **biogenous** and **lithogenous sediments.**

seeding In **mariculture**, when cultured juvenile individuals are released to enrich a natural **stock.**

segmentation The division of the body into similar compartments, or *segments.*

segmented worms (or **annelids**; phylum *Annelida*) The **invertebrates** that display an elongate body with distinct **segmentation** and a digestive tract that lies in a **coelom.**

seismic sea wave *See* **tsunami.**

self-regulating population A **population** with a growth rate that is dependent on its own numbers.

self-shading When **phytoplankton** reduce the amount of light available to phytoplankton that live below.

semidiurnal tide A tidal pattern with two high and two low **tides** each day (Figure 3-33, A).

septa The thin tissue partitions in the **polyp** of **anthozoans.**

sessile An organism that lives attached to the bottom or to a surface.

seta (plural **setae**) The bristles of **polychaetes.**

sex hormone A **hormone** that controls the timing of reproduction and sexual characteristics in **vertebrates.**

sexual reproduction The type of reproduction that involves the union of **gametes.**

shear boundary The boundary between two plates that move past each other on the earth's surface. *See* **fault.**

shelf break The section of the **continental shelf** where the slope abruptly becomes steeper, usually at a depth of 120 to 200 m (400 to 600 feet) (Figure 2-22).

silica (SiO_3) A mineral similar to glass that is the major component of the cell wall, shell, or skeleton of many marine organisms.

siliceous Made of **silica.**

siliceous ooze A type of **biogenous sediment** that consists mostly of the **silica** shell and skeletons of marine organisms. *See* **diatomaceous ooze** and **radiolarian ooze.**

silicoflagellates Unicellular and **eukaryotic** members of the **phytoplankton** that have a star-shaped **silica** skeleton (Figure 5-8).

simple sugar A sugar, such as **glucose,** that cannot be broken down into simpler sugar molecules.

siphon The tubelike extension through which water flows in and out of the **mantle cavity** in **bivalves** and **cephalopods** (where it is also known as **funnel**) and in **tunicates.**

siphonophores **Hydrozoans** that exist as drifting colonies.

sipunculans *See* **peanut worms.**

sirenians or **sea cows** (order *Sirenia*) The marine **mammals** with anterior flippers, no rear limbs, and a paddle-shaped tail (Figure 8-14).

sludge Semiliquid sewage that has been treated and partially decomposed by bacteria.

soft coral (order *Alcyonacea*) A colonial **anthozoan** with no hard skeleton.

solute Any material dissolved in a *solution.*

sonar (*so*und *na*vigation *ra*nging) A technique or equipment used to locate objects underwater by the detection of echoes (Figure 1-14).

Southern Oscillation *See* **El Niño–Southern Oscillation.**

spawning The release of **gametes** or eggs into the water.

species Individuals of the same kind that cannot breed with those belonging to other kinds.

spermaceti organ The **melon** of sperm whales. It contains *spermaceti,* an oil once widely used in candle-making.

spermatophore A packet of sperm in **cephalopods.**

spicule Any of the small **calcareous** or **siliceous** bodies embedded among cells of **sponges** or in the tissues of other **invertebrates.**

spiracle A pair of openings behind the eyes of **cartilaginous fishes.**

spiral valve A spiral portion in the intestine of **cartilaginous fishes.**

sponges (phylum *Porifera*) The **invertebrates** that consist of a complex aggregation of cells, including **collar cells,** and a skeleton of fibers and/or **spicules.**

spongin The resistant fibers of **sponges.**

spore The **asexual,** sometimes resistant structure produced by some **algae.** *See* **zoospore.**

sporophyte The **diploid, spore**-producing generation in many **seaweeds**. *See* **gametophyte.**

spout (or blow) The water vapor and seawater that is observed when whales surface and exhale.

spring tides The **tides** with a large **tidal range;** they occur around the times of full or new moon. *See* **neap tides.**

standing stock The total amount, or **biomass,** of an organism at a given time.

starch A **carbohydrate** that consists of chains of **simple sugars.**

statocyst A fluid-filled cavity provided with sensitive hairs and a small free body that is used to orient animals in regard to gravity.

stenohaline Organism that can tolerate a narrow range of salinities.

stipe The stemlike portion of the **thallus** of seaweeds.

stock The size of a **population.**

stony corals The **anthozoans,** often colonial, that secrete a massive **calcareous** skeleton.

stratification The separation of the **water column** into layers, the densest at the bottom and the less dense at the surface (Figure 3-35). A stratified water column is said to be *stable*. An *unstable* column results when the surface water becomes more dense than the water below.

stromatolites **Blue-green algae (or cyanobacteria)** that form massive **calcareous** skeletons (Figure 5-3).

structural color A color that results when light is reflected by a particular surface.

structural molecule A complex molecule, such as **cellulose,** that provides support and protection.

subduction The downward movement of a plate into the mantle that occurs in **trenches,** which are also known as *subduction zones.*

sublittoral zone *See* **subtidal zone.**

submarine canyon A narrow, deep depression in the **continental shelf** formed by the erosion of rivers or glaciers before the shelf was submerged (Figure 2-23).

subsidence The sinking of a land mass.

substrate The type of bottom or material on or in which an organism lives.

subtidal zone The bottom above the **continental shelf** (Figure 9-22).

succession *See* **ecological succession.**

succulent A fleshy plant that accumulates water.

sulfide One of the minerals that is abundant in the hot water that seeps through **hydrothermal vents.**

surf A **wave** that becomes so high and steep as it approaches the shoreline that it breaks.

surface (or mixed) layer The upper layer of water that is mixed by wind, waves, and currents (Figure 3-37).

surface-to-volume ratio (or S/V ratio) The amount of surface area relative to the total volume of an organism (Figure 4-23).

suspension feeder An animal that feeds on particles suspended in the water (Figure 6-20). *See* **deposit feeder, filter feeder.**

swarming The aggregation of individuals for **spawning** or other purposes.

sweeper tentacle A type of **tentacle** in corals that is used to sting neighboring colonies.

swell A **wave** with a flatter, rounded **crest** and **trough.** Swells are found away from the area where waves are generated by the wind (Figure 3-28).

swim bladder The gas-filled sac in the body cavity of **bony fishes** that is involved in the adjustment of buoyancy (Figure 7-13).

symbiosis A close relationship between two species: a smaller partner, the *symbiont,* and a larger one, the *host.*

system (or organ system) A group of **organs** specialized in one function.

T

tadpole larva The larva of **tunicates.**

tapeworms A group of parasitic **flatworms,** typically consisting of a chain of repeated units.

taste buds Structures in the mouth and other locations of fishes that are sensitive to chemical stimuli.

taxon (plural taxa) A group of organisms that share a common ancestry.

tectonic estuary An **estuary** that results from the sinking of land due to movements of the **crust.**

tentacle A flexible, elongate appendage.

territory A home area that is defended by an animal.

Tethys Sea A shallow sea that once separated the Eurasian and African sections of the "supercontinent" **Pangaea.** It eventually gave rise to the modern Mediterranean sea (Figure 2-21).

tetrapods The **vertebrates** that have two pairs of legs, a group that includes the **cetaceans.**

thallus The complete body of a **seaweed.**

thermal pollution Pollution by heated water.

thermocline   A zone in the **water column** that shows a sudden change in temperature with depth. The *main thermocline* is the zone where the temperature change marks the transition between the warm surface water and the cold deep water (Figures 3-11 and 3-35).

threatened species   A species that occurs in low numbers.

tidal bore   A steep wave generated as high tides move up some estuaries and rivers.

tidal current   A **current** that is generated by **tides.**

tidal energy   The **energy** that can be harnessed as a result of the movement of **tides.**

tidal marsh   *See* **salt marsh.**

tidal range   The difference in water level between successive high and low **tides.**

tide   The periodic, rhythmic rise and fall of the sea surface.

tide pool   A depression that holds seawater at low tide.

tide table   Tables that give the predicted time and height of **tides** for particular points along a coast.

tintinnids   A group of **ciliates** that secrete vaselike cases, or *loricas* (Figure 6-4).

tissue   A group of cells specialized in one function.

tortoiseshell   The polished shell of hawksbill turtles.

trade winds   The steady winds that blow from east to west toward the Equator as a replacement of the hot air that rises at the Equator (Figure 3-21).

transform fault   A large horizontal displacement in the mid-ocean ridge.

transplantation   The intentional introduction of a species.

trematodes   *See* **flukes.**

trench   A narrow, deep depression in the sea floor (Figures 2-14 and 2-15).

trochophore   A planktonic **larva** found in **polychaetes,** some **molluscs,** and **invertebrates** (Figure 14-11, *D*).

trophic level   Each of the steps in a **food chain.**

tsunami (or **seismic sea waves**)   The long, fast waves produced by earthquakes and other seismic disturbances of the sea floor.

tube foot   Any of the external muscular extensions of the **water vascular system** of **echinoderms.**

tubular eye   The specialized eyes of many **midwater** animals that allow upward or downward vision.

tunic   The outer covering of **sea squirts.**

tunicates (subphylum *Urochordata*)   The **chordates** that show the three basic chordate characteristics only in the larva.

turbellarians   The mostly free-living **flatworms** (Figure 6-17).

tusk shells (or scaphopods; class *Scaphopoda*)   The **molluscs** that have an elongate, tapered shell that is open at both ends.

U

upwelling   The process by which colder water richer in **nutrients** rises from a lower to a higher depth. It includes *coastal* and *equatorial upwelling* (Figures 14-37 and 14-40).

urea   A toxic waste product.

urogenital opening   The common opening for urine and **gametes** in **bony fishes** and other animals.

uterus (or **womb**)   In **mammals,** the portion of the female's reproductive tract in which the **embryo** develops.

V

valve   Any of the two shells of **bivalves** and **lamp shells.**

vegetative reproduction   *See* **asexual reproduction.**

veliger   A planktonic **larva** in **gastropods** and **bivalves** (Figure 14-11, A).

ventral   The underside or belly surface of an animal with **bilateral symmetry** (Figure 6-16).

vertebra (plural **vertebrae**)   Each of the bones that make up the backbone.

vertebrates (subphylum *Vertebrata*)   The **chordates** with a backbone.

vestimentiferans   A group of **pogonophorans** that are common at **hydrothermal vents** (Figure 15-39).

viviparous   An animal whose eggs develop inside the female while the **embryo** derives nutrition from the mother.

W

warning coloration   Coloration that allows organisms to escape from predators by advertising something harmful or distasteful.

water column   The vertical column of seawater that extends from the surface to the bottom.

water mass   A body of water that can be identified by its temperature and salinity.

water vascular system   A network of water-filled canals in **echinoderms** used in locomotion and food-gathering.

wave   The undulation that forms as a disturbance moves along the surface of the water. Waves can be described by their *height* (the vertical distance between **crest** and **trough**), *wavelength* (the horizontal distance between adjacent **crests**), and *period* (the time the wave takes to move past a given point) (Figure 3-25).

wave crest The highest part of a **wave** (Figure 3-25).

wave reinforcement The process by which two waves collide and add together to create a higher wave.

wave shock The intensity of the impact of a **wave**.

wave trough The lowest part of a **wave** (Figure 3-25).

weathering The physical and chemical breakdown of rocks.

westerlies The winds that blow from west to east at middle latitudes (Figure 3-21).

world ocean A concept that is used to indicate that all oceans on earth are interconnected.

Y

yolk sac A yolk-containing sac that is attached to the **embryo** of **fishes** and other **vertebrates.**

Z

zonation The presence of organisms within a particular range, as in the *vertical zonation* observed in the **intertidal.**

zooid Each individual member of a colony of **bryozoans** and other colonial **invertebrates.**

zooplankton The **heterotrophic,** animal component of plankton. *See* **phytoplankton.**

zoospore A **spore** provided with one or more **flagella.**

zooxanthellae **Dinoflagellates** that live within the tissues of reef corals and other marine animals (Figure 13-9).

zygote The **diploid** cell that results from **fertilization,** that is, a fertilized egg.

Illustration Credits

Chapter 1

1-1 Courtesy of G.B.R.M.P.A.

1-2 Hans Bertsch.

1-3 Halstead, B.W.: *Poisonous and Venomous Marine Animals of the World*, vol 1: Invertebrates. Department of Defense, 1965. Figures 4 and 5.

1-4 Bishop Museum.

1-5 The J. Paul Getty Museum.

1-6 Viola, H.J. and C. Margolis (eds): *Magnificent Voyagers*, 1985, Smithsonian Institution.

1-7 Painting by George Richmond; Bridgeman/Art Resource.

1-8 Original drawing of barnacles by Charles Darwin, from Darwin, C.: *A Monograph of the Sub-Class Cirripedia*, vol 2, 1852-1854. The Ray Society, British Museum, London, England.

1-9 Redrawn by Bill Ober/Claire Garrison from Menzies, R.J., R.Y. George, and G.T. Rowe: *The Abyssal Environment and Ecology of the World Oceans*, New York, 1973, J.C. Wiley.

1-10 Linklater, Eric: *The Voyage of the Challenger*. New York, 1972, Doubleday and Co., Inc., p. 16.

1-11 Pagecrafters.

1-12 Peter Castro.

1-13 Woods Hole Oceanographic Institute.

1-14 Bill Ober/Claire Garrison.

1-15 Charlie Arneson.

1-16 Mike Huber.

1-17 Harbor Branch Oceanographic Institution/Tom Smoyer.

1-18 George Grice/Woods Hole Oceanographic Institute.

1-19 Woods Hole Oceanographic Institute.

1-20 A,B Mike Huber.

1-21 Robert Ginsberg/RSMAS.

1-22 Peter Castro.

1-23 Charlie Arneson.

1-24 Scott Taylor/Duke Marine Lab.

1-25 Charlie Arneson.

1-26 Pagecrafters.

1-27 NASA.

1-28 NASA.

1-29 Mike Huber.

1-30 Norbert Wu.

1-31 Charlie Arneson.

1-32 P. Curto/The Image Bank.

1-33 Tom and Michele Grimm/International Stock Photo.

1-34 Bill Ober/Claire Garrison.

1-35 Bill Ober/Claire Garrison.

1-A Courtesy Ralph Buchsbaum.

Chapter 2

2-1 R.W. Griggs, Hawaii Volcano Observatory, USGS.

2-2 Pagecrafters.

2-3 Bill Ober/Claire Garrison.

2-4 Bill Ober/Claire Garrison.

2-5 Pagecraftes.

2-6 *Floor of the oceans*, Bruce C. Heezen and Marie Tharp, copyright © by Marie Tharp, 1980.

2-7 Pagecrafters.

2-8 Pagecrafters.

2-9 University of Colorado.

2-10 Deep Sea Drilling Program/Texas A&M University.

2-11 Bill Ober/Claire Garrison.

2-12 Bill Ober/Claire Garrison.

2-13 Pagecrafters.

2-14 Bill Ober/Claire Garrison.

2-15 Bill Ober/Claire Garrison.

2-16 USGS/M.E. Yount.

2-17 Pagecrafters.

2-18 Brent Winebrenner/International Stock Photo.

2-19 Adapted by Bill Ober/Claire Garrison from Anderson, D.L.: "The San Andreas Fault." *Scientific American*, vol. 225 (5), 1971, pp. 52-66.

2-20 Imagery.

2-21 Ziegler, A.M., C.R. Scotese, and S.F. Barrett: "Mesozoic and Cenozoic Paleogeographic Maps" in *Tidal Friction and the Earth's Rotation II*. Figures 2, 3, 4, 6 and 7. Edited by Brosche/Sunderman. Berlin and Heidelberg, 1983, Springer-Verlag.

2-22 Bill Ober/Claire Garrison.

2-23 Pagecrafters.

2-24 Bill Ober/Claire Garrison.

2-25 Peter Castro.

2-26 Scott Taylor/Duke Marine Lab.

2-27 Bill Ober/Claire Garrison.

2-28 Robert D. Ballard/Woods Hole Oceanographic Institute.

2-29 Redrawn by Bill Ober/Claire Garrison from Edmond and von Damm, "Hot Springs on the Ocean Floor." *Scientific American*, vol. 248, 1983, pp. 70-85.

2-A Bill Ober/Claire Garrison.

2-B Frank Huber.

2-C Peter Castro.

Chapter 3

3-1 Mike Huber.

3-2 Bill Ober/Claire Garrison.

3-3 Bill Ober/Claire Garrison.

3-4 Bill Ober/Claire Garrison.

3-5 Mike Huber.

3-6 Bill Ober/Claire Garrison.

3-7 Bill Ober/Claire Garrison.

3-8 Bill Ober/Claire Garrison.

3-9 National Marine Fisheries Service.

3-10 Scott Taylor/Duke Marine Lab.

3-11 Bill Ober/Claire Garrison.

3-12 Norbert Wu.

3-13 (B) Mike Huber; **(A,C)** Scott Taylor/Duke Marine Lab;

3-14 NASA.

3-15 Mike Huber.

3-16 Bill Ober/Claire Garrison.

3-17 Bill Ober/Claire Garrison.

3-18 Norbert Wu.

3-19 Bill Ober/Claire Garrison.

3-20 Bill Ober/Claire Garrison.

3-21 Pagecrafters.

3-22 Bill Ober/Claire Garrison.

3-23 Pagecrafters.

3-24 Pagecrafters.

3-25 Bill Ober/Claire Garrison.

3-26 Bill Ober/Claire Garrison.

3-27 Bill Ober/Claire Garrison.

3-28 Bill Ober/Claire Garrison.

3-29 Mike Huber.

3-30 A,B Bill Ober/Claire Garrison.

3-31 Bill Ober/Claire Garrison.

3-32 Bill Ober/Claire Garrison.

3-33 Bill Ober/Claire Garrison.
3-34 Pagecrafters.
3-35 Bill Ober/Claire Garrison.
3-36 Bill Ober/Claire Garrison.
3-37 Bill Ober/Claire Garrison.
3-A Pagecrafters.
3-B Metropolitan Musem of Art.

Chapter 4

4-1 Charlie Arneson.
4-2 Bill Ober/Claire Garrison.
4-3 Peter Castro.
4-4 Stephen Leatherwood.
4-5 F. Stuart Westmoreland/Tom Stack and Associates.
4-6 Bill Ober/Claire Garrison.
4-7 Mike Huber.
4-8 Bill Ober/Claire Garrison.
4-9 Bill Ober/Claire Garrison.
4-10 Bill Ober/Claire Garrison.
4-11 Reproduced with permission from Hallegraeff, G.M.: *Plankton: A Microscopic World*, 1988, CSIRO/Tobert Brown Associates.
4-12 Bill Ober/Claire Garrison.
4-13 A,B Bill Ober/Claire Garrison.
4-14 Mike Huber.
4-15 Mike Huber.
4-16 Mike Huber.
4-17 Barbara Cousins.
4-18 Barbara Cousins.
4-19 Bill Ober/Claire Garrison.
4-20 Bill Ober/Claire Garrison.
4-21 A Mike Huber.
4-21 B Peter Castro.
4-22 Pagecrafters.
4-23 Bill Ober/Claire Garrison.
4-24 A Mike Huber.
4-24 B Peter Castro.
4-24 C Charlie Arneson.
4-25 Imagery/L.W. Pollock.
4-26 Photo by Imagery/L.W. Pollock. Illustration by Bill Ober/Claire Garrison.
4-27 (A) Courtesy of G.B.R.M.P.A. (B) B. McConnaughey; (C) Frank Huber; (D) Mike Huber.
4-28 Peter Castro.
4-29 Charlie Arneson.
4-30 Bill Ober/Claire Garrison.
4-A Peter Castro.
4-B Jeffrey L. Rotman Photography.

Chapter 5

5-1 Charlie Arneson.
5-2 Courtesy of Dr. H.D. Raj, California State University-Long Beach.
5-3 Charlie Arneson.
5-4 Peter Castro.
5-5 Bill Ober/Claire Garrison.
5-6 Bill Ober/Claire Garrison.
5-7 Courtesy of L. Fritz, National Research Council, Canada.
5-8 Reproduced with permission from Hallegraeff, G.M.: *Plankton: A Microscopic World*, 1988, CSIRO/Robert Brown Associates.
5-9 A Bill Ober/Claire Garrison.
5-9 B Hans Bertsch.
5-10 Peter Castro.
5-11 Charlie Arneson.
5-12 Charlie Arneson.
5-13 Mike Huber.
5-14 Peter Castro.
5-15 Peter Castro.
5-16 Craig Sandgren.
5-17 Charlie Arneson.
5-18 A Bill Ober/Claire Garrison.
5-18 B Peter Castro.
5-19 A,B Charlie Arneson.
5-20 Peter Castro.
5-21 Charles Yarish.
5-22 Dorothy Chappell.
5-23 Charlie Arneson.
5-24 Redrawn by Bill Ober/Claire Garrison from Bold, H.C. and M.J. Wynne: *Introduction to the Algae*, second edition, 1978.
5-25 Dale Glantz.
5-26 Photo by Peter Castro; illustration by Bill Ober/Claire Garrison.
5-27 Peter Castro.
5-28 A Peter Castro.
5-28 B Charlie Arneson.
5-29 Bill Ober/Claire Garrison.
5-A Paul A. Zahl © National Geographic Society.
5-B Japan National Tourist Organization.

Chapter 6

6-1 Peter Castro.
6-2 Charlie Arneson.
6-3 Bill Ober/Claire Garrison.
6-4 Redrawn by Bill Ober/Claire Garrison from Sherman I.W. and U.G. Sherman, *The Invertebrates: Function and Form*, 1976.
6-5 A, B Bill Ober/Claire Garrison.
6-6 Bill Ober/Claire Garrison.
6-7 Bill Ober/Claire Garrison.
6-8 A,B,D Charlie Arneson.
6-8 C Peter Castro.
6-9 Bill Ober/Claire Garrison.
6-10 A,B Charlie Arneson.
6-11 Bill Ober/Claire Garrison.
6-12 A,B Charlie Arneson.
6-13 Charlie Arneson.
6-14 Peter Castro.
6-15 Marsh Youngbluth.
6-16 Bill Ober/Claire Garrison.
6-17 Charlie Arneson.
6-18 Bill Ober/Claire Garrison.
6-19 Bill Ober/Claire Garrison.
6-20 Bill Ober/Claire Garrison.
6-21 A,B Charlie Arneson.
6-21 C Bill Ober.
6-22 A,B Charlie Arneson.
6-23 Bill Ober/Claire Garrison.
6-24 Bill Ober/Claire Garrison.
6-25 Charlie Arneson.
6-26 Bill Ober/Claire Garrison.
6-27 Copyright 1991 M.G. Harasewych.
6-28 A-D Charlie Arneson.
6-28 E Dale Glantz.
6-29 A to D Bill Ober/Claire Garrison.
6-30 A Peter Castro.
6-30 B,C Charlie Arneson.
6-31 A,B,C Bill Ober/Claire Garrison.
6-32 A Scripps Institution of Oceanography, University of California-San Diego/Cindy Clark.
6-32 B Marsh Youngbluth/Harbor Branch Oceanographic Institution.
6-32 C Charlie Arneson.
6-33 Charlie Arneson.
6-34 Peter Castro.
6-35 Bill Ober/Claire Garrison.
6-36 Photo by Peter Castro; Illustration by Bill Ober/Claire Garrison.
6-37 Bill Ober/Claire Garrison.
6-38 Charlie Arneson.
6-39 Charlie Arneson.
6-40 Bill Ober/Claire Garrison.
6-41 Charlie Arneson.
6-42 A,B Courtesy of Dr. Ernest S. Reese, University of Hawaii.
6-43 Mike Huber.
6-44 Mike Huber.
6-45 Bill Ober/Claire Garrison.
6-46 Woods Hole Oceanographic Institute.
6-47 Bill Ober/Claire Garrison.
6-48 A,B,C Bill Ober/Claire Garrison.
6-49 A Charlie Arneson.
6-49 B Charlie Arneson.
6-50 A,B Charlie Arneson.
6-51 A Jeffrey L. Rotman Photography.
6-51 B Peter Castro
6-51 C Charlie Arneson.
6-52 Charlie Arneson.
6-53 Charlie Arneson.
6-54, A Charlie Arneson.
6-54, B Bill Ober/Claire Garrison.
6-55 Charlie Arneson.
6-56 Bill Ober/Claire Garrison.
6-57 Bill Ober/Claire Garrison.
6-A Charlie Arneson.
6-B Bill Ober/Claire Garrison.

Chapter 7

7-1 Charlie Arneson.
7-2 Bill Ober/Claire Garrison.
7-3 Bill Ober/Claire Garrison.
7-4 Peter Castro.
7-5, **A** Charlie Arneson.
7-5, **B** and **C** Bill Ober/Claire Garrison.
7-6 Bill Ober/Claire Garrison.
7-7, **A,B,** and **C** Charlie Arneson.
7-7, **D** and **E** Bill Ober/Claire Garrison.
7-8 Redrawn by Bill Ober/Claire Garrison in part from Bigelow, H.B. and W.C. Shroeder: *Fishes of the Western North Atlantic*, Part 2, 1953.
7-9 Bill Ober/Claire Garrison.
7-10 Redrawn by Bill Ober/Claire Garrison from Bond, C.E.: *Biology of Fishes*, 1979.
7-11 **A** to **F** Charlie Arneson.
7-11 **G** Peter Castro.
7-12 Redrawn by Bill Ober/Claire Garrison from Moyle, P.B. and J.J. Cech: *Fishes: An Introduction to Ichthyology*, 1988.
7-13 Bill Ober/Claire Garrison.
7-14 Redrawn by Bill Ober/Claire Garrison in part from Bond, C.E., *Biology of Fishes*, 1979.
7-15 **A** and **B** Bill Ober/Claire Garrison.
7-16 Bill Ober/Claire Garrison.
7-17 Bill Ober/Claire Garrison. Adapted from Storer, T.I. et al.: *General Zoology*, 1979.
7-18 **A** to **E** Bill Ober/Claire Garrison.
7-19 Bill Ober/Claire Garrison.
7-20 Redrawn by Bill Ober/Claire Garrison from Storer, T.I. et al.: *General Zoology*, 1979.
7-21 Redrawn by Bill Ober/Claire Garrison from Sale, P.: *Proceedings of the Second International Coral Reef Symposium*, 1974.
7-22 Norbert Wu.
7-23 Redrawn by Bill Ober/Claire Garrison from Radakov, D.V.: *Schooling in the Ecology of Fish*, 1962.
7-24 Pagecrafters.
7-25 Frank Huber.
7-26 Bill Ober/Claire Garrison.
7-27 Redrawn by Bill Ober/Claire Garrison from Keenleyside, M.H.A.: *Diversity and Adaptation in Fish Behavior*, 1979, Springer-Verlag, Berlin.
7-28 Bill Ober/Claire Garrison.
7-29 Charlie Arneson.
7-30 **A** and **B** Courtesy of Dr. Ernest S. Reese, University of Hawaii.
7-31 Courtesy of National Marine Fisheries Service.
7-32 Mike Huber.
7-33 A. Kerstitch.
7-34 Bill Ober/Claire Garrison.
7-A Bill Ober/Claire Garrison.

Chapter 8

8-1 Copyright 1991 Sea World, Inc. All rights reserved. Reproduced by permission.
8-2, **A** Mike Huber.
8-2, **B** George H. Balazs.
8-2, **C** Greenpeace/Fretey.
8-3 Dale Glantz.
8-4 Dale Glantz.
8-5 Peter Castro.
8-6, **A** Copyright 1991 Sea World, Inc. All rights reserved. Reproduced by permission.
8-6, **B** Greenpeace.
8-6, **C** Mike Huber.
8-6, **D** Charlie Arneson.
8-6, **E** Peter Castro.
8-7 Redrawn by Bill Ober/Claire Garrison from Lofgren, L.: *Ocean Birds*, 1984.
8-8 Redrawn by Bill Ober/Claire Garrison from Ashmole, N.P.: *Avian Biology*, 1971.
8-9 Bill Ober/Claire Garrison.
8-10, **A** John Heine, Moss Landing Marine Lab.
8-10, **B** George H. Balazs.
8-10, **C** Peter Castro.
8-10, **D** Greenpeace.
8-10, **E** G.L. Kooyman.
8-11 Bill Ober/Claire Garrison.
8-12 B. "Moose" Peterson.
8-13 John Shaw/Tom Stack and Associates.
8-14 Illustration by Bill Ober/Claire Garrison; photo by Pat Rose/Save the Manatee Club.
8-15 Redrawn by Bill Ober/Claire Garrison from Storer, T.I. et al.: *General Zoology*, 1979.
8-16 Bill Ober/Claire Garrison.
8-17 Redrawn by Bill Ober/Claire Garrison from McIntyre, J.: *Mind in the Waters*, 1974.
8-18 Redrawn by Bill Ober/Claire Garrison in part from Slijper, E.J.: *Whales*, 1962.
8-19 Bill Ober/Claire Garrison.
8-20, **A** Greenpeace.
8-20, **B** Mike Huber.
8-21 Bill Ober/Claire Garrison.
8-22 Redrawn by Bill Ober/Claire Garrison from Daugherty, A.E.: *Marine Mammals of California*, 1972.
8-23 Bill Ober/Claire Garrison.
8-24, **A** Copyright 1991 Sea World, Inc. All rights reserved. Reproduced by permission.
8-24, **B** George W. Calef/Government N.W.T.
8-25 Courtesy of National Marine Fisheries Service—La Jolla.
8-26 Center for Coastal Studies.
8-27 **B** and **C** Redrawn by Bill Ober/Claire Garrison from Watson, L.: *Sea Guide to Whales of the World*, 1981; **A** from McIntyre, J.: *Mind in the Waters*, 1974.
8-28 Thomas Kitchin/Tom Stack and Associates.
8-29 Bill Ober/Claire Garrison.
8-30, **A** Copyright 1991 Sea World, Inc., All rights reserved. Reproduced by permission.
8-30, **B** Peter Castro.
8-31 Redrawn by Bill Ober/Claire Garrison from Watson, L.: *Sea Guide to Whales of the World*, 1981.
8-32 Copyright 1991 Sea World, Inc. All rights reserved. Reproduced by permission.
8-33 Bill Ober/Claire Garrison.
8-A Mike Greer/Chicago Zoological Society.

Chapter 9

9-1 Mike Huber.
9-2, **A** and **B** Bill Ober/Claire Garrison.
9-3 Norbert Wu.
9-4 Bill Ober/Claire Garrison.
9-5 Charlie Arneson.
9-6 Peter Castro.
9-7 Bill Ober/Claire Garrison.
9-8 Jeffrey L. Rotman Photography.
9-9 Hans Bertsch.
9-10 Jeff Foott.
9-11 Copyright 1991 John O'Sullivan.
9-12 Charlie Arneson.
9-13 Bill Ober/Claire Garrison.
9-14 Bill Ober/Claire Garrison.
9-15, **A** and **B** Bill Ober/Claire Garrison.
9-16 Bill Ober/Claire Garrison.
9-17 Bill Ober/Claire Garrison.
9-18 NASA.
9-19 Bill Ober/Claire Garrison.
9-20 Bill Ober/Claire Garrison.
9-21 Bill Ober/Claire Garrison.
9-22 Bill Ober/Claire Garrison.
9-A Ralph Lewin/Scripps Institution of Oceanography.

Chapter 10

10-1, **A** Charlie Arneson.
10-1, **B** Charlie Arneson.
10-2 **A**, G. Brad Lewis; **B**, Peter Castro.

10-3 Mike Huber.
10-4 Mike Huber.
10-5 Peter Castro.
10-6 Mike Huber.
10-7 Peter Castro.
10-8 Mike Huber.
10-9 Peter Castro.
10-10 Bill Ober/Claire Garrison.
10-11 Bill Ober/Claire Garrison.
10-12 Bill Ober/Claire Garrison.
10-13 Redrawn by Bill Ober/Claire Garrison from Migdalski, E.C. and G.S. Fishter: *The Fresh and Salt Water Fishes of the World*, 1976, Alfred A. Knopf, New York.
10-14 Redrawn by Bill Ober/Claire Garrison from Koehl, M.A.R.: "The Interaction of Moving Water and Sessile Organisms." *Scientific American*, vol. 247, 1982, pp. 124-134.
10-15 Peter Castro.
10-16 Peter Castro.
10-17 Redrawn by Bill Ober/Claire Garrison from Lobban, et al.: *Physiological Ecology of Seaweeds*, 1985, Cambridge University Press.
10-18 Redrawn by Bill Ober/Claire Garrison from Nybakken, J.W.: *Marine Biology*, 1988, Harper & Row, New York.
10-19 Bill Ober/Claire Garrison.
10-20 Mike Huber.
10-21 Redrawn by Bill Ober/Claire Garrison from Nybakken, J.W.: *Marine Biology*, 1988, Harper & Row, New York.
10-22 Mike Huber.
10-23 Bill Ober/Claire Garrison.
10-24 Bill Ober/Claire Garrison.
10-25 Redrawn by Bill Ober/Claire Garrison from Connell, J.H.: "The Influence of Interspecific Competition and Other Factors on the Distribution of the Barnacle *Cthamalus stellatus*." *Ecology*, vol. 42, 1961, pp. 710-723.
10-26 Peter Castro.
10-27 Peter Castro.
10-28 Bill Ober/Claire Garrison.
10-29 Bill Ober/Claire Garrison.
10-30 Bill Ober/Claire Garrison.
10-31 Bill Ober/Claire Garrison.
10-32 Hans Bertsch.
10-33 Bill Ober/Claire Garrison.
10-34 Bill Ober/Claire Garrison.
10-35 Peter Castro.
10-36 Adapted from Trueman, E.R.: *The Locomotion of Soft-bodied Animals*, Edward Arnold, London, 1975.
10-37 Adapted from Trueman E.R.:

The Locomotion of Soft-bodied Animals, Edward Arnold, London, 1975.
10-38 Mike Huber.
10-39 Bill Ober/Claire Garrison.
10-40 Bill Ober/Claire Garrison.
10-41 Ricketts, E.R. and J. Calvin: *Between Pacific Tides*, 5th edition, 1985, Stanford University Press.
10-42 Bill Ober/Claire Garrison.
10-B Courtesy of Dr. Ernest S. Reese, University of Hawaii.

Chapter 11

11-1 Charlie Arneson.
11-2 Department of the Interior/USGS.
11-3 Department of the Interior/USGS.
11-4 Peter Castro.
11-5 Bill Ober/Claire Garrison.
11-6 **A** and **B** Bill Ober/Claire Garrison.
11-7 Bill Ober/Claire Garrison.
11-8 Bill Ober/Claire Garrison.
11-9 Peter Castro.
11-10 Peter Castro.
11-11 Peter Castro.
11-12 Peter Castro.
11-13 Bill Ober/Claire Garrison.
11-14 Peter Castro.
11-15 **A** Redrawn by Bill Ober/Claire Garrison from Barbour, M.G., et al.: *Coastal Ecology, Bodega Bay*, 1973.
11-15 **B** and **C** Redrawn by Bill Ober/Claire Garrison from Pienkowski, M.W.: *Feeding and Survival Strategies of Estuarine Organisms*, 1981.
11-16 **A** Scott Taylor/Duke Marine Lab.
11-16 **B** Peter Castro.
11-17 Bill Ober/Claire Garrison.
11-18 Scott Taylor/Duke Marine lab.
11-19 Peter Castro.
11-20 Redrawn by Pagecrafters from Chapman, V.J.: *Wet Coastal Ecosystems, Ecosystems of the World*, vol. 1, 1977.
11-21 Charlie Arneson.
11-22 Peter Castro.
11-23 Mike Huber.
11-24 Scott Taylor/Duke Marine Lab.
11-25 Bill Ober/Claire Garrison.
11-B Bill Ober/Claire Garrison.
11-C Bill Ober/Claire Garrison.

Chapter 12

12-1 Norbert Wu.
12-2 Bill Ober/Claire Garrison.
12-3 Charlie Arneson.
12-4 Bill Ober/Claire Garrison.
12-5 Bill Ober/Claire Garrison.
12-6 Bill Ober/Claire Garrison.
12-7 Charlie Arneson.

12-8 Charlie Arneson.
12-9 Charlie Arneson.
12-10 Charlie Arneson.
12-11 Bill Ober/Claire Garrison.
12-12 Bill Ober/Claire Garrison.
12-13 Charlie Arneson.
12-14 Charlie Arneson.
12-15 Scott Taylor/Duke Marine Lab.
12-16 Scott Taylor/Duke Marine Lab.
12-17 Bill Ober/Claire Garrison.
12-18 Bill Ober/Claire Garrison.
12-19 Charlie Arneson.
12-20 Peter Castro.
12-21 Charlie Arneson.
12-22 Redrawn by Bill Ober/Claire Garrison from Berrill, M. and D. Berrill: *A Sierra Club Naturalist's Guide to the North Atlantic Coast*, 1981.
12-23 Pagecrafters.
12-24 Charlie Arneson.
12-25 Dale Glantz.
12-26 Bill Ober/Claire Garrison.
12-27 Bill Ober/Claire Garrison.
12-28 Mia Tegner, Scripps Institution of Oceanography.
12-29 Dale Glantz.
12-B John N. Heine/Moss Landing Marine Laboratory.
12-C Bill Ober/Claire Garrison.

Chapter 13

13-1 Charlie Arneson.
13-2 Bill Ober/Claire Garrison.
13-3, **A** Mike Huber.
13-3, **B** Bill Ober/Claire Garrison.
13-3, **C** Charlie Arneson.
13-4 Peter Parks/Oxford Scientific Films.
13-5 Mike Huber.
13-6 Mike Huber.
13-7 Mike Huber.
13-8 Adapted from Veron, J.E.: *Corals of Australia and the Indo-Pacific*, Angus and Robertson, North Ryde, Australia, 1986.
13-9 Robert Sisson, National Geographic Society.
13-10 Peter Castro.
13-11 Bill Ober/Claire Garrison.
13-12 Mike Huber.
13-13 Pagecrafters.
13-14 Bill Ober/Claire Garrison.
13-15 Charlie Arneson.
13-16 Map by Pagecrafters; photo courtesy of Paul Jokiel.
13-17 Peter Castro.
13-18 Illustration by Bill Ober/Claire Garrison; photo by Peter Castro.
13-19 Peter Castro.
13-20 Peter Castro.
13-21 Bill Ober/Claire Garrison.

13-22 Mike Huber.
13-23 Mike Huber.
13-24 Peter Castro.
13-25 Courtesy of G.B.R.M.P.A.
13-26 Courtesy of G.B.R.M.P.A.
13-27 Jeffrey L. Rotman Photography.
13-28 Illustration by Bill Ober/Claire Garrison; photos by Charlie Arneson and Peter Castro.
13-29 Bill Ober/Claire Garrison.
13-30 Bill Ober/Claire Garrison.
13-31 Peter Castro.
13-32 Bill Ober/Claire Garrison.
13-33 Mike Huber.
13-34 Mike Huber.
13-35 Courtesy of G.B.R.M.P.A.
13-36 Courtesy of Randall Kosaki.
13-37 Charlie Arneson.
13-38 Peter Castro.
13-39 Courtesy of G.B.R.M.P.A.
13-40 Courtesy of G.B.R.M.P.A.
13-41 D.R. Robertson, Smithsonian Tropical Research Institute.
13-42 Peter Castro.
13-43 Charlie Arneson.
13-B 1-3 Courtesy of G.B.R.M.P.A.

Chapter 14

14-1 Charlie Arneson.
14-2, A Mike Huber.
14-2, B Mike Huber.
14-3, A and B Reproduced with permission from Hallegraeff, G.M.: *Plankton: A Microscopic World*, 1988, CSIRO/Robert Brown Associates.
14-3, C N. Hoepffner.
14-4 Reproduced with permission from Hallegraeff, G.M.: *Plankton: A Microscopic World*, 1988, CSIRO/Robert Brown Associates.
14-5 Bill Ober/Claire Garrison.
14-6, A Charlie Arneson.
14-6, B Charlie Arneson.
14-7 Bill Ober/Claire Garrison.
14-8 James M. King; Illustration by Bill Ober.
14-9, A Marsh Youngbluth.
14-9, B Marsh Youngbluth.
14-10 Dale Glantz.
14-11 Bill Ober/Claire Garrison.
14-12 Charlie Arneson.
14-13 Bill Ober/Claire Garrison.
14-14 Bill Ober/Claire Garrison.
14-15 Marsh Youngbluth.
14-16 W. Gregory Brown/Earth Scenes.
14-17 Peter Parks/Oxford Scientific Films.
14-18 Charlie Arneson.
14-19 A. Kerstitch.
14-20 Charlie Arneson.

14-21 Lanna Cheng, Scripps Institution of Oceanography.
14-22 Bill Ober/Claire Garrison.
14-23, A Charlie Arneson.
14-23, B Marsh Youngbluth.
14-23, C Marsh Youngbluth.
14-24 Bill Ober/Claire Garrison.
14-25 Norbert Wu.
14-26 Bill Ober/Claire Garrison.
14-27 Bill Ober/Claire Garrison.
14-28 Bill Ober/Claire Garrison.
14-29 Redrawn from Hardy A.C.: The herring in relation to its animate environment. Pt. 1. *The food and feeding habits of the herring*. Fisheries Investigations, London, Ser. II, vol. 7, 1924, pp. 1-53.
14-30 Courtesy of Dr. H.D. Raj, California State University-Long Beach.
14-31 Redrawn by Pagecrafters from *Atlas of Living Resources of the Sea*, FAO, 1981, and Cushing, D.H.: *Marine Ecology and Fisheries*, Cambridge University Press, 1975.
14-32 Peter Castro.
14-33 Bill Ober/Claire Garrison.
14-34 Redrawn from Sverdrup et al: *The Oceans*, Prentice-Hall, Englewood Cliffs, New Jersey, 1942.
14-35 Bill Ober/Claire Garrison.
14-36 Bill Ober/Claire Garrison.
14-37 Bill Ober/Claire Garrison.
14-38 Pagecrafters.
14-39 NOAA.
14-40 Bill Ober/Claire Garrison.
14-41 Redrawn by Bill Ober/Claire Garrison from Rasmussen, E.M.: El Niño—The Ocean/Atmosphere Connection. *Oceanus*, vol. 27, 1984, pp. 5-12.
14-42 Redrawn by Pagecrafters from Rasmussen, E.M.: El Niño—The Ocean/Atmosphere Connection. *Oceanus*, vol. 27, 1984, pp. 5-12.
14-43 Pagecrafters.
14-B Bill Ober/Claire Garrison.
14-I Redrawn by Pagecrafters from McGowan: *The Biology of the Ocean Pacific* (C.B. Miller, ed.), 1972.

Chapter 15

15-1 Marsh Youngbluth.
15-2 Bill Ober/Claire Garrison.
15-3 Bill Ober/Claire Garrison.
15-4 Redrawn by Bill Ober/Claire Garrison from various sources.
15-5 Bill Ober/Claire Garrison.
15-6 Charlie Arneson.
15-7 Redrawn by Bill Ober/Claire Garrison from Herring, P.J.: Biolumines-

cence of invertebrates other than insects. In P.J. Herring (ed): *Bioluminescence in Action*, Academic Press, 1978, pp. 199-240.
15-8 Photo insert by Institute of Oceanographic Sciences, Deacon Laboratory; illustration redrawn by Bill Ober/Claire Garrison from Young, R.E.: "The systematics and aereal distribution of pelagic cephalopods from the seas off southern California," *Smithsonian Contributions to Zoology*, no. 97, 1972.
15-9 Mike Huber.
15-10 Bill Ober/Claire Garrison.
15-11 Mike Huber.
15-12 Bill Ober/Claire Garrison.
15-13 B, Redrawn by Bill Ober/Claire Garrison from Marshall, N.B.: *Developments in Deep-Sea Biology*, Blandford Press, London, 1979.
15-14 Pagecrafters.
15-15 Norbert Wu.
15-16 Bill Ober.
15-17 Redrawn by Bill Ober/Claire Garrison from Lookett, N.A.: Adaptations to the deep-sea environment. In F. Crescitelli (ed): *Handbook of Sensory Physiology*, vol. VII/5, Springer-Verlag, Berlin, 1977.
15-18 Charlie Arneson.
15-19 Charlie Arneson.
15-20 Norbert Wu.
15-21 Redrawn by Bill Ober/Claire Garrison from Hardy: *The Open Sea: The World of Plankton*, 1970.
15-22 Redrawn by Pagecrafters. From Warner, J.A. et al: "Cryptic Bioluminescence in a Midwater Shrimp." *Science*, vol. 203, 1979, pp. 1109-1110.
15-23 Adapted by Bill Ober/Claire Garrison from Herring, P.J. and N.A. Locket: "The Luminescence and Photophores of Euphausiid Crustaceans." *Journal of Zoology* vol. 186, 1978, pp. 431-436.
15-24 Bill Ober.
15-25 Adapted by Bill Ober from Migdalski, E.C. and G.S. Fishter: *The Fresh and Salt Water Fishes of the World*, 1976, Alfred A. Knopf, New York.
15-26 Bill Ober.
15-27 Redrawn by Bill Ober/Claire Garrison from various sources.
15-28 Robert R. Hessler, Scripps Institution of Oceanography.
15-29 Norbert Wu.
15-30 Norbert Wu.

15-31 Scripps Institution of Oceanography.

15-32 Robert R. Hessler, Scripps Institution of Oceanography.

15-33 Norbert Wu.

15-34 Redrawn by Bill Ober from Smith, K.L. et al: "Free Vehicle Capture of Abyssopelagic Animals." *Deep Sea Research*, vol. 26A, 1979, pp. 57-64.

15-35 Scripps Institution of Oceanography.

15-36 Redrawn by Bill Ober from various sources.

15-37 Bill Ober/Claire Garrison.

15-38 J. Edmond, M.I.T.

15-39 Redrawn by Bill Ober/Claire Garrison from *Scientific American*, vol. 256, no. 5, 1987.

15-40 Redrawn by Bill Ober/Claire Garrison from Childress, J.J., Felbeck H., and Somero G.N.: "Symbiosis in the Deep Sea." *Scientific American* vol. 256, 1987, pp. 114-120.

15-B Peter Castro.

15-C Charlie Arneson.

15-D Mike Huber.

Chapter 16

16-1 Peter Castro.

16-2 Bill Ober/Claire Garrison.

16-3 Mike Huber.

16-4 Bill Ober/Claire Garrison.

16-5 Gregory Heisler/The Image Bank.

16-6 Pagecrafters.

16-7 Norbert Wu.

16-8, **A** and **B** Peter Castro.

16-8, **C** and **D** Peter Castro.

16-9 Bill Ober/Claire Garrison.

16-10 Peter Castro.

16-11 Ocean Farms of Hawaii, Inc.

16-12 Bill Ober/Claire Garrison.

16-13 Bill Ober/Claire Garrison.

16-14 Perú Tourist Office.

16-15 Bill Ober/Claire Garrison.

16-16, **A** Marsh Youngbluth.

16-16, **B** Peter Castro.

16-17 Dale Glantz.

16-18 Southeast Asian Fisheries Development Center.

16-19 **A** Mike Huber.

16-19, **B** Peter Castro.

16-20, **A** and **B** Ocean Farms of Hawaii, Inc.

16-20, **C** Ocean Farms of Hawaii, Inc.

16-21 Mike Huber.

16-22 Courtesy of G.R. Pettit, Arizona State University.

16-23 Frank Huber.

16-24 United States Dept. of the Interior/Bureau of Mines.

16-25 American Petroleum Institute.

16-26 Scripps Institution of Oceanography, University of California-San Diego/Bruce Heezen/Lamont Doherty.

16-27 Charlie Arneson.

16-29 Bill Ober/Claire Garrison.

16-A Courtesy of Jaime Tres.

16-B Courtesy of Jaime Tres.

Chapter 17

17-1 Alaska Chamber of Commerce

17-2 Bill Ober.

17-3 Alaska Chamber of Commerce

17-4 Dale Glantz

17-5 Courtesy of Masahiro Dojiri.

17-6 Ted Streshinsky/Photo 20-20.

17-7 Bill Ober.

17-8 Greenpeace/Larry Lipsky.

17-9 Greenpeace/Van der Veer.

17-10 Redrawn by Bill Ober/Claire Garrison from Gerlach, S.A.: *Marine Pollution*, 1981.

17-11 Courtesy Paul Jokiel.

17-12 Peter Castro.

17-13 Dale Glantz.

17-14 Courtesy of Steve Coles.

17-15 Courtesy of Paul Jokiel.

17-16 Oil painting by Alfred G. Mi-lotte (1972); courtesy Victor B. Schaeffer.

17-17 Peabody Museum of Salem/Photo by Mark Sexton.

17-18 Greenpeace/Morgan.

17-19 John N. Heine/Moss Landing Marine Laboratory.

17-20 Greenpeace/Roger Grace.

17-21 Peter Castro.

17-22 Courtesy Fish & Wildlife, Ashland, Oregon.

17-23, **A** and **B** Courtesy Dr. Bella Galil, Tel Aviv University, Israel.

17-24 Peter Castro.

17-25 Mike Huber.

17-26 Dale Glantz.

17-27 Bill Ober/Claire Garrison.

17-A Bill Ober/Claire Garrison.

17-B Scott Taylor/Duke Marine Lab.

Chapter 18

18-1 Peter Castro.

18-2 Bill Ober/Claire Garrison.

18-3 Bill Ober/Claire Garrison.

18-4 Peter Castro.

18-5 George Ancona/International Stock Photo.

18-6 Peter Castro.

18-7 Bill Ober/Claire Garrison.

18-8 Paul Chesley.

18-9 Peter Castro.

18-10 Peabody Museum/Photo by Mark Sexton.

18-11 Peabody Museum/Photo by Mark Sexton.

18-12 Greenpeace/Visser.

18-13 Peter Castro.

18-14 Peter Castro.

18-15 Metropolitan Museum of Art, New York.

18-16 Peter Castro.

18-17 Peter Castro.

18-A Bill Ober/Claire Garrison.

Index

Dredging
  near coral reefs, 531
  of sea floor, 5
Drift, continental; *see* Continental drift
Drift nets, 498, 537
Drilling, offshore, 510–511, 519
Drowned river valley estuaries,
  323
Drugs from sea, 509
Drupella, 394
DSL; *see* Deep scattering layer
Ducks, 337, 357
Dugongs, 239–240
  threat of extermination of, 538
Dulse, 136
Dunaliella, 100, 101
Dungeness crab, 494
Dynamite fishing, 531

E

Eagle rays, 196
Ear stones in fishes, 213
Eared seals, 239
Earth
  geologic history of, 36
  structure of, 24–26
Earthquakes
  and mid-ocean ridge, 29–30
  and trenches, 34, 35–36
Easterlies, polar, 66, 67
Eastern spinner dolphin, 537
Echineis, 200, 202–203
Echinocardium, 315
Echinoderms, 178–184
Echinoecus, 352
Echinoidea, 181–182
Echinometra, 311, 397
Echiura, 160
Echiurans, 160–161
Echolocation in mammals, 247–250,
  420
Ecklonia, 363, 365
Ecological niche, 268
Ecological succession, 308, 342
Ecology, 264–285
  biological zonation in, 283–285
  of coral reefs, 389–400
  definition of, 264
  flow of energy and materials in,
    271–283
  organization of communities in,
    264–271
Ecosystem, 98
  continental shelf as, 350–368
  estuaries as, 327–344

Ectocarpus, 130
Ectoprocta, 163
Ectotherms, 102, 229
Eelgrass, 138, 342, 356–357, 359, 539
  harvesting of, 551
Eelpout, 475
Eels, 331
  migration of, 218–219
EEZ; *see* Exclusive economic zones
Eggs, 106–107
  of fishes, 222-223
  of sea turtles, 230
Eggshells, pesticide effects on,
  524–525
Egregia, 131, 311, 365
Egrets, 337
Eisenia, 131, 297
Ekman layer, 440
Ekman spiral, 439–440
Ekman transport, 440–441
El Niño, 368, 499
  and coral reefs, 379–380
  and corals, 394
El Niño–Southern Oscillation (ENSO),
  443–447
Electric ray, 4, 196
Electricity, generation of, 513–514
Elephant fish, 196
Elephant seals, 238, 246
  harem of, 256
Elk kelp, 363
Elkhorn coral, 383, 390
Embiotoca, 224
Embryo, 107–108
Embryological development, 108
Emerita, 316, 318, 525
Emersion time, 289–293
Emperor penguin, 233
Emperor seamount chain, 46
Encrusting algae in coral reef building,
  376–378
Encrusting sponges, 148, 149
Endangered species, 531–538, 543
Endocladia, 132
Endolithic algae, 120
Endophytes, 129
Endoplasmic reticulum, 95
Endoskeleton of echinoderms, 178, 179
Endotherms, 102–103, 232
Energy
  flow of, 271–283
    in epipelagic food webs, 428–429
  and ocean conservation, 543
  ocean in production of, 513–514
  pyramid of, 275

Enewetak Atoll, 385, 389
Engineering, genetic, 508
Engraulis, 205, 489, 490, 492, 497,
  498–499
Enhydra, 239, 367, 538
Enrichment, nutrient, 381–382
Ensis, 354
ENSO; *see* El Nino–Southern
  Oscillation
Enteromorpha, 87, 129, 136, 304, 310,
  331, 353, 359
Enteropneusts, 184
Entoprocts, 163
Environment
  conservation of, 541
  marine, human impact on,
    517–543
  restoration of, 541–542
Enzymes, 88
Epibenthic sleds, 470
Epifauna, 288
  on deep sea floor, 471
  subtidal, 350
Epinephelus, 220
Epipelagic, 285, 403–447, 449
  adaptations to life in, 414–427
  food webs of, 427–447
  organisms of, 404–414
Epiphytes, 121
  in seagrass beds, 357
Eptatretus, 191, 475
Equatorial currents, 66
Equatorial upwelling, 441–442
Eretmochelys, 229, 541
Ereunetes, 337
Erikson, Leif, 4
Erolia, 337
Eschrichtius robustus, 13, 244, 255, 257,
  356, 536
Esophagus in fishes, 206
Essential nutrients, cycle of,
  280–283
Estuaries, 322–344
  communities in, 330–342
  feeding interaction among organisms
    in, 342–344
  human effects on, 530
  origins and types of, 322–324
  physical characteristics of,
    324–327
  problems of life in, 327–330
Eubacteria, 118
Eubalaena glacialis, 243, 534
Eucheuma, 135
Eudendrium, 269

Groupers, 220
Growth, population, 266–267
Guano, 236
Guano islands, 498–499
Guitarfish, 194
Gulf Stream, 69
Gulls, 235, 236, 337
Gulper eel, 467, 468
Gunboat diplomacy, 552, 555
Gunnels, 199, 312
Guyots, 44
Gymnothorax, 198
Gyres, 66, 68, 69, 364
  central, 432, 443

H

Habitats, 22, 264
  destruction of, 497–498
  modification of, 529–531
  restoration of, 541–542
  undersea, 9, 10
Hadal zone, 284
Haddock, 492
Hadopelagic zone, 285, 465
Haematopus, 304–305, 337
Hagfishes, 191–192, 475
Hakes, 493, 502
Haldane, J.B.S., 110
Halibuts, 198, 493, 494
Haliclystus, 359
Halicreas, 449
Halimeda, 129–130, 378
Haliotis, 166, 360, 494
Halobates, 419
Halophytes, 139
Hammerhead sharks, 193, 194
Haploid cells, 106
Haploops, 353
Harbor seals, 238
Hard-bottom subtidal communities, 358–368
Harems of pinnipeds, 256
Harp seal, 238
Harpagifer, 223
Hatchetfishes, 455, 456, 461
Hawaiian Islands
  creation of, 46–47
  rocky shore of, 289
Hawaiian monk seal, 238
Hawksbill turtle, 229, 230, 541
Hazardous wastes, 526
Heart
  of fishes, 207
  of molluscs, 170

Heart urchins, 181–182, 315, 353–354
Heat, latent
  of evaporation, 55
  of melting, 54
Heat capacity, 55
Heavy metals, effect of, on marine environments, 526–527
Hemichordates, 184
Hemigrapsus, 296, 340
Hemoglobin
  of fishes, 210
  of marine mammals, 247
Herbivores, 269
  in estuaries, 343
  in seagrass beds, 357
  zooplankton as, 407–408
Heredity, 104–105
Hermaphrodites
  deep sea, 469
  in fishes, 220
  in molluscs, 171
Hermatypic corals, 370
Hermissenda, 166
Hermit crabs, 175
Hermosira, 55
Herons, 337
Herrings, 489, 492
  feeding in, 205
  mouth of, 205
  overfishing of, 497
  schools of, 215
Hesperonoe, 337
Hesperophycus, 306
Heterocentrotus, 191
Heterotrophic bacteria, 119
Heterotrophs, 90, 119, 273
Higgins larvae, 162
Himalaya Mountains, 36
Hippocampus, 199
Hippoglossus, 198, 493
Hirudinea, 159
Histioteuthis, 453, 454, 456, 461
Histrio, 268
Hogfish, 222
Holdfast of seaweeds, 126, 127, 295, 360
Holoplankton, 412
Holothuria, 316
Homarus, 174
Homeotherms, 102–103, 232
Homing behavior in salmon, 218
Homotrema, 145
Hopkins Marine Station, 7, 14

Hormones, 88
  sex, of fishes, 220
Hormuz, strait of, 558
Horse latitudes, 69
Horseshoe crab, 176, 178
Horsetail kelp, 361
Host, 269–271
Hot springs, deep-sea, 45–48, 477–480
Human cultures, oceans and, 549–556
Human population growth, 485
Humans, impact of, on marine environment, 517–543
Humpback whale, 243
  behavior of, 252
  courtship in, 257
  hunting of, 536
  migration of, 254, 255
Hunting
  sonic, 249
  by traditional cultures, 550–551
  of whales, 533–537
Hydranassa, 337
Hydrobia, 165, 333, 337
Hydrocarbons
  chlorinated, 523–525
  effect of, on marine environments, 518–521
Hydrodamalis, 533
Hydrogen bonds in water, 52
Hydrogen sulfide, 315
  in estuaries, 326
  in hydrothermal vents, 478
Hydrolab, 10
Hydrostatic skeleton of nematodes, 157
Hydrothermal vents, 45–48, 477–480
  mining along, 512
Hydrozoans, 151
Hymenamphiastra, 121
Hypothesis
  construction of, 16
  testing, 18–20

I

Ice
  melting of, 54
  molecules in crystals of, 53, 54
  shelf, in Antarctica, 351
Ice ages, 40–41
Idiacanthus, 456
Iguana, marine, 231, 232
Ilyanassa, 337
Inchworm, 316

Mesopelagic, 287, 449, 451–465
    adaptations to, 456–465
    animals of, 452–456
Metals
    heavy, effect of, on marine
        environments, 526–527
    mining of, 511–512
Metamorphosis of larvae
    of corals, 372–373
    of sponges, 148
    of subtidal organisms, 351–352
Methane in greenhouse effect, 532
Methyl mercury, 526
Metridia, 452
Metridium, 312
Microfossils, 40
Microherbivores and coral reefs, 397
Micronesian stick map, 4
Micronesians, 551
Micropogon, 509
Micropogonias, 330
Middens, 3
Middle intertidal zone, 305–310
Mid-ocean ridge, 34
    discovery of, 28–29
    and hydrothermal vents, 45–48
    significance of, 29–32
Midway Island, 46
Migration
    of fishes, 216–219
        in estuaries, 330–331
    of pinnipeds and cetaceans,
        254–255
    of tubenoses, 234–235
    of tunas, 494
    of sea turtles, 229–230
    vertical, 426–427
        mesopelagic, 285, 458–459
    of whales, 254–255, 536
Milk of marine mammals, 257
Milkfish, 504, 505
Miller, Stanley L., 110
Milne-Edwards, Henri, 7
Minamata disease, 527
Minerals
    on sea floor, 511–512
    in seawater, 511–512
Mining, ocean, 511–512
Minke whale, 242, 243, 503
    hunting of, 536
Mirounga, 238, 246, 256
Mitochondrion(a), 96
    and bacteria, similarities in, 270
Mitosis, 105, 133
Mitsukurina, 193

Mixed layer in ocean, 81
Mixed tide, 305
    semidiurnal, 76–78
Mola mola, 412
Mole crabs, 525
Molecules
    organic, 110–111
    structural, 87
    of water, 52
Molluscs, 164–171
    biology of, 169–171
    as food, 494
Molpadia, 354
Molting in arthropods, 171
Monachus, 238, 538
Monera, 113, 118
Monitor, 553
Monk seals, 238
    threat of extermination of, 538
Monkfish, 502
Monoplacophorans, 169
Monsoons, 441
Montastrea, 372
Monterey Canyon, 43
Moon snails, 318, 333, 337, 355–356
Moray eels, 198
Motion
    in ocean, 65–78
        vertical, 78–82
    of water particles in wave, 71
Motor oil and ocean pollution, 543
Mt. Veniaminof, 35
Mucous nets, 412
Mud, 314–318
    in estuaries, 326–327
        adaptation to, 329–330
Mud flat communities, 331–337
Mud shrimps, 316, 334
Mud snails, 165, 315, 333, 337
Mudskippers, 341–342
Mugil, 165
Mullets, 330, 489
Multicellular organization, 96
Murex, 165
Muscle and swimming ability, 423,
    425, 426
Mushroom corals, 372
Mussels, 167, 290, 297, 306, 342
    competition in, 300
    as food, 494
    and hydrothermal vents, 479
    predators, 306–309
    in salt marshes, 339
Mutualism, 271
    on coral reef, 400

Mya, 315, 333, 334, 353, 354
Myctophum, 455, 456
Myoglobin
    in fishes, 210
    in marine mammals, 247
    and swimming ability, 423
Myomeres, 201–202
Mytilus, 51, 97, 167, 290, 297, 300,
    306, 342, 494, 505
Myxine, 191–192

N

Nanaloricus, 162
Nanoplankton, 405
Nassarius, 353
Native species, threat of introduced
    species to, 538–539
Natural products, marine, 509
Natural selection, 109, 267, 531
Nauplius, 176
Nautilus, 169, 472–473
Neanthes, 158, 333, 337, 353
Neap tides, 75, 289
Nekton, 98, 284, 413–414
    in subtidal zone, 347
Nematocysts, 151, 154
    of Portuguese man-of-war, 153
Nematodes, 156–157
Nematolampas, 454
Nemertean worms, 156
Nemichthys, 199
Nereis, 168
Nereocystis, 131, 136, 363
Nerita, 292
Neritic zone, 285, 347, 403
Nerve cells of cnidarians, 154
Nerve cord, 190
    of chordates, 184
Nerve net of cnidarians, 154
Nervous system
    in crustaceans, 175–176
    in echinoderms, 183
    in fishes, 211–213
    in flatworms, 155
    in molluscs, 170–171
Net plankton, 404–405
Nets
    drift, 498, 537
    gill, 494
    for mesopelagic organisms, 455
    mucous, 412
    plankton, 404
Neuston, 418–419
New Zealand fur seal, 238
Niche, ecological, 268